Mathematical Geosciences

Joseph L. Awange · Béla Paláncz ·
Robert H. Lewis · Lajos Völgyesi

Mathematical Geosciences

Hybrid Symbolic-Numeric Methods

Second Edition

 Springer

Joseph L. Awange
Department of Spatial Sciences
School of Earth and Planetary Sciences
Curtin University
Perth, WA, Australia

Robert H. Lewis
Fordham University
New York, NY, USA

Béla Paláncz
Department of Geodesy and Surveying
Faculty of Civil Engineering
Budapest University of Technology
and Economics
Budapest, Hungary

Lajos Völgyesi
Department of Geodesy and Surveying
Faculty of Civil Engineering
Budapest University of Technology
and Economics
Budapest, Hungary

ISBN 978-3-030-92497-3   ISBN 978-3-030-92495-9   (eBook)
https://doi.org/10.1007/978-3-030-92495-9

1st edition: © Springer International Publishing AG 2018
2nd edition: © The Editor(s) (if applicable) and The Author(s), under exclusive license to Springer Nature
Switzerland AG 2023

This Springer imprint is published by the registered company Springer Nature Switzerland AG
The registered company address is: Gewerbestrasse 11, 6330 Cham, Switzerland

Foreword to the First Edition

Hybrid symbolic-numeric computation (HSNC, for short) is a large and growing area at the boundary of mathematics and computer science, devoted to the study and implementation of methods that mix symbolic with numeric computation.

As the title suggests, this is a book about some of the methods and algorithms that benefit from a mix of symbolic and numeric computation. Three major areas of computation are covered herein. The first part discusses methods for computing all solutions to a system of polynomials. Purely symbolic methods, e.g., via Gröbner bases tend to suffer from algorithmic inefficiencies, and purely numeric methods such as Newton iterations have trouble finding all solutions to such systems. One class of hybrid methods blends numerics into the purely algebraic approach, e.g., computing numeric Gröbner bases or Dixon resultants (the latter being extremely efficient, e.g., for elimination of variables). Another mixes symbolic methods into more numerical approaches, e.g., finding initializations for numeric homotopy tracking to obtain all solutions.

The second part goes into the realm of "soft" optimization methods, including genetic methods, simulated annealing, and particle swarm optimization, among others. These are all popular and heavily used, especially in the context of global optimization. While often considered as "numeric" methods, they benefit from symbolic computation in several ways. One is that implementation is typically straightforward when one has access to a language that supports symbolic computation. Updates of state, e.g., to handle mutations and gene crossover, are easily coded. (Indeed, this sort of thing can be so deceptively simple, baked into the language so to speak, that one hardly realizes symbolic computation is happening). Among many applications in this part, there is, again, that of solving systems of equations. Also covered is mixed-integer programming (wherein some variables are discrete-valued and others continuous). This is a natural area for HSNC since it combines aspects of exact and numeric methods in the handling of both discrete and continuous variables.

The third part delves into data modeling. This begins with use of radial basis functions and proceeds to machine learning, e.g., via support vector machine (SVM) methods. Symbolic regression, a methodology that combines numerics with evolutionary programming, is also introduced for the purpose of modeling data. Another area seeing recent interest is that of robust optimization and regression, wherein one seeks results that remain relatively stable with respect to perturbations in input or random parameters used in the optimization. Several hybrid methods are presented to address problems in this realm. Stochastic modeling is also discussed. This is yet another area in which hybrid methods are quite useful.

Symbolic computing languages have seen a recent trend toward ever more high level support for various mathematical abstractions. This appears for example in exact symbolic programming involving probability, geometry, tensors, engineering simulation, and many other areas. Under the hood is a considerable amount of HSNC (I write this as one who has been immersed at the R&D end of hybrid computation for two decades). Naturally, such support makes it all the easier for one to extend hybrid methods; just consider how much less must be built from scratch to support, say, stochastic equation solving, when the language already supports symbolic probability and statistics computations. This book presents to the reader some of the major areas and methods that are being changed, by the authors and others, in furthering this interplay of symbolic and numeric computation. The term hybrid symbolic-numeric computation has been with us for over two decades now. I anticipate the day when it falls into disuse, not because the technology goes out of style, but rather that it is just an integral part of the plumbing of mathematical computation.

Urbana–Champaign, IL, USA   Daniel Lichtblau
July 2017   Ph.D. Mathematics UIUC 1991
                                                      Algebra, Applied Mathematics
                                                      Wolfram Research, Champaign

Preface to the Second Edition

This second edition of Mathematical Geosciences book adds five new topics: **Solution equations with uncertainty**, which proposes two novel methods for solving nonlinear geodetic equations as stochastic variables when the parameters of these equations have uncertainty characterized by probability distribution. The first method, an *algebraic technique*, partly employs symbolic computations and is applicable to polynomial systems having different uncertainty distributions of the parameters. The second method, a *numerical technique*, uses stochastic differential equation in *Ito* form; **Nature Inspired Global Optimization** where meta-heuristic algorithms are based on natural phenomenon such as particle swarm optimization. This approach simulates, e.g., schools of fish or flocks of birds and is extended through discussion of geodetic applications. Black Hole algorithm, which is based on the black hole phenomena, is added, and a new variant of the algorithm code is introduced and illustrated based on examples; **The application of the Gröbner Basis to integer programming** based on numeric symbolic computation is introduced and illustrated by solving some standard problems. An extension of the **applications of integer programming** solving phase ambiguity in Global Navigation Satellite Systems (GNSSs) is considered as a global quadratic mixed-integer programming task, which can be transformed into a pure integer problem with a given digit of accuracy. Three alternative algorithms are suggested, two of which are based on local and global linearization via McCormic Envelopes and **Machine learning techniques** (MLT) that offer effective tools for stochastic process modeling. The stochastic modeling section is extended by the stochastic modeling via MLT and their effectiveness is compared with that of the modeling via stochastic differential equations (SDE). Mixing MLT with SDE also known as frequently neural differential equations is also introduced and illustrated by an image classification via a regression problem.

Joseph L. Awange
Perth, Australia

Béla Paláncz
Budapest, Hungary
January 2022

Robert H. Lewis
New York, USA

Lajos Völgyesi
Budapest, Hungary
January 2022

Acknowledgements The authors wish to express their sincere thanks to *Dr. Daniel Lichtblau* for his helpful comments and for agreeing to write a foreword for our book. *R. Lewis* and *B. Paláncz* highly appreciate and thank to *Prof. Jon Kirby*, Head of the Department of Spatial Sciences, Curtin University, Australia, for his hospitality and his support of their visiting Curtin. *B. Paláncz* wishes also thank the TIGeR Foundation for the partial support of his staying in Perth. This work was funded partially in the project No. K-124286 supported by the Hungarian National Research, Development, and Innovation Office (NKFIH).

Preface to the First Edition

It will surprise no one to hear that digital computers have been used for numerical computations ever since their invention during World War II. Indeed, until around 1990, it was not widely understood that computers could do anything else. For many years, when students of mathematics, engineering, and the sciences used a computer, they wrote a program (typically in Fortran) to implement mathematical algorithms for solving equations in one variable, or systems of linear equations, or differential equations. The input was in so-called float numbers with 8 to 12 significant figures of accuracy. The output was the same type of data, and the program worked entirely with the same type of data. This is *numerical computing*.

By roughly 1990, computer algebra software had become available. Now, it was possible to enter data like $x^2 + 3x + 2$ and receive output like $(x+2)(x+1)$. The computer is doing algebra! More precisely, it is doing *symbolic computing*. The program that accomplishes such computing almost certainly uses no float numbers at all.

What is still not widely understood is that often it is productive to have algorithms that do both kinds of computation. We call these *hybrid symbolic-numeric* methods. Actually, such methods have been considered by some mathematicians and computer scientists since at least 1995 (ref. to ISSAC 1995 conference). In this book, the authors provide a much-needed introduction and reference for applied mathematicians, geoscientists, and other users of sophisticated mathematical software.

No mathematics beyond the undergraduate level is needed to read this book, nor does the reader need any pure mathematics background beyond a first course in linear algebra. All methods discussed here are illustrated with copious examples.

A brief list of topics covered:

- Systems of polynomial equations with resultants and Gröbner bases
- Simulated annealing
- Genetic algorithms
- Particle swarm optimization
- Integer programming
- Approximation with radial basis functions

- Support vector machines
- Symbolic regression
- Quantile regression
- Robust regression
- Stochastic modeling
- Parallel computations

Most of the methods discussed in book will probably be implemented by the reader on a computer algebra system (CAS). The two most fully developed and widely used CAS are *Mathematica* and *Maple*. Some of the polynomial computations here are better done on the specialized system *Fermat*. Other systems worthy of mention are *Singular* and *SageMath*.

The second author is a regular user of *Mathematica*, who carried out the computations; therefore, frequent mention is made of *Mathematica* commands. However, this book is not a reference manual for any system, and we have made an effort to keep the specialized commands to a minimum and to use commands whose syntax makes them as self-explanatory as possible. More complete *Mathematica* programs to implement some of the examples are available online. Similarly, a program written in *Fermat* for the resultant method called Dixon-EDF is available online.

Introduction

Numeric and Symbolic Methods—What Are They?

Basically, a *numeric (or numerical) method* is one that could be done with a simple hand-held calculator, using basic arithmetic, square roots, some trigonometry functions, and a few other functions most people learn about in high school. Depending on the task, one may have to press the calculator buttons thousands (or even millions) of times, but theoretically, a person with a calculator and some paper could implement a numerical method. When finished, the paper would be full of arithmetic.

A *symbolic method* involves algebra. It is a method that if a person implemented, would involve algebraic or higher rational thought. A person implementing a symbolic method will rarely need to reach for a calculator. When finished, there may be some numbers, but the paper would be full of variables like x, y, z.

Students usually meet the topic of quadratic equations in junior high school. Suppose you wanted to solve the equation $x^2 + 3x - 2 = 0$. With a hand-held calculator, one could simply do "intelligent guessing." Let us guess, say, $x = 1$. Plug it in, get a positive result. OK, that is too big. Try $x = 0$; that is too small. Go back and forth; stop when satisfied with the accuracy. It does not take long to get $x = 0.56155$, which might well be considered accurate enough. Furthermore, it is easy to write a computer program to implement this idea. That is a numeric method.

But wait. There is another answer, which the numeric method missed, namely -3.56155. Even worse, if one were to continue this method on many problems, one would soon notice that some equations don't seem to have solutions, such as $x^2 - 2x + 4 = 0$. A great deal of effort could be expended in arithmetic until finally giving up and finding no solution.

The problem is cured by learning algebra and the symbolic method called the quadratic formula. Given $ax^2 + bx + c = 0$ the solution is $x = \frac{-b \pm \sqrt{b^2 - 4ac}}{2a}$. It is now immediately obvious why some problems have no solution: it happens precisely when $b^2 - 4ac < 0$.

In the previous example, $x^2 + 3x - 2 = 0$, we see that the two roots are exactly $(-3 \pm \sqrt{17})/2$. There is no approximation whatever. Should a decimal answer correct to, say, 16 digits be desired, that would be trivially obtained on any modern computer.

There is more. Not only does the symbolic method concisely represent all solutions, it invites the question, can we define a new kind of number in which the negative under the square root may be allowed? The symbolic solution has led to a new concept, that of complex numbers!

Symbolic methods may be hard to develop, and they may be difficult for a computer to implement, but they lead to insight.

Fortunately, we are not forced into a strict either/or dichotomy. There are symbolic-numeric methods, hybrids using the strengths of both ideas.

Numeric Solution

In order to further illustrate numeric, symbolic, and symbolic-numeric solutions, let us consider an algebraic system of polynomial equations. For such systems, there may be no solution, one solution, or many solutions. With numerical solutions, one commonly utilizes iterative techniques starting from an initially guessed value. Let us start with a two variable system of two equations $f(x, y) = 0$ and $g(x, y) = 0$,

$$f = (x - 2)^2 + (y - 3)^2 \,,$$

$$g = \left(x - \frac{1}{2}\right)^2 + \left(y - \frac{3}{4}\right)^2 - 5.$$

This actual problem has two real solutions, see Fig. 1.

Fig. 1. Geometrical representation of a multivariate
polynomial system

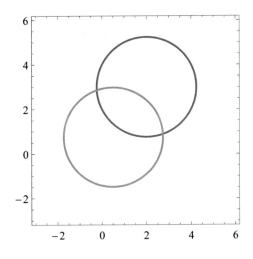

Fig. 2 Local solution with initial guess and iteration steps

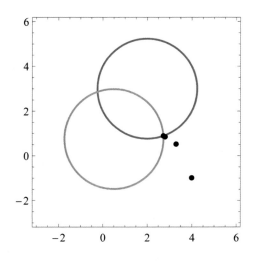

A numeric solution starts with the initial value and proceeds step-by-step locally. Depending on the method, we expect to converge to one of the solutions in an efficient manner. Employing the initial value (4, −1) and a multivariate Newton's method, the solution after seven steps is (2.73186, 0.887092). Let us visualize the iteration steps, see Fig. 2.

However, if the initial guess is not proper, for example (0, 0), then, we may have a problem with the convergence since the Jacobian may become singular.

Symbolic Solution

Let us transform the original system into another one, which has the same solutions, but for which variables can be isolated and solved one-by-one. Employing Gröbner basis, we can reduce one of the equations to a univariate polynomial,

$$gry = 2113 - 3120y + 832y^2$$

$$grxy = -65 + 16x + 24y$$

First, solving the quadratic equation gry, we have

$$y = \frac{1}{104}\left(195 - 2\sqrt{2639}\right),$$

$$y = \frac{1}{104}\left(195 + 2\sqrt{2639}\right).$$

Then employing these roots of y, the corresponding values of x can be computed from the second polynomial of the Gröbner basis as

$$x = \frac{1}{104}\left(130 + 3\sqrt{2639}\right),$$

$$x = \frac{1}{104}\left(130 - 3\sqrt{2639}\right).$$

So, we have computed both solutions with neither guessing nor iteration. In addition, there is no round-off error. Let us visualize the two solutions, see Fig. 3:

Let us summarize the main features of the symbolic and numeric computations:

Numeric computations:

– usually require initial values and iterations. They are sensitive to round-off errors, provide only one local solution,
– can be employed for complex problems, and the computation times are short in general because the steps usually translate directly into computer machine language.

Symbolic computations:

– do not require initial values and iterations. They are not sensitive for round-off errors and provide all solutions,
– often cannot be employed for complex problems, and the computation time is long in general because the steps usually require computer algebra system software.

Ideally, the best strategy is to divide the algorithm into symbolic and numeric parts in order to utilize the advantages of both techniques. Inevitably, numeric computations will always be used to a certain extent. For example, if polynomial

Fig. 3 Global solution—all solutions without initial guess and iteration

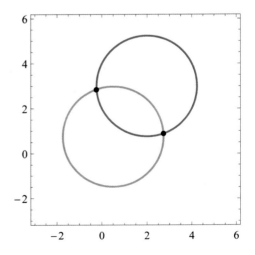

gry above had been degree, say, five, then a numeric univariate root solver would have been necessary.

Hybrid (Symbolic-Numeric) Solution

Sometimes, we can precompute a part of a numerical algorithm in symbolic form. Here is a simple illustrative example.

Consider a third polynomial and add it to our system above:

$$h = \left(x + \frac{1}{2}\right)^2 + \left(y - \frac{7}{4}\right)^3 - 5.$$

In that case, there is no solution, since there is no common point of the three curves representing the three equations, see Fig. 4.

However, we can look for a solution of this overdetermined system in the minimal least squares sense by using the objective function

$$G = f^2 + g^2 + h^2,$$

or

$$G = \left(-5 + (-2+x)^2 + (-3+y)^2\right)^2 +$$
$$\left(-5 + \left(\frac{1}{2}+x\right)^2 + \left(-\frac{7}{4}+y\right)^3\right)^2 + \left(-5 + \left(-\frac{1}{2}+x\right)^2 + \left(-\frac{3}{4}+y\right)^2\right)^2$$

and minimizing it.

Fig. 4 Now, there is no solution of the overdetermined system.

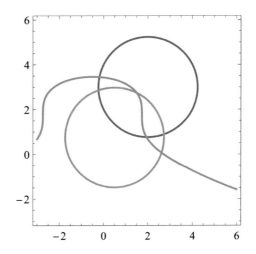

Fig. 5 The solution of the overdetermined system

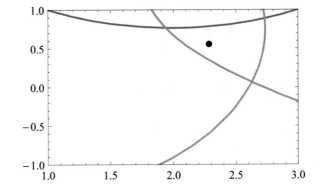

Employing Newton's method, we get

$$x = 2.28181,$$

$$y = 0.556578.$$

The computation time for this was 0.00181778 s. The solution of the overdetermined system can be seen in Fig. 5.

Here, the gradient vector as well as the Hessian matrix is computed in numerical form in every iteration step. But we can compute the gradient in symbolic form:

$$\text{grad} = \left(\frac{1}{32} \left(2x(173 + 192(-2 + x)x) + 216xy - 16(41 + 26x)y^2 \right. \right.$$

$$+ 64(1 + 2x)y^3 + 3(-809 + 740y)) - \frac{137829}{512} + \frac{555x}{8} + \frac{27x^2}{8} + \frac{60527y}{128}$$

$$\left. - 41xy - 13x^2y - \frac{6321y^2}{16} + 6xy^2 + 6x^2y^2 + \frac{767y^3}{4} - \frac{105y^4}{2} + 6y^5 \right).$$

Employing this symbolic form the computation time can be reduced. The running time can be further reduced if the Hessian matrix is also computed symbolically,

$$H = \begin{bmatrix} \frac{173}{16} + 12x(-4 + 3x) + \frac{27y}{4} - 13y^2 + 4y^3 & \frac{555}{8} + y(-41 + 6y) + x\left(\frac{27}{4} + 2y(-13 + 6y)\right) \\ \frac{555}{8} + y(-41 + 6y) + x\left(\frac{27}{4} + 2y(-13 + 6y)\right) & \frac{60527}{128} - 41x - 13x^2 - \frac{6321y}{8} \\ & + 12xy + 12x^2y + \frac{2301y^2}{4} - 210y^3 + 30y^4 \end{bmatrix}.$$

Now, the computation time is less than half of the original one.

So using symbolic forms, the *computation time can be reduced* considerably. This so-called *hybrid computation* has an additional advantage too, namely the symbolic part of the algorithm does *not generate round-off errors.*

Another approach of applying the hybrid computation is to *merge* symbolic evaluation with numeric algorithm. This technique is illustrated using the following example.

Let us consider a linear, non-autonomous differential equation system of n variables in matrix form:

$$\frac{d}{dx}y(x) = A(x)y(x) + b(x),$$

where A is a matrix of $n \times n$ dimensions, $y(x)$ and $b(x)$ are vectors of n dimensions, and x is a scalar independent variable. In the case of a boundary value problem, the values of some dependent variables are not known at the beginning of the integration interval, at $x = x_a$, but they are given at the end of this interval, at $x = x_b$. The usually employed methods need subsequent integration of the system because of their trial-error technique or they require solution of a large linear equation system, in the case of discretization methods. The technique is based on the symbolic evaluation of the well-known Runge–Kutta algorithm. This technique needs only one integration of the differential equation system and a solution of the linear equation system representing the boundary conditions at $x = x_b$.

The well-known fourth-order Runge–Kutta method, in our case, can be represented by the following formulas:

$$R1_i = A(x_i)y(x_i) + b(x_i),$$

$$R2_i = A\left(x_i + \frac{h}{2}\right)\left(y(x_i) + \frac{R1_i h}{2}\right) + b\left(x_i + \frac{h}{2}\right),$$

$$R3_i = A\left(x_i + \frac{h}{2}\right)\left(y(x_i) + \frac{R2_i h}{2}\right) + b\left(x_i + \frac{h}{2}\right),$$

$$R4_i = A(x_i + h)(y(x_i) + R3_i h) + b(x_i + h)$$

and then the new value of $y(x)$ can be computed as:

$$y_{i+1} = y(x_i) + \frac{(R1_i + 2(R2_i + R3_i) + R4_i)h}{6}.$$

A symbolic system like *Mathematica* is able to carry out this algorithm not only with numbers but also with symbols. It means that the unknown elements of $y_a = y(x_a)$ can be considered as unknown symbols. These symbols will appear in every evaluated y_i value, as well as in $y_b = y(x_b)$ too.

Let us consider a simple illustrative example. The differential equation is:

$$\left(\frac{d^2}{dx^2}y(x)\right) - \left(1 - \frac{x}{5}\right)y(x) = x.$$

The given boundary values are:

$$y(1) = 2$$

and

$$y(3) = -1.$$

After introducing new variables, we get a first-order system,

$$y1(x) = y(x)$$

and

$$y2(x) = \frac{d}{dx}y(x)$$

the matrix form of the differential equation is:

$$\left[\frac{d}{dx}y1(x), \; \frac{d}{dx}y2(x)\right] = \begin{bmatrix} 0 & 1 \\ 1 - x/5 & 0 \end{bmatrix}[y1(x), \; y2(x)] + [0, \; x].$$

Employing *Mathematica*'s notation:

```
⇒ A[x_] := {{0, 1}, {1 − 1/5x, 0}};
⇒ b[x_] := {0, x};
⇒ x0 = 1;
⇒ y0 = {2., s}
```

The unknown initial value is *s*. The order of the system $M = 2$. Let us consider the number of the integration steps as $N = 10$, so the step size is $h = 0.2$.

```
⇒ ysol = RKSymbolic[x0, y0, A, b, 2, 10, 0.2];
```

The result is a list of list data structure containing the corresponding (x, y) pairs, where the *y* values depend on *s*.

```
⇒ ysol[[2]][[1]];
⇐ {{1, 2.}, {1.2, 2.05533 + 0.200987 s}, {1.4, 2.22611 + 0.407722 s},
     {1.6, 2.52165 + 0.625515 s}, {1.8, 2.95394 + 0.859296 s},
     {2., 3.53729 + 1.11368 s}, {2.2, 4.28801 + 1.39298 s},
     {2.4, 5.22402 + 1.70123 s}, {2.6, 6.36438 + 2.0421 s},
     {2.8, 7.72874 + 2.41888 s}, {3., 9.33669 + 2.8343 s}}
```

Consequently, we have got a symbolic result using traditional numerical Runge–Kutta algorithm.

In order to compute the proper value of the unknown initial value, s, the boundary condition can be applied at $x = 3$. In our case $y1(3) = -1$.

$\Rightarrow$ eq = ysol$[[1]][[1]]$ == -1
$\Leftarrow$ $9.33669 + 2.8343$ s == -1

Let us solve this equation numerically and assign the solution to the symbol s:

$\Rightarrow$ sol = Solve$[$eq, s$]$
$\Leftarrow$ $\{\{$s $\rightarrow -3.647\}\}$
$\Rightarrow$ s = s$/$.sol
$\Leftarrow$ $\{-3.647\}$
$\Rightarrow$ s = s$[[1]]$
$\Leftarrow$ $\{-3.647\}$

Then we get the numerical solution for the problem:

$\Rightarrow$ ysol$[[2]][[1]]$
$\Leftarrow$ $\{\{1, 2.\}, \{1.2, 1.32234\}, \{1.4, 0.739147\}, \{1.6, 0.240397\},$
$\quad \{1.8, -0.179911\}, \{2., -0.524285\}, \{2.2, -0.792178\}, \{2.4, -0.980351\},$
$\quad \{2.6, -1.08317\}, \{2.8, -1.09291\}, \{3., -1.\}\}$

The truncation error can be decreased by using smaller step size h, and the round-off error can be controlled by the employed number of digits.

Contents

Part I
Solution of Nonlinear Systems

Chapter 1
Solution of Algebraic Polynomial Systems

1.1 Zeros of Polynomial Systems

Let us consider the following polynomial

$$p = 2x + x^3 y^2 + y^2.$$

The monomials are $x^3 y^2$ with coefficient 1, and $x^1 y^0$ with coefficient 2 and $x^0 y^2$ with coefficient 1. The degree of such a monomial is defined as the sum of the exponents of the variables. For example, the second monomial $x^3 y^2$, has degree $3 + 2 = 5$. The *degree of the polynomial* is the maximum degree of its constituent monomials. In this case $\deg(p) = \max(1, 5, 2) = 5$.

Some polynomials contain *parameters* as well as variables. For example, the equation of a circle centered at the origin is $x^2 + y^2 - r^2 = 0$. Only x and y are actual variables; the r is a parameter.

Now consider a polynomial system like

$$g(x, y) = a_1 + a_2 x + a_3 xy + a_4 y,$$
$$h(x, y) = b_1 + b_2 x^2 y + b_3 xy^2.$$

The total degree of the system is defined to be

$$\deg(g) \, \deg(h) = 2 \times 3 = 6.$$

Notice that we do not count the parameters in this computation.

Define the *roots* or *zeros* of a polynomial system to be the set of pairs (r, s) such that $g(r, s) = 0$ and $h(r, s) = 0$.

Supplementary Information The online version contains supplementary material available at https://doi.org/10.1007/978-3-030-92495-9_1.

Bézout's Theorem: Consider two polynomial equations in two unknowns: $g(x, y) = h(x, y) = 0$. If this system has only finitely many zeros $(x, y) \in \mathbb{C}^2$, then the number of zeros is at most $\deg(g)\,\deg(h)$. Here $\deg(g)$ and $\deg(h)$ are the total degree of $g(x, y)$ and $h(x, y)$.

1.2 Resultant Methods

In this section we introduce two different symbolic methods: Sylvester and Dixon resultants see Dickenstein and Emiris (2005). These techniques eliminate variables and yield univariate polynomials, which then can be solved numerically.

1.2.1 Sylvester Resultant

Let us consider the following system

$$p = xy - 1,$$
$$g = x^2 + y^2 - 4.$$

Since linear systems of equations are well known, let's try to convert this into a useful system of linear equations. With x as the "real" variable and y as a "parameter," consider x^0, x^1, and x^2 as three independent symbols. The two equations in the original system give us two linear equations, and we generate a third by multiplying p by x. This yields (Fig. 1.1).

$$M(y) \begin{pmatrix} x^0 \\ x^1 \\ x^2 \end{pmatrix} = \bar{0},$$

where $M(y)$ is

$$\begin{pmatrix} -1 & y & 0 \\ y^2 - 4 & 0 & 1 \\ 0 & -1 & y \end{pmatrix} \begin{pmatrix} x^0 \\ x^1 \\ x^2 \end{pmatrix} = \bar{0}.$$

Since x^0 is really 1, any solution to this homogeneous system must be nontrivial. Thus

$$\det(M(y)) = -1 + 4y^2 - y^4 = 0.$$

Fig. 1.1 Graphical
interpretation of the real roots
of the system

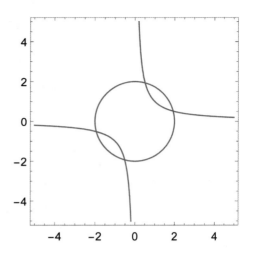

Solving this gives us y; we have *eliminated* x. This function is built into *Mathematica*,

$\Rightarrow$ Resultant$[p, g, x]$
$\Leftarrow 1 - 4y^2 + y^4$

For the other variable

$\Rightarrow$ Resultant$[p, g, y]$
$\Leftarrow 1 - 4x^2 + x^4$

The solutions of these two polynomials are the solutions of the system $\{p(x, y),\ g(x, y)\}$.

$\Rightarrow$ Roots$[-1 + 4y^2 - y^4 == 0, y]$
$\Leftarrow y = \sqrt{2 - \sqrt{3}} || y = -\sqrt{2 - \sqrt{3}} || y = \sqrt{2 + \sqrt{3}} || y = -\sqrt{2 + \sqrt{3}}$
$\Rightarrow$ Roots$[1 - 4x^2 + x^4 == 0, x]$
$\Leftarrow x = \sqrt{2 - \sqrt{3}} || x = -\sqrt{2 - \sqrt{3}} || x = \sqrt{2 + \sqrt{3}} || x = -\sqrt{2 + \sqrt{3}}$

The main handicap of the *Sylvester resultant* is that it can be directly employed only for systems of two polynomials.

1.2.2   Dixon Resultant

Let us introduce a new variable b and define the following polynomial,

$$\delta(x, y, b) = \frac{p(x,y)g(x,b) - p(x,b)g(x,y)}{y - b} = b - 4x + x^3 + y - bxy.$$

(One can show that the numerator above is always divisible by $y - b$). We call this polynomial the *Dixon polynomial*.

It is easy to see that plugging in any common root of $\{p(x,y), \ g(x,y)\}$ forces the Dixon polynomial to be 0, for *any* value of b. The Dixon polynomial can be written as,

$$\delta(x, y, b) = b^0 \left(-4x + x^3 + y\right) + b^1 (1 - xy).$$

Then the following homogeneous linear system should have solutions for every b

$$-4x + x^3 + y = 0,$$
$$1 - xy = 0$$

or, where x is considered as parameter

$$\begin{pmatrix} -4x + x^3 & 1 \\ 1 & -x \end{pmatrix} \begin{pmatrix} y^0 \\ y^1 \end{pmatrix} = 0,$$

therefore

$$\det \left[\begin{pmatrix} -4x + x^3 & 1 \\ 1 & -x \end{pmatrix} \right] = -1 + 4x^2 - x^4,$$

must be zero. The matrix

$$\begin{pmatrix} -4x + x^3 & 1 \\ 1 & -x \end{pmatrix}$$

is called the *Dixon matrix*, and its determinant is called as *Dixon resultant*.

[*Historical note*: the argument above, for two variables, was first used by Bezout.]

Let us employ *Mathematica*

$\Rightarrow$ $<<$ Resultant`Dixon`

$\Rightarrow$ DixonPolynomial$[\{p,g\},\{y\},\{b\}]$

$\Leftarrow$ $b - 4x + x^3 + y - bxy$

$\Rightarrow$ DixonMatrix$[\{p,g\},\{y\},\{b\}]$//MatrixForm

$\Leftarrow$ $\begin{pmatrix} -4x + x^3 & 1 \\ 1 & -x \end{pmatrix}$

$\Rightarrow$ DixonResultant$[\{p,g\},\{y\},\{b\}]$

$\Leftarrow$ $-1 + 4x^2 - x^4$

Similarly for the other variable, we get

$\Rightarrow$ DixonResultant$[\{p,g\},\{x\},\{a\}]$

$\Leftarrow$ $-1 + 4y^2 - y^4$

Here a and b are dummy formal variables (symbolic variables), without assigned values.

The Dixon resultant method can be generalized to polynomial systems of more than two polynomials. For example,

$\Rightarrow$ $P = x + y + z;$

$\Rightarrow$ $G = x - 2y + z^3;$

$\Rightarrow$ $S = x^2 - 2y^3 + z;$

To eliminate variables x and y, we introduce dummy variables X and Y, then

$\Rightarrow$ DixonResultant$[\{P,G,S\},\{x,y\},\{X,Y\}]$//Expand

$\Leftarrow$ $324z + 144z^2 + 24z^3 + 144z^4 - 72z^5 + 36z^6 + 72z^7 - 24z^9$

or

$\Rightarrow$ %/12//Expand

$\Leftarrow$ $27z + 12z^2 + 2z^3 + 12z^4 - 6z^5 + 3z^6 + 6z^7 - 2z^9$

(% refers to the last previous output.)

Remark 1   For other multivariate resultant methods such as Sturmfels' approach, see Awange and Paláncz (2016).

Remark 2   For three or more variables, the discussion above of the Dixon resultant has been simplified. Sometimes the Dixon matrix is not square, and sometimes when it is square the determinant is identically 0. Then the method would seem to fail. *Kappur, Saxena,* and *Yang* (1994) showed how to proceed and define the Dixon resultant using maximal minors, see Exercises 1.6.4 Dixon KSY solution.

Remark 3 If the system contains parameters, then the resultant will contain those parameters.

Remark 4 If there are n equations, the Dixon resultant will eliminate $n - 1$ variables.

1.3   Göbner Basis

This technique introduced by *Buchberger* and named after his PhD supervisor *Gröbner,* is more general and mostly more efficient than the resultant methods, unless parameters are present. To have an idea how this method works, first, let us see how the *greatest common divisor of polynomials* can be defined.

1.3.1   Greatest Common Divisor of Polynomials

Greatest common divisor, GCD, is a familiar concept from arithmetic, as in GCD (12, 30) = 6. The same concept applies to polynomials, of any number of variables.

Let us consider two univariate polynomials $s(x)$ and $v(x)$ with the same variable x

$$\Rightarrow s = 8 + 22x + 21x^2 + 8x^3 + x^4 ;$$
$$\Rightarrow v = 6 + 11x + 6x^2 + x^3 ;$$

The greatest common divisor (GCD) of these polynomials

$$\Rightarrow \text{gcd} = \text{PolynomialGCD}[s, v]$$
$$\Leftarrow 2 + 3x + x^2$$

Let us divide $s(x)$ with this GCD

$$\Rightarrow \{\{Cs\}, Rs\} = \text{PolynomialReduce}[s, \text{gcd}, x]$$
$$\Leftarrow \{\{4 + 5x + x^2\}, 0\}$$

The remainder is zero and we can check

$$\Rightarrow \text{Csgcd}$$
$$\Leftarrow (2 + 3x + x^2)(4 + 5x + x^2)$$
$$\Rightarrow \text{Expand}[\%]$$
$$\Leftarrow 8 + 22x + 21x^2 + 8x^3 + x^4$$

Let us carry out these operations with $v(x)$, too

$\Rightarrow \{\{Cv\}, Rv\} = \text{PolynomialReduce}[v, \text{gcd}, x]$
$\Leftarrow \{\{3+x\}, 0\}$

and

$\Rightarrow \text{Csgcd}$
$\Leftarrow (3+x)(2+3x+x^2)$
$\Rightarrow \text{Expand}[\%]$
$\Leftarrow 6+11x+6x^2+x^3$

This means that the original polynomials $s(x)$ and $v(x)$ can be expressed as the linear combination of the GCD, like

$$s(x) = Cs(x)\,\text{gcd}(x) + 0\,\text{gcd}(x)$$

and

$$v(x) = 0\,\text{gcd}(x) + Cv(x)\,\text{gcd}(x)$$

or

$$\begin{pmatrix} s(x) \\ v(x) \end{pmatrix} = \begin{pmatrix} Cs(x) & 0 \\ 0 & Cv(x) \end{pmatrix} \begin{pmatrix} \text{gcd}(x) \\ \text{gcd}(x) \end{pmatrix}.$$

Since there is only one variable, the roots of $\text{gcd}(x)$, the GCD of these two polynomials $\{s(x), v(x)\}$ are the roots of the polynomial system. This important fact is because any one-variable polynomial can be factored over $\mathbb{C}$ into linear pieces. The roots of $\text{gcd}(x) = 2+3x+x^2 = 0$ are in Fig. 1.2.

Fig. 1.2 Common roots of polynomials

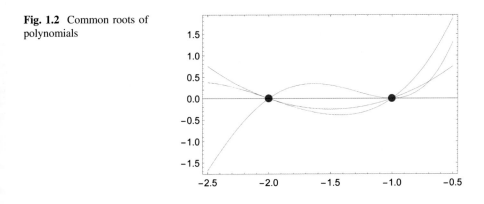

However, we normally have polynomials of two variables (x, y)

$\Rightarrow p$

$\Leftarrow -1 + xy$

$\Rightarrow g$

$\Leftarrow -4 + x^2 + y^2$

In case of more than one variable, the greatest common divisor, though it exists, does not play the role it did in the previous paragraph. That role is filed by the *Gröbner basis* (Buchberger and Winkler 1998).

$\Rightarrow \{G1, G2\} = \text{GroebnerBasis}[\{p, g\}, \{x, y\}]$

$\Leftarrow \{1 - 4y^2 + y^4, x - 4y + y^3\}$

As in the case of univariate polynomial, for the two variables (x, y), the original system can be expressed as the linear combination of the polynomials of the Gröbner basis $\{G1(y), G2(x, y)\}$, where the coefficients are also polynomials. The coefficients are the remainders.

$\Rightarrow \{c1, r1\} = \text{PolynomialReduce}[p, \{G1, G2\}, \{x, y\}]$

$\Leftarrow \{\{-1, y\}, 0\}$

Then $p(x, y)$ can be expressed as

$\Rightarrow \{-1, y\}. \begin{pmatrix} G1 \\ G2 \end{pmatrix} // \text{Simplify}$

$\Leftarrow \{-1 + x\,y\}$

and

$\Rightarrow \{c2, r2\} = \text{PolynomialReduce}[g, \{G1, G2\}, \{x, y\}]$

$\Leftarrow \{\{-4 + y^2, x + 4y - y^3\}, 0\}$

then $g(x, y)$

$\Rightarrow \{-4 + y^2, x + 4y - y^3\}. \begin{pmatrix} G1 \\ G2 \end{pmatrix} // \text{Simplify}$

$\Leftarrow \{-4 + x^2 + y^2\}$

In matrix form

$$\begin{pmatrix} p(x,y) \\ g(x,y) \end{pmatrix} = \begin{pmatrix} -1 & y \\ -4+y^2 & x+4y-y^3 \end{pmatrix} \begin{pmatrix} G1(y) \\ G2(x,y) \end{pmatrix}$$

or

$$\begin{pmatrix} p(x,y) \\ g(x,y) \end{pmatrix} = \begin{pmatrix} -1 \\ -4+y^2 \end{pmatrix} G1(y) + \begin{pmatrix} y \\ x+4y-y^3 \end{pmatrix} G2(x,y).$$

The roots of the system $\{p(x,y), g(x,y)\}$ are the same as the roots of the system $\{G1(y), G2(x,y)\}$. Note that this basis consists of special polynomials, since $G1(y)$ is a univariate polynomial!

Generally speaking, the original polynomial system $\{p(x,y), g(x,y)\}$ can be expressed as a linear combination of the basis polynomials $\{G1(x,y), G2(x,y)\}$.

There are many other basis polynomials too and the set of these basis polynomials is called the *ideal* of the original polynomial. However, the Gröbner basis is a special basis, since one of its polynomials is a univariate one. If the Gröbner basis is 1, the polynomials have no common divisor, consequently they have no common roots.

Remark The theory of Gröbner bases is much more extensive and sophisticated than we can go into here. Our focus is on using Gröbner bases to eliminate variables.

Let us employ the built-in function for the system {P, S, G} considered in previous Sect. 1.2.2,

$\Rightarrow$ GroebnerBasis[{P, S, G}, {x, y, z}]
$\Leftarrow \{-27z - 12z^2 - 2z^3 - 12z^4 + 6z^5 - 3z^6 - 6z^7 + 2z^9, 3y + z - z^3, 3x + 2z + z^3\}$

where the first element of the Gröbner basis is the same provided by the Dixon resultant.

Now, let us compute the Gröbner basis of the following system

$\Rightarrow$ U = $x^2 + y^2 == 1$
$\Leftarrow x^2 + y^2 - 1$
$\Rightarrow$ V = $x^2 + y^2 == 2$
$\Leftarrow x^2 + y^2 - 2$
$\Rightarrow$ GroebnerBasis[{U, V}, {x, y}]
$\Leftarrow \{1\}$

There are no common roots, see Fig. 1.3, however the upper limit of the number of roots is $2 \times 2 = 4$.

Fig. 1.3 No common roots of
polynomials

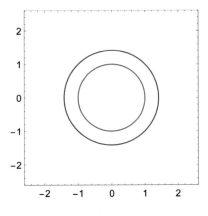

1.3.2   Reduced Gröbner Basis

The *Mathematica* built in function can carry out the elimination process too,
employing the so called reduced Gröbner Basis.

To get the univariate polynomial of x, we should eliminate y and z,

$$\Rightarrow \texttt{grbx} = \texttt{GroebnerBasis}[\{\texttt{P,S,G}\}, \{\texttt{x}\}, \{\texttt{y,z}\}]$$
$$\Leftarrow \{-27x + 18x^2 - 342x^3 + 306x^4 - 186x^5 + 229x^6$$
$$-18x^7 + 12x^8 + 8x^9\}$$

and then for the other variables,

$$\Rightarrow \texttt{grby} = \texttt{GroebnerBasis}[\{\texttt{P,S,G}\}, \{\texttt{y}\}, \{\texttt{x,z}\}]$$
$$\Leftarrow \{-21y - 23y^2 - 30y^3 - 36y^4 - 9y^5 + 6y^6 - 12y^7 + 8y^8\}$$
$$\Rightarrow \texttt{grbz} = \texttt{GroebnerBasis}[\{\texttt{P,S,G}\}, \{\texttt{z}\}, \{\texttt{x,y}\}]$$
$$\Leftarrow \{-27z - 12z^2 - 2z^3 - 12z^4 + 6z^5 - 3z^6 - 6z^7 + 2z^9\}$$

These algebraic methods are very impressive and useful, but they are limited by
the size of the system. In general, systems with more than ten unknowns cannot be
solved this way due to time and space (RAM) limitations.

1.3.3   Polynomials with Inexact Coefficients

Computing Gröbner bases with inexact coefficients is often desired in industrial
applications, but the computation with floating-point numbers is quite unstable if
performed naively (Sasaki 2014). The solution methods of the Gröbner basis are
very sensitive to round off error, therefore sometimes in case of systems that are

over-constrained or have roots with multiplicities, and are given with inexact coefficients, using hybrid symbolic-numeric methods are required (Szanto 2011).

Lichblau (2013) discussed computation of Gröbner bases using approximate arithmetic for coefficients and showed how certain considerations of tolerance, corresponding roughly to accuracy and precision from numeric computation, allow us to obtain good approximate solutions to problems that are overdetermined.

Let us consider the following polynomial system,

$$\Rightarrow \text{polys} = -4 + x^2 - 1.49071xy + y^2, -8 + x^2 - 0.4xz + z^2,$$
$$-4 + t^2 - 0.894427tx + x^2, -4 + y^2 - 1.49071yz + z^2,$$
$$-8 + t^2 - 0.666667ty + y^2, -4 + t^2 - 0.894427tz + z^2;$$

If we try to find the Gröbner basis, we get the trivial answer $\{1.\}$, which means there is no relationship between the polynomials.

$$\Rightarrow \text{sol} = \text{GroebnerBasis}[\text{polys}, \ \{x, y, z, t\}]$$
$$\Leftarrow \{1.\}$$

Even employing rationalization of the coefficients will not solve the problem,

$$\Rightarrow n = 10;$$
$$\Rightarrow \text{polysR} = \text{Map}[\text{Rationalize}[\#, 10^{-n}]\&, \text{polys}]$$
$$\Leftarrow -4 + x^2 - (149071xy)/100000 + y^2, -8 + x^2 - (2xz)/5 + z^2,$$
$$-4 + t^2 - (216200tx)/241719 + x^2, -4 + y^2 - (149071yz)/100000 + z^2,$$
$$-8 + t^2 - (666503ty)/999754 + y^2, -4 + t^2 - (216200tz)/241719 + z^2$$
$$\Rightarrow \text{solR} = \text{GroebnerBasis}[\text{polysR}, \{x, y, z, t\}]$$
$$\Leftarrow \{1.\}$$

However, applying an *approximate hybrid technique*,

$$\Rightarrow \text{solA} = \text{GroebnerBasis}[\text{polys}, x, y, z, t, \text{Tolerance} \rightarrow 10^{(-3)}]$$

yields

$$\Rightarrow \text{solt} = \text{NSolve}[\text{solA}[[1]], t]$$
$$\Rightarrow \ \{\{t \rightarrow -1.00002 - 0.0044912 \ i\}, \{t \rightarrow -1.00002 + 0.0044912 \ i\},$$
$$\{t \rightarrow 1.00002 - 0.0044912 \ i\}, \{t \rightarrow 1.00002 + 0.0044912 \ i\}\}$$

Since we are interested in real solutions,

$$\Rightarrow \text{Map}[\text{Re}[\#[[2]]]\&, \text{Flatten}[\text{solt}]]$$
$$\Leftarrow \{-1.00002, -1.00002, 1.00002, 1.00002\}$$

Then the other variables are

$$\Rightarrow \texttt{solz} = \texttt{NSolve}[\texttt{solA}[[2]]/.\texttt{t} \rightarrow 1.00002, \texttt{z}]$$
$$\Leftarrow \{\{\texttt{z} \rightarrow 2.2361\}\}$$

$$\Rightarrow \texttt{soly} = \texttt{NSolve}[\texttt{solA}[[3]]/.\texttt{t} \rightarrow 1.00002, \texttt{y}]$$
$$\Leftarrow \{\{\texttt{y} \rightarrow 3.00005\}\}$$

$$\Rightarrow \texttt{solx} = \texttt{NSolve}[\texttt{solA}[[4]]/.\texttt{t} \rightarrow 1.00002, \texttt{x}]$$
$$\Leftarrow \{\{\texttt{x} \rightarrow 2.23606\}\}$$

Let us check our result via least squares technique employing global minimization. Our objective function is

$$\Rightarrow \texttt{G} = \texttt{Total}[\texttt{Map}[\#^2\&, \texttt{polys}]]$$
$$\Leftarrow \quad (-4 + t^2 - 0.894427tx + x^2)^2 + (-8 + t^2 - 0.666667ty + y^2)^2$$
$$+ (-4 + x^2 - 1.49071xy + y^2)^2 + (-4 + t^2 - 0.894427tz + z^2)^2$$
$$+ (-8 + x^2 - 0.4xz + z^2)^2 + (-4 + y^2 - 1.49071yz + z^2)^2$$

and

$$\Rightarrow \texttt{NMinimize}[\texttt{G}, \{\texttt{x}, \texttt{y}, \texttt{z}, \texttt{t}\}]$$
$$\Leftarrow 2.10012 \times 10^{-10}, \texttt{x} \rightarrow 2.23607, \texttt{y} \rightarrow 3., \texttt{z} \rightarrow 2.23607, \texttt{t} \rightarrow 1.0023$$

1.4   Using Dixon-EDF for Symbolic Solution of Polynomial Systems

We have discussed the basic idea of a system of polynomial equations in the Introduction. Earlier in this chapter we introduced the ideas of resultants and Gröbner bases and did some examples. In this section we will show some much more difficult problems that reveal the great power of the Dixon resultant as extended with "Early Detection of Factors", or Dixon-EDF.

As before, we have in each case n equations in n variables $x_1, x_2, ..., x_n$ and some parameters. We assume that the system is neither over- nor underdetermined. Usually $3 < n < 15$, though we can work with more variables if the system is sparse enough and does not involve variables with high exponent. In most examples from actual applications, one rarely sees an exponent larger than 2.

Again, by "solve the system" we mean we have eliminated all but one of the variables. We are left with one equation in one variable and the parameters.

If desired, numerical values for the parameters can then be substituted, and the variable obtained by one-variable numerical solvers.

The ideas in this section were developed by Lewis (2008, 2017).

1.4.1   Explanation of Dixon-EDF

The basic idea of the Dixon method is to construct a square matrix M whose determinant D is a multiple of the resultant. Usually M is not unique, it is obtained as a maximal minor, in a larger matrix we shall call M^+, and there are usually many maximal minors—any one of which will do. The entries in M are polynomials in parameters. The factors of D that are not the resultant are called the *spurious factors*, and their product is sometimes referred to as the *spurious factor*.

The naive way to proceed is to compute D, factor it, and separate the spurious factor from the actual resultant. But there are problems. On the one hand, the determinant may be so large as for it to be impractical or even impossible to compute, even though the resultant is relatively small; the spurious factor is huge. On the other hand, the determinant may be so large that factoring it is impractical.

Lewis developed three heuristic methods to overcome these problems. The first may be used on any polynomial system. It uses known factors of D to compute other factors. The second also may be used on any polynomial system and it discovers factors of D so that the complete determinant is never produced. The third applies only when the resultant appears as a factor of D in certain exponential patterns. These methods were discovered by experimentation and may apply to other resultant formulations, such as the Macaulay.

This method is to exploit the observed fact that D has many factors. In other words, we try to turn the existence of spurious factors to our advantage. By elementary row and column manipulations (Gaussian elimination) we discover probable factors of D and extract them from $M_0 = M$. This produces a smaller matrix M_1, still with polynomial entries, and a list of discovered numerators and denominators.

Here is very simple example.

$$M_0 = \begin{pmatrix} 9 & 2 \\ 4 & 4 \end{pmatrix} \qquad \text{numerators:} \qquad \text{denominators:}$$

Suppose we wish to keep the arithmetic very simple, and never work with numbers bigger than 9. We factor a 2 out of the second column, then a 2 from the second row. Thus:

$$M_0 = \begin{pmatrix} 9 & 1 \\ 2 & 1 \end{pmatrix} \qquad \text{numerators: 2, 2} \qquad \text{denominators:}$$

We change the second row by subtracting 2/9 of the first:

$$M_0 = \begin{pmatrix} 9 & 1 \\ 0 & 7/9 \end{pmatrix} \qquad \text{numerators: } 2, 2 \qquad \text{denominators:}$$

We pull out the denominator 9 from the second row, and factor out 9 from the first column:

$$M_0 = \begin{pmatrix} 1 & 1 \\ 0 & 7 \end{pmatrix} \qquad \text{numerators: } 2, 2, 9 \qquad \text{denominators: } 9$$

We "clean up" by dividing out the common factor of 9 from the numerator and denominator lists; any 1 that occurs may be erased and the list compacted. Since the first column is canonically simple, we are finished with one step of the algorithm, and have produced a one-smaller M_1 for the next step.

$$M_1 = (7) \qquad \text{numerators: } 2, 2 \qquad \text{denominators: } 1$$

The algorithm terminates by pulling out the 7:

$$\text{numerators: } 2, 2, 7 \qquad \text{denominators: } 1$$

As expected (since the original matrix contained only integers) the denominator list is empty. The product of all the entries in the numerator list is the determinant, but we never needed to deal with any number larger than 9.

The accelerated Dixon resultant by the Early Discovery Factors (Dixon EDF) algorithm was suggested and implemented in the computer algebra system *Fermat* by Lewis (2008).

The Dixon resultant is a very attractive tool for solving system of multivariate polynomial geodetic equations (see Paláncz et al. 2008). Comparing it to other multipolynomial resultant like Strumfelds's method it has advantages of (i) its small size of the Dixon matrix, (ii) faster computational speed, (iii) being robust.

In the following sections we provide some examples where Dixon EDF method proved to be very effective.

1.4.2 Distance from a Point to a Standard Ellipsoid

Given an ellipsoid $x^2/a^2 + y^2/b^2 + z^2/c^2 - 1 = 0$ and a point (u, v, w), compute the point (x, y, z) on the ellipsoid closest to the point. We have three variables x, y, z. We derive equations using partial derivatives, so we must add two more variables to stand for $\partial z/\partial x$, $\partial z/\partial y$. There are six parameters a, b, c, u, v, w. The new variables representing $\partial z/\partial x$, $\partial z/\partial y$ are artifacts. We don't care about them. We want to

Fig. 1.4 Given u, v, w find x, y, z

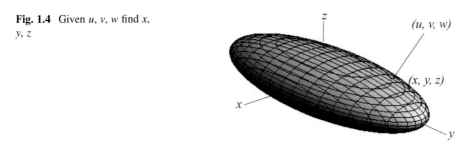

know just x, y, z. One advantage of resultants is that you can't tell a Gröbner basis algorithm not to bother with some of the variables (Fig. 1.4).

This is a fairly easy problem. The resultant is degree 6 in x.

With Dixon: 0.038 s, 22 MB RAM, with Magma: 1 s, 100 MB. (Similar results were obtained with Maple and Mathematica.)

But we can say more. The coefficient of x^6 is

$$b^2c^2 - 2abc^2 + a^2c^2 - 2ab^2c + 4a^2bc - 2a^3c + a^2b^2 - 2a^3b + a^4.$$

This factors into $(a - c)^2(a - b)^2$, so we learn that if $b = a$ or $c = a$ there is a simpler solution. In fact, if $c = a$ it drops to degree 4. As we pointed out in the Introduction, the symbolic method leads to insight!

1.4.3 Distance from a Point to Any 3D Conic

Here is the image for a general ellipsoid, but we could have any 3D conic (Fig. 1.5).
Given

$$ax^2 + by^2 + cz^2 + d\,xy + e\,xz + f\,yz + gx + hy + iz + j = 0$$

Fig. 1.5 Given u, v, w find x, y, z

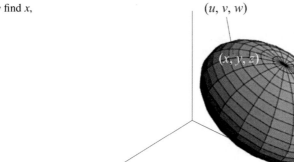

and point (u, v, w), compute point (x, y, z) with shortest distance. We have again three variables x, y, z, but now 13 parameters $a, b, c, \ldots, u, v, w$. At least one artifact variable must be added.

This problem is much harder than the previous but Dixon-EDF handles it well. The coefficient of x^6 now has two factors, one is $af^2 - def + be^2 + cd^2 - 4abc$. If this were 0, the resultant simplifies.

1.4.4   Pose Estimation

Suppose we have a quadrilateral $ABCE$; it does not have to be planar. The distances between each pair of vertices are known. The object moves. We observe it from point P, noting the angles spanned by each pair of vertices. The classic four point pose problem is to deduce the distances X_1, X_2, X_3, X_4 (Fig. 1.6).

It is easy to derive six equations from the law of cosines:

$$X_1^2 + X_2^2 - X_1 X_2\, r - |AB|^2$$
$$X_1^2 + X_3^2 - X_1 X_3\, q - |AC|^2$$
$$X_2^2 + X_3^2 - X_2 X_3\, p - |BC|^2$$
$$X_1^2 + X_4^2 - X_1 X_4\, s - |AE|^2$$
$$X_4^2 + X_3^2 - X_4 X_3\, t - |CE|^2$$
$$X_2^2 + X_4^2 - X_2 X_4\, u - |BE|^2$$

r, p, q, s, t, u, are cosines.

Fig. 1.6 Pose estimation problem

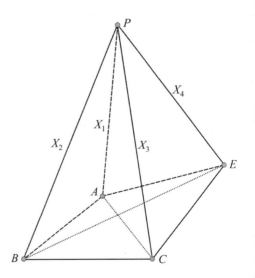

There are four variables X_1, X_2, X_3, X_4. The parameters are the lengths of AB, BC, CE, AE, AC, BE, and the six cosines.

Using any four equations but including the diagonals AC and BE yields an easy system of equations, solvable by many means. Indeed, one could select, say, the first three equations and obtain a complete three variable system, see the exercises at the end of this chapter. But suppose the object could be flexible! Then we must use only the outside edges; diagonal distances might change. This turns out to be a much harder system to solve. As far as we know, it is only solvable by Dixon-EDF.

1.4.5 How to Run Dixon-EDF

As far as we know, Dixon KSY is implemented only in *Mathematica*, no other large multipurpose CAS. It is a package that must be downloaded and installed. However, the package does not implement EDF.

Dixon-EDF is implemented in Fermat as a series of procedures.

1.5 Applications

1.5.1 Common Points of Geometrical Objects

It is well known that the visualization of curves and surfaces is easy and comfortable via parametric explicit equations of the geometrical objects. However, the implicit form of these equations is sometimes needed. For example one would like to decide whether a point is on a curve or surface or not. Finding the common points of two or more geometrical objects is the generalization of this task. Converting explicit to implicit just means eliminating the parameter.

Application 1 Let us compute the implicit equation of a 2D circle.

The form of the explicit equation with the parameter is,

$$x = \cos(\alpha),$$
$$y = \sin(\alpha)$$

and in addition we know that

$$\sin^2(\alpha) + \cos^2(\alpha) = 1.$$

Solution
Therefore, we have the following system of equations with unknowns (x, y, α)

$$x - \cos(\alpha),$$
$$y - \sin(\alpha),$$
$$-1 + \sin^2(\alpha) + \cos^2(\alpha).$$

and we should eliminate the variable α. Let us compute the Gröbner basis for x and y eliminating α

$\Rightarrow$ GroebnerBasis

$$\left[\{ x - \cos[\alpha], y - \sin[\alpha], \sin[\alpha]^2 + \cos[\alpha]^2 - 1 \}, \{x, y\}, \{\alpha, \cos[\alpha], \sin[\alpha]\} \right]$$
$$\Leftarrow \{ -1 + x^2 + y^2 \}$$

This elimination could easily be done with the Dixon resultant. Note that we really have four variables x, y, $\cos(\alpha)$, and $\sin(\alpha)$ and three equations. With three equations we can eliminate any two variables, so we choose the latter two.

Application 2 Now let us compute the common points of a cardioid and a circle. The parametric equation of the cardioid is, see Fig. 1.7,

$$x = 2 \left(1 + \cos(t) \right) \cos(t),$$
$$y = 2 \left(1 + \cos(t) \right) \sin(t).$$

Fig. 1.7 A cardioid curve

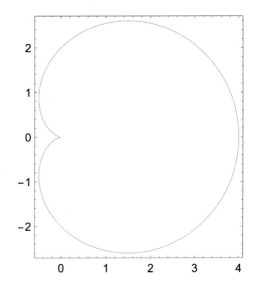

Solution
As a first step, we compute the implicit form of the equation of the cardioid.

$$x - 2\,\cos(t) - 2\,\cos^2(t),$$
$$y - 2\,\sin(t) - 2\,\cos(t)\,\sin(t),$$
$$1 + 2\,\cos^2(t) + \sin^2(t).$$

The Gröbner basis of the system is,

$$\left\{-4x^3 + x^4 - 4y^2 - 4xy^2 + 2x^2y^2 + y^4\right\}.$$

Now let us consider the following circle,

$$x^2 + y^2 - 2 = 0.$$

Then, the two geometrical objects together are as shown in Fig. 1.8.

The next step is the computation of the common points employing these implicit equations. Then the following system should be solved

$$\Rightarrow g1 = -4x^3 + x^4 - 4y^2 - 4xy^2 + 2x^2y^2 + y^4$$
$$\Leftarrow -4x^3 + x^4 - 4y^2 - 4xy^2 + 2x^2y^2 + y^4$$
$$\Rightarrow g2 = x^2 + y^2 - 2$$
$$\Leftarrow -2 + x^2 + y^2$$

Fig. 1.8 The two geometrical objects

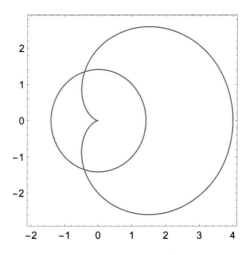

The reduced Gröbner basis for the x coordinate is given as

$\Rightarrow$ GroebnerBasis$[\{g1,g2\},\{x\},\{y\}]$

$\Leftarrow \{-1 - 2x + x^2\}$

Similarly for the y coordinate

$\Rightarrow$ GroebnerBasis$[\{g1,g2\},\{y\},\{x\}]$

$\Leftarrow \{-7 + 2y^2 + y^4\}$

or with the built-in function **Solve**

$\Rightarrow$ solp $= \{x,y\}/.$Solve$[\{g1 == 0, g2 == 0\}, \{x,y\}]//$Simplify

$\Leftarrow \left\{\left\{1-\sqrt{2}, -\sqrt{-1+2\sqrt{2}}\right\}, \left\{1-\sqrt{2}, \sqrt{-1+2\sqrt{2}}\right\}, \left\{1+\sqrt{2}, -i\sqrt{1+2\sqrt{2}}\right\}, \left\{1+\sqrt{2}, i\sqrt{1+2\sqrt{2}}\right\}\right\}$

There are only two real solutions! Let us visualize the common points, see Fig. 1.9.

To solve this problem with the Dixon resultant, just take the three equations defining the cardioid and the one defining the circle. First, eliminate the three variables y, $\cos(t)$, and $\sin(t)$. That yields the equation for x, $-1 - 2x + x^2$. Then repeat, eliminating y, $\cos(t)$, and $\sin(t)$ to get the equation for y.

Fig. 1.9 The common points of the two geometrical objects

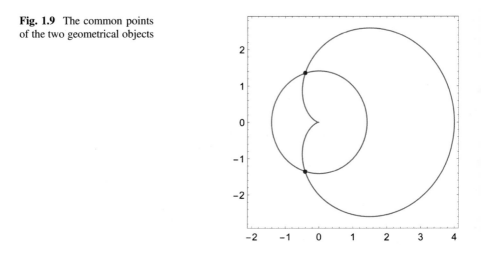

1.5.2   Nonlinear Heat Transfer

The nonlinear dimensionless equation of the steady state heat transfer in 1D is,

$$\frac{d}{dx}\left(\lambda(\theta)\frac{d\theta}{dx}\right) = 0.$$

The boundary conditions are,

$$\theta(0) = 0 \quad \text{and} \quad \theta(1) = 1.$$

The heat transfer coefficient depending on the temperature,

$$\lambda(\theta) = 1 + k\theta.$$

Let us approximate the temperature profile with the following polynomial,

$$\theta(x) = x + c_1\left(x^2 - x\right) + c_2\left(x^3 - x\right),$$

which satisfies the boundary conditions. Let us compute the c_i coefficients.

Solution
Substituting the temperature profile into the differential equation, we get

$$
\begin{aligned}
eq = {}& k + 2c_1 - 2kc_1 + 6kxc_1 + kc_1^2 - 6kxc_1^2 + 6kx^2c_1^2 - 2kc_2 + 6xc_2 + 12kx^2c_2 \\
& + 2kc_1c_2 - 6kxc_1c_2 - 12kx^2c_1c_2 + 20kx^3c_1c_2 + kc_2^2 - 12kx^2c_2^2 + 15kx^4c_2^2.
\end{aligned}
$$

Using the global integral method, the square of the integral should be minimized,

$$
\begin{aligned}
r = \int_0^1 eq^2\,dx = {}& k^2 + 4kc_1 + 2k^2c_1 + 4c_1^2 + 4kc_1^2 + 4k^2c_1^2 + \frac{1}{5}k^2c_1^4 + 6kc_2 + 4k^2c_2 \\
& + 12c_1c_2 + 20kc_1c_2 + 16k^2c_1c_2 + \frac{4}{5}k^2c_1^2c_2 + \frac{6}{5}k^2c_1^3c_2 + 12c_2^2 + 24kc_2^2 \\
& + \frac{84}{5}k^2c_2^2 + \frac{12}{5}k^2c_1c_2^2 + \frac{20}{7}k^2c_1^2c_2^2 + \frac{64}{35}k^2c_2^3 + \frac{111}{35}k^2c_1c_2^3 + \frac{48}{35}k^2c_2^4.
\end{aligned}
$$

Employing the necessary conditions of the minimum, differentiate the integral, we get an algebraic polynomial system for the unknown coefficients

$\Rightarrow$ eq1 $= D[r, c_1]$

$\Leftarrow 4k + 2k^2 + 8c_1 + 8kc_1 + 8k^2c_1 + \dfrac{4}{5}k^2c_1^3 + 12c_2 + 20kc_2 + 16k^2c_2$

$\qquad + \dfrac{8}{5}k^2c_1c_2 + \dfrac{18}{5}k^2c_1^2c_2 + \dfrac{12}{5}k^2c_2^2 + \dfrac{40}{7}k^2c_1c_2^2 + \dfrac{111}{35}k^2c_2^3$

$\Rightarrow$ eq2 $= D[r, c_2]$

$\Leftarrow 6k + 4k^2 + 12c_1 + 20kc_1 + 16k^2c_1 + \dfrac{4}{5}k^2c_1^2 + \dfrac{6}{5}k^2c_1^3 + 24c_2 + 48kc_2 + \dfrac{168k^2c_2}{5}$

$\qquad + \dfrac{24}{5}k^2c_1c_2 + \dfrac{40}{7}k^2c_1^2c_2 + \dfrac{192}{35}k^2c_2^2 + \dfrac{333}{35}k^2c_1c_2^2 + \dfrac{192}{35}k^2c_2^3$

The Gröbner basis for c_1,

$\Rightarrow$ GroebnerBasis[$\{$eq1, eq2$\}, \{c_1\}, \{c_2\}$]

$\Leftarrow \{2222640000k + 25041744000k^2 + 106983636480k^3 + 216207482400k^4 + 217869466458k^5$

$\qquad + 105383544084k^6 + 21747027960k^7 + 982690800k^8 + 4445280000c_1 + 45638208000kc_1$

$\qquad + 172774344960k^2c_1 + 305473573440k^3c_1 + 294315313236k^4c_1 + 170466205476k^5c_1$

$\qquad + 78560129424k^6c_1 + 27948563160k^7c_1 + 4880962800k^8c_1 + 4834771200k^2c_1^2$

$\qquad + 14644375200k^3c_1^2 + 3743455968k^4c_1^2 - 11828581632k^5c_1^2 - 15676868844k^6c_1^2$

$\qquad - 6328977648k^7c_1^2 - 849050160k^8c_1^2 + 398664000k^2c_1^3 - 377496000k^3c_1^3 + 1763997312k^4c_1^3$

$\qquad + 2443968240k^5c_1^3 + 2017898358k^6c_1^3 + 441062580k^7c_1^3 + 61164425k^8c_1^3 + 52698240k^4c_1^4$

$\qquad + 370528368k^5c_1^4 + 23358168k^6c_1^4 + 197374240k^7c_1^4 + 18557000k^8c_1^4 - 4040400k^4c_1^5$

$\qquad - 27938400k^5c_1^5 - 12698784k^6c_1^5 - 6985752k^7c_1^5 + 4109544k^8c_1^5 - 1465920k^6c_1^6 - 716616k^7c_1^6$

$\qquad - 872028k^8c_1^6 + 55600k^6c_1^7 + 18640k^7c_1^7 + 58568k^8c_1^7 - 3120k^8c_1^8 + 100k^8c_1^9\}$

and for c_2, we get similar polynomial.

$\Rightarrow$ GroebnerBasis[$\{$eq1, eq2$\}, \{c_2\}, \{c_1\}$]

$\Leftarrow \{-709927680k^2 - 1419855360k^3 - 904619968k^4 - 194692288k^5 + 4100908k^6 + 1419855360c_2$

$\qquad + 4259566080kc_2 + 5168334976k^2c_2 + 3237393152k^3c_2 + 1556328200k^4c_2 + 647559304k^5c_2$

$\qquad + 151166960k^6c_2 - 92198400c_2^2 - 276595200kc_2^2 + 220266368k^2c_2^2 + 901524736k^3c_2^2$

$\qquad + 983883152k^4c_2^2 + 487021584k^5c_2^2 + 107131248k^6c_2^2 + 16464000c_2^3 + 49392000kc_2^3$

$\qquad + 253787072k^2c_2^3 + 425254144k^3c_2^3 + 384371484k^4c_2^3 + 179976412k^5c_2^3 + 43090110k^6c_2^3$

$\qquad - 6679680k^2c_2^4 - 13359360k^3c_2^4 - 604072k^4c_2^4 + 6075608k^5c_2^4 + 5758480k^6c_2^4 - 1117200k^2c_2^5$

$\qquad - 2234400k^3c_2^5 + 811440k^4c_2^5 + 1928640k^5c_2^5 + 1604652k^6c_2^5 + 312480k^4c_2^6 + 312480k^5c_2^6$

$\qquad + 267470k^6c_2^6 + 16800k^4c_2^7 + 16800k^5c_2^7 + 24612k^6c_2^7 + 1560k^6c_2^8 + 75k^6c_2^9\}$

From a practical point of view, it is more convenient to employ numerical Gröbner basis function, as in *Mathematica* using $k = 1$,

$\Rightarrow$ sol $=$ NSolve[$\{$eq1, eq2$\}$/.kR1, $\{c_1, c_2\}$, Reals]//Flatten

$\Leftarrow \{c_1 \rightarrow -0.6251338312334316, c_2 \rightarrow 0.19045444692157196\}$

Then the temperature profile is

$\Rightarrow$ T $= \theta$/.sol

$\Leftarrow x - 0.625134\,(-x + x^2) + 0.190454\,(-x + x^3)$

Fig. 1.10 The dimensionless temperature profile in case of $k = 1$

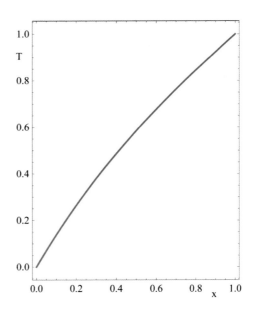

Figure 1.10 shows the dimensionless temperature profile for $k = 1$,
The general function for any $k = \kappa$ can be written as,

$$\Rightarrow \Omega[\kappa\_] := \theta /. \ (\texttt{NSolve}[\{\texttt{eq1}, \texttt{eq2}\} /. \texttt{k} \rightarrow \kappa, \{c_1, c_2\}, \texttt{Reals}] // \texttt{Flatten})$$

Let us test this function for $k = 1$

$$\Rightarrow \Omega[1]$$
$$\Leftarrow x - 0.625134(-x + x^2) + 0.190454(-x + x^3)$$

We utilized the common capability of the Computer Algebra System (CAS) type language providing symbolic computation as well as any size of digits in order to reduce round-off error.

One can realize that this example is a nice illustration of the hybrid computation, since our function is computed partly in numerical and partly in symbolic way.

1.5.3 Helmert Transformation

Let us consider a 2D Helmert transformation with parameters α and β,

$$\begin{pmatrix} X \\ Y \end{pmatrix} = s \begin{pmatrix} \cos(\Omega) & -\sin(\Omega) \\ \sin(\Omega) & \cos(\Omega) \end{pmatrix} \begin{pmatrix} x \\ y \end{pmatrix} = \begin{pmatrix} \alpha & -\beta \\ \beta & \alpha \end{pmatrix} \begin{pmatrix} x \\ y \end{pmatrix}.$$

We have three control points in both systems, namely (Table 1.1).

Table 1.1 Numerical data for the 2D Helmert transformation problem	i	x_i	y_i	X_i	Y_i
	1	0.0	1.0	− 2.1	1.1
	2	1.0	0.0	1.0	2.0
	3	1.0	1.0	− 0.9	2.8

Assuming that these values have errors in both systems (EIV model) let us consider the adjustments as Δx_i and ΔX_i $i = 1, 2, 3$.

In order compute these adjustments the following minimization problem should be solved,

$$F = \sum_{i=1}^{3} \left(\Delta x_i^2 + \Delta X_i^2 \right)$$

with the constraints,

$$eq_1 = \alpha(x_1 + \Delta x_1) - \beta y_1 - (X_1 + \Delta X_1),$$

$$eq_2 = \beta(x_1 + \Delta x_1) + \alpha y_1 - Y_1,$$

$$eq_3 = \alpha(x_2 + \Delta x_2) - \beta y_2 - (X_2 + \Delta X_2),$$

$$eq_4 = \beta(x_2 + \Delta x_2) + \alpha y_2 - Y_2,$$

$$eq_5 = \alpha(x_3 + \Delta x_3) - \beta y_3 - (X_3 + \Delta X_3),$$

$$eq_6 = \beta(x_3 + \Delta x_3) + \alpha y_3 - Y_3.$$

To transform this problem into a minimization without constraints, let us employ Lagrange- multipliers,

$$G = F + \sum_{i=1}^{6} \lambda_i \, eq_i = \Delta x_1^2 + \Delta x_2^2 + \Delta x_3^2 + \Delta X_1^2 + \Delta X_2^2 + \Delta X_3^2$$

$$+ (-X_1 - \beta y_1 + \alpha(x_1 + \Delta x_1) - \Delta X_1)\lambda_1 + (\alpha y_1 - Y_1 + \beta(x_1 + \Delta x_1))\lambda_2$$

$$+ (-X_2 - \beta y_2 + \alpha(x_2 + \Delta x_2) - \Delta X_2)\lambda_3 + (\alpha y_2 - Y_2 + \beta(x_2 + \Delta x_2))\lambda_4$$

$$+ (-X_3 - \beta y_3 + \alpha(x_3 + \Delta x_3) - \Delta X_3)\lambda_5 + (\alpha y_3 - Y_3 + \beta(x_3 + \Delta x_3))\lambda_6.$$

Using the necessary condition, after differentiating the objective, we get the following algebraic polynomial system for the unknowns $\{\Delta x_1, \Delta X_1, \Delta x_2, \Delta X_2, \Delta x_3, \Delta X_3, \alpha, \beta, \lambda_1, \lambda_2, \lambda_3, \lambda_4, \lambda_5, \lambda_6\}$

$$2\Delta x_1 + \alpha\lambda_1 + \beta\lambda_2,$$

$$2\Delta X_1 - \lambda_1,$$

$$2\Delta x_2 + \alpha\lambda_3 + \beta\lambda_4,$$

$$2\Delta X_2 - \lambda_3,$$

$$2\Delta x_3 + \alpha\lambda_5 + \beta\lambda_6,$$

$$2\Delta X_3 - \lambda_5,$$

$$x_1\lambda_1 + \Delta x_1\lambda_1 + y_1\lambda_2 + x_2\lambda_3 + \Delta x_2\lambda_3 + y_2\lambda_4 + x_3\lambda_5 + \Delta x_3\lambda_5 + y_3\lambda_6,$$

$$-y_1\lambda_1 + x_1\lambda_2 + \Delta x_1\lambda_2 - y_2\lambda_3 + x_2\lambda_4 + \Delta x_2\lambda_4 - y_3\lambda_5 + x_3\lambda_6 + \Delta x_3\lambda_6,$$

$$\alpha x_1 - X_1 - \beta y_1 + \alpha\Delta x_1 - \Delta X_1,$$

$$\beta x_1 + \alpha y_1 - Y_1 + \beta\Delta x_1,$$

$$\alpha x_2 - X_2 - \beta y_2 + \alpha\Delta x_2 - \Delta X_2,$$

$$\beta x_2 + \alpha y_2 - Y_2 + \beta\Delta x_2,$$

$$\alpha x_3 - X_3 - \beta y_3 + \alpha\Delta x_3 - \Delta X_3,$$

$$\beta x_3 + \alpha y_3 - Y_3 + \beta\Delta x_3.$$

Substituting the numerical values for $\{x_i, y_i\}$ and $\{X_i, Y_i\}$ from Table 1.1, we get 14 polynomials of the Gröbner basis of the problem. The rows of Table 1.2 show the exponents of the unknown variables in the different polynomials,

The first base is a degree six polynomial for λ_6 ,

$$- 7630949955162482528767108340$$
$$+ 4295983922788968266779304813\lambda_6 - 489181086373271123938589713\lambda_6^2$$
$$+ 10461095486070027991388157780\lambda_6^3 + 104018749323715741160794050000\lambda_6^4$$
$$- 382929968026648328876789062\lambda_6^5 + 349089071788949996689453125\lambda_6^6$$

Which has two real solutions: $\{\lambda_6 \to -2.14502, \quad \lambda_6 \to 0.238268\}$. We consider the positive solution. (The reason will be given later.) Then the solutions can be obtained with successive elimination from the other bases,

$$\{\Delta x_1 \to 0.021889, \quad \Delta X_1 \to 0.165307, \quad \Delta x_2 \to 0.021537, \quad \Delta X_2 \to 0.079912,$$
$$\Delta x_3 \to -0.109804, \quad \Delta X_3 \to -0.116768, \quad \alpha \to 1.057144, \quad \beta \to 1.957834\}.$$

Table 1.2 Numerical data for the 2D Helmert transformation problem

Δx_1	ΔX_1	Δx_2	ΔX_2	Δx_3	ΔX_3	α	β	λ_1	λ_2	λ_3	λ_4	λ_5	λ_6
0	0	0	0	0	0	0	0	0	0	0	0	0	6
0	0	0	0	0	0	0	0	0	0	0	0	1	5
0	0	0	0	0	0	0	0	0	0	0	1	0	5
0	0	0	0	0	0	0	0	0	0	1	0	0	5
0	0	0	0	0	0	0	0	0	1	0	0	0	5
0	0	0	0	0	0	0	0	1	0	0	0	0	5
0	0	0	0	0	0	0	1	0	0	0	0	0	5
0	0	0	0	0	0	1	0	0	0	0	0	0	5
0	0	0	0	0	1	0	0	0	0	0	0	0	5
0	0	0	0	1	0	0	0	0	0	0	0	0	5
0	0	0	1	0	0	0	0	0	0	0	0	0	5
0	0	1	0	0	0	0	0	0	0	0	0	0	5
0	1	0	0	0	0	0	0	0	0	0	0	0	5
1	0	0	0	0	0	0	0	0	0	0	0	0	5

We could solve the problem via direct minimization, too. Employing global minimization method, we can get the same solution. The minimization problem has two local minimums and negative λ_6 refers to the other local minimum, which is not the global one.

1.6   Exercises

1.6.1   Solving a System with Different Techniques

Let us consider the following system,

$$f(x, y, z) = x\,y\,z - 1,$$

$$g(x, y, z) = x^2 + 2y^2 + 4z^2 - 7,$$

$$h(x, y, z) = 2x^2 + y^3 + 6z - 7.$$

We do not know approximate solutions, therefore we have no idea which initial values would be proper to start with in case of numerical (iterative) solutions.

Problem

(a) Estimate the number of common roots
(b) Find common roots via *Sylvester* resultant

(c) Find common roots via *Dixon* resultant
(d) Find the univariate polynomials for the unknowns (x, y, z) via *Gröbner basis*
(e) Compute the roots of these polynomials
(f) Carry out the computation with built-in function **NSolve**
(g) Employ high precision computation

Solution
Considering the degree of the polynomials of the system, the total degree of the system is

$$d = 3 \times 2 \times 3 = 18,$$

Therefore the upper limit of the number of the common roots is 18.

Solution via *Sylvester resultant*

$\Rightarrow f = x\,y\,z - 1$
$\Leftarrow -1 + x\,y\,z$
$\Rightarrow g = x^2 + 2y^2 + 4z^2 - 7$
$\Leftarrow -7 + x^2 + 2y^2 + 4z^2$
$\Rightarrow h = 2x^2 + y^3 + 6z - 7$
$\Leftarrow -7 + 2x^2 + y^3 + 6z$

Eliminate z from $f(x, y, z)$ and $g(x, y, z)$

$\Rightarrow \mathtt{fg} = \mathtt{Resultant}[\mathtt{f, g, z}]$
$\Leftarrow 4 - 7x^2y^2 + x^4y^2 + 2x^2y^4$

Eliminate z from $g(x, y, z)$ and $h(x, y, z)$

$\Rightarrow \mathtt{gh} = \mathtt{Resultant}[\mathtt{g, h, z}]$
$\Leftarrow -56 - 76x^2 + 16x^4 + 72y^2 - 56y^3 + 16x^2y^3 + 4y^6$

Eliminate y from $fg(x, y)$ and $gh(x, y)$

$\Rightarrow \mathtt{fggh} = \mathtt{Resultant}[\mathtt{fg, gh, y}]$
$\Leftarrow 1048576 - 37748736x^2 + 1241776128x^4 - 5052301312x^6$
$\quad + 7358103552x^8 + 11476934656x^{10} - 13391571968x^{12}$
$\quad - 6389601280x^{14} + 5032881408x^{16} - 314205696x^{18} + 369083136x^{20}$
$\quad - 382296064x^{22} + 92680960x^{24} - 3911168x^{26} - 927488x^{28} + 51200x^{30} + 4096x^{32}$

The real roots

$\Rightarrow \mathtt{solx} = \mathtt{NSolve}[\mathtt{fggh} == 0, \mathtt{x, Reals}]//\mathtt{Simplify}$
$\Leftarrow \{\{x \to -1.754\}, \{x \to -1.\}, \{x \to 1.\}, \{x \to 1.754\}\}$

Solution Dixon resultant (Mathematica)

$\Rightarrow$ dr = DixonResultant[{f, g, h}, {y, z}, {Y, Z}]

$\Leftarrow -256\,(64x - 768x^2 + 3456x^3 - 4560x^4 - 696x^5 + 4032x^6 - 2534x^7$
$\quad + 5760x^8 + 1203x^9 - 2256x^{10} + 1409x^{11} + 192x^{12} - 531x^{13} + 25x^{15} + 4x^{17})$

$\Rightarrow$ solx = NSolve[dr[[2]] == 0, x, Reals]//Simplify

$\Leftarrow \{\{x \rightarrow -1.754\}, \{x \rightarrow -1.\}, \{x \rightarrow 0\}\}$

Employing reduced Gröbner basis for *x*

$\Rightarrow$ grx = GroebnerBasis[{f, g, h}, {x, y, z}, {y, z}]

$\Leftarrow \{64 - 768x + 3456x^2 - 4560x^3 - 696x^4 + 4032x^5 - 2534x^6$
$\quad + 5760x^7 + 1203x^8 - 2256x^9 + 1409x^{10} + 192x^{11} - 531x^{12} + 25x^{14} + 4x^{16}\}$

$\Rightarrow$ solx = NSolve[grx == 0, x, Reals]//Simplify

$\Leftarrow \{\{x \rightarrow -1.754\}, \{x \rightarrow -1.\}\}$

Employing built-in function **NSolve** for the system. The number of solutions

$\Rightarrow$ solT = Length[NSolve[{f, g, h}, {x, y, z}]]

$\Leftarrow 16$

The real solutions

$\Rightarrow$ solR = NSolve[{f, g, h}, {x, y, z}, Reals]

$\Leftarrow \{\{x \rightarrow -1.754, y \rightarrow -1.24075, z \rightarrow 0.4595\}, \{x \rightarrow -1., y \rightarrow -1., z \rightarrow 1.\}\}$

Checking the solutions

$\Rightarrow$ Map[{f, g, h}/.#&, solR]

$\Leftarrow \{\{-1.40998 \times 10^{-14}, 1.62981 \times 10^{-13}, 3.61933 \times 10^{-14}\},$
$\quad \{7.99361 \times 10^{-15}, 7.72715 \times 10^{-14}, 2.04281 \times 10^{-14}\}\}$

Employing higher (any) precision

$\Rightarrow$ solR30 = NSolve[{f, g, h}, {x, y, z}, Reals, WorkingPrecisionR30]

$\Leftarrow \{\{x \rightarrow -1.75400475361904621847786941109,$

$\quad y \rightarrow -1.24074650135092991106121800126,$

$\quad z \rightarrow 0.459500697239036497067167646364\},$

$\quad \{x \rightarrow -1.000000000000000000000000000000,$

$\quad y \rightarrow -1.000000000000000000000000000000,$

$\quad z \rightarrow 1.000000000000000000000000000000\}\}$

and checking the solution via back substitution,

⇒ Map[{f,g,h}/.#&,solR30]

⇐ {{0. × 10^{-30}, 0. × 10^{-29}, 0. × 10^{-29}}, {0. × 10^{-30}, 0. × 10^{-29}, 0. × 10^{-29}}}

1.6.2 Planar Ranging

Problem Let us assume that the distances ($t1$, $t2$) of an unknown point (xu, yu) from two different locations ($x1$, $y1$) and ($x2$, $y2$) are known, to compute the unknown coordinates.

Solution
The distance equations for the two known points are,

⇒ eq1 = $(x1 - xu)^2 + (y1 - yu)^2 - t1^2$;

⇒ eq2 = $(x2 - xu)^2 + (y2 - yu)^2 - t2^2$;

Let the coordinate values of the two points be,

⇒ xm = {48177.62, 49600.15};

⇒ ym = {6531.28, 7185.19};

and the measured distances be,

⇒ tm = {611.023, 1529.482};

Applying the *Sylvester resultant*, we eliminate *yu*

⇒ Eqxu = Resultant[eq1, eq2, yu]

⇐ $t1^4 - 2t1^2 t2^2 + t2^4 - 2t1^2 x1^2 + 2t2^2 x1^2 + x1^4 + 2t1^2 x2^2$
$\quad - 2t2^2 x2^2 - 2x1^2 x2^2 + x2^4 + 4t1^2 x1\, xu - 4t2^2 x1\, xu - 4x1^3 xu$
$\quad - 4t1^2 x2\, xu + 4t2^2 x2\, xu + 4x1^2 x2\, xu + 4x1\, x2^2 xu - 4x2^3 xu$
$\quad + 4x1^2 xu^2 - 8x1\, x2\, xu^2 + 4x2^2 xu^2 - 2t1^2 y1^2 - 2t2^2 y1^2 + 2x1^2 y1^2$
$\quad + 2x2^2 y1^2 - 4x1\, xu\, y1^2 - 4x2\, xu\, y1^2 + 4xu^2 y1^2 + y1^4 + 4t1^2 y1\, y2$
$\quad + 4t2^2 y1\, y2 - 4x1^2 y1\, y2 - 4x2^2 y1\, y2 + 8x1\, xu\, y1\, y2 + 8x2\, xu\, y1\, y2$
$\quad - 8xu^2 y1\, y2 - 4y1^3 y2 - 2t1^2 y2^2 - 2t2^2 y2^2 + 2x1^2 y2^2 + 2x2^2 y2^2$
$\quad - 4x1\, xu\, y2^2 - 4x2\, xu\, y2^2 + 4xu^2 y2^2 + 6y1^2 y2^2 - 4y1\, y2^3 + y2^4$

and similarly

$$\Rightarrow \text{Eqyu} = \text{Resultant}[\text{eq1}, \text{eq2}, \text{xu}]$$
$$\Leftarrow \text{t1}^4 - 2\text{t1}^2\text{t2}^2 + \text{t2}^4 - 2\text{t1}^2\text{x1}^2 - 2\text{t2}^2\text{x1}^2 + \text{x1}^4 + 4\text{t1}^2\text{x1}\,\text{x2}$$
$$+ 4\text{t2}^2\text{x1}\,\text{x2} - 4\text{x1}^3\text{x2} - 2\text{t1}^2\text{x2}^2 - 2\text{t2}^2\text{x2}^2 + 6\text{x1}^2\text{x2}^2 - 4\text{x1}\,\text{x2}^3$$
$$+ \text{x2}^4 - 2\text{t1}^2\text{y1}^2 + 2\text{t2}^2\text{y1}^2 + 2\text{x1}^2\text{y1}^2 - 4\text{x1}\,\text{x2}\,\text{y1}^2 + 2\text{x2}^2\text{y1}^2$$
$$+ \text{y1}^4 + 2\text{t1}^2\text{y2}^2 - 2\text{t2}^2\text{y2}^2 + 2\text{x1}^2\text{y2}^2 - 4\text{x1}\,\text{x2}\,\text{y2}^2 + 2\text{x2}^2\text{y2}^2$$
$$- 2\text{y1}^2\text{y2}^2 + \text{y2}^4 + 4\text{t1}^2\text{y1}\,\text{yu} - 4\text{t2}^2\text{y1}\,\text{yu} - 4\text{x1}^2\text{y1}\,\text{yu} + 8\text{x1}\,\text{x2}\,\text{y1}\,\text{yu}$$
$$- 4\text{x2}^2\text{y1}\,\text{yu} - 4\text{y1}^3\text{yu} - 4\text{t1}^2\text{y2}\,\text{yu} + 4\text{t2}^2\text{y2}\,\text{yu} - 4\text{x1}^2\text{y2}\,\text{yu}$$
$$+ 8\text{x1}\,\text{x2}\,\text{y2}\,\text{yu} - 4\text{x2}^2\text{y2}\,\text{yu} + 4\text{y1}^2\text{y2}\,\text{yu} + 4\text{y1}\,\text{y2}^2\text{yu} - 4\text{y2}^3\text{yu}$$
$$+ 4\text{x1}^2\text{yu}^2 - 8\text{x1}\,\text{x2}\,\text{yu}^2 + 4\text{x2}^2\text{yu}^2 + 4\text{y1}^2\text{yu}^2 - 8\text{y1}\,\text{y2}\,\text{yu}^2 + 4\text{y2}^2\text{yu}^2$$

then solving the two univariate polynomial equations separately,

$$\Rightarrow \text{NRoots}[(\text{Eqxu}/.\{\text{x1} \rightarrow \text{xm}[[1]], \text{x2} \rightarrow \text{xm}[[2]], \text{y1} \rightarrow \text{ym}[[1]],$$
$$\text{y2} \rightarrow \text{ym}[[2]], \text{t1} \rightarrow \text{tm}[[1]], \text{t2} \rightarrow \text{tm}[[2]]\}) == 0, \text{xu}]$$
$$\Leftarrow \text{xu} == 48071.6 || \text{xu} == 48565.3$$
$$\Rightarrow \text{NRoots}[(\text{Eqyu}/.\{\text{x1} \rightarrow \text{xm}[[1]], \text{x2} \rightarrow \text{xm}[[2]], \text{y1} \rightarrow \text{ym}[[1]],$$
$$\text{y2} \rightarrow \text{ym}[[2]], \text{t1} \rightarrow \text{tm}[[1]], \text{t2} \rightarrow \text{tm}[[2]]\}) == 0, \text{yu}]$$
$$\Leftarrow \text{yu} == 6058.98 || \text{yu} == 7133.03$$

1.6.3 3D Resection

In case of 3D resection problem, the distances between the points P_i are known $S_{i,j}$ as well as angles $\varphi_{i,j}$ and the length of the edges, S_i should be computed.

Problem The means practically, we should solve the *Grunert equations*,

$$S_{1,2}^2 = S_1^2 + S_2^2 - 2S_1S_2 \cos(\varphi_{1,2}),$$

$$S_{2,3}^2 = S_2^2 + S_3^2 - 2S_2S_3 \cos(\varphi_{2,3}),$$

$$S_{3,1}^2 = S_3^2 + S_1^2 - 2S_3S_1 \cos(\varphi_{3,1}),$$

where the unknowns are s_1, s_2, s_3, see Fig. 1.11.

In case of for a regular tetrahedron $\varphi_{i,j} = \pi/3$, and $2\cos(\pi/3) = 1$, and $S_{i,j} = d$. Introducing $x_i = S_i$, our system is,

$$\Rightarrow \text{p1} = \text{x1}^2 - \text{x1x2} + \text{x2}^2 - \text{d} == 0;$$
$$\Rightarrow \text{p2} = \text{x2}^2 - \text{x2x3} + \text{x3}^2 - \text{d} == 0;$$
$$\Rightarrow \text{p3} = \text{x3}^2 - \text{x1x3} + \text{x1}^2 - \text{d} == 0;$$

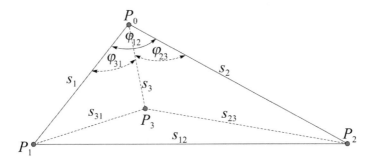

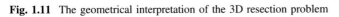

Fig. 1.11 The geometrical interpretation of the 3D resection problem

Let us solve this problem in symbolic way employing Dixon resultant.

Solution

$\Rightarrow$ Clear[X1, X2, X3]

$\Rightarrow$ DixonResultant[{p1[[1]], p2[[1]], p3[[1]]}, {x2, x3}, {X2, X3}]

$\Leftarrow 4(d^3x1^2 - 3d^2x1^4 + 3dx1^6 - x1^8)$

and solve the univariate polynomial equation,

$\Rightarrow$ Solve[% == 0, x1]

$\Leftarrow \{\{x1 \rightarrow 0\}, \{x1 \rightarrow 0\}, \{x1 \rightarrow -\sqrt{d}\}, \{x1 \rightarrow -\sqrt{d}\},$
$\qquad \{x1 \rightarrow -\sqrt{d}\}, \{x1 \rightarrow \sqrt{d}\}, \{x1 \rightarrow \sqrt{d}\}, \{x1 \rightarrow \sqrt{d}\}\}$

Then we eliminate the variables $(x1, x3)$

$\Rightarrow$ DixonResultant[{p1[[1]], p2[[1]], p3[[1]]}, {x1, x3}, {X1, X3}]

$\Leftarrow -4(d^3x2^2 - 3d^2x2^4 + 3dx2^6 - x2^8)$

$\Rightarrow$ Solve[% == 0, x2]

$\Leftarrow \{\{x2 \rightarrow 0\}, \{x2 \rightarrow 0\}, \{x2 \rightarrow -\sqrt{d}\}, \{x2 \rightarrow -\sqrt{d}\},$
$\qquad \{x2 \rightarrow -\sqrt{d}\}, \{x2 \rightarrow \sqrt{d}\}, \{x2 \rightarrow \sqrt{d}\}, \{x2 \rightarrow \sqrt{d}\}\}$

and then let us eliminate the variables $(x1, x2)$

$\Rightarrow$ DixonResultant[{p1[[1]], p2[[1]], p3[[1]]}, {x1, x2}, {X1, X2}]

$\Leftarrow -4(d^3x3^2 - 3d^2x3^4 + 3dx3^6 - x3^8)$

$\Rightarrow$ Solve[% == 0, x3]

$\Leftarrow \{\{x3 \rightarrow 0\}, \{x3 \rightarrow 0\}, \{x3 \rightarrow -\sqrt{d}\}, \{x3 \rightarrow -\sqrt{d}\},$
$\qquad \{x3 \rightarrow -\sqrt{d}\}, \{x3 \rightarrow \sqrt{d}\}, \{x3 \rightarrow \sqrt{d}\}, \{x3 \rightarrow \sqrt{d}\}\}$

Only positive solutions are accepted.

1.6.4 Pose Estimation

Now, we consider a similar but considerably more difficult problem discussed in Sect. 1.4.4. Suppose we have a quadrilateral ABCE; it does not have to be planar. The distances between each pair of vertices are known. The object moves. We observe it from point P, noting the angles spanned by each pair of vertices. The classic four point pose problem is to deduce the distances $X1$, $X2$, $X3$, $X4$.

$\Rightarrow$ eq1 $= X1^2 + X2^2 - X1\,X2\,r - AB$
$\Leftarrow -AB + X1^2 - r\,X1\,X2 + X2^2$

$\Rightarrow$ eq2 $= X1^2 + X3^2 - X1\,X3\,q - AC$
$\Leftarrow -AC + X1^2 - q\,X1\,X3 + X3^2$

$\Rightarrow$ eq3 $= X2^2 + X3^2 - X3\,X2\,p - BC$
$\Leftarrow -BC + X2^2 - p\,X2\,X3 + X3^2$

$\Rightarrow$ eq4 $= X1^2 + X4^2 - X1\,X4\,s - AE$
$\Leftarrow -AE + X1^2 - s\,X1\,X4 + X4^2$

$\Rightarrow$ eq5 $= X4^2 + X3^2 - X4\,X3\,t - CE$
$\Leftarrow -CE + X3^2 - t\,X3\,X4 + X4^2$

$\Rightarrow$ eq6 $= X2^2 + X4^2 - X4\,X2\,u - BE$
$\Leftarrow -BE + X2^2 - u\,X2\,X4 + X4^2$

There are four variables $X1$, $X2$, $X3$, $X4$. The parameters are AB, BC, CE, AE, AC, BE. An overdetermined system results from applying the law of cosines to each triangle having vertex P. Using four equations including the diagonals AC and BE gives an easy system of equations, solvable by many means.

For example in order to reduce the problem let us consider the equations eq3, eq5 and eq6 containing variables ($X2$, $X3$, $X4$). The *Gröbner basis* is,

$\Rightarrow$ AbsoluteTiming[sol = GroebnerBasis[{eq3, eq5, eq6}, {X2, X3, X4}];]
$\Leftarrow$ {2.48481, Null}

The number of the polynomials is

$\Rightarrow$ Length[sol]
$\Leftarrow$ 31

⇒ Map[Exponent[#, {X2, X3, X4}]&, sol]//TableForm

⇐   0   0   8
      0   1   7
      0   1   6
      0   1   7
      0   1   7
      0   1   5
      0   1   7
      0   1   7
      0   1   7
      0   1   7
      0   1   7
      0   1   7
      0   1   7
      0   1   4
      0   1   6
      0   1   6
      0   1   4
      0   1   7
      0   1   7
      0   2   2
      1   1   3
      1   1   3
      1   1   3
      1   1   5
      1   1   2
      1   1   3
      1   1   4
      1   1   3
      1   1   3
      1   1   4
      2   0   2

therefore the problem is solvable, since considering the first polynomials we get an univariate polynomial of degree eight.

$\Rightarrow$ Sol$[[1]]$

$\Leftarrow BC^4 - 4BC^3BE + 6BC^2BE^2 - 4BCBE^3 + BE^4 - 4BC^3CE + 12BC^2BECE - 12BCBE^2CE + 4BE^3CE + 6BC^2CE^2 - 12BCBECE^2$
$+ 6BE^2CE^2 - 4BCCE^3 + 4BECE^3 + CE^4 - 2BC^2BECEp^2 + 4BCBE^2CEp^2 - 2BE^3CEp^2 + 4BCBECE^2p^2 - 4BE^2CE^2p^2$
$- 2BECE^3p^2 + BE^2CE^2p^4 + 8BC^3X4^2 - 24BC^2BEX4^2 + 24BCBE^2X4^2 - 8BE^3X4^2 - 24BC^2CEX4^2 + 48BCBECEX4^2$
$- 24BE^2CEX4^2 + 24BCCE^2X4^2 - 24BECE^2X4^2 - 8CE^3X4^2 + 2BC^2BEp^2X4^2 - 4BCBE^2p^2X4^2 + 2BE^3p^2X4^2 +$
$2BC^2CEp^2X4^2 - 16BCBECEp^2X4^2 + 14BE^2CEp^2X4^2 - 4BCCE^2p^2X4^2 + 14BECE^2p^2X4^2 + 2CE^3p^2X4^2 -$
$2BE^2CEp^4X4^2 - 2BECE^2p^4X4^2 - 2BC^3t^2X4^2 + 6BC^2BEt^2X4^2 - 6BCBE^2t^2X4^2 + 2BE^3t^2X4^2 + 4BC^2CEt^2X4^2 -$
$8BCBECEt^2X4^2 + 4BE^2CEt^2X4^2 - 2BCCE^2t^2X4^2 + 2BECE^2t^2X4^2 - BC^2BEp^2t^2X4^2 + 2BCBE^2p^2t^2X4^2 - BE^3p^2t^2X4^2 -$
$BECE^2p^2t^2X4^2 + BC^3ptuX4^2 - BC^2BEptuX4^2 - BCBE^2ptuX4^2 + BE^3ptuX4^2 - BC^2CEptuX4^2 +$
$6BCBECEptuX4^2 - 5BE^2CEptuX4^2 - BCCE^2ptuX4^2 - 5BECE^2ptuX4^2 + CE^3ptuX4^2 + BCBECEp^3tuX4^2$
$+ BE^2CEp^3tuX4^2 + BECE^2p^3tuX4^2 - 2BC^3u^2X4^2 + 4BC^2BEu^2X4^2 - 2BCBE^2u^2X4^2 + 6BC^2CEu^2X4^2$
$- 8BCBECEu^2X4^2 - 6BCCE^2u^2X4^2 + 2BE^2CEu^2X4^2 + 4BECE^2u^2X4^2 + 2CE^3u^2X4^2 - BC^2CEp^2u^2X4^2 - BE^2CEp^2u^2X4^2$
$+ 2BCCE^2p^2u^2X4^2 - CE^3p^2u^2X4^2 + 24BC^2X4^4 - 48BCBEX4^4 + 24BE^2X4^4 - 48BCCEX4^4 + 48BECEX4^4 + 24CE^2X4^4 -$
$2BC^2p^2X4^4 + 12BCBEp^2X4^4 - 10BE^2p^2X4^4 + 12BCCEp^2X4^4 - 28BECEp^2X4^4 - 10CE^2p^2X4^4 + BE^2p^4X4^4$
$+ 4BECEp^4X4^4 + CE^2p^4X4^4 - 10BC^2t^2X4^4 + 20BCBEt^2X4^4 - 10BE^2t^2X4^4 + 12BCCEt^2X4^4 - 12BECEt^2X4^4$
$- 2CE^2t^2X4^4 + BC^2p^2t^2X4^4 - 4BCBEp^2t^2X4^4 + 3BE^2p^2t^2X4^4 + 2BECEp^2t^2X4^4 + CE^2p^2t^2X4^4 + BC^2t^4X4^4$
$- 2BCBEt^4X4^4 + BE^2t^4X4^4 + 2BC^2ptuX4^4 - 4BCBEptuX4^4 + 2BE^2ptuX4^4 - 4BCCEptuX4^4 + 20BECEptuX4^4 +$
$2CE^2ptuX4^4 - BCBEp^3tuX4^4 - BE^2p^3tuX4^4 - BCCEp^3tuX4^4 - 4BECEp^3tuX4^4 - CE^2p^3tuX4^4 - BC^2pt^3uX4^4$
$+ 2BCBEpt^3uX4^4 - BE^2pt^3uX4^4 - BCCEpt^3uX4^4 - BECEpt^3uX4^4 - 10BC^2u^2X4^4 + 12BCBEu^2X4^4 - 2BE^2u^2X4^4$
$+ 20BCCEu^2X4^4 - 12BECEu^2X4^4 - 10CE^2u^2X4^4 + BC^2p^2u^2X4^4 + BE^2p^2u^2X4^4 - 4BCCEp^2u^2X4^4$
$+ 2BECEp^2u^2X4^4 + 3CE^2p^2u^2X4^4 + 3BC^2t^2u^2X4^4 - 4BCBEt^2u^2X4^4 + BE^2t^2u^2X4^4 - 4BCCEt^2u^2X4^4 +$
$CE^2t^2u^2X4^4 + BCBEp^2t^2u^2X4^4 + BCCEp^2t^2u^2X4^4 + BECEp^2t^2u^2X4^4 - BC^2ptu^3X4^4 - BCBEptu^3X4^4$
$+ 2BCCEptu^3X4^4 - BECEptu^3X4^4 - CE^2ptu^3X4^4 + BC^2u^4X4^4 - 2BCCEu^4X4^4 + CE^2u^4X4^4 + 32BCX4^6$
$- 32BEX4^6 - 32CEX4^6 - 8BCp^2X4^6 + 16BEp^2X4^6 + 16CEp^2X4^6 - 2BEp^4X4^6 - 2CEp^4X4^6 - 16BCt^2X4^6$
$+ 16BEt^2X4^6 + 8CEt^2X4^6 + 2BCp^2t^2X4^6 - 4BEp^2t^2X4^6 - 2CEp^2t^2X4^6 + 2BCt^4X4^6 - 2BEt^4X4^6 + 4BCptuX4^6$
$- 12BEptuX4^6 - 12CEptuX4^6 + BCp^3tuX4^6 + 3BEp^3tuX4^6 + 3CEp^3tuX4^6 - BCpt^3uX4^6 + 3BEpt^3uX4^6$
$+ CEpt^3uX4^6 - 16BCu^2X4^6 + 8BEu^2X4^6 + 16CEu^2X4^6 + 2BCp^2u^2X4^6 - 2BEp^2u^2X4^6 - 4CEp^2u^2X4^6$
$+ 8BCt^2u^2X4^6 - 2BEt^2u^2X4^6 - 2CEt^2u^2X4^6 - 2BCp^2t^2u^2X4^6 - BEp^2t^2u^2X4^6 - CEp^2t^2u^2X4^6 - BCt^4u^2X4^6$
$- BCptu^3X4^6 + BEptu^3X4^6 + 3CEptu^3X4^6 + BCpt^3u^3X4^6 + 2BCu^4X4^6 - 2CEu^4X4^6 - BCt^2u^4X4^6 + 16X4^8$
$- 8p^2X4^8 + p^4X4^8 - 8t^2X4^8 + 2p^2t^2X4^8 + t^4X4^8 + 8ptuX4^8 - 2p^3tuX4^8 - 2pt^3uX4^8 - 8u^2X4^8 + 2p^2u^2X4^8$
$+ 2t^2u^2X4^8 + p^2t^2u^2X4^8 - 2ptu^3X4^8 + u^4X4^8$

or using *Dixon KSY*

$\Rightarrow$ < <Resultant$'$Dixon$'$

$\Leftarrow$ Clear$[x2, x3]$

$\Rightarrow$AbsoluteTiming$[$soldix $=$

DixonMatrix$[\{$eq3, eq5, eq6$\}, \{X2, X3\}, \{x2, x3\}]//$Simplify; $]$

$\Leftarrow \{0.021031, \text{Null}\}$

$\Rightarrow$ Dimensions$[$soldix$]$

$\Leftarrow \{6, 6\}$

$\Rightarrow$ MatrixRank$[$soldix$]$

$\Leftarrow 5$

The matrix is not full rank therefore the maximal minor is used,

$\Rightarrow$AbsoluteTiming[solX4 =

    Apply[Plus, Flatten[Minors[soldix, 5]]]//Simplify;]

$\Leftarrow \{1.00794, \text{Null}\}$

$\Rightarrow$ solX4

$\Leftarrow \mathrm{p}(\mathrm{tX4}(\mathrm{tu}^2\mathrm{X4}^3(-\mathrm{BE}+\mathrm{X4}^2)(-\mathrm{CE}+\mathrm{X4}^2)$
$-\mathrm{X4}(\mathrm{BCt}-\mathrm{BEt}+\mathrm{CEpu}+\mathrm{tX4}^2-\mathrm{puX4}^2)(-(-\mathrm{BC}+\mathrm{BE}+\mathrm{CE}-2\mathrm{X4}^2)^2+\mathrm{t}^2\mathrm{X4}^2(\mathrm{BC}-\mathrm{BE}+\mathrm{X4}^2))$
$-(\mathrm{CEp}+(-\mathrm{p}+\mathrm{tu})\mathrm{X4}^2)(-\mathrm{BCt}^2\mathrm{uX4}^3+\mathrm{uX4}(\mathrm{BC}-\mathrm{CE}+\mathrm{X4}^2)(\mathrm{BC}-\mathrm{BE}-\mathrm{CE}+2\mathrm{X4}^2)))$
$+\mathrm{uX4}(\mathrm{X4}(\mathrm{BE}-\mathrm{X4}^2)(\mathrm{BCt}-\mathrm{BEt}+\mathrm{CEpu}+\mathrm{tX4}^2-\mathrm{puX4}^2)(-\mathrm{BCp}+\mathrm{BEp}+\mathrm{CEp}-2\mathrm{pX4}^2+\mathrm{pt}^2\mathrm{X4}^2+\mathrm{tuX4}^2)$
$+(\mathrm{BC}-\mathrm{BE}-\mathrm{CE}+2\mathrm{X4}^2-\mathrm{ptuX4}^2)(-\mathrm{BCt}^2\mathrm{uX4}^3+\mathrm{uX4}(\mathrm{BC}-\mathrm{CE}+\mathrm{X4}^2)(\mathrm{BC}-\mathrm{BE}-\mathrm{CE}+2\mathrm{X4}^2))$
$+\mathrm{u}^2\mathrm{X4}^3(-(\mathrm{BC}-\mathrm{CE}+\mathrm{X4}^2)(\mathrm{BEpt}+\mathrm{BCu}-\mathrm{CEu}-\mathrm{ptX4}^2+\mathrm{uX4}^2)+\mathrm{BCt}(\mathrm{BEp}+(-\mathrm{p}+\mathrm{tu})\mathrm{X4}^2)))$
$+(\mathrm{BC}-\mathrm{BE}-\mathrm{CE}+2\mathrm{X4}^2)(-(\mathrm{BE}-\mathrm{X4}^2)(-\mathrm{BCp}+\mathrm{BEp}+\mathrm{CEp}-2\mathrm{pX4}^2+\mathrm{pt}^2\mathrm{X4}^2+\mathrm{tuX4}^2)(\mathrm{CEp}+(-\mathrm{p}+\mathrm{tu})\mathrm{X4}^2)$
$+(\mathrm{BC}-\mathrm{BE}-\mathrm{CE}+2\mathrm{X4}^2-\mathrm{ptuX4}^2)(-(-\mathrm{BC}+\mathrm{BE}+\mathrm{CE}-2\mathrm{X4}^2)^2+\mathrm{t}^2\mathrm{X4}^2(\mathrm{BC}-\mathrm{BE}+\mathrm{X4}^2))$
$-\mathrm{uX4}(\mathrm{X4}(\mathrm{BC}-\mathrm{BE}-\mathrm{CE}+2\mathrm{X4}^2)(-\mathrm{BEpt}-\mathrm{BCu}+\mathrm{CEu}+\mathrm{ptX4}^2-\mathrm{uX4}^2)$
$+\mathrm{tX4}(\mathrm{BC}-\mathrm{BE}+\mathrm{X4}^2)(\mathrm{BEp}+(-\mathrm{p}+\mathrm{tu})\mathrm{X4}^2)))+\mathrm{pX4}(\mathrm{tuX4}(-\mathrm{BE}+\mathrm{X4}^2)(-\mathrm{CE}+\mathrm{X4}^2)$
$(-\mathrm{BC}+\mathrm{BE}+\mathrm{CE}-2\mathrm{X4}^2+\mathrm{ptuX4}^2)-\mathrm{uX4}(\mathrm{CEp}+(-\mathrm{p}+\mathrm{tu})\mathrm{X4}^2)(-(\mathrm{BC}-\mathrm{CE}+\mathrm{X4}^2)$
$(\mathrm{BEpt}+\mathrm{BCu}-\mathrm{CEu}-\mathrm{ptX4}^2+\mathrm{uX4}^2)+\mathrm{BCt}(\mathrm{BEp}+(-\mathrm{p}+\mathrm{tu})\mathrm{X4}^2))+(\mathrm{BCt}-\mathrm{BEt}+\mathrm{CEpu}+\mathrm{tX4}^2-\mathrm{puX4}^2)$
$(\mathrm{X4}(\mathrm{BC}-\mathrm{BE}-\mathrm{CE}+2\mathrm{X4}^2)(-\mathrm{BEpt}-\mathrm{BCu}+\mathrm{CEu}+\mathrm{ptX4}^2-\mathrm{uX4}^2)+\mathrm{tX4}(\mathrm{BC}-\mathrm{BE}+\mathrm{X4}^2)(\mathrm{BEp}+(-\mathrm{p}+\mathrm{tu})\mathrm{X4}^2)))$
$+\mathrm{p}(\mathrm{CE}-\mathrm{X4}^2)(\mathrm{BE}^3\mathrm{p}-\mathrm{CE}^2(\mathrm{p}+\mathrm{tu}-\mathrm{pu}^2)\mathrm{X4}^2-\mathrm{CE}(\mathrm{p}^3-2\mathrm{tu}-\mathrm{p}^2\mathrm{tu}+\mathrm{p}(-4+\mathrm{t}^2+2\mathrm{u}^2))\mathrm{X4}^4$
$+\mathrm{p}(-4+\mathrm{p}^2+\mathrm{t}^2-\mathrm{ptu}+\mathrm{u}^2)\mathrm{X4}^6+\mathrm{BE}^2(-\mathrm{CEp}(-2+\mathrm{p}^2)+(-5\mathrm{p}+\mathrm{p}^3+\mathrm{tu})\mathrm{X4}^2)+\mathrm{BC}^2(\mathrm{BEp}-(\mathrm{p}+\mathrm{tu}-\mathrm{pu}^2)\mathrm{X4}^2)$
$+\mathrm{BE}(\mathrm{CE}^2\mathrm{p}+\mathrm{CEp}(-6+2\mathrm{p}^2+\mathrm{t}^2-\mathrm{ptu})\mathrm{X4}^2+(8\mathrm{p}-2\mathrm{p}^3-\mathrm{pt}^2-2\mathrm{tu}+\mathrm{p}^2\mathrm{tu})\mathrm{X4}^4)$
$+\mathrm{BC}(-2\mathrm{BE}^2\mathrm{p}+2\mathrm{CE}(\mathrm{p}+\mathrm{tu}-\mathrm{pu}^2)\mathrm{X4}^2+(\mathrm{p}^2\mathrm{tu}+\mathrm{t}(-2+\mathrm{t}^2)\mathrm{u}-\mathrm{p}(4+(-2+\mathrm{t}^2)\mathrm{u}^2))\mathrm{X4}^4$
$-\mathrm{BEp}(2\mathrm{CE}+(-6+\mathrm{ptu})\mathrm{X4}^2)))+\mathrm{ptX4}(-\mathrm{X4}(\mathrm{BE}-\mathrm{X4}^2)(\mathrm{CE}^2\mathrm{pt}+\mathrm{CE}^2\mathrm{u}+\mathrm{BECE}(2+\mathrm{p}^2)\mathrm{u}+\mathrm{BC}^2(-\mathrm{pt}+\mathrm{u})$
$+\mathrm{BE}^2(-\mathrm{pt}+\mathrm{u})-2\mathrm{CEptX4}^2-4\mathrm{CEuX4}^2-\mathrm{CEp}^2\mathrm{uX4}^2+\mathrm{CEt}^2\mathrm{uX4}^2$
$+\mathrm{CEu}^3\mathrm{X4}^2-\mathrm{BE}(-2\mathrm{pt}+4\mathrm{u}+\mathrm{p}^2\mathrm{u})\mathrm{X4}^2+4\mathrm{uX4}^4+\mathrm{p}^2\mathrm{uX4}^4-\mathrm{t}^2\mathrm{uX4}^4-\mathrm{u}^3\mathrm{X4}^4+\mathrm{BC}(2\mathrm{BE}(\mathrm{pt}-\mathrm{u})-2\mathrm{CEu}$
$+(-\mathrm{u}(-4+\mathrm{u}^2)+\mathrm{pt}(-2+\mathrm{u}^2))\mathrm{X4}^2))+\mathrm{uX4}(\mathrm{ptuX4}^2(\mathrm{BE}-\mathrm{X4}^2)(-\mathrm{CE}+\mathrm{X4}^2)$
$+(\mathrm{BC}-\mathrm{BE}-\mathrm{CE}+2\mathrm{X4}^2)(-(-\mathrm{BC}+\mathrm{BE}+\mathrm{CE}-2\mathrm{X4}^2)^2+\mathrm{t}^2\mathrm{X4}^2(\mathrm{BC}-\mathrm{BE}+\mathrm{X4}^2))$
$+\mathrm{uX4}(-\mathrm{BCt}^2\mathrm{uX4}^3+\mathrm{uX4}(\mathrm{BC}-\mathrm{CE}+\mathrm{X4}^2)(\mathrm{BC}-\mathrm{BE}-\mathrm{CE}+2\mathrm{X4}^2)))))$

$\Rightarrow$ Exponent[solX4, {X2, X3, X4}]

$\Leftarrow \{0, 0, 8\}$

References

Awange JL, Paláncz B (2016) Geospatial algebraic computations, theory and applications. Springer, Berlin, Heidelberg

Buchberger B, Winkler F (1998) Groebner bases and applications. London mathematical society lecture note series 251. Cambridge University Press, Cambridge

Dickenstein A, Emiris IZ (eds) (2005) Solving polynomial equations, foundations, algorithms and applications (algorithms and computation in mathematics), 14. Springer

Kapur D, Saxena T, Yang L (1994) Algebraic and geometric reasoning using Dixon resultants. In: Proceedings of the international symposium on symbolic and algebraic computation. A.C.M. Press

Lewis RH. Computer algebra system fermat. http://home.bway.net/lewis/

Lewis RH (2008) Heuristics to accelerate the Dixon resultant. Math Comput Simul 77(4):400–407

Lewis RH (2017) Dixon-EDF: the premier method for solution of parametric polynomial systems. In: Kotsireas IS, Martinez-Moro E (eds) Applications of computer algebra, Kalamata, Greece, July 20–23, 2015. Springer proceedings in mathematics & statistics, vol 198

Lichtblau D (2013) Approximate Gröbner bases, overdetermined polynomial systems and approximate GCDs. Hindawi Publishing Corporation, ISRN computational mathematics, vol 2013, Article ID 352806, pp 1–12. https://doi.org/10.11155/2013/352806

Paláncz B, Zaletnyik P, Awange JL, Grafarend EW (2008) Dixon resultant's solution of systems of geodetic polynomial equations. J Geodesy 82(8):505–511

Sasaki T (2014) A practical method for floating-point Gröbner basis computation. In: Feng R, Lee W, Sato Y (eds) Computer mathematics. Springer, Berlin, Heidelberg, pp 109–124

Szanto A (2011) Hybrid symbolic-numeric methods for the solution of polynomial systems: tutorial overview. In: Proceeding ISSAC'11 proceedings of the 36th international symposium on symbolic and algebraic computation, San Jose, California, USA, June 08–11, 2011 ACM New York, NY, USA ©2011, pp 9–10

Chapter 2
Homotopy Solution of Nonlinear Systems

2.1 The Concept of Homotopy

The continuous deformation of an object to another object is known as *homotopy*. Let us consider a simple geometric example that defines the homotopy between a circle and a square. The parametric equations of the circle (Fig. 2.1) is given by,

$$x = R\cos(\alpha),$$
$$y = R\sin(\alpha).$$

while the parametric equations of the square are (Fig. 2.2).

$$x = f(\alpha)R\cos(\alpha),$$
$$y = f(\alpha)R\sin(\alpha),$$

where

$$f(\alpha) = \frac{1}{\max(|\sin(\alpha)|, |\cos(\alpha)|)}.$$

Now, the homotopy function is given by

$$H(\alpha, \lambda) = \lambda \begin{pmatrix} R\cos(\alpha) \\ R\sin(\alpha) \end{pmatrix} + (1 - \lambda) \begin{pmatrix} f(\alpha)R\cos(\alpha) \\ f(\alpha)R\sin(\alpha) \end{pmatrix}.$$

In geometrical terms, the homotopy H provides us a continuous, smooth deformation from a *square*—which is obtained for $\lambda = 0$ by $H(\alpha, 0)$—to a circle—which is obtained for $\lambda = 1$ by $H(\alpha, 1)$. One moment of the animation of the

Supplementary Information The online version contains supplementary material available at https://doi.org/10.1007/978-3-030-92495-9_2.

Fig. 2.1 A circle ($R = 1$)

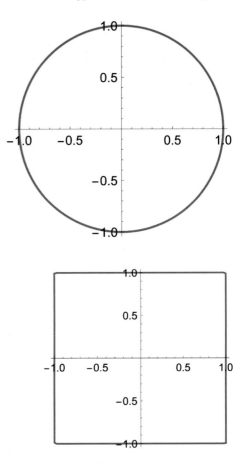

Fig. 2.2 A square

homotopy can be seen in Fig. 2.3 for $\lambda \in [0, 1]$. We call it linear homotopy because H is a linear function of the variable λ.

Lines are also geometric objects. Let us consider two Bezier splines with four control points each,

$\Rightarrow \mathtt{pts1} = \{\{0, -1\}, \{2, 1\}, \{4, 2\}, \{6, 2\}\};$
$\Rightarrow \mathtt{pts2} = \{\{2, -1\}, \{3, 1\}, \{4, -1\}, \{6, 0\}\};$

and a homotopy between them (see Fig. 2.4),

$\Rightarrow \mathtt{BezierCurve}[(1-\lambda)\mathtt{pts1} + \lambda\,\mathtt{pts2}]$

Fig. 2.3 Homotopy between
a square and a circle

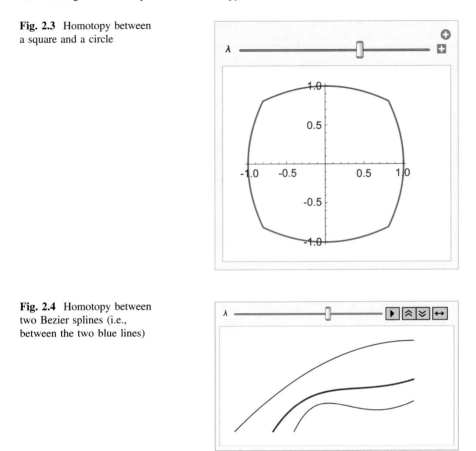

Fig. 2.4 Homotopy between
two Bezier splines (i.e.,
between the two blue lines)

2.2 Solving Nonlinear Equation via Homotopy

A homotopy continuation method deforms the known roots of the *start system* into
the roots of the *target system* (Kotsireas 2001). Now, let us look at how homotopy
can be used to solve a simple polynomial equation. Consider the polynomial
equation of degree two,

$$q(x) = x^2 + 8x - 9 = 0.$$

By deleting the middle term, we can get a simpler equation, which can be solved
easily by inspection,

$$p(x) = x^2 - 9 = 0.$$

This equation also has two roots and will be considered the start system for the target system. The linear homotopy can be defined as follows

$$H(x, \lambda) = (1 - \lambda)p(x) + \lambda q(x) = -9 + x^2 + 8x\lambda.$$

Let us plot the homotopy H for the polynomials $p(x)$ and $q(x)$ for different values of λ (Fig. 2.5).

The homotopy continuation method deforms $p(x) = 0$, the known roots of the start system, into $q(x) = 0$, the roots of the target system. Let us solve the equation $H(x, \lambda) = 0$ for different values of λ. First let $\lambda_1 = 0.2$, and consider $x_0 = 3$, one of the solutions of $p(x) = 0$, as the initial guess value. Solving $H(x, \lambda_1) = 0$ using the Newton–Raphson method, we get

$$\Rightarrow \texttt{x}_0 = 3; \lambda_1 = 0.2; \texttt{x}_1 = \texttt{x/. FindRoot}[\texttt{H(x,}\lambda_1) == 0, (\texttt{x}, \texttt{x}_0)]$$
$$\Leftarrow 2.30483$$

Now use this result as the guess value for the next solution step

$$\Rightarrow \lambda_2 = 0.4; \texttt{x}_2 = \texttt{x/. FindRoot}[\texttt{H(x,}\lambda_2) == 0, (\texttt{x}, \texttt{x}_1)]$$
$$\Leftarrow 1.8$$

and so on,

$$\Rightarrow \lambda_3 = 0.6; \texttt{x}_3 = \texttt{x/. FindRoot}[\texttt{H(x,}\lambda_3) == 0, (\texttt{x}, \texttt{x}_2)]$$
$$\Leftarrow 1.44187$$
$$\Rightarrow \lambda_4 = 0.8; \texttt{x}_4 = \texttt{x/. FindRoot}[\texttt{H(x,} \lambda_4) == 0, (\texttt{x}, \texttt{x}_3)]$$
$$\Leftarrow 1.18634$$
$$\Rightarrow \lambda_5 = 1; \texttt{x}_5 = \texttt{x/. FindRoot}[\texttt{H(x,}\lambda_5) == 0, (\texttt{x}, \texttt{x}_4)]$$
$$\Leftarrow 1.$$

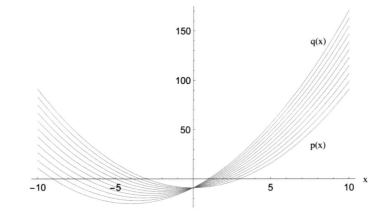

Fig. 2.5 Deformation of the function H from $p(\text{x})$ to $q(\text{x})$ as function of parameter λ

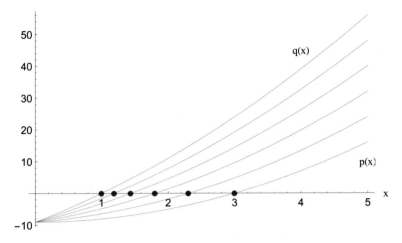

Fig. 2.6 The transition of the root from $x = 3$ to $x = 1$ during the deformation of the function H

Let us display the transition of a root of the polynomial $p(x)$ into a root of the polynomial $q(x)$ (Fig. 2.6),

The homotopy path is the function $x = x(\lambda)$, and x_i is the root of $H(x, \lambda_i)$. Figure 2.7 below shows the path of homotopy transition of the root of $p(x)$ into the root of $q(x)$,

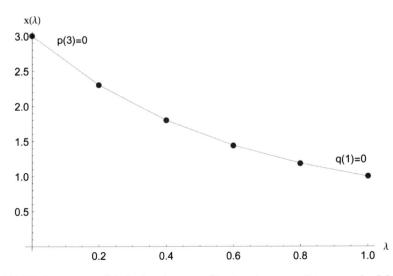

Fig. 2.7 The homotopy path is the function $x = x(\lambda)$, where in every point, at every lambda value $H \equiv 0$

2.3   Tracing Homotopy Path as Initial Value Problem

Comparing homotopy solution with the traditional Newton–Raphson solution, it is clear that if $\Delta\lambda$ is small enough, the convergence may be ensured in every step. Consequently one can consider the root tracing procedure with an infinitesimal step size as an initial value problem of an ordinary differential equation.

Since $H(x, \lambda) = 0$ for every $\lambda \in [0, 1]$, therefore

$$dH(x, \lambda) = \frac{\partial H}{\partial x}dx + \frac{\partial H}{\partial \lambda}d\lambda \equiv 0 \quad \lambda \in [0, 1].$$

Then the initial value problem is

$$H_x\frac{dx(\lambda)}{d\lambda} + H_\lambda = 0$$

with

$$x(0) = x_0.$$

It goes without saying that for systems of equations, the Jacobian should be used.

Here H_x is the Jacobian of H with respect to x_i, $i = 1,\dots, n$, in case of n nonlinear equations with n variables.

In our single variable case, the two partial derivatives of the homotopy function are

$\Rightarrow$ dHdλ$=$ D[H[x, λ], λ]

$\Leftarrow$ 8 x

$\Rightarrow$ dHdx $=$ D[H[x, λ], x]

$\Leftarrow$ 2x(1−λ) + (8 + 2x)λ

Then the right hand side of the differential equation to be solved is

$\Rightarrow$ deqrhs $= -\dfrac{\text{dHdλ}}{\text{dHdx}}/.x \to x[\lambda]$

$\Leftarrow -\dfrac{8x[1]}{2(1-1)x[1] + 1(8+2x[1])}$

The differential equation,

$\Rightarrow$ deq $=$ D[x[λ],λ] $==$ deqrhs

$\Leftarrow$ x′[1]$= -\dfrac{8x[1]}{2(1-1)x[1] + 1(8+2x[1])}$

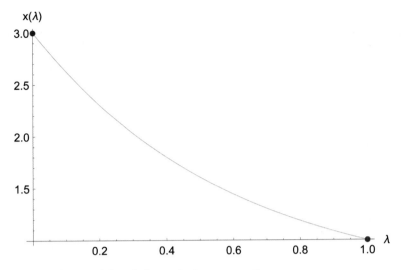

Fig. 2.8 The trajectory of the solution as the homotopy path

The initial value is

$\Rightarrow x0 = x_0$

$\Leftarrow 3$

The numerical solution of this initial value problem is

$\Rightarrow sol = NDSolve[\{deq, x[0] == x0\}, \{x[\lambda]\}, \{\lambda, 0, 1\}];$

The resulting trajectory is the homotopy path (Fig. 2.8),
The value of the corresponding root of $q(x)$ is $x(\lambda)$ at $\lambda = 1$,

$\Rightarrow First[x[\lambda]/.sol/.\lambda \rightarrow 1]$

$\Leftarrow 1.$

2.4   Types of Linear Homotopy

Here we consider the most popular five different types of the homotopy.

2.4.1   General Linear Homotopy

As we have seen, the start system can be constructed intuitively, reducing the
original system (target system) to a more simple system (start system), which roots

can be easily computed. In order to get all of the roots of the target system, the start system should have so many roots as many as the target system has.

The start system can be constructed in different ways however there are two typical techniques for generating the start systems, which are usually employed.

2.4.2 Fixed-Point Homotopy

The starting system can be considered as

$$p(x) = x - x_0,$$

where x_0 is a guess value for the root of the target system $q(x) = 0$.

In that case the homotopy function is

$$H(x, \lambda) = (1 - \lambda)(x - x_0) + \lambda q(x).$$

2.4.3 Newton Homotopy

Another construction for the start system, when we consider $p(x)$ as

$$p(x) = q(x) - q(x_0),$$

in that case the homotopy function is

$$H(x, \lambda) = (1 - \lambda)(q(x) - q(x_0)) + \lambda q(x),$$

or

$$H(x, \lambda) = q(x) - (1 - \lambda)q(x_0).$$

2.4.4 Affine Homotopy

This type of homotopy suggested by Jalali and Seader (2000) requires the first derivative of the target function $q'(x)$, then

$$H(x, \lambda) = (1 - \lambda)q'(x_0)(x - x_0) + \lambda q(x).$$

Undoubtably, this can be effectively employed when the derivative can be given in analytical form. In case of polynomials it is the case. It goes without saying that for a system of equations the Jacobian should be used.

2.4.5 Mixed Homotopy

Rahimian et al. (2011) are concerned with the use of the homotopy function,

$$H(x, \lambda) = (1 - \lambda)(q(x) - q(x_0) + (x - x_0)) + \lambda q(x)$$

to track the approximate solution. Here the start system is a linear combination of the fixed point and the Newton homotopy.

There are other methods, too to construct start system for linear homotopy. In a later section we shall see how one can define the start system for polynomial systems automatically.

It is known that local methods, like Newton–Raphson method, require initial guess of a root, which is close to the intended root for the particular application. Global methods, like homotopy continuation, can find solution from a start guess that is far from the solution.

2.5 Regularization of the Homotopy Function

Sometimes even this form of homotopy method may fail, because of a singularity resulting in diverging paths.

In order to avoid singularities in the field of real numbers, we consider a modified complex homotopy function,

$$H(x, \lambda) = \gamma(1 - \lambda)(x - x_0) + \lambda q(x),$$

where γ is a complex number. For almost all choices of a complex constant γ, all solution paths defined by the homotopy above are regular, i.e., for all $\lambda \in [0, 1]$, the Jacobian matrix of $H (x, \lambda)$ is regular and no diverging path occurs.

2.6 Start System in Case of Algebraic Polynomial Systems

How can we automatically find the proper start system, which will provide all of the solutions of the target system? This problem can be solved if the nonlinear system is a system of algebraic polynomial equations.

Let us consider the case where we are looking for the homotopy solution of $f(x) = 0$, where $f(x)$ is a *polynomial system*, $f(x) \colon \mathbb{R}^n \to \mathbb{R}^n$. To get all of the solutions, one should find out a proper polynomial system, as a start system, $g(x) = 0$, where $g(x) \colon \mathbb{R}^n \to \mathbb{R}^n$ with known or easily computable solutions.

An appropriate start system can be generated in the following way.

Let $f_i(x_1, \ldots, x_n)$, $i = 1, \ldots, n$ be a system of n polynomials. We are interested in the common zeros of the system namely, $f = (f_1(x), \ldots, f_n(x)) = 0$.

Let d_j denote the degree of the jth polynomial—that is, the degree of the highest order monomial in the equation. Then the system will be a proper starting system

$$g_j(x) = e^{i\phi_j}\left(x_j^{d_j} - \left(e^{i\vartheta_j}\right)^{d_j}\right) = 0, \qquad j = 1, \ldots, n,$$

where ϕ_j and θ_j are random real numbers in the interval $[0, 2\pi]$. The equation above has the obvious particular solution $x_j = e^{i\vartheta_j}$ and the complete set of the starting solutions for $j = 1, \ldots, n$ are given by

$$e^{\left(i\vartheta_j + \frac{2\pi ik}{d_j}\right)}, \qquad k = 0, 1, \ldots, d_j - 1.$$

Bezout's theorem states that the number of isolated roots of such a system is bounded by the total degree of the system,

$$\prod_{i=1}^{n} d_i = d_1 \cdot d_2 \cdot \ldots \cdot d_n.$$

Example Let us consider the following system,

$$f_1(x, y) = x^2 + y^2 - 1,$$
$$f_2(x, y) = x^3 + y^3 - 1.$$

The degrees of the polynomials are

$\Rightarrow$ d1 = 2; d2 = 3;

Indeed, this system has the following six roots ($d_1 \, d_2 = 2 * 3 = 6$), as expected.

$\Rightarrow$ NSolve[{f1[x,y] == 0, f2[x,y] == 0}, {x,y}]
$\Rightarrow$ {{x → −1. + 0.707107 i, y → −1. −0.707107 i},
  {x → −1. −0.707107 i, y → −1. + 0.707107 i}, {x → 1., y → 0.},
  {x → 1., y → 0.}, {x → 0., y → 1.}, {x → 0., y → 1.}}
$\Rightarrow$ Length[%]
$\Leftarrow$ 6

Now, we compute the start system. We generate random real numbers in the interval $[0, 2\pi]$ as it follows

$\Rightarrow \phi 1 = \mathtt{Random}[\mathtt{Real}, \{0, 2\pi\}]$

$\Leftarrow 4.46668$

$\Rightarrow \phi 2 = \mathtt{Random}[\mathtt{Real}, \{0, 2\pi\}]$

$\Leftarrow 1.97211$

$\Rightarrow \theta 1 = \mathtt{Random}[\mathtt{Real}, \{0, 2\pi\}]$

$\Leftarrow 2.71370$

$\Rightarrow \theta 2 = \mathtt{Random}[\mathtt{Real}, \{0, 2\pi\}]$

$\Leftarrow 0.163568$

Then the start system is,

$\Rightarrow \mathtt{g1}[\mathtt{x\_, y\_}] := e^{i\phi 1}\left(x^{d1} - \left(e^{i\theta 1}\right)^{d1}\right)$

$\Rightarrow \mathtt{g2}[\mathtt{x\_, y\_}] := e^{i\phi 2}\left(x^{d2} - \left(e^{i\theta 2}\right)^{d2}\right)$

$\Rightarrow \mathtt{g1}[\mathtt{x, y}]$

$\Leftarrow (-0.243245 - 0.969965\,\mathrm{i})((-0.655628 + 0.755084\,\mathrm{i}) + x^2)$

$\Rightarrow \mathtt{g2}[\mathtt{x, y}]$

$\Leftarrow (-0.390631 + 0.920548\,\mathrm{i})((-0.882001 - 0.471247\,\mathrm{i}) + y^3)$

Then the complete set of the starting solutions are,

$\Rightarrow \mathtt{Xi} = \mathtt{Table}[\mathtt{Exp}[\mathtt{i\theta 1 + 2\pi i k/d1}], \{k, 0, d1 - 1\}]$

$\Leftarrow (-0.390631 + 0.920548\,\mathrm{i})((-0.882001 - 0.471247\,\mathrm{i}) + y^3)$

and

$\Rightarrow \mathtt{Yi} = \mathtt{Table}[\mathtt{Exp}[\mathtt{i\,\theta 2 + 2\pi i k/d2}], \{k, 0, d2 - 1\}]$

$\Leftarrow \{0.986653 + 0.16284\mathrm{i}, -0.63435 + 0.773046\,\mathrm{i}, -0.352303 - 0.935886\,\mathrm{i}\}$

We need all of the combination of the initial values $\{X_i, Y_j\}$, $i = 1, 2, j = 1, 2$

$\Rightarrow \mathtt{X0} = \mathtt{Tuples}[\{\mathtt{Xi}, \mathtt{Yi}\}]$

$\Leftarrow \{\{-0.909843 + 0.414953\,\mathrm{i}, 0.986653 + 0.16284\,\mathrm{i}\},$
$\{-0.909843 + 0.414953\,\mathrm{i}, -0.63435 + 0.773046\,\mathrm{i}\},$
$\{-0.909843 + 0.414953\,\mathrm{i}, -0.352303 - 0.935886\,\mathrm{i}\},$
$\{0.909843 - 0.414953\,\mathrm{i}, 0.986653 + 0.16284\,\mathrm{i}\},$
$\{0.909843 - 0.414953\,\mathrm{i}, -0.63435 + 0.773046\,\mathrm{i}\},$
$\{0.909843 - 0.414953\,\mathrm{i}, -0.352303 - 0.935886\,\mathrm{i}\}\}$

These values satisfy the start system.

So we have six initial values for solving our homotopy equations. These initial values will provide the start point of the six homotopy paths. The end points of these paths are the six solutions of the target system.

2.7   Homotopy Methods in *Mathematica*

The algorithm discussed above have been implemented in *Mathematica*,

$\Rightarrow$ $<<$Homotopy$'$LinearHomotopy$'$

There are the following functions

$\Rightarrow$?LinearHomotopyFR

```
Computes the homotopy paths with direct path tracing.
Input parameters:
F - list of functions of the target system,
G - list of functions of the start system,
X - list of variables,
X0 - list of initial values,
γ - list of complex weights,
n - number of subintervals,
λ - dummy variable
Output variables:
sol[[1]] - list of the solutions,
sol[[2]] - list of homotopy paths
```

$\Rightarrow$?LinearHomotopyNDS02

```
Computes the homotopy paths with integration using numeric inverse.
Input parameters:
X - list of variables,
F - list of functions of the target system,
G - list of functions of the start system,
X0 - list of initial values,
γ - list of complex weights,
p - standard precision - > 0; higher precision - > 1,
λ - dummy variable
```

```
Output variables:
sol[[1]] - list of the solutions,
sol[[2]] - list of homotopy paths
```

⇒ ?Paths

```
Display homotopy paths.
Input parameters:
X - list of variables,
sol - list of homotopy paths,
X0 - list of initial values,
λ - dummy variable
```

⇒ ?StartingSystem

```
Computes the start system for polynomial system
Input parameters:
 F - list of functions of the target system,
 X - list of variables,
 d - list of orders of the equations
Output parameters:
 G - list of functions of the start system,
 X0 - solution of the start system, the initial values for homotopy
```

Now let us solve the system considered in the last example. The system is,

$$\Rightarrow F = \{f1[x,y], f2[x,y]\}$$
$$\Leftarrow \{-1+x^2+y^2, -1+x^3+y^3\}$$

The list of variables is

$$\Rightarrow X = \{x,y\}$$
$$\Leftarrow \{x,y\}$$

The degree of the polynomials

$\Rightarrow$ d = {d1, d2}
$\Leftarrow$ {2, 3}

Let us generate the starting system

$\Rightarrow$ sol = StartingSystem[F, X, d];

Now our starting system is

$\Rightarrow$ G = sol[[1]]
$\Leftarrow$ {(0.981373 − 0.192112 i)((0.971733 − 0.236081 i) + x^2),
   (−0.555004 − 0.831847 i)((−0.998626 − 0.0523973 i) + y^3)}

The initial values

$\Rightarrow$ X0 = sol[[2]]
$\Leftarrow$ {{−0.118884 − 0.992908 i, 0.999847 + 0.0174729 i},
   {−0.118884 − 0.992908 i, −0.515056 + 0.857157 i},
   {−0.118884 − 0.992908 i, −0.484792 − 0.87463 i},
   {0.118884 + 0.992908 i, 0.999847 + 0.0174729 i},
   {0.118884 + 0.992908 i, −0.515056 + 0.857157 i},
   {0.118884 + 0.992908 i, −0.484792 − 0.87463 i}}

Let us compute the solution. Since the system of polynomials and the starting system are complex, we do not need the gamma "trick", therefore let,

$\Rightarrow$ γ= {1, 1};

First we employ the direct path tracing

$\Rightarrow$ sol = LinearHomotopyFR[F, G, X, X0, γ, 100, λ];

The roots are

$\Rightarrow$ sol1FR = Chop[sol[[1]], 10^{-8}]
$\Leftarrow$ {{0, 1.}, {−1. − 0.707107 i, −1. + 0.707107 i},
   {1., 0}, {0, 1.}, {1., 0}, {−1. + 0.707107 i, −1. − 0.707107 i}}

The trajectories of the solutions on the complex plane are

$\Rightarrow$ pFRS = Paths[X, sol[[2]], X0, λ]

Now, let us employ the method based on integration of the differential equation system, which employ numerical inverse (Fig. 2.9)

⇒ sol = LinearHomotopyNDS02[X, F, G, X0,γ, 1,λ][[1]]
⇐ {{0, 1.}, {−1. − 0.707107 i − 1. + 0.707107 i},
  {1., 0}, {0, 1.}, {1., 0}, {−1. + 0.707107 i, −1. − 0.707107 i}}

2.8 Parallel Computation

The tracing of the different paths are independent of each other, therefore parallel computation can be employed. Let us carry out the computation in non-parallel way

⇒ AbsoluteTiming[sol = LinearHomotopyFR[F, G, X, X0,γ, 5000,λ];]
⇐ {14.9039, Null}

Without parallel execution, the processor performance is about 12%. The result is

⇒ sol1FR = Chop[sol[[1]], 10^{-8}]
⇐ {{0, 1.}, {−1. − 0.707107 i, −1. + 0.707107 i},
  {1., 0}, {0, 1.}, {1., 0}, {−1. + 0.707107 i, −1. − 0.707107 i}}

The preparation of the initial conditions for the parallel computation

⇒ X0P = Map[{#}&, X0]
⇐ {{{−0.118884 −0.992908 i, 0.999847 +0.0174729 i}},
  {{−0.118884 −0.992908 i, −0.515056 +0.857157 i}},
  {{−0.118884 −0.992908 i, −0.484792 −0.87463 i}},
  {{0.118884 +0.992908 i, 0.999847 +0.0174729 i}},
  {{0.118884 +0.992908 i, −0.515056 +0.857157 i}},
  {{0.118884 +0.992908 i, −0.484792 −0.87463 i}}}

Our computer has four cores with two threads each,

⇒ LaunchKernels[2 $ProcessorCount]//Quiet;
⇐" Updating from Wolfram Research server...".

We should distribute the information for the 8 threads

⇒ DistributeDefinitions[LinearHomotopyFR, F, G, X,γ, X0P];

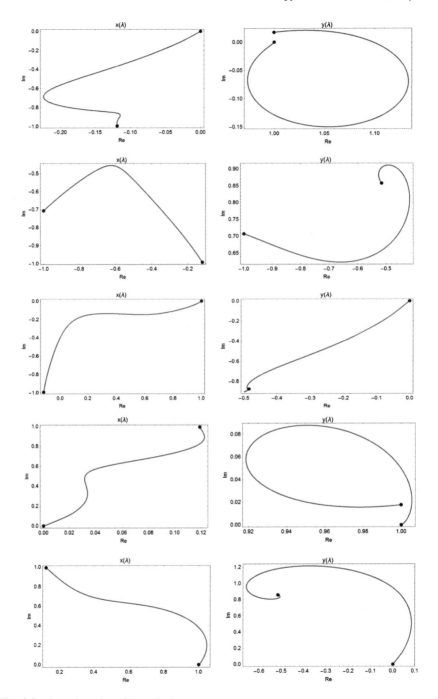

Fig. 2.9 The trajectories of the solutions

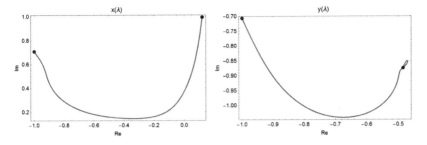

Fig. 2.9 (continued)

Then the parallel computation can be carried out

```
⇒ AbsoluteTiming[sol =
  ParallelMap[LinearHomotopyFR[F,G,X,#,γ,5000,λ][[1]]&,X0P]; ]
⇐ {4.2528,Null}
⇒ Chop[sol,10⁻⁸]
⇐ {{{0,1.}},{{−1. − 0.707107i,−1. + 0.707107 i}},{{1.,0}},
   {{0,1.}},{{1.,0}},{{−1. + 0.707107 i,−1. − 0.707107 i}}}
```

Now, the processor performance is about 100%. It means that using parallel evaluation the computation time can be reduced to one-third of the non-parallel case, 4.25 < 14.9 [s].

2.9   General Nonlinear System

The homotopy method is not restricted to algebraic systems. Now, we consider the following system of equations

$$\Rightarrow eq1 = x^2 + y^2 - 1;$$
$$\Rightarrow eq2 = Sin[x] - y;$$

The system cannot be solved via none of the symbolic methods discussed in Chap. 1, however local numerical solution requiring initial guess values is possible (Fig. 2.10).

$$\Rightarrow solF = FindRoot[\{eq1, eq2\}, \{\{x, -0.5\}, \{y, -0.75\}\}]$$
$$\Leftarrow \{x \rightarrow -0.739085, y \rightarrow -0.673612\}$$

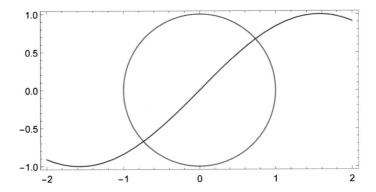

Fig. 2.10 Geometric representation of the non-polynomial system

Let us try to employ homotopy

$\Rightarrow F = \{eq1, eq2\}$
$\Leftarrow \{-1 + x^2 + y^2, -y + Sin[x]\}$
$\Rightarrow X = \{x, y\};$

Considering the starting system as

$\Rightarrow g1 = eq1$
$\Leftarrow -1 + x^2 + y^2$

Employing the linearized form of eq2 at $x = 0$

$\Rightarrow g2 = x - y$
$\Leftarrow x - y$

The solutions of this system are

$\Rightarrow X0 = \{x, y\}/.Solve[\{g1 == 0, g2 == 0\}, \{x, y\}]$
$\Leftarrow \left\{ \left\{ -\dfrac{1}{\sqrt{2}}, -\dfrac{1}{\sqrt{2}} \right\}, \left\{ \dfrac{1}{\sqrt{2}}, \dfrac{1}{\sqrt{2}} \right\} \right\}$

The starting system

$\Rightarrow G = \{g1, g2\}$
$\Leftarrow \{-1 + x^2 + y^2, x - y\}$

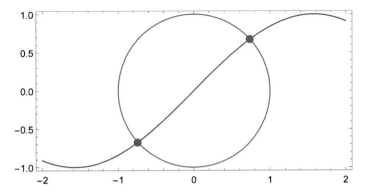

Fig. 2.11 Homotopy paths in case of non-polynomial system

Now, let us employ gamma trick

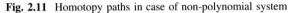

```
⇒ γ= {1+i,i};
⇒ sol = LinearHomotopyFR[F,G,X,X0,γ,100,λ];
⇒ sol[[1]]
⇐ {{−0.739085, −0.673612},{0.739085, 0.673612}}
```

Let us visualize these solutions, see Fig. 2.11

2.10   Nonlinear Homotopy

The idea of this nonlinear homotopy function comes from the construction of the Bezier splines.

2.10.1   Quadratic Bezier Homotopy Function

Bezier curves are used to draw smooth curves along points on a path. In case of two points (P_0, P_1) the point Q_0 is running from P_0 to P_1 while the parameter λ is changing from 0 to 1, see Fig. 2.12

$$\overrightarrow{Q_0} = (1 - \lambda)\overrightarrow{P_0} + \lambda\overrightarrow{P_1}, \quad \lambda \in [0, 1].$$

In case of three points (P_0, P_1, P_2) the point Q_0 is ruining from P_0 to P_1, while point Q_1 is running from P_1 to P_2, see Fig. 2.13

Fig. 2.12 Linear Bezier
spline

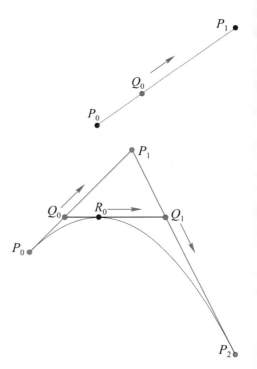

Fig. 2.13 Quadratic Bezier
spline

$$\overrightarrow{Q_1} = (1 - \lambda)\,\overrightarrow{P_1} + \lambda\,\overrightarrow{P_2}\,, \quad \lambda \in [0, 1]$$

and point R_0 is running along a smooth path from P_0 to P_2 in a way,

$$\overrightarrow{R_0} = (1 - \lambda)\,\overrightarrow{Q_0} + \lambda\,\overrightarrow{Q_1}\,, \quad \lambda \in [0, 1]$$

or

$$\overrightarrow{R_0} = (1 - \lambda)^2\,\overrightarrow{P_0} + 2(1 - \lambda)\lambda\,\overrightarrow{P_1} + \lambda^2\,\overrightarrow{P_2}\,, \quad \lambda \in [0, 1]$$

Considering analogy between this Bezier curve construction and the quadratic
homotopy function Nor et al. (2013) suggested the following casting,

$$P_0 \sim G(x)$$

$$P_1 \sim H_1(x, \lambda)$$

$$P_2 \sim F(x)$$

and

$$Q_0 \sim A(x, \lambda)$$
$$Q_1 \sim B(x, \lambda)$$
$$R_0 \sim H_2(x, \lambda)$$

where $H_1(x, \lambda)$ is the linear homotopy function,

$$H_1(x, \lambda) = (1 - \lambda)G(x) + \lambda F(x)$$

and $H_2(x, \lambda)$ is the quadratic Bezier homotopy function. Applying the analogy

$$A(x, \lambda) = (1 - \lambda)G(x) + \lambda H_1(x, \lambda)$$
$$B(x, \lambda) = (1 - \lambda)H_1(x, \lambda) + \lambda F(x)$$

then

$$H_2(x, \lambda) = (1 - \lambda)^2 G(x) + 2\,\lambda(1 - \lambda)H_1(x, \lambda) + \lambda^2 F(x)$$
$$= (1 - \lambda)^2 G(x) + 2\,\lambda(1 - \lambda)((1 - \lambda)G(x) + \lambda F(x)) + \lambda^2 F(x)$$

The coefficients of $H_2(x, \lambda)$ can be computed also as the coefficients of Bezier function. Let us employ the variables

$$H_0 = G(x) \quad \text{for } \lambda = 0$$
$$H_1 = H_1(x, \lambda) \quad \text{for } \lambda \in (0, 1)$$
$$H_2 = F(x) \quad \text{for } \lambda = 1$$

then

$$H_2(x, \lambda) = H_0 B_0^2(\lambda) + H_1 B_1^2(\lambda) + H_2 B_2^2(\lambda) = \sum_{i=0}^{2} H_i B_i^2(\lambda),$$

where $B_i^n(\lambda)$ represents the ith Bernstein basis function of degree n at λ,

$$B_i^n(\lambda) = \binom{n}{i}(1 - i)^{n-i},$$

where $i = 0, 1, 2, \ldots, n$. In *Mathematica* has a built in function for these basis functions, however it is only a numerical one, `BernsteinBasis[n,i, λ]`.

For linear homotopy

$$\sum_{i=0}^{2} B_i^1(\lambda) = (1 - \lambda) + \lambda = 1,$$

$\Rightarrow$ Plot$\left[\sum_{i=0}^{2}$ BernsteinBasis$[1, i, 1], \{\lambda, 0, 1\}\right]$

For quadratic homotopy (Fig. 2.14)

$$\sum_{i=0}^{2} B_i^2(\lambda) = (1 - \lambda)^2 + 2\lambda(1 - \lambda) + \lambda^2 = 1,$$

$\Rightarrow$ Plot$\left[\sum_{i=0}^{2}$ BernsteinBasis$[2, i, 1], \{\lambda, 0, 1\}\right]$

It goes without saying that (Fig. 2.15)

$$H_2(x, 0) = G(x)$$
$$H_2(x, 1) = F(x)$$

Remark A recursive construction of the quadratic homotopy function can be given by the *De Casteljau's* algorithm. In our case see Fig. 2.16.

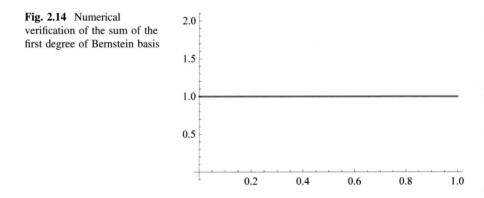

Fig. 2.14 Numerical verification of the sum of the first degree of Bernstein basis

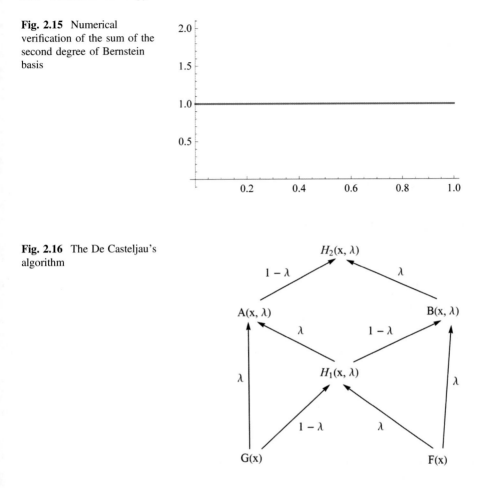

Fig. 2.15 Numerical verification of the sum of the second degree of Bernstein basis

Fig. 2.16 The De Casteljau's algorithm

2.10.2 Implementation in Mathematica

The implementation of this quadratic homotopy function is easy; only one line should be changed,

```
⇒ NonLinearHomotopyFR[F_, G_, X_, X0_, γ_, n_] :=
  Module[{H, X0L, λ0, i, β, m, R, RR, j, k, sol},
   Off[FindRoot :: "lstol"]; λ0 = Table[i 1/n, {i, 0, n}];
   H = Flatten[(1−β)²Thread[Gγ] +
    2 β(1−β)((1−β)Thread[G γ] + βF) + β²F];
   m = Length[X]; k = Length[X0]; RR = {};
   Do[
    X0L = {X0[[j]]};
    Do[AppendTo[X0L,
     Map[#[[2]]&, FindRoot[H/.βλ0[[i + 1]],
      MapThread[{#1, #2}&, {X, X0L[[i]]}]]]]], {i, 1, n}];
    R = {};
    Do[AppendTo[R, Interpolation[
     MapThread[{#1, #2[[i]]}&, {λ0, X0L}]]]], {i, 1, m}];
    AppendTo[RR, {Map[Chop[N[#[[1]]]]&, R], R}],
    {j, 1, k}];
   sol = Transpose[RR];
   {sol[[1]],
    Table[MapThread[#1[λ] − > #2[λ]&, {X, sol[[2, i]]}],
     {i, 1, Length[X0]}]}]
```

2.10.3 *Comparing Linear and Quadratic Homotopy*

Let us consider this simple polynomial system

```
⇒ Clear[f1, f2]
⇒ f1[x_, y_] := x²−2x − y + 1/2
⇒ f2[x_, y_] := x² + 4y²−4
```

First we solve it with numerical Gröbner basis

```
⇒ AbsoluteTiming[sol = NSolve[{f1[x, y], f2[x, y]}, {x, y}]; ]
⇐ {0.0195764, Null}
```

The solution

⇒ sol

⇐ $\{\{x \to 1.16077 - 0.654492i, y \to -0.902513 - 0.210444i\},$
   $\{x \to 1.16077 + 0.654492i, y \to -0.902513 + 0.210444i\},$
   $\{x \to -0.222215, y \to 0.993808\}, \{x \to 1.90068, y \to 0.311219\}\}$

Let us substitute back the solutions into the system and compute the mean of the error norms

⇒ Mean[Map[Norm[#]&, {f1[x, y], f1[x, y]}/.sol]]

⇐ 4.71401×10^{-15}

Now let us carry out the computation with linear homotopy. The system is

⇒ F = {f1[x, y], f2[x, y]}

⇐ $\{1/2 - 2x + x^2 - y, -4 + x^2 + 4y^2\}$

The variables

⇒ X = {x, y}

⇐ {x, y}

The degree of the polynomials

⇒ d = {2, 2}

⇐ {2, 2}

We generate a start system

⇒ sol = StartingSystem[F, X, d];

⇒ G = sol[[1]]

⇐ $(-0.419162 - 0.907912i)((-0.657419 + 0.753525i) + x^2),$
   $(0.609013 - 0.79316i)((-0.278254 + 0.960507i) + y^2)$

with its solutions

⇒ X0 = sol[[2]]

⇐ $\{\{-0.910335 + 0.413873\,i, 0.799454 - 0.600727\,i\},$
   $\{-0.910335 + 0.413873\,i, -0.799454 + 0.600727\,i\},$
   $\{0.910335 - 0.413873\,i, 0.799454 - 0.600727\,i\},$
   $\{0.910335 - 0.413873\,i, -0.799454 + 0.600727\,i\}\}$

Now, we do not need γ trick,

$\Rightarrow \gamma = \{1, 1\};$
$\Rightarrow \text{AbsoluteTiming}[\text{sol10L} = \text{LinearHomotopyFR}[F, G, X, X0, \gamma, 5];\,]$
$\Leftarrow \{0.0210878, \text{Null}\}$

The solution

$\Rightarrow \text{sol10L}[[1]]$
$\Leftarrow \{\{-0.222215, 0.993808\}, \{1.16077 + 0.654492i, -0.902513 + 0.210444i\},$
$\quad \{-0.222215, 0.993808\}, \{1.16077 - 0.654492i, -0.902513 - 0.210444i\}\}$

Back substitution

$\Rightarrow \text{error10L} = \text{Map}[F/.\{x \rightarrow \#[[1]], y \rightarrow \#[[2]]\}\&, \text{sol10L}[[1]]]$
$\Leftarrow \{\{0., 4.44089 \times 10^{-16}\},$
$\quad \{2.22045 \times 10^{-16} - 4.44089 \times 10^{-16}i, 0. - 4.44089 \times 10^{-16}i\},$
$\quad \{0., 4.44089 \times 10^{-16}\}, \{0. + 0.i, 0. + 2.22045 \times 10^{-16}i\}\}$

The average of error norms

$\Rightarrow \text{Mean}[\text{Map}[\text{Norm}[\#]\&, \text{error10L}]]$
$\Leftarrow 4.44089 \times 10^{-16}$

Now we try quadratic homotopy

$\Rightarrow \text{AbsoluteTiming}[\text{sol10NL} = \text{NonLinearHomotopyFR}[F, G, X, X0, \gamma, 5];\,]$
$\Leftarrow \{0.0304908, \text{Null}\}$

The solution

$\Rightarrow \text{sol10NL}[[1]]$
$\Leftarrow \{\{-0.222215, 0.993808\}, \{1.16077 + 0.654492i, -0.902513 + 0.210444i\},$
$\quad \{-0.222215, 0.993808\}, \{1.16077 - 0.654492i, -0.902513 - 0.210444i\}\}$

Back substitution

$\Rightarrow \text{error10NL} = \text{Map}[F/.\{x \rightarrow \#[[1]], y \rightarrow \#[[2]]\}\&, \text{sol10NL}[[1]]]$
$\Leftarrow 0.0, -4.44089 \times 10^{-16}, 0.0 + 0.0i, -4.44089 \times 10^{-16} + 0.0i,$
$\quad 1.11022 \times 10^{-16}, -4.44089 \times 10^{-16}, 0.0 + 0.0i, 0.0 + 2.22045 \times 10^{-16}i$

The average of error norms

$\Rightarrow \text{Mean}[\text{Map}[\text{Norm}[\#]\&, \text{error10NL}]]$
$\Leftarrow 3.91995 \times 10^{-16}$

In order to improve the result of the linear homotopy we need smaller step size, which means more steps,

$\Rightarrow$ AbsoluteTiming[sol10L = LinearHomotopyFR[F, G, X, X0, γ, 300];]
$\Leftarrow$ {0.53911, Null}

The solution

$\Rightarrow$ sol10L[[1]]
$\Leftarrow$ {{1.90068, 0.311219}, {1.16077 + 0.654492i, −0.902513 + 0.210444i},
   {−0.222215, 0.993808}, {1.16077 − 0.654492i, −0.902513 − 0.210444i}}

Back substitution

$\Rightarrow$ error10L = Map[F/.{x $\rightarrow$ #[[1]], y $\rightarrow$ #[[2]]}&, sol10L[[1]]]
$\Leftarrow$ {{−4.44089 × 10^{-16}, −7.21645 × 10^{-16}}, {0.0 + 0.0i, 0.0 + 0.0i, }
   {0.0, 4.44089 × 10^{-16}}, {0. + 0.0i, 0. + 0.0i}}

Now the average of error norms is smaller but the running time longer

$\Rightarrow$ Mean[Map[Norm[#]&, error10L]]
$\Leftarrow$ 3.22858 × 10^{-16}

In this example the quadratic homotopy gives higher precision than the linear one, or linear homotopy requires longer computation time to reach the same error limit.

2.11   Applications

2.11.1   Nonlinear Heat Conduction

Let us consider again a 1D heat conduction problem

$$\frac{d}{dx}\left(\lambda(\theta)\frac{d\theta}{dx}\right) = 0.$$

The boundary conditions are

$$\theta(0) = 0 \quad \text{and} \quad \theta(1) = 1.$$

The heat conduction coefficient depends on the temperature

$$\lambda(\theta) = 1 + k\theta.$$

Substituting $\lambda(\theta)$ into the differential equation and expanding it, we get

⇒ D[(1 + k θ[x]) D[θ[x], x], x]//Expand
⇐ k θ'[x]² + θ"[x] + k θ[x]θ"[x]

Let us employ finite difference method. The equation for the i-th node

$$\Rightarrow \mathtt{eq} = k\left(\frac{\theta_{i+1}-\theta_i}{\Delta}\right)^2 + (1+k\theta_i)\frac{\theta_{i+1}-2\theta_i+\theta_{i-1}}{\Delta^2} \; ;$$

Considering six nodes and numbering them with 1, ..., 6, the boundary conditions are

⇒ θ₁= 0; θ₆= 1;

Let $k = 70$ and the length of the interval $\Delta x = 1/(6-1) = 1/5$

⇒ data = {k → 70,Δ → 1/5};

Then the equations for the internal nodes are

⇒ eqs = Table[eq, {i, 2, 5}]/.data//Expand
⇐ {−50 θ₂−1750 θ₂² + 250θ₃−1750 θ₂θ₃ + 1750 θ₃²,
    25 θ₂−50 θ₃ + 1750 θ₂θ₃−1750 θ₃² + 25 θ₄−1750 θ₃θ₄ + 1750 θ₄²,
    25 θ₃−50 θ₄ + 1750 θ₃θ₄−1750 θ₄² + 25 θ₅−1750 θ₄θ₅ + 1750 θ₅²,
    1775 + 25 θ₄−1800 θ₅ + 1750 θ₄θ₅−1750 θ₅²}

If we employ linear homotopy, then the equations are

⇒ F = eqs;

The variables

⇒ X = Table[[θᵢ, {i, 2, 5}]
⇐ {θ₂,θ₃,θ₄,θ₅}

The degree of the equations

⇒ d = {2, 2, 2, 2}
The starting system

⇒ sol = StartingSystem[F, X, d];
⇒ G = sol[[1]];

The initial values of the starting system

⇒ X0 = sol[[2]];

Since the starting system is complex

⇒ γ= {1, 1, 1, 1};

Let us employ the direct path tracing method

⇒ sol = LinearHomotopyFR[F, G, X, X0, γ, 100, λ];

Cutting imaginary tails,

⇒ sol1FR = Chop[sol[[1]], 10^{-8}];

Considering the positive solutions

⇒ sol2 = Select[
sol1FR, (#[[1]] > 0) ∧ (#[[2]] > 0) ∧ (#[[3]] > 0) ∧ (#[[4]] > 0)&]
⇐ {{0.348022, 0.5655, 0.734005, 0.875685}}

Let us visualize the temperature profile (Fig. 2.17).
We can find a natural starting system in this case, namely using the linear heat conduction problem, $k = 0$. The equations of the starting system are,

⇒ eqsL = Table[eq, {i, 2, 5}]/.{k → 0, Δ → 1/5}//Expand
⇐ {−50 θ$_2$ + 25 θ$_3$, 25 θ$_2$−50 θ$_3$ + 25 θ$_4$, 25 θ$_3$−50 θ$_4$ + 25 θ$_5$, 25 + 25 θ$_4$−50 θ$_5$}

The initial conditions are the solutions of this linear system

⇒ solL = NSolve[eqsL, X]//Flatten
⇐ {θ$_2$ → 0.2, θ$_3$ → 0.4, θ$_4$ → 0.6, θ$_5$ → 0.8}

The degree of the original system is $d = 2^4 = 16$. Formally therefore, we need to create 16 initial values for the 16 paths,

⇒ X0 = Table[X/.solL, {i, 1, 16}];
⇒ sol = LinearHomotopyFR[F, G, X, X0, γ, 100, λ];.
The solutions are

Fig. 2.17 Temperature
profile

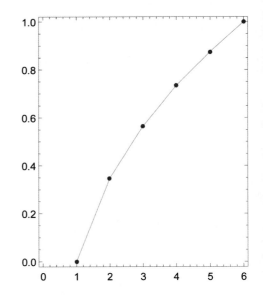

$\Leftarrow \{0.348022, 0.565500, 0.7340052, 0.875685\}$

The result using complex starting system was the same.

2.11.2 Local Coordinates via GNSS

Throughout history, position determination has been one of the most important tasks of mountaineers, pilots, sailors, civil engineers etc. In modern times, Global Navigation Satellite Systems (GPS) provide an ultimate method to accomplish this task. If one has a hand held GPS receiver, which measures the travel time of the signal transmitted from the satellites, the distance traveled by the signal from the satellites to the receiver can be computed by multiplying the measured time by the speed of light in vacuum. The distance of the receiver from the i-th GPS satellite, the pseudo-range observation d_i, is related to the unknown position of the receiver, X, Y, Z by

$$d_i = \sqrt{(a_i - X)^2 + (b_i - Y)^2 + (c_i - Z)^2} + \xi \quad i = 0, 1, 2, 3,$$

where $\{a_i, b_i, c_i\}$ are the coordinates of the satellite, and ξ is the adjustment of the satellite clock as an additional unknown variable. Let us determine the coordinates of a location on the Earth employing Global Positioning System (GNSS) $\{X, Y, Z\}$.

Now let us employ the variables: x_1, x_2, x_3 and x_4 representing the 4 unknowns $\{X, Y, Z, \xi\}$.

$$f1 = x_1^2 + x_2^2 + x_3^2 - x_4^2 - 2x_1a_0 + a_0^2 - 2x_2b_0 + b_0^2 - 2x_3c_0 + c_0^2 + 2x_4d_0 - d_0^2,$$

$$f2 = x_1^2 + x_2^2 + x_3^2 - x_4^2 - 2x_1a_1 + a_1^2 - 2x_2b_1 + b_1^2 - 2x_3c_1 + c_1^2 + 2x_4d_1 - d_1^2,$$

$$f3 = x_1^2 + x_2^2 + x_3^2 - x_4^2 - 2x_1a_2 + a_2^2 - 2x_2b_2 + b_2^2 - 2x_3c_2 + c_2^2 + 2x_4d_2 - d_2^2,$$

$$f4 = x_1^2 + x_2^2 + x_3^2 - x_4^2 - 2x_1a_3 + a_3^2 - 2x_2b_3 + b_3^2 - 2x_3c_3 + c_3^2 + 2x_4d_3 - d_3^2.$$

We construct the starting system keeping only one of the unknown variables from each equation, in the following way:

$$g1 = x_1^2 - 2x_1a_0 + a_0^2 + b_0^2 + c_0^2 - d_0^2,$$
$$g2 = x_2^2 + a_1^2 - 2x_2b_1 + b_1^2 + c_1^2 - d_1^2,$$
$$g3 = x_3^2 + a_2^2 + b_2^2 - 2x_3c_2 + c_2^2 - d_2^2,$$
$$g4 = -x_4^2 + a_3^2 + b_3^2 + c_3^2 + 2x_4d_3 - d_3^2.$$

Let us use the following numerical data, see *Awange* and *Palácz* (2016) page 180, Table 12.1. Then the system to be solved is,

```
⇒ F = {f1, f2, f3, f4}/.data
⇐ {1.0305 × 10¹⁴ − 2.9665 × 10⁷x1 + x1² + 4.0933 × 10⁷x2 + x2² + 1.4857 × 10⁷x3 + x3²
   + 4.8621 × 10⁷x4 − x4²,
   1.9504 × 10¹⁴ + 3.16000 × 10⁷x1 + x1² + 2.6602 × 10⁷x2 + x2² − 3.4268 × 10⁷x3 + x3²
   + 4.5829 × 10⁷x4 − x4²,
   2.8173 × 10¹⁴ − 396964.82x1 + x1² + 2.3735 × 10⁷x2 + x2² − 4.7434 × 10⁷x3 + x3²
   + 4.1258 × 10⁷x4 − x4²,
   1.6464 × 10¹⁴ + 2.4960 × 10⁷x1 + x1² + 4.6765 × 10⁷x2 + x2² − 6556945.4x3 + x3²
   + 4.6845 × 10⁷x4 − x4²}
```

The starting system

```
⇒ G = {g1, g2, g3, g4}/.data
⇐ {1.03055 × 10¹⁴ − 2.96646 × 10⁷x1 + x1², 1.95045 × 10¹⁴
   + 2.66023 × 10⁷x2 + x2²,
   2.81726 × 10¹⁴ − 4.74338 × 10⁷x3 + x3², 1.64642 × 10¹⁴
   + 4.68448 × 10⁷x4 − x4²}
```

The initial values for the starting system can be computed via solving the independent equations of the starting system

```
⇒ X = {x1, x2, x3, x4};
⇒ Transpose[Partition[Map[
   #[[2]]&, Flatten[MapThread[Nsolve[#1 == 0, #2]&, {G, X}]]], 2]]
⇐ {{4.01833 × 10⁶, −1.33011 × 10⁷ − 4.25733 × 10⁶i, 6.96083 ∗ 10⁶, −3.28436 × 10⁶},
   {2.56463 × 10⁷, −1.33011 × 10⁷ + 4.25733 × 10⁶i, 4.0473 × 10⁷, 5.01291 × 10⁷}}
```

Now we use the γ trick

$\Rightarrow \gamma = i\{1,1,1,1\};$

Employing linear homotopy

$\Rightarrow \text{sol} = \texttt{LinearHomotopyFR}[F, G, X, X0, \gamma, 100, \lambda];$

The solution,

$\Leftarrow \{1.11159 \times 10^6, -4.34826 \times 10^6, 4.52735 \times 10^6, 100.001\}$

using more digits

$\Rightarrow \texttt{NumberForm}[\%, 16]$
$\Leftarrow \{1.111590459962204 \times 10^6, -4.348258630909095 \times 10^6,$
$\quad 4.527351820245796 \times 10^6, 100.000550673451\}$

Let us compute the solution via Gröbner basis,

$\Rightarrow \texttt{NSolve}[F, X]$
$\Leftarrow \{\{x1 \rightarrow -2.89212 \times 10^6, x2 \rightarrow 7.56878 \times 10^6, x3 \rightarrow -7.20951 \times 10^6,$
$\quad x4 \rightarrow 5.74799 \times 10^7\},$
$\quad \{x1 \rightarrow 1.11159 \times 10^6, x2 \rightarrow -4.34826 \times 10^6, x3 \rightarrow 4.52735 \times 10^6,$
$\quad x4 \rightarrow 100.001\}\}$

using 16 digits,

$\Leftarrow \{x1 \rightarrow 1.111590459962204 \times 10^6, x2 \rightarrow -4.348258630909095 \times 10^6,$
$\quad x3 \rightarrow 4.527351820245798 \times 10^6, x4 \rightarrow 100.0005506748347\}$

we have the same result.

2.12 Applications

2.12.1 GNSS Positioning N-Point Problem

Problem In case of more than 4 satellites, $m > 4$, let us consider two distance representations (Paláncz 2006).. The algebraic one is,

$$f_i = (x_1 - a_i)^2 + (x_2 - b_i)^2 + (x_3 - c_i)^2 - (x_4 - d_i)^2,$$

and the geometrical one,

$$g_i = d_i - \sqrt{(x_1 - a_i)^2 + (x_2 - b_i)^2 + (x_3 - c_i)^2} - x_4,$$

The results will be not equivalent in least square sense, namely

$$\min_{x_1, x_2, x_3, x_4} \sum_{i=1}^{m} f_i \neq \min_{x_1, x_2, x_3, x_4} \sum_{i=1}^{m} g_i^2,$$

Let us consider six satellites with the following numerical values,

$\Rightarrow$ datan $= \{a_0 \rightarrow 14177553.47, a_1 \rightarrow 15097199.81, a_2 \rightarrow 23460342.33,$
$a_3 \rightarrow -8206488.95, a_4 \rightarrow 1399988.07, a_5 \rightarrow 6995655.48,$
$b_0 \rightarrow -18814768.09, b_1 \rightarrow -4636088.67, b_2 \rightarrow -9433518.58,$
$b_3 \rightarrow -18217989.14, b_4 \rightarrow -17563734.90, b_5 \rightarrow -23537808.26,$
$c_0 \rightarrow 12243866.38, c_1 \rightarrow 21326706.55, c_2 \rightarrow 8174941.25,$
$c_3 \rightarrow 17605231.99, c_4 \rightarrow 19705591.18, c_5 \rightarrow -9927906.48,$
$d_0 \rightarrow 21119278.32, d_1 \rightarrow 22527064.18, d_2 \rightarrow 23674159.88,$
$d_3 \rightarrow 20951647.38, d_4 \rightarrow 20155401.42, d_5 \rightarrow 24222110.91\};$

Solution
The number of the equations,

$\Rightarrow$ m = 6;

The prototype form of the error of the i-th equation is,

$\Rightarrow$ en $= d_i - \sqrt{(x1 - a_i)^2 + (x2 - b_i)^2 + (x3 - c_i)^2} - x4;$

then

$\Rightarrow$ Map[D[en^2,#]&,x1,x2,x3,x4]

$$\Leftarrow \left\{ -\frac{2(x1 - a_i)\left(-x4 - \sqrt{(x1 - a_i)^2 + (x2 - b_i)^2 + (x3 - c_i)^2} + d_i\right)}{\sqrt{(x1 - a_i)^2 + (x2 - b_i)^2 + (x3 - c_i)^2}}, \right.$$

$$-\frac{2(x2 - b_i)\left(-x4 - \sqrt{(x1 - a_i)^2 + (x2 - b_i)^2 + (x3 - c_i)^2} + d_i\right)}{\sqrt{(x1 - a_i)^2 + (x2 - b_i)^2 + (x3 - c_i)^2}},$$

$$-\frac{2(x3 - c_i)\left(-x4 - \sqrt{(x1 - a_i)^2 + (x2 - b_i)^2 + (x3 - c_i)^2} + d_i\right)}{\sqrt{(x1 - a_i)^2 + (x2 - b_i)^2 + (x3 - c_i)^2}},$$

$$\left. -2\left(-x4 - \sqrt{(x1 - a_i)^2 + (x2 - b_i)^2 + (x3 - c_i)^2} + d_i\right)\right\}$$

in general form,

$$\Rightarrow v = \text{Map}\left[\sum_{i=0}^{m-1} \#\&, \% \right]$$

$$\Leftarrow \left\{ \sum_{i=0}^{-1+m} -\frac{2(x1-a_i)\left(-x4-\sqrt{(x1-a_i)^2 + (x2-b_i)^2 + (x3-c_i)^2} + d_i\right)}{\sqrt{(x1-a_i)^2 + (x2-b_i)^2 + (x3-c_i)^2}}, \right.$$

$$quad \sum_{i=0}^{-1+m} -\frac{2(x2-b_i)\left(-x4 - \sqrt{(x1-a_i)^2 + (x2-b_i)^2 + (x3-c_i)^2} + d_i\right)}{\sqrt{(x1-a_i)^2 + (x2-b_i)^2 + (x3-c_i)^2}},$$

$$quad \sum_{i=0}^{-1+m} -\frac{2(x3-c_i)\left(-x4-\sqrt{(x1-a_i)^2 + (x2-b_i)^2 + (x3-c_i)^2} + d_i\right)}{\sqrt{(x1-a_i)^2 + (x2-b_i)^2 + (x3-c_i)^2}},$$

$$quad \sum_{i=0}^{-1+m} -2\left(-x4-\sqrt{(x1-a_i)^2 + (x2-b_i)^2 + (x3-c_i)^2} + d_i\right)\right\}$$

In our case $m = 6$, therefore the numeric form of the equations,

$\Rightarrow$ vn $= v/.\, m \rightarrow m/.\text{datan}//\text{Expand};$
For example the first equation

$\Rightarrow$ vn[[1]]

$\Leftarrow -1.05849 \times 10^8 + 12\text{x}1$

$+ 6.801919 \times 10^{14} / \left(\sqrt{\left((-1.50972 \times 10^7 + \text{x}1)^2 + (463609.67 + \text{x}2)^2 + (-2.13267 \times 10^7 + \text{x}3)^2 \right)} \right)$

$- (4.50541 \times 10^7 \text{x}1) / \left(\sqrt{\left((-1.50972 \times 10^7 + \text{x}1)^2 + (463609.67 + \text{x}2)^2 + (-2.13267 \times 10^7 + \text{x}3)^2 \right)} \right)$

$+ 5.64346 \times 10^{13} / \left(\sqrt{\left((-139999.07 + \text{x}1)^2 + (1.75637 \times 10^7 + \text{x}2)^2 + (-1.97056 \times 10^7 + \text{x}3)^2 \right)} \right)$

$- (4.03108 \times 10^7 \text{x}1) / \left(\sqrt{\left((-139999.07 + \text{x}1)^2 + (1.75637 \times 10^7 + \text{x}2)^2 + (-1.97056 \times 10^7 + \text{x}3)^2 \right)} \right)$

$- 3.43879 \times 10^{14} / \left(\sqrt{\left((820649.95 + \text{x}1)^2 + (1.8218 \times 10^7 + \text{x}2)^2 + (-1.76052 \times 10^7 + \text{x}3)^2 \right)} \right)$

$- (4.19033 \times 10^7 \text{x}1) / \left(\sqrt{\left((820649.95 + \text{x}1)^2 + (1.8218 \times 10^7 + \text{x}2)^2 + (-1.76052 \times 10^7 + \text{x}3)^2 \right)} \right)$

$+ 5.98839 \times 10^{14} / \left(\sqrt{\left((-1.41776 \times 10^7 + \text{x}1)^2 + (1.88148 \times 10^7 + \text{x}2)^2 + (-1.22439 \times 10^7 + \text{x}3)^2 \right)} \right)$

$- (4.22386 \times 10^7 \text{x}1) / \left(\sqrt{\left((-1.41776 \times 10^7 + \text{x}1)^2 + (1.88148 \times 10^7 + \text{x}2)^2 + (-1.22439 \times 10^7 + \text{x}3)^2 \right)} \right)$

$+ 1.11081 \times 10^{15} / \left(\sqrt{\left((-2.34603 \times 10^7 + \text{x}1)^2 + (9.43352 \times 10^6 + \text{x}2)^2 + (-8.17494 \times 10^6 + \text{x}3)^2 \right)} \right)$

$- (4.73483 \times 10^7 \text{x}1) / \left(\sqrt{\left((-2.34603 \times 10^7 + \text{x}1)^2 + (9.43352 \times 10^6 + \text{x}2)^2 + (-8.17494 \times 10^6 + \text{x}3)^2 \right)} \right)$

$+ 3.38899 \times 10^{14} / \left(\sqrt{\left((-6.99566 \times 10^6 + \text{x}1)^2 + (2.35378 \times 10^7 + \text{x}2)^2 + (9.92791 \times 10^6 + \text{x}3)^2 \right)} \right)$

$- (4.84442 \times 10^7 \text{x}1) / \left(\sqrt{\left((-6.99566 \times 10^6 + \text{x}1)^2 + (2.35378 \times 10^7 + \text{x}2)^2 + (9.92791 \times 10^6 + \text{x}3)^2 \right)} \right)$

$- (3.01944 \times 10^7 \text{x}4) / \left(\sqrt{\left((-1.50972 \times 10^7 + \text{x}1)^2 + (4.63609 \times 10^6 + \text{x}2)^2 + (-2.13267 \times 10^7 + \text{x}3)^2 \right)} \right)$

$+ (2\text{x}1\text{x}4) / \left(\sqrt{\left((-1.50972 \times 10^7 + \text{x}1)^2 + (4.63609 \times 10^6 + \text{x}2)^2 + (-2.13267 \times 10^7 + \text{x}3)^2 \right)} \right)$

$- (2.79998 \times 10^6 \text{x}4) / \left(\sqrt{\left((-1.39999 \times 10^6 + \text{x}1)^2 + (1.75637 \times 10^7 + \text{x}2)^2 + (-1.97056 \times 10^7 + \text{x}3)^2 \right)} \right)$

$+ (2\text{x}1\text{x}4) / \left(\sqrt{\left((-1.39999 \times 10^6 + \text{x}1)^2 + (1.75637 \times 10^7 + \text{x}2)^2 + (-1.97056 \times 10^7 + \text{x}3)^2 \right)} \right)$

$+ (1.6413 \times 10^7 \text{x}4) / \left(\sqrt{\left((8.20649 \times 10^6 + \text{x}1)^2 + (1.8218 \times 10^7 + \text{x}2)^2 + (-1.76052 \times 10^7 + \text{x}3)^2 \right)} \right)$

$+ (2\text{x}1\text{x}4) / \left(\sqrt{\left((8.20649 \times 10^6 + \text{x}1)^2 + (1.8218 \times 10^7 + \text{x}2)^2 + (-1.76052 \times 10^7 + \text{x}3)^2 \right)} \right)$

$- (2.83551 \times 10^7 \text{x}4) / \left(\sqrt{\left((-1.41776 \times 10^7 + \text{x}1)^2 + (1.88148 \times 10^7 + \text{x}2)^2 + (-1.22439 \times 10^7 + \text{x}3)^2 \right)} \right)$

$+ (2\text{x}1\text{x}4) / \left(\sqrt{\left((-1.41776 \times 10^7 + \text{x}1)^2 + (1.88148 \times 10^7 + \text{x}2)^2 + (-1.22439 \times 10^7 + \text{x}3)^2 \right)} \right)$

$- (4.69207 \times 10^7 \text{x}4) / \left(\sqrt{\left((-2.34603 \times 10^7 + \text{x}1)^2 + (9.43352 \times 10^6 + \text{x}2)^2 + (-8.17494 \times 10^6 + \text{x}3)^2 \right)} \right)$

$+ (2\text{x}1\text{x}4) / \left(\sqrt{\left((-2.34603 \times 10^7 + \text{x}1)^2 + (9.43352 \times 10^6 + \text{x}2)^2 + (-8.17494 \times 10^6 + \text{x}3)^2 \right)} \right)$

$- (1.39913 \times 10^7 \text{x}4) / \left(\sqrt{\left((-6.99566 \times 10^6 + \text{x}1)^2 + (2.35378 \times 10^7 + \text{x}2)^2 + (9.92791 \times 10^6 + \text{x}3)^2 \right)} \right)$

$+ (2\text{x}1\text{x}4) / \left(\sqrt{\left((-6.99566 \times 10^6 + \text{x}1)^2 + (2.35378 \times 10^7 + \text{x}2)^2 + (9.92791 \times 10^6 + \text{x}3)^2 \right)} \right)$

We pick up a fairly wrong initial guess provided by the solution of a triplet subset,

$\Rightarrow$ X0 = {{596951.53, -4.8527796×10^6, 4.0887586×10^6, 3510.400237}}

$\Leftarrow$ {{596952.0, -4.85278×10^6, 4.08876×10^6, 3510.4}}

The variables,

$\Leftarrow$ V = {x1, x2, x3, x4};

The start system using fixed point homotopy,

$\Rightarrow$ gF = V $-$ First[X0]

$\Leftarrow$ {$-596952. + \text{x}1$, $4.85278 \times 10^6 + \text{x}2$, $-4.08876 \times 10^6 + \text{x}3$, $-3510.4 + \text{x}4$}

To avoid singularity of the homotopy function, let

$\Rightarrow$ γ= i{1,1,1,1};

$\Rightarrow$ AbsoluteTiming[solH = LinearHomotopyFR[vn, gF, V, X0, γ, 10];]

$\Leftarrow$ {0.837211, Null}

$\Rightarrow$ solH[[1]]

$\Leftarrow$ {{596930., $-4.84785 \times 10^6, 4.08823 \times 10^6, 15.5181$}}

$\Rightarrow$ NumberForm[%, 15]

$\Leftarrow$ 596929.65, -4.84785155×10^6, 4.0882268×10^6, 15.5180507

Displaying the homotopy paths,

$\Rightarrow$ paH = PathsV[V, solH[[2]], X0];

$\Rightarrow$ paH = Flatten[paH]

Now we try to use nonlinear homotopy (Fig. 2.18),

$\Rightarrow$ AbsoluteTiming[solH = NonLinearHomotopyFR[vn, gF, V, X0, γ, 3];]

$\Leftarrow$ {0.0333693, Null}

$\Rightarrow$ solH[[1]]

$\Leftarrow$ {{596930., $-4.84785 \times 10^6, 4.08823 \times 10^6, 15.5181$}}

$\Rightarrow$ NumberForm[%, 15]

$\Leftarrow$ 596929.65, -4.84785155×10^6, 4.0882268×10^6, 15.5180507

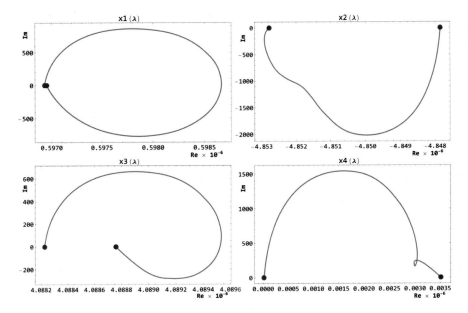

Fig. 2.18 The homotopy paths in case of linear homotopy

Displaying the homotopy paths,

⇒ paH = PathsV[V, solH[[2]], X0];

⇒ paH = Flatten[paH]

Now the running time of the nonlinear homotopy is shorter (Fig. 2.19).

As demonstrated by these examples, the linear and nonlinear homotopy methods proved to be powerful solution tools in solving nonlinear geodetic problems, especially if it is difficult to find proper initial values to ensure convergence of local solution methods. These global numerical methods can be successful when symbolic computation based on Groebner basis or Dixon resultant fail because of the size and complexity of the problem. Since the computations along the different paths are independent, parallel computation can be successfully employed. In general, nonlinear homotopy gives more precise result or required less iteration steps and needs less computation time at a fixed precision than its linear counterpart.

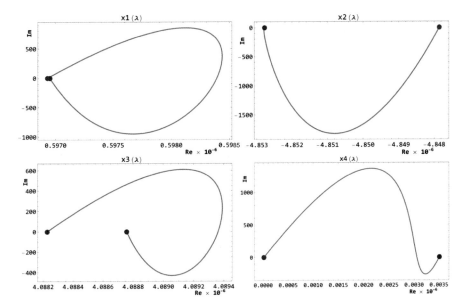

Fig. 2.19 The homotopy paths in case of nonlinear homotopy

References

Awange JL, Paláncz B (2016) Geospatial algebraic computations, theory and applications, 3rd edn. Springer, Heidelberg, New York

Jalali F, Seader JD (2000) Use of homotopy-continuation method in stability analysis of multiphase reacting systems. Comput Chem Eng 24:1997–2000

Kotsireas IS (2001) Homotopies and polynomial systems solving I: basic principles. ACM SIGSAM Bulletin 35(1):19–32

Nor HM, Ismail AIM, Majid AA (2013) Quadratic bezier homotopy function for solving system of polynomial equations. Matematika 29(2):159–171

Paláncz B (2006) GPS N-points problem. Math Educ Res 11(2):153–177

Rahimian SK, Farhang J, Seader JD, White RE (2011) A new homotopy for seeking all real roots of a nonlinear equation. Comput Chem Eng 35(3):403–411

Chapter 3
Overdetermined and Underdetermined Systems

3.1 Concept of the Over and Underdetermined Systems

As we know, usually overdetermined systems have no solution while underdetermined systems have infinitely many solutions. From an engineering point of view we can consider the solution of an overdetermined system to be the minimum of the sum of the squares of the errors, while in the case of an underdetermined system, we look for the solution having the minimal norm.

3.1.1 Overdetermined Systems

Let us consider the following system

$$f_1 = x^2 + 2x - 1,$$
$$f_2 = 5xy - 1 + x^3,$$
$$f_3 = y^2 - 12 + 10\sin(x).$$

Now, we have more equations than unknowns. These equations have pairwise solutions, ((f_1, f_2): solid line, (f_2, f_3): dashed line, (f_1, f_3): dotted line), but there is no solution for the three equations simultaneously, see Fig. 3.1.

Supplementary Information The online version contains supplementary material available at https://doi.org/10.1007/978-3-030-92495-9_3.

Fig. 3.1 Overdetermined
system

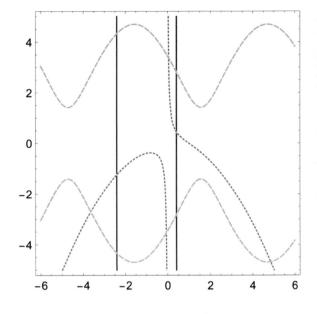

We may consider the solution in the sense of the least squares, namely

$$\sum_{i=1}^{3} f_i^2(x, y) \rightarrow \min_{x,y} .$$

In our case the minimization results in a single local minimum, see Fig. 3.2.

$\Rightarrow$ sol $=$ NMinimize$\left[f_1^2 + f_2^2 + f_3^2, \{x, y\} \right]$
$\Leftarrow \{15.0617, \{x \rightarrow 1.16201, y \rightarrow -0.117504\}\}$

3.1.2   Underdetermined Systems

In this case, we have less equations than unknown variables, i.e.,

$$f_1 = x^2 + 2x + 2yz^2 - 5,$$
$$f_2 = 5xy - 1 + x^3 + xz.$$

Let us visualize the system for different z values, see Fig. 3.3.

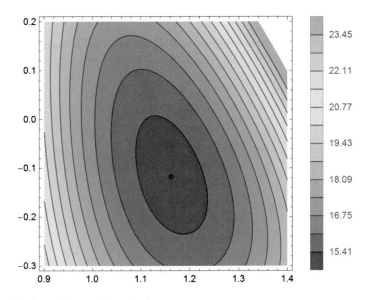

Fig. 3.2 Solution of the overdetermined system

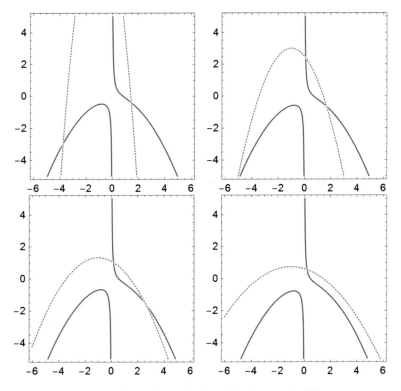

Fig. 3.3 The graphs of the solutions of the underdetermined system for different constant z values, $z = (0.5,\ 1.0,\ 1.5,\ 2.0)$

It is clear that there are infinitely many solutions. We consider the solution, which has minimal norm,

$$\|\{x, y, z\}\| \rightarrow \min_{x,y,z}$$

under the constraint

$$f_i(x, y, z) = 0, \quad i = 1, 2, \ldots n.$$

Namely, in our case the solution of this constrained minimization problem is

$\Rightarrow$ sol $=$ NMinimize[$\{$Norm[$\{$x, y, z$\}$], f$_1 =$ 0, f$_2 =$ 0$\}$,$\{$x, y, z$\}$,

  Method $\rightarrow$ " SimulatedAnnealing"]

$\Leftarrow \{1.80538, \{$x $\rightarrow$ 0.288172, y $\rightarrow$ 0.975707, z $\rightarrow$ $-1.49142\}\}$

We employed here a stochastic global minimization technique, Simulated Annealing, which will be discussed later in the second part of the book (Fig. 3.4).

The solution of the over or underdetermined systems can be computed via minimization or constrained minimization, respectively. Here, we compute always the global minimum therefore the solutions of such systems are unique.

Fig. 3.4 Solution of the underdetermined system

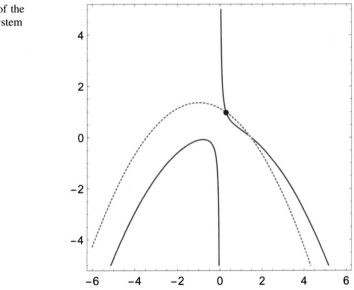

3.2 Gauss–Jacobi Combinatorial Solution

This technique can be applied to overdetermined systems. Let us suppose that we have m equations with n unknowns, $m > n$. Then we can select $\binom{m}{n}$ number of determined subsystems size of $(n \times n)$. In case of linear systems, the sum of the weighted solutions of these subsystems will give exactly the solution of the overdetermined system.

Let us illustrate the algorithm with a small linear system,

$$a_1 x + b_1 y = c_1,$$
$$a_2 x + b_2 y = c_2,$$
$$a_3 x + b_3 y = c_3.$$

The possible combinations of these equations are: $\{1, 2\}$, $\{1, 3\}$ and $\{2, 3\}$. Let us suppose that the solutions of these subsystems are (x_{12}, y_{12}), (x_{13}, y_{13}), and (x_{23}, y_{23}). Then the solution of the overdetermined linear system is

$$x = \frac{\pi_{12} x_{12} + \pi_{13} x_{13} + \pi_{23} x_{23}}{\pi_{12} + \pi_{13} + \pi_{23}}$$

and

$$y = \frac{\pi_{12} y_{12} + \pi_{13} y_{13} + \pi_{23} y_{23}}{\pi_{12} + \pi_{13} + \pi_{23}},$$

where the $\pi_{i,j}$ weights can be computed from the square of the sub-determinants of the coefficient matrix of the overdetermined system as follows:

the transpose of the coefficient matrix is

$$\begin{pmatrix} a_1 & b_1 \\ a_2 & b_2 \\ a_3 & b_3 \end{pmatrix}^T = \begin{pmatrix} a_1 & a_2 & a_3 \\ b_1 & b_2 & b_3 \end{pmatrix},$$

the squares of the different sub-determinants are,

$$\pi_{12} = \left\{ \begin{matrix} a_1 & a_2 \\ b_1 & b_2 \end{matrix} \right\}^2 = (a_1 b_2 - a_2 b_1)^2,$$

$$\pi_{13} = \left\{ \begin{matrix} a_1 & a_3 \\ b_1 & b_3 \end{matrix} \right\}^2 = (a_1 b_3 - a_3 b_1)^2,$$

$$\pi_{23} = \left\{ \begin{matrix} a_2 & a_3 \\ b_2 & b_3 \end{matrix} \right\}^2 = (a_2 b_3 - a_3 b_2)^2.$$

Let us see a numerical example.

$\Rightarrow$ eq1 = x + 2y = 3;
$\Rightarrow$ eq2 = 4x + 7y = 6;
$\Rightarrow$ eq3 = 5x + 8y = 9;
$\Rightarrow$ n = 2; m = 3;

The number of the combinations

$\Rightarrow$ mn = Binomial[m, n]
$\Leftarrow$ 3

The actual combinations are

$\Rightarrow$ R = Subsets[Range[m], {n}]
$\Leftarrow$ {{1, 2}, {1, 3}, {2, 3}}

The solutions of the determinant subsystems are,

$\Rightarrow$ sol12 = LinearSolve$\left[\begin{pmatrix} 1 & 2 \\ 4 & 7 \end{pmatrix}, \begin{pmatrix} 3 \\ 6 \end{pmatrix} \right]$; MatrixForm[sol12]

$\Leftarrow \begin{pmatrix} -9 \\ 6 \end{pmatrix}$

$\Rightarrow$ sol13 = LinearSolve$\left[\begin{pmatrix} 1 & 2 \\ 5 & 8 \end{pmatrix}, \begin{pmatrix} 3 \\ 9 \end{pmatrix} \right]$; MatrixForm[sol13]

$\Leftarrow \begin{pmatrix} -3 \\ 3 \end{pmatrix}$

$\Rightarrow$ sol23 = LinearSolve$\left[\begin{pmatrix} 4 & 7 \\ 5 & 8 \end{pmatrix}, \begin{pmatrix} 6 \\ 9 \end{pmatrix} \right]$; MatrixForm[sol23]

$\Leftarrow \begin{pmatrix} 5 \\ -2 \end{pmatrix}$

Let us compute the weights. The transpose of the coefficient matrix is

$\Rightarrow$ Transpose$\left[\begin{pmatrix} 1 & 2 \\ 4 & 7 \\ 5 & 8 \end{pmatrix} \right]$ //MatrixForm

$\Leftarrow \begin{pmatrix} 1 & 4 & 5 \\ 2 & 7 & 8 \end{pmatrix}$

then the weights are,

$$\Rightarrow \Pi_{12}=(1*7-4*2)^2$$
$$\Leftarrow 1$$
$$\Rightarrow \Pi_{13}=(1*8-5*2)^2$$
$$\Leftarrow 4$$
$$\Rightarrow \Pi_{23}=(4*8-5*7)^2$$
$$\Leftarrow 9$$

Therefore the solution of the overdetermined system can be computed as

$$\Rightarrow x = (\Pi_{12}\,\text{sol}12[[1]]+\Pi_{13}\,\text{sol}13[[1]]+\Pi_{23}\,\text{sol}123[[1]])/(\Pi_{12}+\Pi_{13}+\Pi_{23})$$
$$\Leftarrow \left\{\frac{12}{7}\right\}$$
$$\Rightarrow y = (\Pi_{12}\,\text{sol}12[[2]]+\Pi_{13}\,\text{sol}13[[2]]+\Pi_{23}\,\text{sol}123[[2]])/(\Pi_{12}+\Pi_{13}+\Pi_{23})$$
$$\Leftarrow \{0\}$$

Now, let us compare this result to the solution computed in sense of least squares. The system in matrix form is

$$\Rightarrow A = \begin{pmatrix} 1 & 2 \\ 4 & 7 \\ 5 & 8 \end{pmatrix}; \ b = \begin{pmatrix} 3 \\ 6 \\ 9 \end{pmatrix};$$

The rank of the coefficient matrix,

$$\Rightarrow \texttt{MatrixRank}\left[\begin{pmatrix} 1 & 2 \\ 4 & 7 \\ 5 & 8 \end{pmatrix}\right]$$
$$\Leftarrow 2$$

The rank of the coefficient matrix extended with right hand side vector (b)

$$\Rightarrow \texttt{MatrixRank}\left[\begin{pmatrix} 1 & 2 & 3 \\ 4 & 7 & 6 \\ 5 & 8 & 9 \end{pmatrix}\right]$$
$$\Leftarrow 3$$

These ranks are different; therefore there is no traditional solution. The solution can be computed in sense of least squares via pseudoinverse, $\mathbf{P} = \left(\mathbf{A}^T\mathbf{A}\right)^{-1}\mathbf{A}^T$,

$\Rightarrow$ P = Inverse[Transpose[A].A].Transpose[A]; MatrixForm[P]]

$$\Leftarrow \begin{pmatrix} -\frac{23}{14} & -\frac{11}{7} & \frac{25}{14} \\ 1 & 1 & -1 \end{pmatrix}$$

then the solution is

$\Rightarrow$ x = P.b; MatrixForm[x]

$$\Leftarrow \begin{pmatrix} \frac{12}{7} \\ 0 \end{pmatrix}$$

One can also use a built-in function of the pseudoinverse directly,

$\Rightarrow$ PseudoInverse[A]//MatrixForm

$$\Leftarrow \begin{pmatrix} -\frac{23}{14} & -\frac{11}{7} & \frac{25}{14} \\ 1 & 1 & -1 \end{pmatrix}$$

$\Rightarrow$ %.b//MatrixForm

$$\Leftarrow \begin{pmatrix} \frac{12}{7} \\ 0 \end{pmatrix}$$

3.3   Gauss–Jacobi Solution in Case of Nonlinear Systems

The Gauss-Jacobi method can be extended to nonlinear systems. Let us consider a simple example $n = 2$ and $m = 3$.

$$f_1(x, y) = 0,$$
$$f_2(x, y) = 0,$$
$$f_3(x, y) = 0.$$

In this case, the algorithm is the following:

Let us compute the solutions of the nonlinear subsystems via *resultant, Gröbner basis* or *homotopy* methods, namely $\eta_{12} = (x_{12}, y_{12})$, $\eta_{13} = (x_{13}, y_{13})$, and $\eta_{23} = (x_{23}, y_{23})$.

Compute the means of these solutions, $\eta = (x_s, y_s)$

$$\eta = \frac{1}{3}(\eta_{12} + \eta_{13} + \eta_{23}).$$

Let us linearize the overdetermined system at this point.

The linearized form of the j-th nonlinear equation at $\eta = (x_s, y_s)$, for $j = 1,..., 3$

$$f_j(\eta) + \frac{\partial f_j}{\partial x}(\eta)(x - x_s) + \frac{\partial f_j}{\partial y}(\eta)(y - y_s) = 0$$

or

$$\frac{\partial f_j}{\partial x}(\eta)x + \frac{\partial f_j}{\partial y}(\eta)y + \left(f_j(\eta) - \frac{\partial f_j}{\partial x}(\eta)x_s - \frac{\partial f_j}{\partial y}(\eta)y_s \right) = 0.$$

Let us recall the computation of the weight of the $\{1, 2\}$ subsystem in linear case $(n = 2$ and $m = 3)$

$$\pi_{12} = \left\{ \begin{matrix} a_{11} & a_{12} \\ a_{21} & a_{22} \end{matrix} \right\}^2 = (a_{11}a_{22} - a_{21}a_{12})^2.$$

Similarly, the weight of the solution of the linearized subsystem $\{1, 2\}$ is

$$\pi_{12}(\eta) = \left\{ \begin{matrix} \frac{\partial f_1}{\partial x}(\eta) & \frac{\partial f_1}{\partial y}(\eta) \\ \frac{\partial f_2}{\partial x}(\eta) & \frac{\partial f_2}{\partial y}(\eta) \end{matrix} \right\}^2 = \left(\frac{\partial f_1}{\partial x}(\eta)\frac{\partial f_2}{\partial y}(\eta) - \frac{\partial f_2}{\partial x}(\eta)\frac{\partial f_1}{\partial y}(\eta) \right)^2.$$

Then computing the other two weights similarly, the approximate Gauss–Jacobi solution of the overdetermined nonlinear system is

$$x_{GJ} = \frac{\pi_{12}x_{12} + \pi_{13}x_{13} + \pi_{23}x_{23}}{\pi_{12} + \pi_{13} + \pi_{23}}$$

and

$$y_{GJ} = \frac{\pi_{12}y_{12} + \pi_{13}y_{13} + \pi_{23}y_{23}}{\pi_{12} + \pi_{13} + \pi_{23}}.$$

The precision of the solution can be increased by further linearization of the system at (x_{GJ}, y_{GJ}) and using a pseudoinverse solution.

Remarks Keep in mind that we did not need initial values!

It goes without saying that after computing the average of the solutions of the subsystems, one may employ standard iteration techniques; however there is no guarantee for convergence.

Let us illustrate the method via solving the following system,

$$f_1(x, y) = x^2 - xy + y^2 + 2x,$$
$$f_2(x, y) = x^2 - 2xy + y^2 + x - y - 2,$$
$$f_3(x, y) = xy + y^2 - x + 2y - 3.$$

We are looking for the real solution in least square sense

$$\sum_{i=1}^{3}\left(f_i^2(x,y)\right)^2 \to \min_{x,y}.$$

$\Rightarrow$ f1 = x^2 − xy + y^2 + 2x;
$\Rightarrow$ f2 = x^2 − 2xy + y^2 + x − y − 2;
$\Rightarrow$ f3 = xy + y^2 − x + 2y − 3;

The solutions of the subsystems {1, 2} is

$\Rightarrow$ sol12 = NSolve[{f1, f2}, {x, y}]
$\Leftarrow$ {{x → −2., y → 0.}, {x → −2., y → 0.},
        {x → −0.5 + 0.866025 i, y → −1.5 + 0.866025 i},
        {x → −0.5 − 0.866025 i, y → −1.5 − 0.866025 i}}

The real solution,

$\Rightarrow$ xy12 = sol12[[1]]
$\Leftarrow$ {x → −2., y → 0.}

The subsystem {1, 3} has two real solutions

$\Rightarrow$ sol13 = NSolve[{f1, f3}, {x, y}]
$\Leftarrow$ {{x → −2.43426, y → −0.565741}, {x → −0.5 + 0.866025 i, y → 1.},
        {x → −0.5 − 0.866025 i, y → 1.}, {x → −1.23241, y → −1.76759}}
$\Rightarrow$ xy13a = sol13[[1]]
$\Leftarrow$ {x → −2.43426, y → −0.565741}
$\Rightarrow$ xy13b = sol13[[4]]
$\Leftarrow$ {x → −1.23241, y → −1.76759}

We choose the solution which gives the smaller residual, i.e.,

$\Rightarrow$ obj = f1^2 + f2^2 + f3^2
$\Leftarrow$ $(-2+x+x^2-y-2xy+y^2)^2 + (2x+x^2-xy+y^2)^2 + (-3-x+2y+xy+y^2)^2$

and

$\Rightarrow$ obj/.xy13a
$\Leftarrow$ 0.14225
$\Rightarrow$ obj/.xy13b
$\Leftarrow$ 1.38861

Therefore the first solution $\{x \to -2.43426, y \to -0.565741\}$ is chosen. Considering the subsystem $\{2, 3\}$, we get

$\Rightarrow$ sol23 = NSolve[{f2, f3}, {x, y}]
$\Leftarrow \{\{x \to -2.5, y \to -0.5\}, \{x \to -1., y \to 1.\},$
$\quad \{x \to 2., y \to 1.\}, \{x \to -1., y \to -2.\}\}$

Now, we have four real solutions, the residuals are

$\Rightarrow$ Map[obj/.#&, sol23]
$\Leftarrow \{0.0625, 1., 49., 1.\}$

Therefore we choose again the first solution since it has the smallest residual corresponding with the global minimum,

$\Rightarrow$ sol23a = sol23[[1]]
$\Leftarrow \{x \to -2.5, y \to -0.5\}$

Let us compute the average of the solutions of the subsystems,

$\Rightarrow \{xs, ys\} = \frac{1}{3}$Apply[Plus, Map[{x, y}/.#&, {xy12, xy13a, sol23a}]]
$\Leftarrow \{-2.31142, -0.355247\}$

or

$\Rightarrow$ Mean[Map[{x, y}/.#&, {xy12, xy13a, sol23a}]]
$\Leftarrow \{-2.31142, -0.355247\}$

Now we can compute the weights at $\{xs, ys\}$,

$\Rightarrow$ n12 = det$\left[\left(\begin{matrix} D[f1, x] & D[f1, y] \\ D[f2, x] & D[f2, y] \end{matrix}\right)\right]^2$/.$\{x \to xs, y \to ys\}$

$\Leftarrow 3.76967$

$\Rightarrow$ n13 = det$\left[\left(\begin{matrix} D[f1, x] & D[f1, y] \\ D[f3, x] & D[f3, y] \end{matrix}\right)\right]^2$/.$\{x \to xs, y \to ys\}$

$\Leftarrow 20.1326$

$\Rightarrow$ n23 = det$\left[\left(\begin{matrix} D[f2, x] & D[f2, y] \\ D[f3, x] & D[f3, y] \end{matrix}\right)\right]^2$/.$\{x \to xs, y \to ys\}$

$\Leftarrow 47.9295$

Then the weighted solution is

$$\Rightarrow \{xy12, xy13, xy23\} = Map[\{x, y\}/.\#\&, \{xy12, xy13a, sol23a\}]$$
$$\Leftarrow \{\{-2., 0.\}, \{-2.43426, -0.565741\}, \{-2.5, -0.5\}\}$$
$$\Rightarrow \{xGJ, yGJ\} = \frac{n12xy12 + n13xy13 + n23xy23}{n12 + n13 + n23}$$
$$\Leftarrow \{-2.45533, -0.492186\}$$

Again, we did not need to guess an initial value, and no iteration was necessary! However, we may repeat (this is not a traditional numerical iteration) the algorithm.

Let us employ one step iteration, in order to improve the solution. The linear form of $f(x, y)$ at $(x, y) = (x0, y0)$ is

$$f(x, y) = f(x0, y0) + \frac{\partial f(x0, y0)}{\partial x}(x - x0) + \frac{\partial f(x0, y0)}{\partial y}(y - y0).$$

Therefore the linearized system is

$$\Rightarrow A = \begin{pmatrix} D[f1, x] & D[f1, y] \\ D[f2, x] & D[f2, y] \\ D[f3, x] & D[f3, y] \end{pmatrix} /.\{x \rightarrow xGJ, y \rightarrow yGJ\}; MatrixForm[A]$$

$$\Leftarrow \begin{pmatrix} -2.41848 & 1.47096 \\ -2.9263 & 2.9263 \\ -1.49219 & -1.43971 \end{pmatrix}$$

$$\Rightarrow b = \begin{pmatrix} D[f1, x]x + D[f1, y]y - f1 \\ D[f2, x]x + D[f2, y]y - f2 \\ D[f3, x]x + D[f3, y]y - f3 \end{pmatrix} /.\{x \rightarrow xGJ, y \rightarrow yGJ\}; MatrixForm[b]$$

$$\Leftarrow \begin{pmatrix} 5.06243 \\ 5.85395 \\ 4.45073 \end{pmatrix}$$

Having A and b, the solution of the linearized system is

$$\Rightarrow \{xss, yss\} = Flatten[LeastSquares[A, b]]$$
$$\Leftarrow \{-2.4646, -0.500688\}$$

Then the residual

$$\Rightarrow obj/.\{x \rightarrow xss, y \rightarrow yss\}$$
$$\Leftarrow 0.0403161$$

Let us check the result via direct minimization of $f = f_1^2 + f_2^2 + f_3^2$.

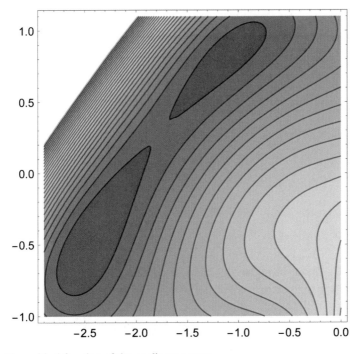

Fig. 3.5 The residual function of the nonlinear system

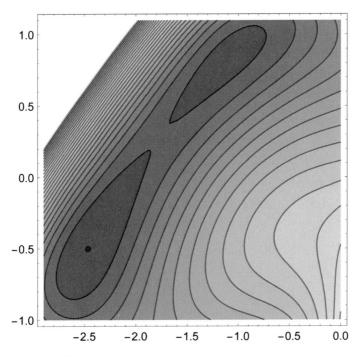

Fig. 3.6 The global minimum

There are two local minimums, see Fig. 3.5. We are looking for the global one, see Fig. 3.6.

> NMinimize[obj, {x, y}, Method → {RandomSearch, SearchPoints → 1000}]
⇐ {0.0403159, {x → −2.46443, y → −0.500547}}

This is really the global one, since the other minimum is

> FindMinimum[obj, {x, −1}, {y, 0.75}]
⇐ {0.631509, {x → −1.09675, y → 0.835606}}

Comparing the global minimum to the Gaussian solution with one iteration step,

> {xss, yss}
⇐ {−2.4646, −0.500688}

Remark The main handicap of the Gauss–Jacobi method is that frequently, we have cases where $m \gg n$, which will result in too many subsystems (combinatorics runaway!).

The advantage is that the solutions of the subsystems can be computed in parallel.

3.4   Transforming Overdetermined System into a Determined System

Let us consider again our overdetermined system. We are looking for the solution in the least squares sense. Therefore, the solution should minimize the following objective function,

$$G(x, y) = f_1(x, y)^2 + f_2(x, y)^2 + f_3(x, y)^2.$$

The necessary condition of the optimum is

$$F_1(x, y) = \frac{\partial G}{\partial x}(x, y) = 0,$$
$$F_2(x, y) = \frac{\partial G}{\partial y}(x, y) = 0.$$

This is already a determined system.

$\Rightarrow$ F1 = D[obj,x]

$\Leftarrow 2(1+2x-2y)(-2+x+x^2-y-2xy+y^2)$

$\quad +2(2+2x-y)(2x+x^2-xy+y^2)+2(-1+y)(-3-x+2y+xy+y^2)$

$\Rightarrow$ F2 = D[obj,y]

$\Leftarrow 2(-1-2x+2y)(-2+x+x^2-y-2xy+y^2)+$

$\quad 2(-x+2y)(2x+x^2-xy+y^2)+2(2+x+2y)(-3-x+2y+xy+y^2)$

Let us compute the real solutions of this algebraic system,

$\Rightarrow$ sol = NSolve[{F1,F2},{x,y}, Reals]

$\Leftarrow \{\{x \rightarrow -2.46443, y \rightarrow -0.500547\},$

$\quad \{x \rightarrow -1.76765, y \rightarrow 0.282464\}, \{x \rightarrow -1.09675, y \rightarrow 0.835606\},$

$\quad \{x \rightarrow -1.8227, y \rightarrow -1.28262\}, \{x \rightarrow -1.06936, y \rightarrow -1.8973\}\}$

The number of the real solutions is

$\Rightarrow$ Length[sol]

$\Leftarrow 5$

We choose the solution that provides the smallest residual

$\Rightarrow$ Map[obj/.#&, sol]

$\Leftarrow \{0.0403159, 1.23314, 0.631509, 6.14509, 0.577472\},$

which is the first one,

$\Rightarrow$ sol[[1]]

$\Leftarrow \{x \rightarrow -2.46443, y \rightarrow -0.500547\}$

3.5   Extended Newton–Raphson Method

The Newton–Raphson method solves a linear system by iteration. The coefficient matrix is the Jacobian of the linearized system, which is a square matrix with full rank when the nonlinear system is determined. Let us consider the nonlinear system $f(x)$, then, in the i-th iteration step, we have the linearization as

$$f(x_{i+1}) = f(x_i) + J(x_i)(x_{i+1} - x_i).$$

Then the linear system to be solved is $f(x_{i+1}) = 0$, which means we should compute x_{i+1} from the linear system,

$$J(x_i)\, x_{i+1} = J(x_i)x_i - f(x_i).$$

When the nonlinear system is over- or underdetermined, this linear system is over- or underdetermined, too. Therefore one can generalize or extend the Newton–Raphson method for over- or underdetermined systems by employing a pseudoinverse solution.

There is a *Mathematica* function to do that

⇒ < <Newton 'NewtonExtended'

⇒ ?NewtonExtended

Computes the solution of an overdetermined nonlinear system
Input parameters:
f - list of functions of the system.
x - list of variables,
x0 - list of the initial values,
eps - error limit for the iteration, default value: 10^-12
n - maximum number of the iterations, default value: 100
Output:
list of the iterative solutions

Let us employ this method to solve our overdetermined problem. Since this method is a local technique, we generate ten initial values in the region $[-2, 0] \times [-1, 1]$, randomly

⇒ iv = {RandomReal[{−2,0}, 10], RandomReal[{−1,1}, 10]}//Transpose
⇐ {{−0.211648, −0.843957}, {−0.458807, −0.27978},
    {−0.907141, −0.160848}, {−1.34021, −0.596869},
    {−0.806718, 0.114513}, {−1.60056, 0.930948}, {−0.825661, −0.738885},
    {−0.819798, −0.16371}, {−0.893981, 0.26915}, {−1.53862, −0.121982}}

Then the solutions are,

⇒ soli = Map[Last[NewtonExtended[{f1, f2, f3}, {x, y}, #]]&, iv]
⇐ {{−1.06936, −1.8973}, {−1.09675, 0.835606},
 {−1.09675, 0.835606}, {−1.09675, 0.835606},
 {−1.09675, 0.835606}, {−1.09675, 0.835606}, {−1.06936, −1.8973},
 {−1.09675, 0.835606}, {−1.09675, 0.835606}, {−2.46443, −0.500547}}

We choose the solution, which provide the smallest residual,

⇒ ob = Map[obj/.{x → #[[1]], y → #[[2]]}&, soli]
⇐ {0.577472, 0.631509, 0.631509, 0.631509, 0.631509,
 0.631509, 0.577472, 0.631509, 0.631509, 0.0403159}

The position of this minimum in the list is

⇒ First[Position[ob, Min[ob]]//Flatten]
⇐ 10.

Then the corresponding solution is

⇒ soli[[%]]
⇐ {−2.46443, −0.500547}

Alternatively, we can employ the average of the solutions of the subsystems (see Gauss-Jacobi solution) as an initial value,

⇒ {xs, ys}
⇐ {−2.31142, −0.355247}

namely,

⇒ soli = NewtonExtended[{f1, f2, f3}, {x, y}, {xs, ys}]//Last
⇐ {−2.46443, −0.500547}

Remark The solutions belong to different initial values and can be computed in parallel.

3.6 Solution of Underdetermined Systems

We have seen that this problem can be considered as a constrained minimization problem, which can be solved in different ways. Our system represents such a constraint. For example

$$\Rightarrow \text{f1} = 3x^2 + 2xy + 2y^2 + z + 3u + xv - 6;$$
$$\Rightarrow \text{f2} = 2x^2 + x + y^2 + 10z + 2u + yv - 23;$$
$$\Rightarrow \text{f3} = 3x^2 + xy + 2y^2 + 2z + 9u + zv - 9;$$
$$\Rightarrow \text{f4} = x^2 + 3y^2 + 2z + 3u + uv - 5;$$

and the objective function to be minimized is the norm of the solution,

$$\Rightarrow G = x^2 + y^2 + z^2 + u^2 + v^2$$

3.6.1 Direct Minimization

We employ global constrained minimization. The solutions is

$$\Rightarrow \text{sol1} = \text{NMinimize}[\{G, \text{f1} == 0, \text{f2} == 0, \text{f3} == 0, \text{f4} == 0\}, \{x, y, z, u, v\}]$$
$$\Leftarrow \{5.96532, \{x \rightarrow 0.975755, y \rightarrow 0.0701982,$$
$$z \rightarrow 2.00317, u \rightarrow 0.00669931, v \rightarrow 0.997781\}\}$$

This is a solution satisfying the equations to a certain error,

$$\Rightarrow \{\text{f1}, \text{f2}, \text{f3}, \text{f4}\}/.\text{sol1}[[2]]$$
$$\Leftarrow \{6.28357 \times 10^{-8}, -1.39885 \times 10^{-7}, -1.15531 \times 10^{-7}, 4.31944 \times 10^{-8}\}$$

3.6.2 Method of Lagrange Multipliers

Introducing new variables λ_i, we can transform the constrained problem into an unconstrained one,

$$\Rightarrow G = G + \lambda 1\,\text{f1} + \lambda 2\,\text{f2} + \lambda 3\,\text{f3} + \lambda 4\,\text{f4}$$
$$\Leftarrow u^2 + v^2 + x^2 + y^2 + z^2 + \left(-6 + 3u + vx + 3x^2 + 2xy + 2y^2 + z\right)\lambda 1$$
$$+ \left(-23 + 2u + x + 2x^2 + vy + y^2 + 10z\right)\lambda 2$$
$$+ \left(-9 + 9u + 3x^2 + xy + 2y^2 + 2z + vz\right)\lambda 3 + \left(-5 + 3u + uv + x^2 + 3y^2 + 2z\right)\lambda 4$$

The necessary conditions of the optimum is that every partial derivative is zero,

$\Rightarrow$ eqs = Map[D[G, #]&,{x, y, z, u, v,λ1,λ2,λ3,λ4}]
$\Leftarrow$ {2x + (v + 6x + 2y)λ1 + (1 + 4x)λ2 + (6x + y)λ3 + 2xλ4, 2y + (2x + 4y)λ1 + (v + 2y)λ2
+ (x + 4y)λ3 + 6yλ4, 2z + λ1 + 10λ2 + (2 + v)λ3 + 2λ4, 2u + 3λ1 + 2λ2 + 9λ3 + (3 + v)λ4,
2v + xλ1 + yλ2 + zλ3 + uλ4, $-6 + 3u + vx + 3x^2 + 2xy + 2y^2 + z, -23 + 2u + x + 2x^2 + vy + y^2$
$+ 10z, -9 + 9u + 3x^2 + xy + 2y^2 + 2z + vz, -5 + 3u + uv + x^2 + 3y^2 + 2z$}

This algebraic system can be solved via numerical Gröbner basis,

$\Rightarrow$ sol = NSolve[eqs,{x, y, z, u, v,λ1,λ2,λ3,λ4}];

Let us select the real solutions

$\Rightarrow$ sol2 =
Select[sol,
(Im[#[[1, 2]]] == 0 $\wedge$ Im[#[[2, 2]]] == 0 $\wedge$ Im[#[[3, 2]]] == 0$\wedge$
Im[#[[4, 2]]] == 0 $\wedge$ Im[#[[5, 2]]] == 0)&]
$\Leftarrow$ {{x $\rightarrow$ 0.0532491,y $\rightarrow$ -2.15786,z $\rightarrow$ 3.14204,u $\rightarrow$ -2.1505,v $\rightarrow$ 4.09418,
λ1 $\rightarrow$ -5.62931,λ2 $\rightarrow$ -0.553225,λ3 $\rightarrow$ -0.395869,λ4 $\rightarrow$ 3.64499},
{x $\rightarrow$ -0.237014, y $\rightarrow$ -0.137118, z $\rightarrow$ 2.07153, u $\rightarrow$ 1.03871, v $\rightarrow$ -2.28339,
λ1 $\rightarrow$ -0.607563, λ2 $\rightarrow$ -1.21588, λ3 $\rightarrow$ -0.100285, λ4 $\rightarrow$ 4.29743},
{x $\rightarrow$ 0.975687,y $\rightarrow$ 0.0703823, z $\rightarrow$ 2.00318, u $\rightarrow$ 0.00670792,
v $\rightarrow$ 0.997813,λ1 $\rightarrow$ 0.416967,λ2 $\rightarrow$ -0.619598,λ3 $\rightarrow$ -1.18648,
λ4 $\rightarrow$ 2.66475}, {x $\rightarrow$ -1.38903,y $\rightarrow$ -0.0438894,
z $\rightarrow$ 2.11802, u $\rightarrow$ -0.308493, v $\rightarrow$ 0.796651,λ1 $\rightarrow$ -0.0577799,
λ2 $\rightarrow$ -0.630616,λ3 $\rightarrow$ -0.538566,λ4 $\rightarrow$ 1.81704}}

Finding the corresponding minimal norm can remove the extraneous solutions corresponding to local minimums. The lengths of the norm of the different real solutions are

$\Rightarrow$ ss = Map[Norm[{x, y, z, u, v}]/.#&, %]
$\Leftarrow$ {5.99321, 3.26482, 2.4424, 2.67342}

Then we choose the solution providing the smallest norm

$\Rightarrow$ sol2[[Position[ss, Min[ss]]//Flatten//First]]
$\Leftarrow$ {x $\rightarrow$ 0.975687,y $\rightarrow$ 0.0703823,z $\rightarrow$ 2.00318,
u $\rightarrow$ 0.00670792,v $\rightarrow$ 0.997813, λ1 $\rightarrow$ 0.416967,
λ2 $\rightarrow$ -0.619598,λ3 $\rightarrow$ -1.18648,λ4 $\rightarrow$ 2.66475}

This is a solution indeed since substituting back, we get

$\Rightarrow \{f1, f2, f3, f4\}/.\%$
$\Leftarrow \{4.44089 \times 10^{-16}, 0., 8.88178 \times 10^{-16}, 8.88178 \times 10^{-16}\}$

The solution resulting from the direct global minimization above was the same, see Sect. 3.6.1.

$\Rightarrow sol1$
$\Leftarrow \{5.96532, \{x \rightarrow 0.975755, y \rightarrow 0.0701982, z \rightarrow 2.00317,$
$\quad u \rightarrow 0.00669931, v \rightarrow 0.997781\}\}$

Remark In case of non-algebraic system, we may employ the homotopy method discussed in the earlier chapter.

3.6.3 Method of Penalty Function

Let us introduce a parameter K as a weight of the norm of the constraints. Then our objective function is

$\Rightarrow F(K\_) := u^2 + v^2 + x^2 + y^2 + z^2 + K(f1^2 + f2^2 + f3^2 + f4^2)$

Employing an initial value for the parameter $K = 10$, we get

$\Rightarrow sol = NMinimize[F[10], \{x, y, z, u, v\}]$
$\Leftarrow \{5.75969, \{x \rightarrow 0.998177, y \rightarrow 0.0639928, z \rightarrow 1.9847,$
$\quad u \rightarrow 0.037631, v \rightarrow 0.79637\}\}$

The constraints are not satisfied correctly,

$\Rightarrow \{f1, f2, f3, f4\}/.sol[[2]]$
$\Leftarrow \{0.0175328, -0.0317418, -0.0502123, 0.120914\}$

Let us increase the value of $K = 500$,

$\Rightarrow sol = NMinimize[F(500), \{x, y, z, u, v\},$
$\quad Method \rightarrow \{"RandomSearch", "SearchPoints" \rightarrow 500\}]$
$\Leftarrow \{5.96079, \{x \rightarrow 0.976156, y \rightarrow 0.0702523, z \rightarrow 2.0028,$
$\quad u \rightarrow 0.00735433, v \rightarrow 0.993576\}\}$

This is a better, acceptable solution,

⇒ {f1, f2, f3, f4}/.sol[[2]]

⇐ {0.000415708, −0.000620085, −0.00118282, 0.0026602}

3.6.4 Extended Newton–Raphson

Let us guess five initial values randomly between {− 1, 1},

⇒ x0 = Table[1 − 2 RandomReal[], {5}]

⇐ {0.89982, 0.676421, 0.376708, −0.842843, −0.867964}

Then the Extended Newton–Raphson can be employed again,

⇒ sol = NewtonExtended[{f1, f2, f3, f4}, {x, y, z, u, v}, x0]//Last

⇐ {1.01847, 0.673925, 2.03846, 1.02869, −4.43554}

⇒ Norm[sol]

⇐ 5.13605

⇒ %^2

⇐ 5.98607

Substituting the results back in to the equation,

⇒ {f1, f2, f3, f4}/.MapThread[(#1 − > #2)&, {{x, y, z, u, v}, sol}]

⇐ {1.77636×10^{-15}, 3.552716×10^{-15}, -1.77636×10^{-15}, 8.88178×10^{-16}}

Here parallel computation can be employed again.

3.7 Applications

3.7.1 Geodetic Application—The Minimum Distance Problem

The problem definition: Given a point in space P (X, Y, Z) and an ellipsoid, find the closest ellipsoid point $p(x, y, z)$ to this P, see Fig. 3.7. We will assume the ellipsoid has a circular cross section on the xy-plane. (See also Sects. 1.4.2 and 1.4.3 where this problem is solved symbolically.)

Fig. 3.7 The closest point
problem

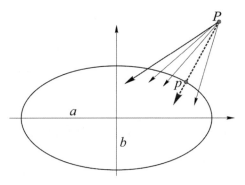

The distance to be minimized is

$$\Rightarrow d = (X - x)^2 + (Y - y)^2 + (Z - z)^2 ;$$

The point p is on the surface of the ellipsoid, therefore the coordinates of point p
(x, y, z) should satisfy the following constraint,

$$\Rightarrow c = \tfrac{x^2 + y^2}{a^2} + \tfrac{z^2}{b^2} - 1 ;$$

This optimization problem can be represented as an underdetermined problem,
where the norm of the solution vector is the distance to be minimized. Let us solve
this problem using the *Extended Newton–Raphson* method. Introducing new variables (α, β, γ)

$$\Rightarrow \text{denoting} = \{x \to X{-}\alpha, y \to Y{-}\beta, z \to Z{-}\gamma\} ;$$

Then the constraint is,

$$\Rightarrow \text{eq} = c/.\text{denoting}$$
$$\Leftarrow \frac{(X{-}\alpha)^2 + (Y{-}\beta)^2}{a^2} + \frac{(Z{-}\gamma)^2}{b^2} - 1$$

or

$$\Rightarrow \text{eq} = \text{eq}\,a^2 b^2 //\text{Expand}$$
$$\Leftarrow -a^2 b^2 + a^2 \gamma^2 + a^2 z^2 - 2a^2 \gamma z + \alpha^2 b^2 + b^2 \beta^2 + b^2 X^2 - 2\alpha b^2 X + b^2 Y^2 - 2b^2 \beta Y$$

Using actual numerical data, let the values of the coordinates of P be

$\Rightarrow \text{data} = \{X \rightarrow 3770667.9989, Y \rightarrow 446076.4896,$
$\quad Z \rightarrow 5107686.2085, a \rightarrow 6378136.602, b \rightarrow 6356751.860\};$
$\Rightarrow \text{eqn} = \text{eq}/.\text{data}$
$\Leftarrow 4.04083 \times 10^{13}\alpha^2 - 3.04733 \times 10^{20}\alpha + 4.04083 \times 10^{13}\beta^2$
$\quad -3.60504 \times 10^{19}\beta + 4.06806 \times 10^{13}\gamma^2 - 4.15568 \times 10^{20}\gamma + 2.33091 \times 10^{22}$

Normalizing the equation,

$\Rightarrow \text{eqn} = \text{eqn}/\text{eqn}[[1]]//\text{Expand}$
$\Leftarrow 1. -0.0130735\,\alpha + 1.73358 \times 10^{-9}\alpha^2 - 0.001546622\,\beta$
$\quad + 1.73358 \times 10^{-9}\beta^2 - 0.0178286\,\gamma + 1.74527 \times 10^{-9}\gamma^2$

Now, we have one equation with three unknowns (α, β, γ). So we have an underdetermined system of one equation with three unknowns. The norm of the solution vector (α, β, γ) should be minimized,

$$d = \alpha^2 + \beta^2 + \gamma^2$$

A reasonable initial guess can be $(\alpha, \beta, \gamma) = \{0, 0, 0\}$,

$\Rightarrow \text{sol} = \text{Last}[\text{NewtonExtended}(\{\text{eqn}\}, \{\alpha, \beta, \gamma\}, \{0., 0., 0.\})]$
$\Leftarrow \{26.6174, 3.14888, 36.2985\}$

Therefore, the coordinates of point $p(x, y, z)$ are

$\Rightarrow \text{sol} = \{X-\alpha, Y-\beta, Z-\gamma\}/.\text{data}/.$
$\quad \{\alpha \rightarrow \text{sol}[[1]], \beta \rightarrow \text{sol}[[2]], \gamma \rightarrow \text{sol}[[3]]\}$
$\Leftarrow \{3.77064 \times 10^6, 446073., 5.10765 \times 10^6\}$

Displayed in more digits as

$\Rightarrow \text{NumberForm}[\%, 15]$
$\Leftarrow \{3.77064138151124 \times 10^6, 446073.34071727, 5.10764991001691 \times 10^6\}$

The solution cab now be visualized in Fig. 3.8.

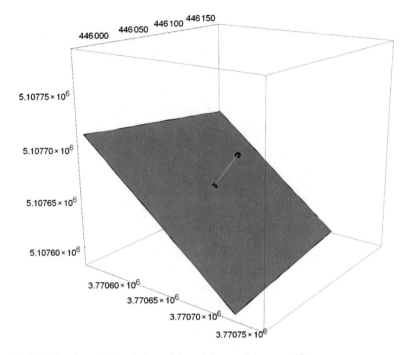

Fig. 3.8 Visualization of the solution of the minimum distance problem

3.7.2   Global Navigation Satellite System (GNSS) Application

Let us consider the satellite global positioning problem discussed and solved via homotopy method in Sect. 2.11.2. Recall that we need to determine the coordinates $\{X, Y, Z\}$ of a location on Earth employing the Global Navigation Satellite System (GNSS). The equations relating the satellite position and the unknown point to be positioned are given in terms of distances as

$$d_i = \sqrt{(a_i - X)^2 + (b_i - Y)^2 + (c_i - Z)^2} + \xi \quad i = 0, \ldots n > 3,$$

where $\{a_i, b_i, c_i\}$ are the coordinates of the satellite and ξ the adjustment of the satellite clock as additional unknown variable.

We introduce new variables: x_1, x_2, x_3 and x_4 for the 4 unknowns $\{X, Y, Z, \xi\}$. Since the number of the satellites is $n = 6 > 3$, we have an overdetermined system.

The numerical data are

$\Rightarrow$ datan $= \{a_0 \rightarrow 14177553.47, a_1 \rightarrow 15097199.81, a_2 \rightarrow 23460342.33,$
 $a_3 \rightarrow -8206488.95, a_4 \rightarrow 1399988.07, a_5 \rightarrow 6995655.48,$
 $b_0 \rightarrow -18814768.09, b_1 \rightarrow -4636088.67, b_2 \rightarrow -9433518.58,$
 $b_3 \rightarrow -18217989.14, b_4 \rightarrow -17563734.90, b_5 \rightarrow -23537808.26,$
 $c_0 \rightarrow 12243866.38, c_1 \rightarrow 21326706.55, c_2 \rightarrow 8174941.25,$
 $c_3 \rightarrow 17605231.99, c_4 \rightarrow 19705591.18, c_5 \rightarrow -9927906.48,$
 $d_0 \rightarrow 21119278.32, d_1 \rightarrow 22527064.18, d_2 \rightarrow 23674159.88,$
 $d_3 \rightarrow 20951647.38, d_4 \rightarrow 20155401.42, d_5 \rightarrow 24222110.91\};$

The number of satellites

$\Rightarrow$ m $= 6$;

The i-th equation is,

$\Rightarrow$ e $=(x1 - a_i)^2 + (x2 - b_i)^2 + (x3 - c_i)^2 - (x4 - d_i)^2;$

Let us choose the first 4 equations as a determined subsystem,

$\Rightarrow$ eqs $=$ Table$[e, \{i, 1, 4\}]/$.datan
$\Leftarrow \{(-1.50972 \times 10^7 + x1)^2 + (4.63609 \times 10^6 + x2)^2$
 $+ (-2.13267 \times 10^7 + x3)^2 - (-2.25271 \times 10^7 + x4)^2, (-2.34603 \times 10^7 + x1)^2$
 $+ (9.43352 \times 10^6 + x2)^2 + (-8.17494 \times 10^6 + x3)^2 - (-2.36742 \times 10^7 + x4)^2,$
 $(8.20649 \times 10^6 + x1)^2 + (1.8218 \times 10^7 + x2)^2 + (-1.76052 \times 10^7 + x3)^2$
 $-(-2.09516 \times 10^7 + x4)^2, (-1.39999 \times 10^6 + x1)^2$
 $+ (1.75637 \times 10^7 + x2)^2 + (-1.97056 \times 10^7 + x3)^2 - (-2.01554 \times 10^7 + x4)^2\}$

and solve them via numerical Gröbner basis

$\Rightarrow$ solin $=$ NSolve$[$eqs$, \{x1, x2, x3, x4\}]$
$\Leftarrow \{\{x1 \rightarrow -2.68254 \times 10^6, x2 \rightarrow 1.16034 \times 10^7,$
 $x3 \rightarrow -9.36553 \times 10^6, x4 \rightarrow 6.1538 \times 10^7\},$
 $\{x1 \rightarrow 597005., x2 \rightarrow -4.84797 \times 10^6, x3 \rightarrow 4.0883 \times 10^6, x4 \rightarrow 120.59\}\}$

We shall use the second solution as an initial guess for the Extended Newton–Raphson method because it has a physical reality. The overdetermined system with the six equations is given as,

⇒ eqT = Table[e, {i, 0, 5}]/.datan

$\Leftarrow \{(x1 - 1.41776 \times 10^7)^2 + (x2 + 1.88148 \times 10^7)^2 + (x3 - 1.22439 \times 10^7)^2$
$\quad -(x4 - 2.11193 \times 10^7)^2, (x1 - 1.50972 \times 10^7)^2 + (x2 + 4.63609 \times 10^6)^2$
$\quad + (x3 - 2.13267 \times 10^7)^2 - (x4 - 2.25271 \times 10^7)^2, (x1 - 2.34603 \times 10^7)^2$
$\quad + (x2 + 9.43352 \times 10^6)^2 + (x3 - 8.17494 \times 10^6)^2 - (x4 - 2.36742 \times 10^7)^2,$
$\quad (x1 + 8.20649 \times 10^6)^2 + (x2 + 1.8218 \times 10^7)^2 + (x3 - 1.76052 \times 10^7)^2$
$\quad -(x4 - 2.09516 \times 10^7)^2, (x1 - 1.39999 \times 10^6)^2 + (x2 + 1.75637 \times 10^7)^2$
$\quad + (x3 - 1.97056 \times 10^7)^2 - (x4 - 2.01554 \times 10^7)^2, (x1 - 6.99566 \times 10^6)^2$
$\quad + (x2 + 2.35378 \times 10^7)^2 + (x3 + 9.92791 \times 10^6)^2 - (x4 - 2.42221 \times 10^7)^2\}$

Let us employ Newton's method

⇒ sol = NewtonExtended
    [eqT, {x1, x2, x3, x4}, {x1, x2, x3, x4}/.solin[[2]]]//Last
$\Leftarrow \{596929., -4.84785 \times 10^6, 4.08822 \times 10^6, 13.4526\}$
⇒ NumberForm[%, 12]
$\Leftarrow \{596928.910449, -4.84784931442 \times 10^6, 4.08822444717 \times 10^6, 13.4525758581\}$

Now, let us check the results via direct minimization. The function to be minimized is

⇒ f = Apply[Plus, Table[e²/.datan, i, 0, m − 1]]//Simplify

$\Leftarrow ((-6.99566 \times 10^6 + x1)^2 + (2.35378 \times 10^7 + x2)^2 + (9.92791 \times 10^6 + x3)^2$
$\quad -1.(-2.42221 \times 10^7 + x4)^2)^2 + ((-2.34603 \times 10^7 + x1)^2 + (9.43352 \times 10^6 + x2)^2$
$\quad + (-8.17494 \times 10^6 + x3)^2 - 1.(-2.36742 \times 10^7 + x4)^2)^2 + ((-1.50972 \times 10^7 + x1)^2$
$\quad + (4.63609 \times 10^6 + x2)^2 + (-2.13267 \times 10^7 + x3)^2 - 1.(-2.25271 \times 10^7 + x4)^2)^2$
$\quad + ((-1.41776 \times 10^7 + x1)^2 + (1.88148 \times 10^7 + x2)^2 + (-1.22439 \times 10^7 + x3)^2$
$\quad -1.(-2.11193 \times 10^7 + x4)^2)^2 + ((8.20649 \times 10^6 + x1)^2 + (1.8218\ 10^7 + x2)^2$
$\quad + (-1.76052 \times 10^7 + x3)^2 - 1.(-2.09516 \times 10^7 + x4)^2)^2 + ((-1.39999 \times 10^6 + x1)^2$
$\quad + (1.75637 \times 10^7 + x2)^2 + (-1.97056 \times 10^7 + x3)^2 - 1.(-2.01554 \times 10^7 + x4)^2)^2$

Then, the global minimization via the **NMinimize** becomes,

⇒ solN = NMinimize[f, {x1, x2, x3, x4}]//AbsoluteTiming
$\Leftarrow \{0.0700622, \{2.21338 \times 10^{18},$
$\quad \{x1 \rightarrow 59693., x2 \rightarrow -4.84785 \times 10^6, x3 \rightarrow 4.0882 \times 10^6, x4 \rightarrow 13.4526\}\}\}$

or

⇒ NumberForm[%, 12]
$\Leftarrow \{0.0700621592115, \{2.21337940327 \times 10^{18},$
$\quad \{x1 \rightarrow 596928.910449, x2 \rightarrow -4.84784931442 \times 10^6,$
$\quad x3 \rightarrow 4.08822444717 \times 10^6, x4 \rightarrow 13.4525758564\}\}\}$

3.7.3 Geometric Application

The infrared depth camera of Microsoft Kinect XBOX (see, Fig. 3.9) can provides 2D RGB color image and the RGB data representing the depth—the distance of the object—in 11 bits (0.2048).

This low resolution causes discontinuities, which can be as much as 10 cm above 4 m object distance, and is about 2 cm when the object distance is less than 2.5 m. In our case, a spherical object having a radius $R = 0.152$ m was placed in the real world position $x = 0$, $y = 0$ and the object distance $z = 3$ m. Since the intensity threshold resolution is low, one may have quantized levels caused by round-off processes. Figure 3.10 represents the quantized depth data points simulated as synthetic data for illustration.

These points are those of the measured points of the sphere, but they have such layer type form because of the low depth resolution, see Fig. 3.11.

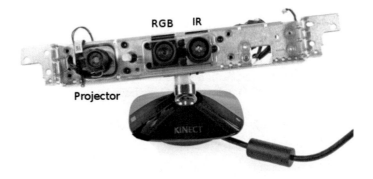

Fig. 3.9 Microsoft Kinect XBOX

Fig. 3.10 Quantized depth data points

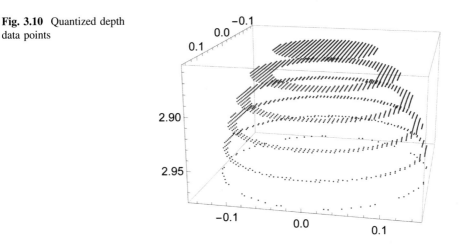

Fig. 3.11 The sphere with
the measured points

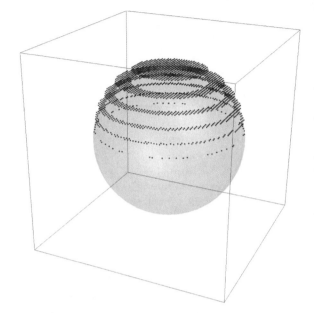

Our task is to compute the position—coordinates of the center point (a, b, c)—
and the radius of the sphere (r) from the measured data, the coordinates of the
points.

In this case, we want to fit the given n points to a sphere such that the sum of the
squared *algebraic distance* is minimized. The local error is defined as

$$\delta_i = (x_i - a)^2 + (y_i - b)^2 + (z_i - c)^2 - R^2,$$

where (x_i, y_i, z_i) are the coordinates of the measured points. First, let us suppose that
the position as well as the radius of the sphere is unknown.

The parameters to be estimated are R, a, b and c. Let us introduce a new
parameter,

$$d = a^2 + b^2 + c^2 - R^2.$$

Then δ_i can be expressed as

$$\delta_i = (-2x_i, -2y_i, -2z_i, 1) \begin{pmatrix} a \\ b \\ c \\ d \end{pmatrix} + x_i^2 + y_i^2 + z_i^2.$$

Indeed

$$\Rightarrow \{-2x_i, -2y_i, -2z_i, 1\}. \begin{pmatrix} a \\ b \\ c \\ a^2+b^2+c^2-R^2 \end{pmatrix} + x_i^2 + y_i^2 + z_i^2$$

$$== \{(x_i-a)^2 + (y_i-b)^2 + (z_i-c)^2 - R^2\} // \texttt{Simplify}$$

$\Leftarrow$ True

Consequently the parameter estimation problem is linear, namely one should solve a linear overdetermined system. In matrix form

$$M\,p = h,$$

where

$$M = \begin{pmatrix} -2x_1 & -2y_1 & -2z_1 & 1 \\ \cdots & \cdots & \cdots & \cdots \\ -2x_i & -2y_i & -2z_i & 1 \\ \cdots & \cdots & \cdots & \cdots \\ -2x_n & -2y_n & -2z_n & 1 \end{pmatrix}, \quad p = \begin{pmatrix} a \\ b \\ c \\ d \end{pmatrix} \quad \text{and} \quad h = \begin{pmatrix} -(x_1^2+y_1^2+z_1^2) \\ \cdots \\ -(x_i^2+y_i^2+z_i^2) \\ \cdots \\ -(x_n^2+y_n^2+z_n^2) \end{pmatrix}.$$

Then the solution in a least squares sense is

$$\begin{pmatrix} a \\ b \\ c \\ d \end{pmatrix} = M^{-1} h,$$

where M^{-1} is the pseudoinverse of M.

The attractive feature of the algebraic parameter estimation is that its formulation results in a linear least squares problem, which has a non-iterative closed-form solution. Therefore the algebraic distance based fitting can be used as a quick way to calculate approximate parameter values without requiring an initial guess. First, we create the M matrix from data measured,

Then let us compute a, b, c and d

$\Rightarrow$ AbsoluteTiming[p = {a, b, c, d} = PseudoInverse[M].h]

$\Leftarrow$ {0.0386268, {1.54043 $\times 10^{-15}$, -2.07993×10^{-15}, 2.99766, 8.96339}}

and the estimated radius of the sphere becomes

$\Rightarrow$ R $= \sqrt{a^2 + b^2 + c^2 - d}$

$\Leftarrow$ 0.150345

Now, let us consider the *geometrical distance error*, which is the difference of the distance and not the square of the distance,

$$\Delta_i = \sqrt{(x_i - a)^2 + (y_i - b)^2 + (z_i - c)^2} - R \qquad i = 1, 2, \ldots n.$$

The equations can be easily generated. The prototype of the equation is,

$$\Rightarrow \mathsf{G} = \sqrt{(\mathsf{x} - \mathsf{a})^2 + (\mathsf{y} - \mathsf{b})^2 + (\mathsf{z} - \mathsf{c})^2} - \mathsf{R};$$

Then our system is

$\Rightarrow \mathtt{eqs} = \mathtt{Map[G/.\{x \rightarrow \#[[1]], y \rightarrow \#[[2]], z \rightarrow \#[[3]]\}\&, dataQ]}$; and so we have 2241 nonlinear equations. For example, the first one is,

$\Rightarrow \mathtt{eqs[[1]]}$

$$\Leftarrow \sqrt{(-0.14634 - \mathsf{a})^2 + (-0.0337707 - \mathsf{b})^2 + (2.97744 - \mathsf{c})^2} - \mathsf{R}$$

$\Rightarrow \mathtt{Length[eqs]}$

$\Leftarrow \mathtt{2241}$

Now we employ Newton's method. The previous result employing algebraic distance can be used as an initial guess,

$\Rightarrow \mathtt{solNE = AbsoluteTiming[}$
$\quad \mathtt{NewtonExtended[eqs, \{a, b, c, R\}, \{0, 0, 2.99766, 0.150345\}];]}$
$\Leftarrow \mathtt{\{1.44983, Null\}}$

The solution is

$\Rightarrow \mathtt{solNE//Last}$
$\Leftarrow \mathtt{\{-1.71339 \times 10^{-19}, 1.02352 \times 10^{-18}, 3.0002, 0.152028\}}$

The number of the iterations

$\Rightarrow \mathtt{Length[solNE]}$
$\Leftarrow \mathtt{7}$

One may employ direct minimization, too. The objective function is

$$\rho = \sum_{i=1}^{n} \Delta_i^2 = \sum_{i=1}^{n} \left(\sqrt{(x_i - a)^2 + (y_i - b)^2 + (z_i - c)^2} - R \right)^2.$$

Here we use local method to find the minimum

> solLM = AbsoluteTiming[FindMinimum[Apply[Plus,
  Map[#²&, eqs]], a, 0, b, 0, c, 2.99766, R, 0.150345]]
< {0.605015, {0.0496067,
  {a → 5.93913 × 10⁻¹², b → −1.3497 × 10⁻¹⁹, c → 3.0002, R → 0.152028}}}

Alternatively, we can employ global minimization, which does not require initial guess, however we need constrain of the region of R,

> solGM = AbsoluteTiming[NMinimize[
  {Apply[Plus, Map[#²&, eqs]], 0 < R < 0.5}, a, b, c, R]]
< {6.92959, {0.0496067,
  {a → 5.93913 × 10⁻¹², b → 2.7251 × 10⁻¹¹, c → 3.0002, R → 0.152028}}}

3.8   Exercises

3.8.1   Solution of Overdetermined System

Let us solve the following overdetermined system, transformation into determined system

$$f_1 = x^2 - xy + y^2 - 3,$$
$$f_2 = x^2 - 2xy + x + y^2 - y,$$
$$f_3 = xy - x + y^2 + 2y - 9.$$

via direct minimization, Gauss-Jacobi method and Extended Newton–Raphson

(a) The objective function for direct minimization is,

> G = f1² + f2² + f3²
< $(x + x^2 - y - 2xy + y^2)^2 + (-3 + x^2 - xy + y^2)^2 + (-9 - x + 2y + xy + y^2)^2$

Let us employ global minimization,

> solxy = NMinimize[G, {x, y}]
< 3.9443 × 10⁻³⁰, x → 1., y → 2.

It means that there is a solution, which practically satisfies all of the three equations,

$\Rightarrow \{f1, f2, f3\}/.solxy[[2]]$
$\Leftarrow \{0., 8.88178 \times 10^{-16}, -1.77636 \times 10^{-15}\}$

(b) Now as second technique, we use Gauss–Jacobi combinatorial approach. The solutions of the subsystems (1, 2) is

$\Rightarrow sol12 = NSolve[\{f1, f2\}, \{x, y\}]$
$\Leftarrow \{\{x \rightarrow -2., y \rightarrow -1.\}, \{x \rightarrow 1.73205, y \rightarrow 1.73205\},$
$\quad \{x \rightarrow -1.73205, y \rightarrow -1.73205\}, \{x \rightarrow 1., y \rightarrow 2.\}\}$

We choose the solution, which gives the smaller residual. The residual is

$\Rightarrow G/.sol12$
$\Leftarrow \{36., 1.6077, 22.3923, 4.48862 \times 10^{-28}\}$

This is the last one,

$\Rightarrow sol12op = sol12//Last$
$\Leftarrow \{x \rightarrow 1., y \rightarrow 2.\}$

The subsystem (1, 3) has two real solutions

$\Rightarrow sol13 = NSolve[\{f1, f3\}, \{x, y\}, Reals]$
$\Leftarrow \{\{x \rightarrow 1., y \rightarrow 2.\}, \{x \rightarrow 1.3553, y \rightarrow 1.95137\}\}$

Similar way one can select the proper solution of the subsystem

$\Rightarrow G/.sol13$
$\Leftarrow \{2.68213 \times 10^{-29}, 0.0579703\}$

Therefore we pick up the first solution,

$\Rightarrow sol12op = sol13//First$
$\Leftarrow \{x \rightarrow 1., y \rightarrow 2.\}$

Then for subsystem (2, 3) we get

$\Rightarrow sol23 = NSolve[\{f2, f3\}, \{x, y\}, Reals]$
$\Leftarrow \{\{x \rightarrow -3., y \rightarrow -2.\}, \{x \rightarrow -2.386, y \rightarrow -2.386\},$
$\quad \{x \rightarrow 1.886, y \rightarrow 1.886\}, \{x \rightarrow 1., y \rightarrow 2.\}\}$

Selection of the proper solution for subsystem (2, 3)

$\Rightarrow$ G/.sol23
$\Leftarrow \{16., 7.25225, 0.310248, 2.88709 \times 10^{-26}\}$

So we get the last solution again,

$\Rightarrow$ sol23op = sol23//Last
$\Leftarrow \{x \rightarrow 1., y \rightarrow 2.\}$

Now it is clear, that the solution of the overdetermined system is (1, 2).

(c) As last technique, let us employ the Extended Newton Method,

$\Rightarrow$ < <Newton'NewtonExtended'

Let us employ this method to solve our overdetermined problem. Since this method is a local technique, we generate ten initial values in the region $[-2, 2] \times [-3, 3]$, randomly

$\Rightarrow$ iv = {RandomReal[{-2, 2}, 10], RandomReal[{-3, 3}, 10]}//Transpose
$\Leftarrow \{\{-1.22005, 0.629526\}, \{-0.870999, 0.544002\}, \{0.662122, -2.56815\},$
$\{-1.03441, 0.169312\}, \{1.86959, 1.69326\}, \{1.52015, -2.34583\},$
$\{0.181326, 1.03311\}, \{-1.36694, -2.74561\}, \{-0.331234, -1.9184\},$
$\{-1.81949, -0.354369\}\}$

Then the solutions are

$\Rightarrow$ soli = Map[Last[NewtonExtended[{f1, f2, f3}, {x, y}, #]]&, iv]
$\Leftarrow \{\{1., 2.\}, \{1., 2.\}, \{-2.10295, -2.35517\}, \{1., 2.\}, \{1., 2.\},$
$\{-1.67642, -2.67781\}, \{1., 2.\}, \{-2.19274, -2.34771\},$
$\{-2.29053, -2.23825\}, \{-2.21021, -2.2809\}\}$

One can use here parallel computation. We choose the solution, which provide the smallest residual,

$\Rightarrow$ ob = Map[G/.{x $\rightarrow$ #[[1]], y $\rightarrow$ #[[2]]}&, soli]
$\Leftarrow \{0., 0., 5.39282, 0., 0., 11.2655, 0., 5.46, 5.63858, 5.42019\}$

There are more candidates, but they are the same. The position of this minimum in the list is

$\Rightarrow$ First[Position[ob, Min[ob]]//Flatten]
$\Leftarrow 1$

the corresponding solution is

$\Rightarrow$ soli[[%]]

$\Leftarrow \{1., 2.\}$

3.8.2 Solution of Underdetermined System

Let us solve the following underdetermined system

$$f_1 = x^2 + y^2 + z - 10,$$
$$f_2 = x^2 + x + y^2 - 5$$

via direct minimization, penalty function method, method of Lagrange multipliers and Extended Newton–Raphson.

(a) Using direct minimization means, we select the minimum norm solution from the infinite solutions. Therefore our objective will be the norm of the solution and the constrains are the equations, namely

$\Rightarrow$ sol = NMinimize[{Norm[{x, y}], f1 == 0, f2 == 0}, {x, y}]

$\Leftarrow \{1.79214, \{x \to 1.78823, y \to 0.118427\}\}$

(b) Let us introduce a parameter κ as a weight of the norm of the constrains, then our objective function is

$\Rightarrow$ F[κ_] := Norm[x, y] + κ(f1^2 + f2^2)

Employing an initial value for the parameter $\kappa = 10$

$\Rightarrow$ sol = NMinimize[F[10], {x, y}]

$\Leftarrow \{1.79088, \{x \to 1.78603, y \to .113296\}\}$

The constrains are not satisfied correctly

$\Rightarrow$ {f1, f2}/.sol[[2]]

$\Leftarrow \{0.00039999, -0.0112166\}$

Let us increase the value of $\kappa = 500$

$\Rightarrow$ sol = NMinimize[F[500], {x, y}, Method $\to$ {"RandomSearch",
          "SearchPoints" $\to$ 500}]

$\Leftarrow \{1.79212, \{x \to 1.78818, y \to 0.118325\}\}$

This is a better, acceptable solution,

$\Rightarrow \{\mathtt{f1}, \mathtt{f2}\}/.\mathtt{sol}[[2]]$
$\Leftarrow \{0.000415708, -0.000620085, -0.00118282, 0.0026602\}$

(c) Introducing new variables, we can transform the constrained problem into an unconstrained one. In addition to make the numerical solution easier, just use a little bit different objective for the norm,

$\Rightarrow \mathtt{G} = \mathtt{x}^2 + \mathtt{y}^2 + \lambda 1 \mathtt{f1} + \lambda 2 \mathtt{f2}$
$\Leftarrow \mathtt{x}^2 + \mathtt{y}^2 + (-3 + \mathtt{x}^2 - \mathtt{xy} + \mathtt{y}^2)\lambda 1 + (-5 + \mathtt{x} + \mathtt{x}^2 + \mathtt{y}^2)\lambda 2$

Now let us minimize the objective,

$\Rightarrow \mathtt{solxy} = \mathtt{NMinimize}[\mathtt{G}, \{\mathtt{x}, \mathtt{y}, \lambda 1, \lambda 2\}, \mathtt{Method} \rightarrow \{\text{"}\mathtt{RandomSearch}\text{"},$
   $\text{"}\mathtt{SearchPoints}\text{"} \rightarrow 500\}]$

> Message Template[NMinimize , cvdiv ,
>
> Failed to converge to a solution. The function may be unbounded. ,
>
> 2 , 49 , 8 , 16640943049945555241 , Local]

$\Leftarrow -7.529408327549746 \times 10^{622}, \mathtt{x} \rightarrow 2.75378 * 10^{207}, \mathtt{y} \rightarrow -2.75378 * 10^{207},$
   $\lambda 1 \rightarrow -2.29129 * 10^{207}, \lambda 2 \rightarrow -1.52752 * 10^{207}$

The global optimization technique may fail. Let us try an algebraic way. The necessary conditions of the optimum is

$\Rightarrow \mathtt{eqs} = \mathtt{Map}[\mathtt{D}[\mathtt{G}, \#]\&, \{\mathtt{x}, \mathtt{y}, \lambda 1, \lambda 2\}]$
$\Leftarrow \{2\mathtt{x} + (2\mathtt{x} - \mathtt{y})\lambda 1 + (1 + 2\mathtt{x})\lambda 2, \ 2\mathtt{y} + (-\mathtt{x} + 2\mathtt{y})\lambda 1 + 2\mathtt{y}\lambda 2,$
   $-3 + \mathtt{x}^2 - \mathtt{xy} + \mathtt{y}^2, \ -5 + \mathtt{x} + \mathtt{x}^2 + \mathtt{y}^2\}$

This algebraic system can be solved via numerical Gröbner basis,

$\Rightarrow \mathtt{sol} = \mathtt{NSolve}[\mathtt{eqs}, \{\mathtt{x}, \mathtt{y}, \lambda 1, \lambda 2\}, \mathtt{Reals}]$
$\Leftarrow \{\{\mathtt{x} \rightarrow 0.673918, \mathtt{y} \rightarrow 1.96772, \lambda 1 \rightarrow -0.389762, \lambda 2 \rightarrow -0.676982\},$
   $\{\mathtt{x} \rightarrow 1.78823, \mathtt{y} \rightarrow 0.118427, \lambda 1 \rightarrow 0.0299103, \lambda 2 \rightarrow -0.804091\}\}$

since the second solution has the smaller norm

$\Rightarrow$ Norm[{x, y}]/.sol
$\Leftarrow$ {2.07992, 1.79214}

we choose this one as a solution.

(d) The Extended Newton method can be employed as last technique. Let us try
ten random initial conditions in the region [0, 2] × [0, 1]

$\Rightarrow$ iv = {RandomReal[{0, 2}, 10], RandomReal[{0, 1}, 10]}//Transpose
$\Leftarrow$ {{0.538531, 0.101471}, {1.04811, 0.903041}, {1.7324, 0.413993},
    {1.54789, 0.829299}, {0.565043, 0.591834}, {1.35753, 0.492477},
    {1.32906, 0.67617}, {0.0835006, 0.148338}, {1.93989, 0.0707861},
    {0.18167, 0.155675}}

Then the solutions are

$\Rightarrow$ soli = Map[Last[NewtonExtended[{f1, f2}, {x, y}, #]]&, iv]
$\Leftarrow$ {{1.78823, 0.118427}, {0.673918, 1.96772}, {1.78823, 0.118427},
    {1.78823, 0.118427}, {0.673918, 1.96772}, {1.78823, 0.118427},
    {1.78823, 0.118427}, {0.673918, 1.96772}, {1.78823, 0.118427},
    {0.673918, 1.96772}}

One can use here parallel computation. We choose the solution, which provide
the smallest residual,

$\Rightarrow$ ob = Map[Norm[{x, y}]/.{x → #[[1]], y → #[[2]]}&, soli]
$\Leftarrow$ {1.79214, 2.07992, 1.79214, 1.79214, 2.07992,
    1.79214, 1.79214, 2.07992, 1.79214, 2.07992}

There are more candidates, but they are the same. The position of this minimum
in the list is

$\Rightarrow$ First[Position[ob, Min[ob]]//Flatten]
$\Leftarrow$ 1

the corresponding solution

$\Rightarrow$ soli[[%]]
$\Leftarrow$ {1.78823, 0.118427}

Chapter 4
Nonlinear Geodetic Equations with Uncertainties: Algebraic-Numeric Solutions

4.1 Introductory Remarks

Geodetic parameters are normally estimated based on functional and stochastic models (Awange 2012, 2018; Hoffman-Wellenhof 2008), where the former expresses the geometrical relationship between the observations and the estimates while the latter is invoked to account for the fact that the observations/measurements often consist of doubtful quality usually introduced through uncertain components (e.g., Niemeier and Tengen 2017). Uncertain components are related to the dispersion of the measurements through variances and standard deviations (see, e.g., Grafarend and Awange 2012). Examples of geodetic estimations incorporating stochasticity include, e.g., Global Navigation Satellite Systems (GNSS; Hoffman-Wellenhof 2008); the seven-parameter datum transformation problem [conformal group $C_7(3)$; Reinking (2001), Koch (2001) Lenzmann and Lenzmann (2001a, b), Grafarend and Awange (2003)]; and in laser scanning, (e.g., Paláncz et al. 2017).

With the advancement of measuring sensors that invoke push button capabilities such as laser scanners that generate millions of point cloud observations (e.g. Paláncz et al. 2017), uncertainties of observations are becoming more complex to manage. More often, users are not familiar with the internal mechanisms of these observing sensors to analyze the repeated observations to obtain adequate measure of the dispersion of measurements (Niemeier and Tengen 2017). In GNSS measurements for example, positions, navigation and time (PNT) information are derived from distance measurements between the receivers and the satellites that deliver multi-constellation measurements of ranges (e.g., Awange 2012, 2018; Tian et al. 2020). These measured distances suffer from various sources of errors ranging

Supplementary Information The online version contains supplementary material available at https://doi.org/10.1007/978-3-030-92495-9_4.

J. L. Awange et al., *Mathematical Geosciences*,
https://doi.org/10.1007/978-3-030-92495-9_4

from orbital, atmospheric to the receiver dynamics that introduce bias (Montenbruck et al. 2017; Khodabandeh and Teunissen 2016; Prange et al. 2017), which degrade the quality of the PNT solutions obtained.

To address the uncertainty in geodetic problems, Niemeier and Tengen (2017) extend the classical concept of geodetic network adjustment by introducing a new method for uncertainty assessment that analyses the raw data and possible influencing factors using uncertainty modeling according to GUM (Guidelines to the Expression of Uncertainty in Measurements). GUM has attracted wide use in the field of meteorology, but rarely adapted within Geodesy. Other techniques that have been employed to tackle uncertainty in geodetic observations include, e.g., the total least squares (TLS; Awange et al. 2016), weighted total least squares (WTLS; Malissiovas et al. 2020), Procrustes (Grafaend and Awange 2003), collocation (Moritz 1973) and Autoregressive moving average (ARMA; Schubert et al. 2020). These methods work well when the functions are linear and simple. In many geodetic problems, however, the underlying processes are rather complex, with some having covariance functions that are often oscillating making it difficult to find corresponding analytical functions for modelling (Schubert et al. 2020). The situation is further complicated when the functional models are nonlinear as is often the case in most geodetic systems of equations such as the case of the seven-parameter datum transformation problem (see e.g., Grafarend and Awange 2003).

This contribution proposes two novel methods for computing the solutions of nonlinear geodetic equations whose stochastic variables have parameters that have uncertainty characterized by a probability distribution. The first method is algebraic, which partly exploits the symbolic computation structure of the system, while the second method is numeric and uses homotopy via stochastic differential equations of the Ito form (Awange et al. 2018).

4.2   Nonlinear System of Equations with Uncertainties

Nonlinear equations whose coefficients or parameters are uncertain can be represented by a probability distribution, mostly a normal distribution, namely a parameter can be characterized as $p + \delta$, where $\delta = N(0, \sigma)$ is a normal distribution. The solution is considered as a stochastic variable represented by its probability density function (*pdf*). It is here emphasized that the interest is in the *uncertainty* of the solution, since its deterministic value, i.e., the mean of its distribution can be frequently computed in traditional way!

4.2.1   Problem Definition

Let us consider the following simple polynomial equation,

$$q = x^2 + (8 + \delta)x - 9,$$

where the coefficient of term x has uncertainty characterized by $\delta = N(0, \sigma)$ for the desire is to find the solution of $q(x) = 0$ as a stochastic variable. Note that our method is also applicable for other distributions that differ from the Gaussian.

4.2.1.1   Algebraic Solution

The most common way is to solve the equation symbolically, and transform the result containing δ into a density function, i.e.,

$$x_1 = \frac{1}{2}\left(-8 - \delta - \sqrt{100 + 16\delta + \delta^2}\right), x_2 = \frac{1}{2}\left(-8 - \delta + \sqrt{100 + 16\delta + \delta^2}\right)$$

The first root's *pdf* is,

$$pdf(x) = pdf\left(\frac{1}{2}\left(-8 - \delta - \sqrt{100 + 16\delta + \delta^2}\right)\right).$$

Assuming $\sigma = 0.2$, then the *pdf* of $\delta = N(0, 0.2)$ is as shown in Fig. 4.1a. The actual values of sigma can be estimated on the basis of measurement or of practical guess. Throughout this article the sigma values are just illustrative values.

Now we need the *pdf(x)*. In general the *pdf(x)* of a transformed distribution can easily be computed in *Mathematica*. Let us transform this distribution according to $x_1 = f(\delta)$, then density function of the first root is (Fig. 4.1b).

The mean value and the standard deviation can be computed per definition

$$m_1 = \int_{-\infty}^{+\infty} \xi pdf(\xi)\, d\xi = -9.00036$$

and

$$\mu_1 = \sqrt{\int_{-\infty}^{+\infty} (\xi - m_1)^2 pdf(\xi)\, d\xi} = 0.18$$

Similarly, for the other root, $m_2 = 1.00036$ and $\mu_2 = 0.0200$.

It can be clearly seen, that the standard deviation of the input (σ) is transformed differently in case of algebraic solution the two different roots ($\mu_2 \ll \mu_1 < \sigma$).

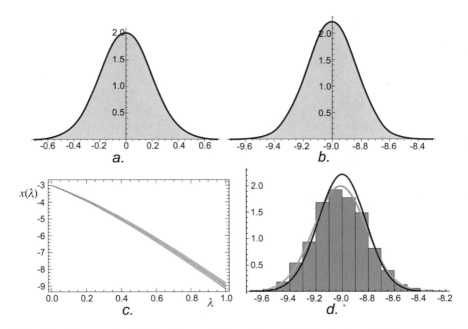

Fig. 4.1 a The error distribution of the coefficient as input; **b** the distribution of the first root; **c** the trajectories of the mean values (red) and the standard deviations (blue); **d** the *pdf* of the stochastic homotopy (green) and that of the algebraic solution (red)

Next, we consider a numerical solution, which can be employed even in cases when algebraic solutions are not possible.

4.2.1.2 Employing Stochastic Homotopy

The basic idea is to employ linear homotopy via stochastic differential equations represented by *Ito* process. Now our *target system* is

$$q(x) = x^2 + (8 + \delta)x - 9.$$

We need a *start system,* which can define in many different ways, see i.e. Awange and Paláncz (2016). Here we consider it as

$$p(x) = x^2 - 9,$$

which has two solutions. Now let us consider the negative one,

$$x_0 = -3.$$

Then the linear homotopy function is

$$H(x, \lambda) = (1 - \lambda)p(x) + \lambda q(x).$$

To get the differential equation form, we compute the partial derivatives of the homotopy function,

$$\frac{\partial H(x, \lambda)}{\partial x} = 2x(1 - \lambda) + (8 + 2x + \delta)\lambda,$$

and

$$\frac{\partial H(x, \lambda)}{\partial \lambda} = x(8 + \delta).$$

Then the right hand side of differential equation is

$$rhs = -\frac{\frac{\partial H(x, \lambda)}{\partial \lambda}}{\frac{\partial H(x, \lambda)}{\partial x}} = -\frac{x(8 + \delta)}{2x(1 - \lambda) + (8 + 2x + \delta)\lambda}.$$

We then linearize this term in δ at $\delta = 0$, in order to get *Ito stochastic differential equation form*, see Awange et al. (2018), and consider that $x = x(\lambda)$, then we get for *rhs*

$$-\frac{\delta x(\lambda)^2}{2(4\lambda + x(\lambda))^2} - \frac{4x(\lambda)}{4\lambda + x(\lambda)}.$$

Therefore, the *Ito form* is

$$\delta x(\lambda) = -\frac{4x(\lambda)}{x(\lambda) + 4\lambda} d\lambda - \frac{x(\lambda)^2}{2(x(\lambda) + 4\lambda)^2} dw(\lambda),$$

where $w(\lambda)$ is a *Wiener process*.

This stochastic differential equation should be solved with the initial condition $x(0) = -3$ up to $x(1)$ in numerical way. During the solution, we generated 2000 trajectories. The stochastic *Runge–Kutta method* is employed with step size 0.00025. The two important parameters of the numerical solution are the step size and the number of the generated trajectories. The proper choice these values can ensure a stable solution.

The mean value and the standard deviation of the trajectory are presented in Fig. 4.1c.

The numerical result for the mean and the standard deviation of the resulting distribution at $x(1)$ are, $m_1 = -9.01052$ and $\mu_1 = 0.199157$. These values are somewhat different—especially in case of the standard deviation—from the result given by the algebraic method. There are two reasons for these deviations; the linearization in δ and round off error of the numerical solution.

The density function can be constructed via generation of numerical random data from the $x(1)$ distribution (Fig. 4.1d) shows the histogram of the generated data and the fitted *pdf* (green) as well as the *pdf* of the corresponding algebraic solution (red).

It should be pointed out that in neither algebraic nor stochastic techniques was the Gaussian distribution assumed. It should also be emphasized that the stochastic homotopy method requires more computation power and seems less precise than the algebraic method. However, there are certain situations when the algebraic technique is not applicable as will be shown later in Sect. 4.3, see 3D resection.

In the next section, we show how to employ these techniques for system of equations.

4.2.2 Systems of Equations

Here we demonstrate how to solve problems with more variables. Two important facts will turn out:

(1) It is recommended to employ Gröbner Basis or Dixon Resultant to reduce polynomial system to a polynomial with a single variable as target system in order to avoid ill conditioned homotopy matrix of the Ito differential equation system, see Awange et al. (2018).
(2) In case the target system can be solved as algebraic equation symbolically, the problem can be solved in algebraic way. Otherwise one should employ stochastic homotopy.

As an illustration, let us consider the following polynomial system,

$$f_1(x,y) = 2x^2 - y^3,$$
$$f_2(x,y) = x^2 + \left(\frac{1}{8} + \delta\right)y^2 - 1.$$

The real solutions in error free case can be seen in Fig. 4.2a.

With the system having uncertainty, the first step is to reduce the system via *Gröbner basis*. Using reduced Gröbner basis to eliminate variable x we get,

$$GroebnerBasis(\{g_1, g_2\}, x) = \{-8 + y^2 + 4y^3 + 8y^2\delta, -8 + 8x^2 + y^2 + 8y^2\delta\}.$$

To compute the *algebraic solution*, we need to find the symbolic solution of the polynomial,

$$-8 + y^2 + 4y^3 + 8y^2\delta = 0,$$

which in error free case ($\delta = 0$) results one real solution, see Fig. 4.2b.

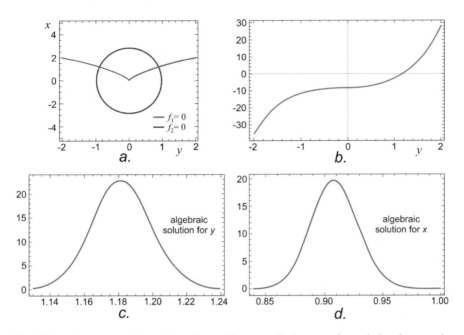

Fig. 4.2 a The contour plot of the polynomial system; **b** the error free solution for y; **c** the pdf(y) probability density function for the y variable; **d** The pdf(x) probability density function for the x variable

4.2.2.1 Algebraic Solution

The symbolic solution for this real root containing the error term δ is,

$$
y = \frac{1}{12}(-1 - 8\delta)
$$

$$
+ \frac{(1+8\delta)^2}{12\left(1727 - 24\delta - 192\delta^2 - 512\delta^3 + 24\sqrt{6}\sqrt{863 - 24\delta - 192\delta^2 - 512\delta^3}\right)^{1/3}}
$$

$$
+ \frac{1}{12}\left(1727 - 24\delta - 192\delta^2 - 512\delta^3 + 24\sqrt{6}\sqrt{863 - 24\delta - 192\delta^2 - 512\delta^3}\right)^{1/3}
$$

Assuming $\sigma = 0.03$, the pdf(y) can be computed via distribution transformation of $y(\delta)$, see Fig. 4.2c.

The mean and standard deviation can be computed as earlier using numerical integration, $m_y = 1.18213$ and $\sigma_y = 0.0175$.

The computation of pdf(x) is somewhat more complicated. First we solve the $\{g_1(y, \delta) = 0, g_2(x, y, \delta) = 0\}$ system symbolically. The error free ($\delta = 0$) form of this system has two real solutions, see Fig. 4.2a. We consider now the positive one, $x(\delta)$. However the symbolic solution is too complex for distribution transform, therefore Taylor expansion at $\delta = 0$ is used,

$$x(\delta) = 0.908516 - 0.673711\ \delta + 0.436908\ \delta^2 - 0.218715\ \delta^3.$$

Then distribution transform can be applied to get $pdf(x)$, see Fig. 4.2d.

The mean and the standard deviation are $m_x = 0.90891$ and $\sigma_x = 0.02023$. For details of the computation, we refer to the corresponding *Electronic Supplement*.

4.2.2.2   Stochastic Homotopy

Our target system for y is the first polynomial from the Gröbner basis,

$$q(y) = -8 + y^2 + 4y^3 + 8y^2\delta.$$

We shall employ here Newton linear homotopy, therefore our homotopy function is,

$$H(y, \lambda) = (1 - \lambda)(q(y) - qy_0)) + \lambda q(y).$$

where y_0 is a close root to the solution for the y, see Fig. 4.2b, let say, $y_0 = 1$.

Therefore the start system is

$$p(y) = q(y) - q(y_0) = -5 + y^2 + 4y^3 - 8\ \delta + 8y^2\delta.$$

Then the homotopy function is

$$H(y, \lambda, \delta) = -5 + y^2 + 4y^3 - 8\ \delta + 8y^2\delta - 3\ \lambda + 8\ \delta\ \lambda.$$

To get the differential equation form, we compute the partial derivatives of the homotopy function then the right hand side of the differential equation is

$$rhs = -\frac{-3 + 8\delta}{2y + 12y^2 + 16y\delta}$$

after linearization in δ at $\delta = 0$, in order to get Ito stochastic differential equation form

$$dy(\lambda) = \frac{3}{2y(\lambda) + 12y(\lambda)^2}\,d\lambda - \frac{8(2 + 3y(\lambda))}{y(\lambda)(1 + 6y(\lambda))^2}\,dw(\lambda),$$

with initial condition $y(0) = 1$. The trajectories of the mean value and the standard deviation can be seen in Fig. 4.3a.

The numerical result for the mean value and for the standard deviation of the distribution resulted at $y(1)$ are, $m_y = 1.19125$ and $\sigma_y = 0.174751$.

The density function can be constructed via generation of numerical random data from the $y(1)$ distribution. Figure 4.3b shows the histogram of the generated data

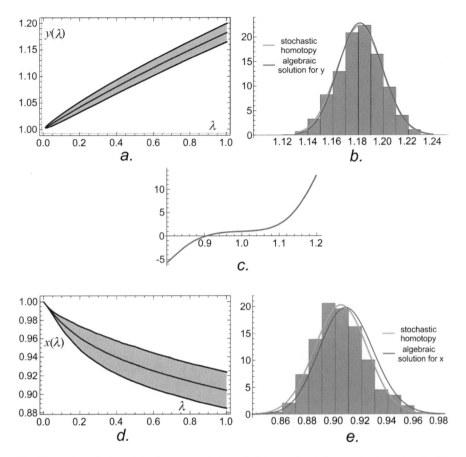

Fig. 4.3 a The solution of the Ito system, **b** the *pdf* of the stochastic homotopy (green) and of the algebraic solution (blue), **c** the positive real root of the error free case, **d** the solution of the *Ito* system, **e** comparison of the two methods

and the fitted *pdf* (green) as well as the *pdf* of the corresponding algebraic solution (blue).

The other variable *pdf*(x) can be similarly computed. Now we eliminate y then the reduced Gröbner basis,

$GroebnerBasis\left(g_1, g_2, g_3\right) =$

$$\{-128 + 384x^2 - 383x^4 + 128x^6 + 24x^4\delta + 192x^4\delta^2 + 512x^4\delta^3,$$
$$- 128 + 255x^2 - 128x^4 + 4y - 24x^2\delta + 64y\delta - 192x^2\delta^2 + 256y\delta^2 - 512x^2\delta^3,$$
$$x^2 - 4y + 4x^2y + 8x^2\delta, \ -8 + 8x^2 + y^2 + 8y^2\delta, \ -2x^2 + y^3\}$$

Let us select the first basis as target system,

$$q(x, \delta) = -128 + 384x^2 - 383x^4 + 128x^6 + 24x^4\delta + 192x^4\delta^2 + 512x^4\delta^3$$

We are looking for the positive real root, see Fig. 4.3c.

Therefore we select $x(0) = 1$, as initial condition. The start system employing *Newton linear homotopy*,

$$p(x, \delta) = q(x, \delta) - q(x_0, \delta) = -129. + 384x^2 - 383x^4 + 128x^6$$
$$- 24.\delta + 24x^4\delta - 192.\delta^2 + 192x^4\delta^2 - 512.\delta^3 + 512x^4\delta^3$$

Using the standard steps as before, we get for the differential equation,

$$dx(\lambda) = -\frac{1}{768x(\lambda) - 1532x(\lambda)^3 + 768x(\lambda)^5} d\lambda$$
$$- \frac{1152\left(1 - 2x(\lambda)^2 + x(\lambda)^4\right)}{x(\lambda)\left(192 - 383x(\lambda)^2 + 192x(\lambda)^4\right)^2} dw(\lambda)$$

with initial condition $x(0) = 1$. The trajectories of the mean value and the standard deviation can be seen in Fig. 4.3d.

The numerical result for the mean value and for the standard deviation of the distribution resulted at $y(1)$ are, $m_x = 0.9042$ and $\sigma_x = 0.01965$.

The density function can be constructed via generation of numerical random data from the $x(1)$ distribution. Figure 4.3e shows the histogram of the generated data and the fitted *pdf* (green) as well as the *pdf* of the corresponding algebraic solution (blue).

The details of the computation can be found in the corresponding *Mathematica* notebook.

4.2.2.3 Solution of Equations Simultaneously

In certain simple cases one can employ this technique to solve equations simultaneously too. Let us illustrate this technique in case of stochastic homotopy.

- **Polynomial system**

We consider same polynomial system. First we define the target system,

$$\Rightarrow \text{F1}[\text{x\_,y\_}] :\, = \text{f1}$$
$$\Rightarrow \text{F2}[\text{x\_,y\_}] :\, = \text{f2}$$

The start system should have solution close to the real solutions,

$\Rightarrow$ g1[x_,y_] : = $x^2 - 1$
$\Rightarrow$ g2[x_,y_] : = $y^2 - 1$

Now our homotopy function is

$$\Rightarrow \texttt{H[x\_,y\_,}\lambda\texttt{\_]} : = \texttt{Flatten}\left[(1 - \lambda)\begin{pmatrix} \texttt{g1[x,y]} \\ \texttt{g2[x,y]} \end{pmatrix} + \lambda \begin{pmatrix} \texttt{F1[x,y]} \\ \texttt{F2[x,y]} \end{pmatrix}\right]$$

or

$\Rightarrow$ H[x,y,λ]
$\Leftarrow \{(-1+x^2)(1-\lambda)+(2x^2-y^3)\lambda, (-1+y^2)(1-\lambda)+(-1+x^2+y^2(\frac{1}{8}+\delta))\lambda\}$

Considering that $H(x, y, \lambda) = 0$ should be valid for every $\lambda \in [0, 1]$, therefore,

$$dH(x, y, \lambda) = \frac{\partial H}{\partial x}dx + \frac{\partial H}{\partial y}dy + \frac{\partial H}{\partial \lambda}d\lambda \equiv 0, \quad \lambda \in [0, 1].$$

Then the initial value problem can be written as,

$$H_x \frac{dx(\lambda)}{d\lambda} + H_y \frac{dy(\lambda)}{d\lambda} + H_\lambda = 0$$

with $x(0) = x_0$ and $y(0) = y_0$.

Considering $\xi = \begin{pmatrix} x \\ y \end{pmatrix}$ is the Jacobian of H_ξ with respect to $\xi_i, i = 1, \ldots, n$, in case of n nonlinear equations with ξ_i variables, in general. So in multivariable case one should compute the inverse of the Jacobian in order to get explicit form, see Awange and Paláncz (2016).

First let us define a function for the Jacobi computation as it follows,

$\Rightarrow$ Jacobi[F_,X_] : = Outer[D,F,X]]

Then the total Jacobian, $\left(\frac{\partial H}{\partial x} \frac{\partial H}{\partial y} \frac{\partial H}{\partial \lambda}\right)$ is,

$\Rightarrow$ Jxyλ = Jacobi[H[x,y,λ],{x,y,λ}]; MatrixForm[Jxyλ]
$$\Leftarrow \begin{pmatrix} 2x(1-\lambda)+4x\lambda & -3y^2\lambda & 1+x^2-y^3 \\ 2x\lambda & 2y(1-\lambda)+2y(\frac{1}{8}+\delta)\lambda & x^2-y^2+y^2(\frac{1}{8}+\delta) \end{pmatrix}$$

The Jacobian respecting to x and y

$\Rightarrow$ Jxy = Take[Jxyλ,{1,2},{1,2}]; MatrixForm[Jxy]
$$\Leftarrow \begin{pmatrix} 2x(1-\lambda)+4x\lambda & -3y^2\lambda \\ 2x\lambda & 2y(1-\lambda)+2y(\frac{1}{8}+\delta)\lambda \end{pmatrix} \quad \text{and}$$

$\Rightarrow$ Jxλ = Jacobi[H[x,y,λ],{x,y,λ}]; MatrixForm[Jxλ]

$$\Leftarrow \begin{pmatrix} 2x(1-\lambda)+4x\lambda & -3y^2\lambda & 1+x^2-y^3 \\ 2x\lambda & 2y(1-\lambda)+2y(\frac{1}{8}+\delta)\lambda & x^2-y^2+y^2(\frac{1}{8}+\delta) \end{pmatrix}$$

$\Rightarrow$ Jx = Take[Jxλ,{1,2},{1,2}]; MatrixForm[Jx]

$$\Leftarrow \begin{pmatrix} 2x(1-\lambda)+4x\lambda & -3y^2\lambda \\ 2x\lambda & 2y(1-\lambda)+2y(\frac{1}{8}+\delta)\lambda \end{pmatrix}$$

Now, we set up the differential equation system. Explicit form of the differential equation system should be expressed.

The right hand side of the linear system,

$\Rightarrow$ b = $-$Take[Jxλ,{1,2},{2+1}]; MatrixForm[b]

$$\Leftarrow \begin{pmatrix} -1-x^2+y^3 \\ -x^2+y^2-y^2(\frac{1}{8}+\delta) \end{pmatrix}$$

Solving the linear system, we get the right hand side,

$\Rightarrow$ rhs = (LinearSolve[Jx,b]//Simplify)/.Map[# $\rightarrow$ #[λ]&,{x,y}]//
　　Flatten]

$\Leftarrow$ {$(-2(8-7\lambda+8\delta\lambda)+(16+7\lambda-8\delta\lambda)y[\lambda]^3-2x[\lambda]^2(8+\lambda(-7+8\delta+12y[\lambda])))$

$/(4x[\lambda](8+\lambda+8\delta\lambda+\lambda^2(-7+8\delta+12y[\lambda]))),$

$-\dfrac{-8\lambda+8x[\lambda]^2+(-7+8\delta)(1+\lambda)y[\lambda]^2+8\lambda y[\lambda]^3}{2y[\lambda](8+\lambda+8\delta\lambda+\lambda^2(-7+8\delta+12y[\lambda]))}$}

To get Ito form we need to linearize at $\delta = 0$ as usual

$\Rightarrow$ rhsδ = Map[(Series[#,{δ,0,1}]//Normal)&,rhs]

$\Leftarrow$ {$\dfrac{24\delta(-2\lambda^3y[\lambda]+2\lambda^2x[\lambda]^2y[\lambda]-2\lambda y[\lambda]^3-2\lambda^2y[\lambda]^3-\lambda^3y[\lambda]^4)}{x[\lambda](8+\lambda-7\lambda^2+12\lambda^2y[\lambda])^2}$

$+\dfrac{-2(8-7\lambda)+(16+7\lambda)y[\lambda]^3-2x[\lambda]^2(8+\lambda(-7+12y[\lambda]))}{4x[\lambda](8+\lambda+\lambda^2(-7+12y[\lambda]))},$

$-\dfrac{-8\lambda+8x[\lambda]^2-7(1+\lambda)y[\lambda]^2+8\lambda y[\lambda]^3}{2y[\lambda](8+\lambda+\lambda^2(-7+12y[\lambda]))}$

$+\dfrac{\delta\dfrac{8(\lambda+\lambda^2)(-8\lambda+8x[\lambda]^2-7(1+\lambda)y[\lambda]^2+8\lambda y[\lambda]^3)}{(8+\lambda-7\lambda^2+12\lambda^2y[\lambda])^2}}{2y[\lambda]}$

$-\dfrac{\delta\dfrac{8(1+\lambda)y[\lambda]^2}{8+\lambda+\lambda^2(-7+12y[\lambda])}}{2y[\lambda]}$}

Then the terms of the Ito form are,

$\Rightarrow$ pλx = rhsδ[[1]][[2]]//Simplify

$\Leftarrow \dfrac{2(-8+7\lambda)+(16+7\lambda)y[\lambda]^3 - 2x[\lambda]^2(8-7\lambda+12\lambda y[\lambda])}{4x[\lambda](8+\lambda+\lambda^2(-7+12y[\lambda]))}$

$\Rightarrow$ pwx = Coefficient[rhsδ[[1]][[1]],δ]//Simplify

$\Leftarrow -\dfrac{24\lambda y[\lambda](2\lambda^2 - 2\lambda x[\lambda]^2 + 2(1+\lambda)y[\lambda]^2 + \lambda^2 y[\lambda]^3)}{x[\lambda](8+\lambda-7\lambda^2+12\lambda^2 y[\lambda])^2}$

$\Rightarrow$ pλy = rhsδ[[2]][[1]]//Simplify

$\Leftarrow -\dfrac{-8\lambda+8x[\lambda]^2 - 7(1+\lambda)y[\lambda]^2 + 8\lambda y[\lambda]^3}{2y[\lambda](8+\lambda+\lambda^2(-7+12y[\lambda]))}$

$\Rightarrow$ pwy = Coefficient[rhsδ[[2]][[2]],δ]//Simplify

$\Leftarrow -\dfrac{16(1+\lambda)(2\lambda^2 - 2\lambda x[\lambda]^2 + 2(1+\lambda)y[\lambda]^2 + \lambda^2 y[\lambda]^3)}{y[\lambda](8+\lambda-7\lambda^2+12\lambda^2 y[\lambda])^2}$

The initial conditions are

$\Rightarrow$ x0 = 1; y0 = 1;

The Ito form of the system is

$\Rightarrow$ proc = ItoProcess[{dx[λ] = pλx dλ + pwx dw[λ],

dy[λ] = pλy dλ + pwy dw[λ]},{x[λ],y[λ]},{{x,y},{x0,y0}},

λ,w $\approx$ WienerProcess[0,σ]]]

$\Leftarrow$ ItoProcess[{{ $\dfrac{2(-8+7\lambda)+(16+7\lambda)y[\lambda]^3 - 2x[\lambda]^2(8-7\lambda+12\lambda y[\lambda])}{4x[\lambda](8+\lambda+\lambda^2(-7+12y[\lambda]))}$,

$-\dfrac{-8\lambda+8x[\lambda]^2 - 7(1+\lambda)y[\lambda]^2 + 8\lambda y[\lambda]^3}{2y[\lambda](8+\lambda+\lambda^2(-7+12y[\lambda]))}$ },

{{$-\dfrac{0.72\lambda y[\lambda](2\lambda^2 - 2\lambda x[\lambda]^2 + 2(1+\lambda)y[\lambda]^2 + \lambda^2 y[\lambda]^3)}{x[\lambda](8+\lambda-7\lambda^2+12\lambda^2 y[\lambda])^2}$ },

{$-\dfrac{0.48(1+\lambda)(2\lambda^2 - 2\lambda x[\lambda]^2 + 2(1+\lambda)y[\lambda]^2 + \lambda^2 y[\lambda]^3)}{y[\lambda](8+\lambda-7\lambda^2+12\lambda^2 y[\lambda])^2}$ }},

{x[λ],y[λ]}},{{x,y},{1,1}},{λ,0}]

Generating 2000 trajectories with step size 0.001,

$\Rightarrow$ psolxy = RandomFunction[proc,{0,1.,0.001},2000]

The mean values for x

$\Rightarrow$ Mean[psolxy[1]][[1]]

$\Leftarrow$ 0.907979

and for y

⇒ Mean[psolxy[1]][[2]]

⇐ 1.18162

The standard deviations for x

⇒ StandardDeviation[psolxy[1]][[1]]

⇐ 0.0206019

and for y

⇒ StandardDeviation[psolxy[1]][[2]]

⇐ 0.0178747

Here the standard deviation of y is somewhat different from that provided by the other two methods: $0.0178747 > (0.0175422, 0.0169806)$.

- **Non polynomial system**

In non-polynomial case it is rare when decomposition is possible therefore the demand to solve system of equations simultaneously is very frequent. Now let us consider the following problem as illustration.

$$f_1(x, y) = x - \sin((2 + \delta)x + 3y) + \cos((3 + \delta)x - 5y).$$
$$f_2(x, y) = y - \sin((1 + \delta)x - 2y) + \cos((1 + \delta)x + 3y).$$

where the error δ effects only the coefficients of the variable x. We generate the contour plot of the equations in the region of $[-2, 2]^2$, see Fig. 4.4.

Fig. 4.4 Non polynomial system

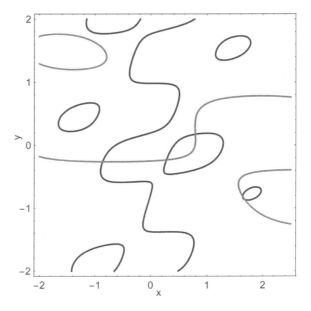

$\Rightarrow$ f1 = x $-$ Sin[(2+δ)x+3y]+Cos[(3+δ)x $-$ 5y];

$\Rightarrow$ f2 = y $-$ Sin[(1+δ)x $-$ 2y]+Cos[(1+δ)x+3y];

$\Leftarrow$ ContourPlot[{(f1/.{$\delta \to 0$}) == 0,(f2/.{$\delta \to 0$}) == 0},

  {x, $-$ 2,2.5},{y, $-$ 2,2},FrameLabel $\to$ {"x","y"}]

This system cannot be solved in an algebraic way and it even cannot be reduced to independent equations with a single local variable. The only way to find the density function of the roots is stochastic homotopy solving simultaneously for both equations as a system. It goes without saying that only a local solution is possible. Let us aim the following root, which is close to ($x \approx 1.2$, $y \approx 0.2$).

$\Rightarrow$ FindRoot[{(f1/.$\delta \to 0$) = 0,(f2/.$\delta \to 0$) = 0},{{x,1.2},{y,0.2}}]

$\Leftarrow$ {x $\to$ 0.79114,y $\to$ 0.174395}

Then let us consider the initial conditions for the Ito system as,

$\Rightarrow$ x0 = 0.8;y0 = 0.18;

First we define the target system is,

$\Rightarrow$ F1[x_,y_] : = f1

$\Rightarrow$ F2[x_,y_] : = f2

Our start system has solution close to the root,

$\Rightarrow$ g1[x_,y_] : = x $-$ x0

$\Rightarrow$ g2[x_,y_] : = y $-$ y0

Now our homotopy function is

$\Rightarrow$ H[x_,y_,λ_] : = Flatten$\left[(1-\lambda) \begin{pmatrix} g1[x,y] \\ g2[x,y] \end{pmatrix} + \lambda \begin{pmatrix} F1[x,y] \\ F2[x,y] \end{pmatrix} \right]$

or

$\Rightarrow$ H[x,y,λ]

$\Rightarrow$ {($-$ 0.8+x)(1 $-$ λ)+λ(x+Cos[5y $-$ x(3+δ)] $-$ Sin[3y+x(2+δ)]),

  ($-$ 0.18+y)(1 $-$ λ)+λ(y+Cos[3y+x(1+δ)]+Sin[2y $-$ x(1+δ)])}

Then the total Jacobian,

$\Rightarrow$ Jxyλ = Jacobi[H[x,y,λ],{x,y,λ}];

The Jacobian respecting to x and y

$\Rightarrow$ Jxy = Take[Jxyλ,{1,2},{1,2}];

and

$\Rightarrow$ Jxλ = Jacobi[H[x,y,λ],{x,y,λ}];

$\Rightarrow$ Jx = Take[Jxλ,{1,2},{1,2}];

The right hand side of the linear system,

$\Rightarrow$ b $= -$Take$[$Jx$\lambda,\{1,2\},\{2+1\}]$;

Solving the linear system, we get the right hand side,

$\Rightarrow$ rhs $= ($LinearSolve$[$Jx,b$]$//Simplify$)$/.Map$[\#\to\#[\lambda]\&,\{$x,y$\}]$
//Flatten;

To get Ito form we need to linearize at $\delta = 0$ as usual

$\Rightarrow$ rhs$\delta =$ Map$[($Series$[\#,\{\delta,0,1\}]$//Normal$)\&,$rhs$]$;

Then the terms of the Ito form are,

$\Rightarrow$ pλx $=$ rhs$\delta[[1]][[1]]$//Simplify;
$\Rightarrow$ pwx $=$ Coefficient$[$rhs$\delta[[1]][[2]],\delta]$//Simplify;
$\Rightarrow$ pλy $=$ rhs$\delta[[2]][[1]]$//Simplify;
$\Rightarrow$ pwy $=$ Coefficient$[$rhs$\delta[[2]][[2]],\delta]$//Simplify;

The Ito form the system is

$\Rightarrow$ proc $=$ ItoProcess$[\{$dx$[\lambda] =$ pλx d$\lambda +$ pwx dw$[\lambda]$,
dy$[\lambda] =$ pλy d$\lambda +$ pwy dw$[\lambda]\},\{$x$[\lambda]$,y$[\lambda]\},\{\{$x,y$\},\{$x0,y0$\}\},\lambda$,
w $\approx$ WienerProcess$[0,\sigma]]$;

Generating 1000 trajectories with step size 0.0005,

$\Rightarrow$ psolxy $=$ RandomFunction$[$proc,$\{0,1.,0.0005\}$,1000$]$

The mean values for x

$\Rightarrow$ Mean$[$psolxy$[1]][[1]]$
$\Leftarrow$ 0.791797

and for y

$\Rightarrow$ Mean$[$psolxy$[1]][[2]]$
$\Leftarrow$ 0.174356

The standard deviations for x

$\Rightarrow$ StandardDeviation$[$psolxy$[1]][[1]]$
$\Leftarrow$ 0.0229247

and for y

$\Rightarrow$ StandardDeviation$[$psolxy$[1]][[2]]$
$\Leftarrow$ 0.00173606

As was expected the error has effect mainly on the x coordinate of the root.

4.2.3 Special Cases

4.2.3.1 Different Types of Uncertainties

Let us consider the *Minimum Distance* Mapping Problem where the measured coordinates X, Y and Z are not error free, and their error distributions are different for the different coordinates, namely δX, δY and δZ.

In order to relate a point $P(X, Y, Z)$ to a surface point $p(x, y, z)$ of an ellipsoid, one works with a bundle of half-straight lines—so called projection lines—that depart from P and intersect the ellipsoid. There is one projection line that is at minimum distance relating P to p (see Fig. 4.5a).

The distance to be minimized is,

$$d = (X - x)^2 + (Y - y)^2 + (Z - z)^2.$$

The constrain below represents that the point p should be an element of the ellipsoid-of revolution,

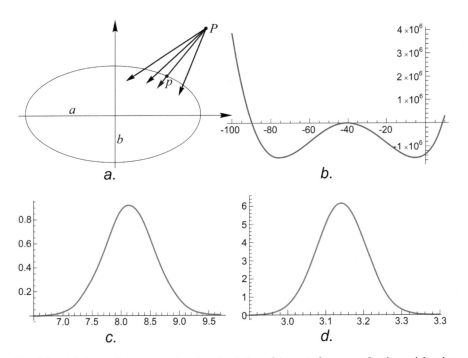

Fig. 4.5 **a** Minimum distance mapping, **b** real solution of the error free $gr\mu = 0$ polynomial, **c** the density function of $D\mu$, **d** the density function of Dx

$$c = \frac{(x^2 + y^2)}{a^2} + \frac{z^2}{b^2} - 1.$$

In order to transform this constrained optimization problem into an unconstrained one, the Lagrange multiplier form is employed,

$$obj = d + \mu c = (-x + X)^2 + (-y + Y)^2 + (-z + Z)^2 + (-1 + \frac{x^2 + y^2}{a^2} + \frac{z^2}{b^2})\mu,$$

where μ is the Lagrange multiplier. Then the necessary conditions of the minimum that $f_i = 0$, for $i = 1, 2, 3, 4$

$$f_1 = \frac{\partial obj(x, y, z, \mu)}{\partial x} = -2(-x + X) + \frac{2x\mu}{a^2} = 0$$

$$f_2 = \frac{\partial obj(x, y, z, \mu)}{\partial y} = -2(-y + Y) + \frac{2y\mu}{a^2} = 0$$

$$f_3 = \frac{\partial obj(x, y, z, \mu)}{\partial z} = -2(-z + Z) + \frac{2z\mu}{a^2} = 0$$

$$f_4 = \frac{\partial obj(x, y, z, \mu)}{\partial \mu} = -1 + \frac{x^2 + y^2}{a^2} + \frac{z^2}{b^2} = 0$$

Employing reduced Gröbner Basis for μ

$$\begin{aligned}
\mathrm{gr}\,\mu = {}& \mathrm{GroebnerBasis}(\{f_1, f_2, f_3, f_4\}, \mu, \{x, y, z\}) \\
= {}& \{a^4 b^4 - a^2 b^4 X^2 - a^2 b^4 Y^2 - a^4 b^2 Z^2 + 2a^4 b^2 \mu + 2a^2 b^4 \mu \\
& - 2a^2 b^2 X^2 \mu - 2a^2 b^2 Y^2 \mu - 2a^2 b^2 Z^2 \mu + a^4 \mu^2 + 4a^2 b^2 \mu^2 \\
& + b^4 \mu^2 - a^2 X^2 \mu^2 - a^2 Y^2 \mu^2 - b^2 Z^2 \mu^2 + 2a^2 \mu^3 + 2b^2 \mu^3 + \mu^4\}
\end{aligned}$$

Let us assume that the coordinates X, Y and Z are not error free, but having errors of different type of distributions, δ_X, δ_Y and δ_Z. To avoid round off errors we use rational representation of the numerical data, namely

$$\left\{\begin{aligned}
&X = \frac{3770667}{1000000} + \delta_X, \quad Y = \frac{44607}{10000} + \delta_Y, \quad Z = \frac{10215373}{2000000} + \delta_Z, \\
&a = \frac{797267}{125000}, \quad b = \frac{25427}{4000}\end{aligned}\right\}$$

In order to identify the proper root, for us the real positive one, we solve $\mathrm{gr}\mu = 0$ polynomial equation with error free form of these data, $\delta_X = 0$, $\delta_Y = 0$ and $\delta_Z = 0$, see Fig. 4.5b.

This is the last (4th) solution of $\mathrm{gr}\mu = 0$. Therefore we shall consider the last symbolic solution, now with non-zero error terms. This solution is considerably complex for distribution transform. Therefore, we use series expansion,

$$\mu(\delta_X, \delta_Y, \delta_Z) = 8.14151 + \delta_Y(3.60064 - 0.306851\delta_Z) + \delta_X(3.04365$$
$$+ \delta_Y(-0.222679 + 0.057093\delta_Z) - 0.259384\delta_Z) + 4.14414\delta_Z$$

where δ_i, i = X, Y, Y is Gaussian distributions with zero means and σ_i, X, Y, Y. Let us assume σ_X = 0.07, σ_Y = 0.08 and σ_Z = 0.06. Then, employing distribution transform for $\mu(\delta_X, \delta_Y, \delta_Z)$, we get the stochastic process for μ as $D\mu$. Now we cannot evaluate $pdf(D\mu)$ directly due to the different distributions of the errors. We therefore assume that $D\mu$ is a Gaussian distribution. Its mean value and standard deviation can be computed from $D\mu$, i.e., m_μ = 8.14151 and σ_μ = 0.436116. The corresponding density function can be seen in Fig. 4.5c.

Note that this is the solution for the Lagrange multiplier! We need the solution for the coordinates! The coordinates can be computed from the equations of the necessary conditions

$$x = \frac{635634669289\left(\frac{3770667}{1000000} + \delta_X\right)}{15625000000\left(\frac{635634669289}{15625000000} + \mu\right)}$$

Employing distribution transformation the x as stochastic variable can be computed. However $x(\delta_X, \mu)$ cannot be handled with the standard transformation, therefore we need to employ Taylor expansion, namely

$$x(\delta_X, \mu) = 3.14167 + (-8.14477 + \mu)(-0.0643449 - 0.0170646\,\delta_X)$$
$$+ (-8.14477 + \mu)^3(-0.0000269912 - 7.15821 \times 10^{-6}\delta_X)$$
$$+ (-8.14477 + \mu)^2(0.00131786 + 0.000349503\,\delta_X) + 0.833186\,\delta_X$$

As done in case of μ, assuming Dx is a Gaussian distribution, its mean value and standard deviation can be computed from Dx as m_x = 3.14192 and σ_x = 0.0647324. The corresponding density function can be seen in Fig. 4.5d.

Likewise, the other coordinates y and z can be similarly computed.

4.2.3.2 Transcendental Equations

Let us consider the following system

$$f_1(x, y) = \mathrm{Exp}(x) + \mathrm{Log}(y) - (2 + \delta)$$
$$f_2(x, y) = \mathrm{Sin}(x) + \mathrm{Cos}(y) - (1 + \delta)$$

where the additional constant terms have error with distribution δ. The Fig. 4.6a shows the contour plot of the system.

The error free numerical solution is $\{x = 0.624295, y = 1.14233\}$. In order to solve the uncertainty problem we reduce the system to a single variable problem via symbolic computation

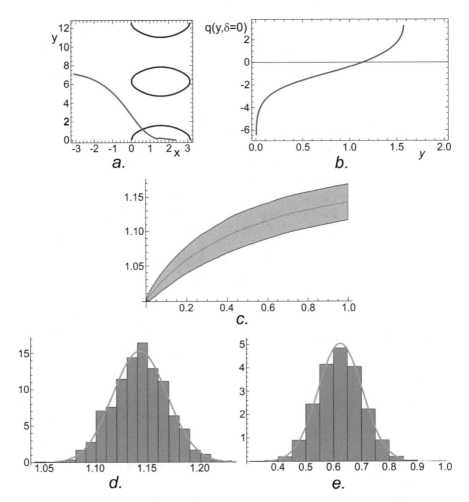

Fig. 4.6 a The contour plot of the transcendent system, **b** the contour plot of the transcendent system, **c** the homotopy trajectories of the mean value and the standard deviation, **d** the *pdf*(y), density function of the stochastic variable y, **e** the *pdf*(y), density function of the stochastic variable x

$$x = \text{ArcSin}(1 + \delta - \text{Cos}(y)).$$

Remarks Note that this reduction of the transcendental system cannot be done in general. Here, we just want to give an example for non-polynomial system. More realistic examples can be found in https://www.researchgate.net/project/Mathematical-Geosciences-A-hybrid-algebraic-numerical-solution.

So we can eliminate the variable x from $f_1(x, y)$ therefore our target system is

$$q(y, \delta) = -2 + e^{\text{ArcSin}(1 + \delta - \text{Cos}(y))} - \delta + \text{Log}(y).$$

Displaying the error free target, see Fig. 4.6b.

Therefore let us consider fixed point homotopy with $y_0 = 1$, so our start system is

$$p(y) = y - y_0 = -1 + y.$$

Then the homotopy function can be written as

$$H(\lambda, y, \delta) = (-1 + y)(1 - \lambda) + \lambda\left(-2 + e^{\text{ArcSin}(1 + \delta - \text{Cos}(y))} - \delta + \text{Log}(y)\right).$$

To get the differential equation form, we compute the partial derivatives of the homotopy function. The resulting *Ito* formula is very complex and as such we refer to the corresponding *Mathematica* notebook.

Generating 1000 trajectories with step size 0.001 in case of $\sigma = 0.06$, the homotopy trajectories of the mean value and the standard deviation can be seen in Fig. 4.6c.

The numerical result for the mean and the standard deviation of the resulting distribution at $y(1)$ are, $m_y = 1.14227$ and $\sigma_y = 0.0263679$.

The density function is constructed via generation of numerical random data from the $y(1)$ distribution. Figure 4.6d shows the histogram of the generated data and the fitted *pdf*(y).

Next, we compute the distribution of the x variable. We have seen that,

$$x(\delta) = \text{ArcSin}(1 + \delta - \text{Cos}(y(\delta))),$$

where δ is linearized to

$$x(\delta) = 0.624224 + 1.12098(-1.14227 + y(\delta)) \\ + (1.23241 + 0.995106(-1.14227 + y(\delta)))\delta$$

Transforming the distribution of δ according to this expression leads to the distribution function *pdf*(x), see Fig. 4.6e.

The numerical result for the mean and the standard deviation of the resulting distribution are, $m_x = 0.624224$ and $\sigma_x = 0.079649$. Such type of problems cannot be solved via algebraic way!

4.2.3.3 Solution of Equations Simultaneously

Now let us consider the following non-polynomial problem,

$$f_1(x, y) = x^2 + y^3 - (2 + \delta_X).$$
$$f_2(x, y) = \sin(x) + \cos(y) + (1 + \delta_Y).$$

where the error δ_X effects the coefficients of the variable x and δ_Y effects the coefficients of the variable y.

$\Rightarrow$ f1 = x^2 + y^3 - (2 + δ_X);
$\Rightarrow$ f2 = Sin[x] + Cos[y] - (1 + δ_Y);

We generate the contour plot of these equations in the region of $[-2, 2]^2$ in error free case, see Fig. 4.7,

$\Rightarrow$ ContourPlot[{(f1/.{$\delta_X \to 0, \delta_Y \to 0$})== 0,
(f2/.{$\delta_X \to 0, \delta_Y \to 0$})== 0},{x, - Pi,Pi},{y, - 2, 2},
FrameLabel $\to$ { "x"," y"}]

This system cannot be solved in an algebraic way and it even cannot be reduced to independent equations with single variable. The only way is to employ stochastic homotopy using simultaneously for both equations as a system. Of course, only a local solution is possible. Let us aim the root which is close to $x \approx 1$, $y \approx -1$. The error free solution is,

Fig. 4.7 The non-polynomial system

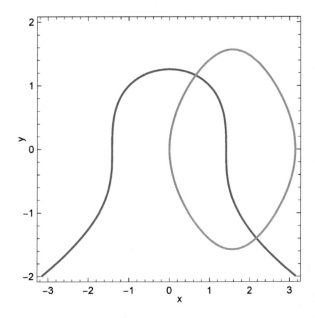

$\Rightarrow$ FindRoot$[\{(f1/.\{\delta_X \rightarrow 0, \delta_Y \rightarrow 0\}) == 0,$
$(f2/.\{\delta_X \rightarrow 0, \delta_Y \rightarrow 0\}) == 0\}, \{\{x, 1\}, \{y, -1\}\}]$
$\Leftarrow \{x \rightarrow 2.17128, y \rightarrow -1.39496\}$

Let us consider the initial conditions close to this root,

$\Rightarrow x0 = 2 ; y0 = -1.4 ;$

First we define the target system,

$\Rightarrow$ F1$[x\_,y\_]$:=f1
$\Rightarrow$ F2$[x\_,y\_]$:=f2

Our start system has solution close to the real solutions,

$\Rightarrow$ g1$[x\_,y\_]$: $= x - x0$
$\Rightarrow$ g2$[x\_,y\_]$: $= y - y0$

Now our homotopy function can be written as,

$$\Rightarrow H[x\_,y\_,\lambda\_] := \text{Flatten}\left[(1-\lambda)\begin{pmatrix} g1[x,y] \\ g2[x,y] \end{pmatrix} + \lambda\begin{pmatrix} F1[x,y] \\ F2[x,y] \end{pmatrix}\right]$$

or

$\Rightarrow H[x,y,\lambda]$
$\Leftarrow \{\{(-2+x)(1-\lambda)+\lambda(-2+x^2+y^3-\delta_x),(1.4+y)(1-\lambda)$
$+\lambda(-1+\text{Cos}[y]+\text{Sin}[x]-\delta_Y)\}$

The Jacobian is defined as earlier,

$\Rightarrow$ Jacobi$[F\_,X\_]$:=Outer$[D,F,X]$

and the total Jacobian,

$\Rightarrow$ Jxyλ=Jacobi$[H[x,y,\lambda],\{x,y,\lambda\}]]$
$\Leftarrow \{\{1-\lambda+2x\lambda, 3y^2\lambda, -x+x^2+y^3-\delta_x\},$
$\{\lambda\text{Cos}[x], 1-\lambda-\lambda\text{Sin}[y], -2.4-y+\text{Cos}[y]+\text{Sin}[x]-\delta_Y\}\}$

The Jacobian respecting to x and y

$\Rightarrow$ Jxy $=$ Take$[$Jxy$\lambda,\{1,2\},\{1,2\}]$
$\Leftarrow \{\{1-\lambda+2x\lambda, 3y^2\lambda\}, \{\lambda\text{Cos}[x], 1-\lambda-\lambda\text{Sin}[y]\}\}$

and

$\Rightarrow$ Jxλ=Jacobi$[H[x,y,\lambda],\{x,y,\lambda\}]$
$\Leftarrow \{\{1-\lambda+2x\lambda, 3y^2\lambda, -x+x^2+y^3-\delta_x\},$
$\{\lambda\text{Cos}[x], 1-\lambda-\lambda\text{Sin}[y], -2.4-y+\text{Cos}[y]+\text{Sin}[x]-\delta_Y\}\}$
$\Rightarrow$ Jx $=$ Take$[$Jx$\lambda,\{1,2\}, \{1,2\}]$
$\Leftarrow \{\{1-\lambda+2x\lambda, 3y^2\lambda\}, \{\lambda\text{Cos}[x], 1-\lambda-\lambda\text{Sin}[y]\}\}$

Now, we set up the differential equation system. The right hand side of the linear system,

$\Rightarrow$ b $= -$Take$[$Jx$\lambda,\{1,2\},\{2+1\}]$

$\Leftarrow \{\{x - x^2 - y^3 + \delta_x\}, \{2.4 + y - \text{Cos}[y] - \text{Sin}[x] + \delta_Y\}\}$

Solving the linear system, we get the right hand side of the differential equations,

$\Rightarrow$ rhs $=$ (LinearSolve[Jx,b]//Simplify) /.Map[# $\rightarrow$ #[λ]&,{x,y}]
//Flatten//Simplify

$\Leftarrow \{($Sec$[x[\lambda]](-0.5 + 0.5\lambda + 0.5\lambda \,Sin[y[\lambda]])\delta_x +$
Sec$[x[\lambda]](-0.5 + 0.5\lambda + 0.5\lambda$Sin$[y[\lambda]])x[\lambda] +$
Sec$[x[\lambda]](0.5 - 0.5\lambda - 0.5\lambda \,Sin[y[\lambda]])x[\lambda]^2 +$
$y[\lambda]^2 (\lambda(3.6(-1.5\,Cos[y[\lambda]])Sec[x[\lambda]] + 1.5\lambda \,Sec[x[\lambda]]\delta_Y -$
1.5λTan$[x[\lambda]] + Sec[x[\lambda]](0.5 + 1.\lambda - 0.5\lambda \,Sin[y[\lambda]])y[\lambda]))/$
$($Sec$[x[\lambda]](-0.5(1. - 1.\lambda)^2 + (0.5 - 0.5\lambda)\lambda \,Sin[y[\lambda]]) +$
λSec$[x[\lambda]](-1. + 1.\lambda + 1.\lambda \,Sin[y[\lambda]])x[\lambda] + 1.5\lambda^2 y[\lambda]^2),$
$(1.(-2.(0.5 - 0.5\lambda + 1.\lambda x[\lambda])(2.4 - 1.Cos[y[\lambda]] - 1.Sin[x[\lambda]] +$
$\delta_Y + y[\lambda]) + \lambda \,Cos[x[\lambda]](\delta_x + x[\lambda] - 1.x[\lambda]^2 - 1.y[\lambda]^3)))/$
$(2.(-1. + 1.\lambda + 1.\lambda \,Sin[y[\lambda]])(0.5 - 0.5\lambda + 1.\lambda x[\lambda]) +$
$3.\lambda^2$Cos$[x[\lambda]]y[\lambda]^2)\}$

To get Ito form we need to linearize at $\delta = 0$ as usual

$\Rightarrow$ rhs$\delta =$ Map$[($Series$[\#,\{\delta_x,0,1\},\{\delta_Y,0,1\}]//$Normal)&,rhs]//Simplify

$\Leftarrow \{($Sec$[x[\lambda]](-0.5 + 0.5\lambda + 0.5\lambda \,Sin[y[\lambda]])\delta_x +$
Sec$[x[\lambda]](-0.5 + 0.5\lambda + 0.5\lambda \,Sin[y[\lambda]])x[\lambda] +$
Sec$[x[\lambda]](0.5 - 0.5\lambda - 0.5\lambda \,Sin[y[\lambda]])x[\lambda]^2 +$
$y[\lambda]^2 (\lambda(3.6(-1.5\,Cos[y[\lambda]])Sec[x[\lambda]] + 1.5\lambda$Sec$[x[\lambda]]\delta_Y -$
1.5λTan$[x[\lambda]] + Sec[x[\lambda]](0.5 + 1.0\lambda - 0.5\lambda \,Sin[y[\lambda]])y[\lambda]))/$
$($Sec$[x[\lambda]](-0.5(1.0 - 1.0\lambda)^2 + (0.5 - 0.5\lambda)\lambda \,Sin[y[\lambda]]) +$
λSec$[x[\lambda]](-1.0 + 1.0\lambda + 1.0\lambda \,Sin[y[\lambda]])x[\lambda] + 1.5\lambda^2 y[\lambda]^2),$
$(-1.2 + 1.2\lambda + 0.5\,Cos[y[\lambda]] - 0.5\lambda \,Cos[y[\lambda]] + 0.5\,Sin[x[\lambda]] -$
$0.5\lambda \,$Sin$[x[\lambda]] + 0.5\lambda \,Cos[x[\lambda]]\delta_x - 2.4\lambda x[\lambda] +$
$0.5\lambda \,$Cos$[x[\lambda]]x[\lambda] + 1.\lambda \,Cos[y[\lambda]]x[\lambda] + 1.\lambda \,Sin[x[\lambda]]x[\lambda] -$
$0.5\lambda \,$Cos$[x[\lambda]]x[\lambda]^2 + \delta_Y(-0.5 + 0.5\lambda - 1.\lambda x[\lambda]) - 0.5y[\lambda] +$
$0.5\lambda y[\lambda] - 1.\lambda x[\lambda]y[\lambda] - 0.5\lambda \,Cos[x[\lambda]]y[\lambda]^3)/$
$(-0.5 + 1.\lambda - 0.5\lambda^2 + 0.5\lambda \,Sin[y[\lambda]] - 0.5\lambda^2 \,Sin[y[\lambda]] +$
$\lambda(-1. + 1.\lambda + 1.\lambda \,Sin[y[\lambda]])x[\lambda] + 1.5\lambda^2 \,Cos[x[\lambda]]y[\lambda]^2)\}$

Then the terms of the Ito form are,

$\Rightarrow \text{pwx1} = \text{Coefficient}[\text{rhs}\delta[[1]],\delta_x]//\text{Simplify}$

$\Leftarrow (\text{Sec}[x[\lambda]](-0.5 + 0.5\lambda + 0.5\lambda\text{Sin}[y[\lambda]]))/$
  $(\text{Sec}[x[\lambda]](-0.5(1.0 - 1.0\lambda)^2 + (0.5 - 0.5\lambda)\lambda\text{Sin}[y[\lambda]]) +$
  $\lambda\text{Sec}[x[\lambda]](-1.0 + 1.0\lambda + 1.0\lambda\text{Sin}[y[\lambda]])x[\lambda] + 1.5\lambda^2 y[\lambda]^2)$

$\Rightarrow \text{pwx2} = \text{Coefficient}[\text{rhs}\delta[[1]],\delta_Y]//\text{Simplify}$

$\Leftarrow (1.5\lambda\text{Sec}[x[\lambda]]y[\lambda]^2)/$
  $(\text{Sec}[x[\lambda]](-0.5(1.0 - 1.0\lambda)^2 + (0.5 - 0.5\lambda)\lambda\text{Sin}[y[\lambda]]) +$
  $\lambda\text{Sec}[x[\lambda]](-1.0 + 1.0\lambda + 1.0\lambda\text{Sin}[y[\lambda]])x[\lambda] + 1.5\lambda^2 y[\lambda]^2)$

$\Rightarrow \text{p}\lambda\text{x} = (\text{rhs}\delta[[1]] - \text{pwx1}\delta_X - \text{pwx2}\delta_Y)//\text{Simplify}$

$\Leftarrow (\text{Sec}[x[\lambda]](-0.5 + 0.5\lambda + 0.5\lambda\text{Sin}[y[\lambda]])x[\lambda] +$
  $\text{Sec}[x[\lambda]](0.5 - 0.5\lambda - 0.5\lambda\text{Sin}[y[\lambda]])x[\lambda]^2 +$
  $y[\lambda]^2(\lambda(3.6 - 1.5\text{Cos}[y[\lambda]])\text{Sec}[x[\lambda]] - 1.5\lambda\text{Tan}[x[\lambda]] +$
  $\text{Sec}[x[\lambda]](0.5 + 1.\lambda - 0.5\lambda\text{Sin}[y[\lambda]])y[\lambda]))/$
  $(\text{Sec}[x[\lambda]](-0.5(1.0 - 1.0\lambda)^2 + (0.5 - 0.5\lambda)\lambda\text{Sin}[y[\lambda]]) +$
  $\lambda\text{Sec}[x[\lambda]](-1.0 + 1.0\lambda + 1.0\lambda\text{Sin}[y[\lambda]])x[\lambda] + 1.5\lambda^2 y[\lambda]^2)$

$\Rightarrow \text{pwy1} = \text{Coefficient}[\text{rhs}\delta[[2]],\delta_x]//\text{Simplify}$

$\Leftarrow (0.5\lambda\text{Cos}[x[\lambda]])/(-0.5 + 1.0\lambda - 0.5\lambda^2 + 0.5\lambda\text{Sin}[y[\lambda]] -$
  $0.5\lambda^2\text{Sin}[y[\lambda]] + \lambda(-1.0 + 1.0\lambda + 1.0\lambda\text{Sin}[y[\lambda]])x[\lambda] +$
  $1.5\lambda^2\text{Cos}[x[\lambda]]y[\lambda]^2)$

$\Rightarrow \text{pwy2} = \text{Coefficient}[\text{rhs}\delta[[2]],\delta_Y]//\text{Simplify}$

$\Leftarrow (-0.5 + 0.5\lambda - 1.0\lambda x[\lambda])/(-0.5 + 1.0\lambda - 0.5\lambda^2 + 0.5\lambda\text{Sin}[y[\lambda]] -$
  $0.5\lambda^2\text{Sin}[y[\lambda]] + \lambda(-1. + 1.\lambda + 1.\lambda\text{Sin}[y[\lambda]])x[\lambda] + 1.5\lambda^2\text{Cos}[x[\lambda]]y[\lambda]^2)$

$\Rightarrow \text{p}\lambda\text{y} = (\text{rhs}\delta[[2]] - \text{pwy1}\delta_X - \text{pwy2}\delta_Y)//\text{Simplify}$

$\Leftarrow (-1.2 + 1.2\lambda + 0.5\text{Cos}[y[\lambda]] - 0.5\lambda\text{Cos}[y[\lambda]] +$
  $0.5\text{Sin}[x[\lambda]] - 0.5\lambda\text{Sin}[x[\lambda]] - 0.5\lambda\text{Cos}[x[\lambda]]x[\lambda]^2 +$
  $\lambda x[\lambda](-2.4 + 0.5\text{Cos}[x[\lambda]] + 1.0\text{Cos}[y[\lambda]] + 1.0\text{Sin}[x[\lambda]] - 1.0y[\lambda]) +$
  $(-0.5 + 0.5\lambda)y[\lambda] - 0.5\lambda\text{Cos}[x[\lambda]]y[\lambda]^3)/$
  $(-0.5 + 1.\lambda - 0.5\lambda^2 + 0.5\lambda\text{Sin}[y[\lambda]] - 0.5\lambda^2\text{Sin}[y[\lambda]] +$
  $\lambda(-1. + 1.\lambda + 1.\lambda\text{Sin}[y[\lambda]])x[\lambda] + 1.5\lambda^2\text{Cos}[x[\lambda]]y[\lambda]^2)$

The errors are

$\Rightarrow \sigma_X = 0.06\,; \sigma_Y = 0.1\,;$

Then the Ito form the system is

$\Rightarrow \text{proc} = \text{ItoProcess}[\{dx[\lambda] = \text{p}\lambda\text{x}d\lambda + \text{pwx1}dw1[\lambda] + \text{pwx2}dw2[\lambda],$
  $dy[\lambda] = \text{p}\lambda\text{y}d\lambda + \text{pwy1}dw1[\lambda] + \text{pwy2}dw2[\lambda]\},\{x[\lambda],y[\lambda]\},$
  $\{\{x,y\},\{x0,y0\}\},\lambda,\{w1 \approx \text{WienerProcess}[0,\sigma_X],$
  $w2 \approx \text{WienerProcess}[0,\sigma_Y]\}];$

Generating 1000 trajectories with step size 0.0005,

$\Rightarrow$ `psolxy = RandomFunction[proc,{0,1.,0.0005},500]`

The mean values for x

$\Rightarrow$ `Mean[psolxy[1]][[1]]`

$\Leftarrow$ `2.17685`

and for y

$\Rightarrow$ `Mean[psolxy[1]][[2]]`

$\Leftarrow$ `-1.39984`

The standard deviations for x

$\Rightarrow$ `StandardDeviation[psolxy[1]][[1]]`

$\Leftarrow$ `0.0769091`

and for y

$\Rightarrow$ `StandardDeviation[psolxy[1]][[2]]`

$\Leftarrow$ `0.0563888`

4.3   Geodetic Examples

4.3.1   Planar Ranging

As opposed to GNSS ranging where the targets being observed are satellites in space and in motion, local positioning systems' target are fixed on surface of the Earth.

Consider two distances $\{s_1, s_2\}$ measured from an unknown station $P(x, y)$ two known stations $P_1(x_1, y_1)$ and $P_2(x_2, y_2)$. The two dimensional distance ranging problems involve the determination of the planar coordinates (x, y) from the measured distances and known positions P_1 and P_2.

The nonlinear distance equations relating the given values above with the coordinates of unknown station can be expressed as

$$(x_1 - x)^2 + (y_1 - y)^2 - s_1^2 = 0$$
$$(x_2 - x)^2 + (y_2 - y)^2 - s_2^2 = 0$$

Given the observations and the unknown positions as,

$$x_1 = 328.76, \quad y_1 = 1207.85, \quad s_1 = 294.33 + \delta$$
$$x_2 = 925.04, \quad y_2 = 954.33, \quad s_2 = 506.42 + \delta$$

let us suppose that both distance values (s_i, $i = 1, 2$) have the same error, $\delta = N$ (0, 0.1). Then our equations are,

$$f_1 = (328.76 - x)^2 + (1207.85 - y)^2 - (294.33 + \delta)^2$$
$$f_2 = (925.04 - x)^2 + (954.33 - y)^2 - (506.42 + \delta)^2$$

To get the reduced target system we employ Gröbner bases eliminating x,

$gr = GroebnerBasis\{f_1, f_2\}$
$$\{1.24041 \times 10^6 - 2264.74y + 1.0y^2 - 315.44\delta - 0.256152y\delta - 0.73976\delta^2,$$
$$-24.8448 + 1.0x - 0.425169y + 0.355689\delta\}$$

4.3.1.1   Algebraic Solution

Then considering the first Gröbner basis as target system for y,

$$q(y, \delta) = 1.24041 \times 10^6 - 2264.74y + y^2 - 315.44\delta - 0.256152y\delta - 0.73976\delta^2.$$

The error free $q(y, \delta = 0) = 0$ solutions are displayed in Fig. 4.8a.
Let us select the first symbolic solution of the target system $q(y, \delta)$ as

$$y = 1132.37 + 0.128076\delta$$
$$- 1.0883 \times 10^{-23} \sqrt{3.53368 \times 10^{50} + 5.11268 \times 10^{48}\delta + 6.3849 \times 10^{45}\delta^2}$$

Employing linearization for δ,

$$y(\delta) = 927.797 - 1.35184\delta.$$

Then the $pdf(y(\delta))$ can be computed via distribution transform. The density function of $y(\delta)$ can be seen in Fig. 4.8b.

The mean value and the standard deviation are: $m_y = 927.797$ and $\sigma_y = 0.135184$.

The x as stochastic variable can computed from the second Gröbner basis,

$$-24.8448 + x(\delta) - 0.425169\, y(\delta) + 0.355689\, \delta = 0$$

employing the linearized expression of y we get for x

$$x(\delta) = 419.316 - 0.93045\, \delta.$$

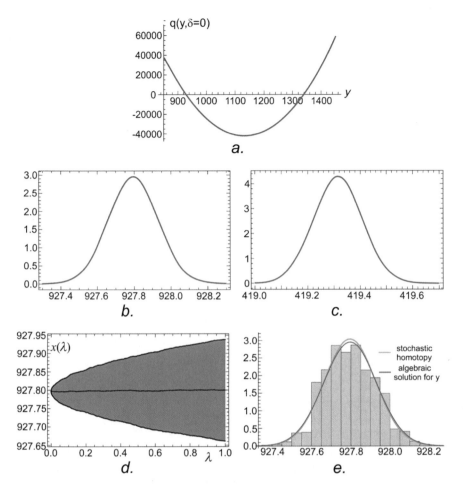

Fig. 4.8 **a** Error free solutions of the target system, **b** the *pdf* of *y* as stochastic variable, **c** the *pdf* of *x* as stochastic variable, **d** the solution of the Ito form, **e** the density functions of the variable *y* computed with different methods

Then the *pdf*(*x*(δ)) can be computed via distribution transform. The density function of *x*(δ) can be seen in Fig. 4.8c.

The mean value and the standard deviation are: $m_x = 419.316$ and $\sigma_x = 0.09304$.

4.3.1.2 Stochastic Homotopy

Now let us consider the start system constructed via fixed point homotopy. Let our target system,

$$Q(y,\delta) = 1.24041 \times 10^6 - 2264.74\,y + y^2 + (-315.44 - 0.256152y)\delta,$$

which is the linearized form of $q(y,\delta)$ in δ. This has the same error free solution as q but this form is more convenient from the point of view of the numerical solution of the Ito differential equation. Now let us consider the start system letting $\delta = 0$ in Q,

$$p = 1.24041 \times 10^6 - 2264.74\,y + y^2,$$

which has two real positive solutions, we consider the smaller one, so

$$y_0 = 927.797.$$

Then the homotopy function is,

$$H = 1.24041 \times 10^6 + y^2 - 315.44\,\delta\lambda + y(-2264.74 - 0.256152\,\delta\lambda)$$

Let $\sigma = 0.1$, then our Ito form is

$$dy(\lambda) = \frac{-315.44 - 0.256152y(\lambda)}{-2264.74 + 2y(\lambda)}\,dw(\lambda).$$

The usual trajectories can be seen in Fig. 4.8d,

The mean and standard deviation are $m_y = 927.8$ and $\sigma_y = 0.13839$. The density function $pdf(y)$ can be seen in Fig. 4.8e.

The solution for the other variable x can be computed similarly, see *Mathematica notebook*.

4.3.2 3D Resection

The Grunert's distance equation, see Fig. 4.9 are given as,

$$S_{1,2}^2 = S_1^2 + S_2^2 - 2S_1S_2\cos(\varphi_{1,2})$$
$$S_{2,3}^2 = S_2^2 + S_3^2 - 2S_2S_3\cos(\varphi_{2,3})$$
$$S_{3,1}^2 = S_3^2 + S_1^2 - 2S_3S_1\cos(\varphi_{3,1})$$

(See also Sect. 1.4.4, Pose Estimation.) The system should be solved for S_1, S_2 and $S_3 \in \mathbf{R}^+$ which are positive, real values. Let us introduce new variables, namely $f_{i,j} = cos(\varphi_{i,j})$, $x_i = S_i$ and $d_{i,j} = S_{i,j}^2$. Then

Fig. 4.9 Tetrahedron: 3D distance and space angle observations

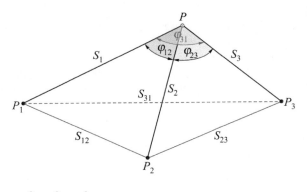

$$p1 = x_1^2 + x_2^2 - d_{1,2}^2 - 2x_1 x_2 f_{1,2}$$
$$p2 = x_2^2 + x_3^2 - d_{2,3}^2 - 2x_2 x_3 f_{2,3}$$
$$p3 = x_1^2 + x_3^2 - d_{1,3}^2 - 2x_1 x_3 f_{1,3}$$

Given the numerical (observation) values as

$$\text{data} = \{d_{1,2} = 1.5603302 + \delta,\ d_{2,3} = 1.358681 + \delta,\ d_{1,3} = 1.718109 + \delta,$$
$$f_{1,2} = \text{Cos}(1.84362),\ f_{2,3} = \text{Cos}(1.768989),\ f_{1,3} = \text{Cos}(2.664537)\}$$

where $d_{i,j}$'s have error δ.

Let us employ now *Dixon resultant* (Lewis 2021) to eliminate x_2 from the second and third equations,

$$dx_1 x_2 = -x_1^4 + 2x_1^2 x_2^2 - x_2^4 + 2x_1^2 d_{1,3}^2 - 2x_2^2 d_{1,3}^2 - d_{1,3}^4 - 2x_1^2 d_{2,3}^2 + 2x_2^2 d_{2,3}^2$$
$$+ 2d_{1,3}^2 d_{2,3}^2 - d_{2,3}^4 - 4x_1^2 x_2^2 f_{1,3}^2 + 4x_1^2 d_{2,3}^2 f_{1,3}^2 + 4x_1^3 x_2 f_{1,3} f_{2,3} + 4x_1 x_2^3 f_{1,3} f_{2,3}$$
$$- 4x_1 x_2 d_{1,3}^2 f_{1,3} f_{2,3} - 4x_1 x_2 d_{2,3}^2 f_{1,3} f_{2,3} - 4x_1^2 x_2^2 f_{2,3}^2 + 4x_2^2 d_{1,3}^2 f_{2,3}^2$$

Then eliminate x_2 from the first equation and from the new polynomial computed above, let us compute the Dixon matrix. The determinant of this matrix provides a polynomial with a single variable of x_1. Substituting the numerical data, we get our target system,

$$q = 130.434 + 558.746\delta + 1039.79\delta^2 + 1097.619\delta^3 + 718.684\delta^4$$
$$+ 298.828\delta^5 + 77.0411\delta^6 + 11.2582\delta^7 + 0.713899\delta^8 - 221.989x_1^2$$
$$- 724.194\delta x_1^2 - 976.74\delta^2 x_1^2 - 696.79\delta^3 x_1^2 - 277.181\delta^4 x_1^2$$
$$- 58.2796\delta^5 x_1^2 - 5.06003\delta^6 x_1^2 + 90.9153x_1^4 + 203.479\delta x_1^4$$
$$+ 170.244\delta^2 x_1^4 + 63.158\delta^3 x_1^4 + 8.77843\delta^4 x_1^4 - 0.371684x_1^6$$
$$- 0.410258\delta x_1^6 - 0.110863\delta^2 x_1^6 + 0.000432337x_1^8$$

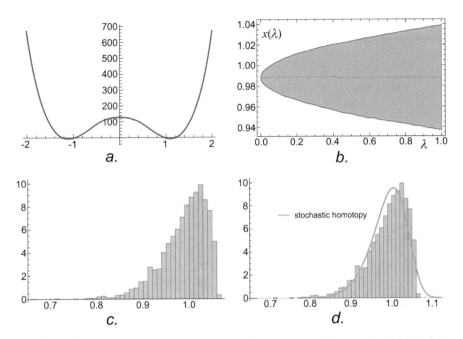

Fig. 4.10 **a** The real roots of the error free target, **b** the trajectories of the solution of the Ito form, **c** the randomly generated data from the $x(1)$ distribution, **d** Fitted Weibull distribution W $(\alpha,\beta,\mu) = W(9.0178, 0.352215, 0.656926)$ for x_1

In this case, symbolic solution is not possible, so we will employ stochastic homotopy. We display the real solutions of the error free case, see Fig. 4.10a.

These roots are

$$\{\{x_1 = -21.0344 - 3.71756\,\mathrm{i}\}, \{x_1 = -21.0344 + 3.71756\,\mathrm{i}\},$$
$$\{x_1 = -1.21812\}, \{x_1 = -0.988277\}, \{x_1 = 0.988277\}, \{x_1 = 1.21812\},$$
$$\{x_1 = 21.0344 - 3.71756\,\mathrm{i}\}, \{x_1 = 21.0344 + 3.71756\,\mathrm{i}\}\}$$

Let us prefer the positive real solution as a proper one, $x = 0.988277$. So we will compute the error distribution of this root. In order to define a corresponding start system, we neglect the last 3 terms of the target system, $\delta = 0$,

$$p = 130.434 - 221.989x_1^2 + 90.9153x_1^4 - 0.371684x_1^6.$$

This start system has a very close root to our preferred root, which will be the initial value for the Ito system $x_0 = 0.988272$. Then the linear homotopy function is (here we switch $x_1 \rightarrow x$)

$$H = (1 - \lambda)p + \lambda q = (130.434 - 221.989x^2 + 90.9153x^4 - 0.371684x^6)(1 - \lambda)$$
$$+ (130.434 - 221.989x^2 + 90.9153x^4 - 0.37168x^6 + 0.000432337x^8$$
$$+ 558.746\delta - 724.194x^2\delta + 203.479x^4\delta - 0.410258x^6\delta + 1039.79\delta^2$$
$$- 976.74x^2\delta^2 + 170.244x^4\delta^2 - 0.110863x^6\delta^2 + 1097.61\delta^3 - 696.79x^2\delta^3$$
$$+ 63.158x^4\delta^3 + 718.684\delta^4 - 277.181x^2\delta^4 + 8.7784x^4\delta^4 + 298.828\delta^5$$
$$- 58.2796x^2\delta^5 + 77.0411\delta^6 - 5.06003x^2\delta^6 + 11.2582\delta^7 + 0.713899\delta^8)\lambda$$

The corresponding Ito form is

$$dx(\lambda) = p\lambda(\lambda, x(\lambda))d\lambda + pw(\lambda, x(\lambda))dw(\lambda),$$

where

$$p\lambda(\lambda, x(\lambda)) = -\frac{0.125\,x(\lambda)^7}{-128365.5 + 105144.1\,x(\lambda)^2 - 644.782\,x(\lambda)^4 + 1.0\,\lambda x(\lambda)^6}$$

and

$$pw(\lambda, x(\lambda)) = \Big(2.07372 \times 10^{10} - 4.38635 \times 10^{10}x(\lambda)^2$$
$$+ 2.96715 \times 10^{10}x(\lambda)^4 + (-6.33598 \times 10^9 - 161548.1\lambda)\,x(\lambda)^6$$
$$+ (5.0405 \times 10^7 + 157037.7\lambda)\,x(\lambda)^8 + (-76481.7 - 29415.6\lambda)\,x(\lambda)^{10}$$
$$+ 29.6541\,\lambda x(\lambda)^{12}\Big)/\Big(x(\lambda)(128365.5 - 105144.1\,x(\lambda)^2$$
$$+ 644.782\,x(\lambda)^4 - 1.0\,\lambda x(\lambda)^6)^2\Big)$$

Let $\sigma = 0.1$ then generating 3000 trajectories with step size 0.003, the trajectories can be seen in Fig. 4.10b.

The mean value and standard deviation of the solution $x(1)$ is $m_x = 987{,}913$ and $\sigma_x = 0.0509105$. Generating random values, we get a definitely non-Gaussian distribution, see Fig. 4.10c.

Let us fit a Weibull distribution, where probability density for value x in a Weibull distribution with location parameter μ is proportional to $(x - \mu)^{\alpha-1}$ $\exp(-((x - \mu)/\beta)^\alpha)$ for $x > \mu$, and is zero for $x \leq \mu$. In our case see Fig. 4.10d.

It goes without saying that this technique is feasible for the other variables, x_2 and x_3, too.

4.3.3 *Ranging by Global Navigation Satellite Systems (GNSS)*

4.3.3.1 Observation Equations

(See also Sect. 2.11.2) Throughout history, position determination has been one of the most important tasks of mountaineers, pilots, sailors, civil engineers etc. In modern times, Global Navigation Satellite Systems (GNSS) provide an ultimate method to accomplish this task. If one has a hand held GNSS receiver, which measures the travel time of the signal transmitted from the satellites, the distance travelled by the signal from the satellites to the receiver can be computed by multiplying the measured time by the speed of light in vacuum. The distance of the receiver from the i-th GNSS satellite, the pseudo-range observations, d_i is related to the unknown position of the receiver, $\{x_1, x_2, x_3\}$ by

$$d_i = \sqrt{(x_1 - a_i)^2 + (x_2 - b_i)^2 + (x_3 - c_i)^2} + x_4,$$

where $\{a_i, b_i, c_i\}$, $i = 0, 1, 2, 3$ are the coordinates of the i-th satellite.

The distance is influenced also by the satellite and receiver' clock biases. The satellite clock bias can be modeled while the receiver' clock bias has to be considered as an unknown variable, x_4. This means, we have four unknowns, consequently we need four satellites to provide a minimum observation. The general form of the equation for the i-th satellite is

$$f_i = (x_1 - a_i)^2 + (x_2 - b_i)^2 + (x_3 - c_i)^2 - (x_4 - d_i)^2.$$

The residual of this type of equation represents the error implicitly. However in geodesy the explicit distance error definition is usual, namely,

$$g_i = d_i - \sqrt{(x_1 - a_i)^2 + (x_2 - b_i)^2 + (x_3 - c_i)^2} - x_4.$$

The relation between the two expressions,

$$g_i = d_i - \sqrt{f_i + (x_4 - d_i)^2} - x_4$$

implies that if $f_i = 0$ then $g_i = 0$ and vice versa. Therefore, in case of four observations, determined system, we employ the first expression, which is easy to handle as a polynomial.

The observation equations for four minimum satellites required are,

$$e_1 = (x_1 - a_0)^2 + (x_2 - b_0)^2 + (x_3 - c_0)^2 - (x_4 - d_0)^2$$
$$e_2 = (x_1 - a_1)^2 + (x_2 - b_1)^2 + (x_3 - c_1)^2 - (x_4 - d_1)^2$$
$$e_3 = (x_1 - a_2)^2 + (x_2 - b_2)^2 + (x_3 - c_2)^2 - (x_4 - d_2)^2$$
$$e_4 = (x_1 - a_3)^2 + (x_2 - b_3)^2 + (x_3 - c_3)^2 - (x_4 - d_3)^2$$

This system can be solved symbolically using Dixon-EDF. The resultant for x1, say, is quadratic and has about 15,000 terms. It involves only x1 and the parameters. But let's continue with numerical data.

Given the observation data as,

$$\text{data} = \{a_0 = 1.483230866 \times 10^7 + \delta, \quad a_1 = -1.579985405 \times 10^7,$$
$$a_2 = 1.98481891 \times 10^6, \quad a_3 = -1.248027319 \times 10^7,$$
$$b_0 = -2.046671589 \times 10^7 + \delta, \quad b_1 = -1.330112917 \times 10^7,$$
$$b_2 = -1.186767296 \times 10^7, \quad b_3 = -2.338256053 \times 10^7,$$
$$c_0 = -7.42863475 \times 10^6 + \delta, \quad c_1 = 1.713383824 \times 10^7,$$
$$c_2 = 2.371692013 \times 10^7, \quad c_3 = 3.27847268 \times 10^6,$$
$$d_0 = 2.4310764064 \times 10^7 + \delta, \quad d_1 = 2.2914600784 \times 10^7,$$
$$d_2 = 2.0628809405 \times 10^7, \quad d_3 = 2.3422377972 \times 10^7\}$$

where the errors δ's are assumed to be in the first coordinates with index 0.

First, this system of polynomials, will be transformed into a system of linear equations and a quadratic equation. Let us expand and multiply by minus one, and arrange the original equations as,

$$\text{eqsL} =$$

$$-x_1^2 - x_2^2 - x_3^2 + x_4^2 + 2x_1 a_0 - a_0^2 + 2x_2 b_0 - b_0^2 + 2x_3 c_0 - c_0^2 - 2x_4 d_0 + d_0^2,$$
$$-x_1^2 - x_2^2 - x_3^2 + x_4^2 + 2x_1 a_1 - a_1^2 + 2x_2 b_1 - b_1^2 + 2x_3 c_1 - c_1^2 - 2x_4 d_1 + d_1^2,$$
$$-x_1^2 - x_2^2 - x_3^2 + x_4^2 + 2x_1 a_2 - a_2^2 + 2x_2 b_2 - b_2^2 + 2x_3 c_2 - c_2^2 - 2x_4 d_2 + d_2^2,$$
$$-x_1^2 - x_2^2 - x_3^2 + x_4^2 + 2x_1 a_3 - a_3^2 + 2x_2 b_3 - b_3^2 + 2x_3 c_3 - c_3^2 - 2x_4 d_3 + d_3^2.$$

Subtract the fourth equation from the other three leads to a system of three linear equations

$$g_1 = a_{0,3} x_1 + b_{0,3} x_2 + c_{0,3} x_3 + d_{0,3} x_4 + e_{0,3}$$
$$g_2 = a_{1,3} x_1 + b_{1,3} x_2 + c_{1,3} x_3 + d_{13} x_4 + e_{1,3}$$
$$g_3 = a_{2,3} x_1 + b_{2,3} x_2 + c_{2,3} x_3 + d_{2,3} x_4 + e_{2,3}$$

The coefficients $\{a_{i,3}, b_{i,3}, c_{i,3}, d_{i,3}, e_{i,3}\}$, $i = 0,1,2$, can be determined as,

$$\{a_{0,3} = 2(a_0 - a_3), \quad b_{0,3} = 2(b_0 - b_3), \quad c_{0,3} = 2(c_0 - c_3), \quad d_{0,3} = -2(d_0 - d_3),$$
$$e_{0,3} = -a_0^2 + a_3^2 - b_0^2 + b_3^2 - c_0^2 + c_3^2 + d_0^2 - d_3^2,$$
$$a_{1,3} = 2(a_1 - a_3), \quad b_{1,3} = 2(b_1 - b_3), \quad c_{1,3} = 2(c_1 - c_3), \quad d_{1,3} = -2(d_1 - d_3),$$
$$e_{1,3} = -a_1^2 + a_3^2 - b_1^2 + b_3^2 - c_1^2 + c_3^2 + d_1^2 - d_3^2,$$
$$a_{2,3} = 2(a_2 - a_3), \quad b_{2,3} = 2(b_2 - b_3), \quad c_{2,3} = 2(c_2 - c_3), \quad d_{2,3} = -2(d_2 - d_3),$$
$$e_{2,3} = -a_2^2 + a_3^2 - b_2^2 + b_3^2 - c_2^2 + c_3^2 + d_2^2 - d_3^2\}$$

In addition, we take one of the nonlinear equations, let say the fourth one,

$$e_4 = (x_1 - a_3)^2 + (x_2 - b_3)^2 + (x_3 - c_3)^2 - (x_4 - d_3)^2$$

Now, we shall solve the linear system for the variables $\{x_1, x_2, x_3\}$, with x_4 as parameter. It means, the relations $x_1 = g_1(x_4)$, $x_2 = g_2(x_4)$, $x_3 = g_3(x_4)$ will be computed. To do that, different elimination methods can be employed.

$$gbx_1 = GroebnerBasis[\{g_1, g_2, g_3\}, \{x_1, x_2, x_3, x_4\}, \{x_2, x_3\}] =$$
$$- x_1 a_{2,3} b_{1,3} c_{0,3} + x_1 a_{1,3} b_{2,3} c_{0,3} + x_1 a_{2,3} b_{0,3} c_{1,3} - x_1 a_{0,3} b_{2,3} c_{1,3}$$
$$- x_1 a_{1,3} b_{0,3} c_{2,3} + x_1 a_{0,3} b_{1,3} c_{2,3} - x_4 b_{2,3} c_{1,3} d_{0,3} + x_4 b_{1,3} c_{2,3} d_{0,3}$$
$$+ x_4 b_{2,3} c_{0,3} d_{1,3} - x_4 b_{0,3} c_{2,3} d_{1,3} - x_4 b_{1,3} c_{0,3} d_{2,3} + x_4 b_{0,3} c_{1,3} d_{2,3}$$
$$- b_{2,3} c_{1,3} e_{0,3} + b_{1,3} c_{2,3} e_{0,3} + b_{2,3} c_{0,3} e_{1,3} - b_{0,3} c_{2,3} e_{1,3} - b_{1,3} c_{0,3} e_{2,3}$$
$$+ b_{0,3} c_{1,3} e_{2,3}$$

in which only x_1 and x_4 can be found. Therefore $x_1 = g(x_4)$ can be computed directly,

$$x_1 = (b_{2,3}(c_{1,3}(x_4 d_{0,3} + e_{0,3}) - c_{0,3}(x_4 d_{1,3} + e_{1,3})) + b_{1,3}(-c_{2,3}(x_4 d_{0,3} + e_{0,3})$$
$$+ c_{0,3}(x_4 d_{2,3} + e_{2,3})) + b_{0,3}(c_{2,3}(x_4 d_{1,3} + e_{1,3}) - c_{1,3}(x_4 d_{2,3} + e_{2,3})))$$
$$/(a_{2,3}(-b_{1,3} c_{0,3} + b_{0,3} c_{1,3}) + a_{1,3}(b_{2,3} c_{0,3} - b_{0,3} c_{2,3}) + a_{0,3}(-b_{2,3} c_{1,3} + b_{1,3} c_{2,3}))$$

Similarly, in the other cases, eliminating x_1 and x_3, and get $x_2 = g(x_4)$,

$$x_2 = (a_{2,3}(b_{1,3}(x_4 d_{0,3} + e_{0,3}) - b_{0,3}(x_4 d_{1,3} + e_{1,3})) + a_{1,3}(-b_{2,3}(x_4 d_{0,3} + e_{0,3})$$
$$+ b_{0,3}(x_4 d_{2,3} + e_{2,3})) + a_{0,3}(b_{2,3}(x_4 d_{1,3} + e_{1,3}) - b_{1,3}(x_4 d_{2,3} + e_{2,3})))$$
$$/(a_{2,3}(-b_{1,3} c_{0,3} + b_{0,3} c_{1,3}) + a_{1,3}(b_{2,3} c_{0,3} - b_{0,3} c_{2,3}) + a_{0,3}(-b_{2,3} c_{1,3} + b_{1,3} c_{2,3}))$$

and $x_3 = g(x_4)$,

$$x_3 = (a_{2,3}(b_{1,3}(x_4 d_{0,3} + e_{0,3}) - b_{0,3}(x_4 d_{1,3} + e_{1,3})) + a_{1,3}(-b_{2,3}(x_4 d_{0,3} + e_{0,3})$$
$$+ b_{0,3}(x_4 d_{2,3} + e_{2,3})) + a_{0,3}(b_{2,3}(x_4 d_{1,3} + e_{1,3}) - b_{1,3}(x_4 d_{2,3} + e_{2,3})))$$
$$/(a_{2,3}(-b_{1,3}c_{0,3} + b_{0,3}c_{1,3}) + a_{1,3}(b_{2,3}c_{0,3} - b_{0,3}c_{2,3}) + a_{0,3}(-b_{2,3}c_{1,3} + b_{1,3}c_{2,3})).$$

4.3.3.2 Algebraic Solution

The solution for x_4 in error free case ($\delta = 0$) is

$$-5.23341 \times 10^9 + 5.23339 \times 10^7 x_4 - 0.91047 x_4^2 = 0$$
$$\{x_4 = 100.001, \ x_4 = 5.74799 \times 10^7\}$$

We shall consider the first solution in the symbolic solution ($\delta > 0$). The solution with error via symbolic computation

$$x_4 = (0.0555556(-2.4074 \times 10^{37} - 2.11252 \times 10^{30}\delta + 1.89369 \times 10^{23}\delta^2$$
$$- 7.21623 \times 10^{15}\delta^3 + \sqrt{(5.79568 \times 10^{74} + 2.0116 \times 10^{67}\delta + 1.4642 \times 10^{60}\delta^2}$$
$$+ 4.1296 \times 10^{52}\delta^3 + 3.1424 \times 10^{45}\delta^4 + 2.881 \times 10^{38}\delta^5 + 5.8762 \times 10^{30}\delta^6)))$$
$$/(-4.65366 \times 10^{28} - 5.86202 \times 10^{20}\delta + 4.18046 \times 10^{14}\delta^2)$$

Employing Taylor series—first order is sufficient,

$$x_4(\delta) = 100.001 + 2.02317\delta.$$

Let the standard deviation of the error distribution $\sigma = 0.3$, the density function for the x_4 variable,

$$\mathrm{NormalDistribution}(100.001, 0.606952)$$

Generating random data and using distribution fitting, we get nearly the same result, see Fig. 4.11a,

4.3.3.3 Stochastic Homotopy

The target system—switching x_4 to x,

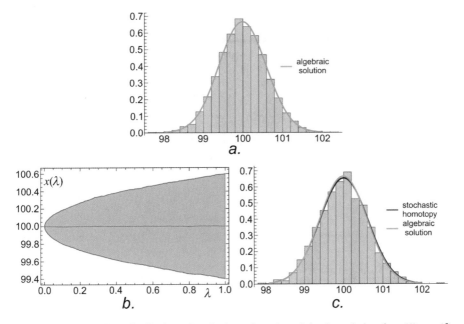

Fig. 4.11 **a** Probability distribution of x_4, **b** the trajectories of the Ito solution for $x(1) = x_4(\delta)$ stochastic variable, **c** the probability distribution of $x_4(\delta)$ generated by the two different methods

$$q(x, \delta) = (-5.42107 \times 10^{54} - 1.09677 \times 10^{53}\delta + 9.60655 \times 10^{45}\delta^2$$
$$- 4.42356 \times 10^{38}\delta^3 + 6.91233 \times 10^{30}\delta^4$$
$$+ x^2(-9.4312 \times 10^{44} - 1.18801 \times 10^{37}\delta + 8.4722 \times 10^{30}\delta^2)$$
$$+ x(5.42105 \times 10^{52} + 4.75697 \times 10^{45}\delta - 4.26422 \times 10^{38}\delta^2$$
$$+ 1.62495 \times 10^{31}\delta^3))/(3.21848 \times 10^{22} + 1.04575 \times 10^{15}\delta)^2$$

The initial value for the Ito differential equation is $x_0 = 100.001$.

Now we employ *affine homotopy*. This type of homotopy suggested by Jalali and Seader (2000) requires the first derivative of the target function,

$$H(x, \lambda) = (1 - \lambda)q'(x_0)(x - x_0) + \lambda q(x)$$

Now we compute the homotopy function as

$$p = x - x_0 = x - 100.001$$

and derivative at $x = x_0$.

$$dqdx = 5.42105 \times 10^{52} + 4.75697 \times 10^{45}\delta - 4.26422 \times 10^{38}\delta^2 + 1.62495 \times 10^{31}\delta^3$$
$$+ 200.001(-9.4312 \times 10^{44} - 1.18801 \times 10^{37}\delta + 8.4722 \times 10^{30}\delta^2))$$
$$/(3.21848 \times 10^{22} + 1.04575 \times 10^{15}\delta)^2$$

Therefore homotopy function,

$$H = (1 - \lambda)dqdx\,p + \lambda q = ((-100.000 + x)(5.42105 \times 10^{52} + 4.75697 \times 10^{45}\delta$$
$$- 4.26422 \times 10^{38}\delta^2 + 1.62495 \times 10^{31}\delta^3 + 200.001(-9.4312 \times 10^{44}$$
$$- 1.18801 \times 10^{37}\delta + 8.4722 \times 10^{30}\delta^2))(1 - \lambda))/(3.21848 \times 10^{22}$$
$$+ 1.0457 \times 10^{15}\delta)^2 + ((-5.42107 \times 10^{54} - 1.09677 \times 10^{53}\delta + 9.60655 \times 10^{45}\delta^2$$
$$- 4.42356 \times 10^{38}\delta^3 + 6.9123 \times 10^{30}\delta^4 + x^2(-9.4312 \times 10^{44} - 1.18801 \times 10^{37}\delta$$
$$+ 8.4722 \times 10^{30}\delta^2) + x(5.42105 \times 10^{52} + 4.75697 \times 10^{45}\delta - 4.26422 \times 10^{38}\delta^2$$
$$+ 1.62495 \times 10^{31}\delta^3))\lambda)/(3.21848 \times 10^{22} + 1.04575 \times 10^{15}\delta)^2$$

Then the in usual way we get the Ito form

$$dx(\lambda) = p\lambda\,d\lambda + pw\,d(\lambda)$$

where

$$p\lambda = \frac{9.4313 \times 10^{48} - 1.88625 \times 10^{47}x(\lambda) + 9.4312 \times 10^{44}x(\lambda)^2}{5.42103 \times 10^{52} + \lambda(1.88625 \times 10^{47} - 1.88624 \times 10^{45}x(\lambda))}$$

and

$$pw = \Big(5.94562 \times 10^{105} + 2.06878 \times 10^{100}\lambda + 7.68478 \times 10^{92}x(\lambda)$$
$$-2.06877 \times 10^{98}\lambda x(\lambda) - 3.84237 \times 10^{90}x(\lambda)^2 + 1.19958 \times 10^{74}\lambda x(\lambda)^2\Big)$$
$$/\Big(5.42103 \times 10^{52} + 1.88625 \times 10^{47}\lambda - 1.88624 \times 10^{45}\lambda x(\lambda)^2\Big)$$

Generating 2000 trajectories with step size 0.001, the result can be seen in Fig. 4.11b.

The mean value of the solution is $m_{x4} = 100.001$ $m_{x4} = 100.001$ and standard deviation $\sigma_{x4} = 0.599631$.

Generating random values, the distribution is fitted to this histogram. This leads to some different result, namely a mixture of distributions with weights $w1$ and $w2$,

$$w1\ \text{NormalDistribution}\ (99.7582, 0.623957)$$
$$+ w2\ \text{GammaDistribution}\ (34345.0, 0.00291818)$$

but this is basically dominated by the Gaussian distribution. Figure 4.11c shows the $pdf(x_4(\delta))$.

The distributions of the other variables can easily computed employing the result of x_4.

4.4   Overdetermined Systems

In geodesy, in most cases, overdetermined systems need to be solved. In this regards, therefore, the methods above are extended to carter for these situations. Overdetermined system with uncertainties may be solved in two ways:

(1) via brutal force—employing the equation system of the minimum conditions of the residual,
(2) via Gauss-Jacobi technique.

In this section we demonstrate these techniques for polynomial systems.

4.4.1   Problem Definitions

Let us consider the following system of equations (3 equations with 2 unknowns)

$$f_1(x, y) = x^2 - xy + y^2 + (2 + \delta)x$$
$$f_2(x, y) = x^2 - 2xy + (1 + \delta)y^2 + x - y - 2$$
$$f_3(x, y) = xy + y^2 - x + 2y - (3 + \delta)$$

We are looking for the real solution in least square sense

$$\sum_{i=1}^{3} (f_i(x, y, \delta))^2 \rightarrow \min_{x,y},$$

where x and y are stochastic variables.

First, let us consider the error free solution in Fig. 4.12a. There exist two local minimums shown in Fig. 4.12b, with the sought global minimum indicated by a circle having the coordinates $\{x = -2.46443, y = -0.500547\}$.

4.4.2   Transforming an Overdetermined System into a Determined One

Let us consider again our overdetermined system. We are looking for the solution in least square sense that should minimize the following objective function

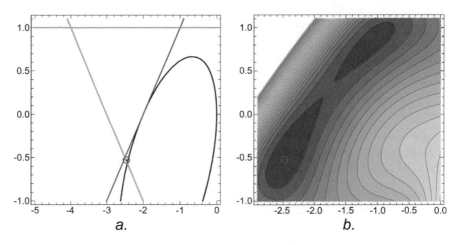

Fig. 4.12 **a** Solution of the overdetermined system in error free case, **b** two local minimums, with the sought global minimum indicated by a circle having the coordinates {x = − 2.46443, y = − 0.500547}

$$G(x, y, \delta) = f_1(x, y, \delta)^2 + f_2(x, y, \delta)^2 + f_3(x, y, \delta)^2.$$

The necessary condition of the optimum is

$$F_1(x, y, \delta) = \frac{\partial G}{\partial x}(x, y, \delta) = 0,$$

$$F_2(x, y, \delta) = \frac{\partial G}{\partial y}(x, y, \delta) = 0.$$

This is already a determined system. First, let us consider our problem in error free case. This system has 5 real solutions, with the first solution providing the smallest residual {x = − 2.46443, y = − 0.500547}.

Now, let us consider non-error free objective function

$$
\begin{aligned}
G = (f_1)^2 + (f_2)^2 + (f_3)^2 &= (-3 - x + 2y + xy + y^2 - \delta)^2 \\
&+ \left(-2 + x + x^2 - y - 2xy + y^2(1 + \delta)\right)^2 + \left(x^2 - xy + y^2 + x(2 + \delta)\right)^2
\end{aligned}
$$

whose necessary conditions are

$$
\begin{aligned}
F_1 = {}& 2 + 4x + 18x^2 + 8x^3 - 4y - 24xy - 18x^2y + 12y^2 + 20xy^2 - 4y^3 \\
& + 2\delta + 8x\delta + 6x^2\delta - 2y\delta - 4xy\delta + 4y^2\delta + 4xy^2\delta - 4y^3\delta + 2x\delta^2 \\
F_2 = {}& -8 - 4x - 12x^2 - 6x^3 - 10y + 24xy + 20x^2y + 6y^2 - 12xy^2 + 12y^3 - 4\delta \\
& - 2x\delta - 2x^2\delta - 12y\delta + 8xy\delta + 4x^2y\delta - 6y^2\delta - 12xy^2\delta + 8y^3\delta + 4y^3\delta^2
\end{aligned}
$$

This is already a determined system. Now, in order to eliminate one of the variable (y) we employ Gröbner basis,

$$\text{grx}(x, d) = \text{GroebnerBasis}(\{F_1, F_2\}, \{y\}) =$$

$$= 17025 + 6967x + 128499x^2 + 95028x^3 + 94732x^4 + 347616x^5 + 548042x^6$$
$$+ 388141x^7 + 128550x^8 + 16250x^9 + 43835\delta + 199014x\delta + 410729x^2\delta$$
$$+ 433318x^3\delta + 312614x^4\delta + 322431x^5\delta + 307523x^6\delta + 142087x^7\delta$$
$$+ 22146x^8\delta - 600x^9\delta + 4654\delta^2 + 247004x\delta^2 + 515886x^2\delta^2 + 544296x^3\delta^2$$
$$+ 226751x^4\delta^2 - 30753x^5\delta^2 - 26795x^6\delta^2 + 26591x^7\delta^2 + 21948x^8\delta^2$$
$$+ 4644x^9\delta^2 + 27580\delta^3 + 197446x\delta^3 + 463917x^2\delta^3 + 545318x^3\delta^3$$
$$+ 248088x^4\delta^3 - 46447x^5\delta^3 - 84161x^6\delta^3 - 33159x^7\delta^3 - 8880x^8\delta^3$$
$$- 1752x^9\delta^3 + 10957\delta^4 + 117024x\delta^4 + 335461x^2\delta^4 + 426416x^3\delta^4$$
$$+ 225736x^4\delta^4 - 5915x^5\delta^4 - 62623x^6\delta^4 - 27320x^7\delta^4 - 3234x^8\delta^4$$
$$+ 426x^9\delta^4 + 3803\delta^5 + 53852x\delta^5 + 186816x^2\delta^5 + 249198x^3\delta^5$$
$$+ 150908x^4\delta^5 + 31324x^5\delta^5 - 7168x^6\delta^5 - 3876x^7\delta^5 - 762x^8\delta^5$$
$$- 240x^9\delta^5 + 1359\delta^6 + 19842x\delta^6 + 79319x^2\delta^6 + 11505x^3\delta^6 + 85857x^4\delta^6$$
$$+ 37120x^5\delta^6 + 9532x^6\delta^6 + 1164x^7\delta^6 + 48x^8\delta^6 + 64x^9\delta^6 + 386\delta^7$$
$$+ 5952x\delta^7 + 26854x^2\delta^7 + 44688x^3\delta^7 + 40008x^4\delta^7 + 21096x^5\delta^7$$
$$+ 6826x^6\delta^7 + 1508x^7\delta^7 + 240x^8\delta^7 + 62\delta^8 + 1352x\delta^8 + 7345x^2\delta^8$$
$$+ 14474x^3\delta^8 + 14152x^4\delta^8 + 7580x^5\delta^8 + 2340x^6\delta^8 + 368x^7\delta^8 + 4\delta^9$$
$$+ 192x\delta^9 + 1508x^2\delta^9 + 3622x^3\delta^9 + 3502x^4\delta^9 + 1584x^5\delta^9 + 300x^6\delta^9$$
$$+ 12x\delta^{10} + 198x^2\delta^{10} + 640x^3\delta^{10} + 528x^4\delta^{10} + 140x^5\delta^{10} + 12x^2\delta^{11}$$
$$+ 72x^3\delta^{11} + 36x^4\delta^{11} + 4x^3\delta^{12}$$

The other variable x can similarly be eliminated.

4.4.2.1 Algebraic Solution

Further on we are looking for the $x(\delta)$ stochastic solution. In order to solve the *grx* $(x, \delta) = 0$ equation symbolically, we employ Taylor expansion to reduce the degree of x and δ at the error free solution $xP = -2.46443$ and $\delta = 0$,

$$x\delta(x, \delta) = 2.23517 \times 10^{-8} + (2.46443 + x)^3(871283 - 1.57129 \times 10^7\delta)$$
$$+ (2.46443 + x)(580979.4 - 3.25277 \times 10^6\delta) + 374414\,\delta$$
$$+ (2.46443 + x)^2(-3.49287 \times 10^6 + 1.00899 \times 10^7\delta)$$
$$+ (2.46443 + x)^4(-1.17976 \times 10^7 + 1.39755 \times 10^7\delta)$$

So we shall consider the first symbolic solution of $x\delta(x,\delta) = 0$ for $x(\delta)$. To keep our result simple, we employ Taylor expansion in δ and throughout imaginary small terms and considering that δ is real,

$$x(\delta) = -2.46443 - 0.644452\delta - 1.11123\delta^2 - 0.809624\delta^3$$

Now, considering $\sigma = 0.1$, we apply distribution transform of δ to get $pdf(x(\delta))$. The probability distribution of $pdf(x(\delta))$ seems different from a Gaussian one, see Fig. 4.13a.

The mean value and the standard deviation of $x(1)$ solution per definition are $m_x = -2.47554$ and $\sigma_x = 0.06872$.

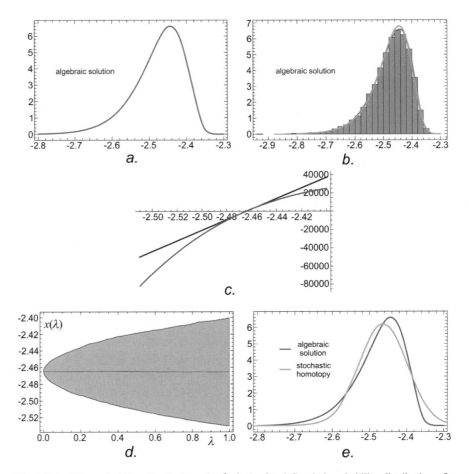

Fig. 4.13 **a** The probability distribution of $x(\delta)$, **b** the fitted Gumbel probability distribution of x (δ), **c** the linearized target function, **d** the homotopy trajectories of the mean and the standard deviations, **e** the density functions of $x(\delta)$ resulting from the two different methods

To find a suitable distribution we generate random data and fit probability distribution to the histogram, see Fig. 4.13b.

The probability density for value x in a Gumbel distribution is

$$\text{Gumbel}(\alpha, \beta) = e^{-e^{(x-\alpha)/\beta} + (x-\alpha)/\beta}.$$

The same technique is feasible for the other variable y, too.

4.4.2.2 Stochastic Homotopy

Our target system is the linearized system $grx(x,\delta)$ at $x = xP$ in order to avoid round off errors, during the solution of the Ito form,

$$\begin{aligned}
q(x, \delta) = {} & 2.2352 \times 10^{-8} + 374413\ \delta - 809069\ \delta^2 + 814976\ \delta^3 - 532678\ \delta^4 \\
& + 296877\ \delta^5 - 151786\ \delta^6 + 61172.8\ \delta^7 - 21286.1\ \delta^8 + 6869.42\ \delta^9 \\
& - 1656.78\ \delta^{10} + 323.13\ \delta^{11} - 59.87\ \delta^{12} + (2.46443 + x)(580979.4 \\
& - 3.25277 \times 10^6\delta + 4.61666 \times 10^6\delta^2 - 3.90172 \times 10^6\delta^3 + 2.4409 \times 10^6\delta^4 \\
& - 1.32622 \times 10^6\delta^5 + 629145\ \delta^6 - 234951\ \delta^7 + 80383.3\ \delta^8 - 22398.7\ \delta^9 \\
& + 4906.12\ \delta^{10} - 902.609\ \delta^{11} + 72.881\ \delta^{12})
\end{aligned}$$

The error free real solution of the linearized system and the original one is presented in Fig. 4.13c where the blue line is the nonlinear target function and the red line is the linearized system in the vicinity of the error free solution xP. Parameters of the applied numeric solution ensuring stable trajectories are 2500 trajectories with step size 0.0002.

Now we employ affine homotopy.

$$H(x, \lambda) = (1 - \lambda)q'(x_0)(x - x_0) + \lambda q(x)$$

where $x_0 = xP$. The homotopy function is,

$$\begin{aligned}
H(x(\lambda), \lambda, \delta) = {} & 1.43178 \times 10^6 + 580979x - 8.01622 \times 10^6\delta - 3.25277 \times 10^6x\delta \\
& + 1.13774 \times 10^7\delta^2 + 4.61666 \times 10^6x\delta^2 - 9.6155 \times 10^6\delta^3 - 3.9017 \times 10^6x\delta^3 \\
& + 6.01548 \times 10^6\delta^4 + 2.4409 \times 10^6x\delta^4 - 3.26837 \times 10^6\delta^5 - 1.3262 \times 10^6x\delta^5 \\
& + 1.55048 \times 10^6\delta^6 + 629145\ x\delta^6 - 579020\ \delta^7 - 234951\ x\delta^7 + 198099\ \delta^8 \\
& + 80383.3\ x\delta^8 - 55200.1\ \delta^9 - 22398.7\ x\delta^9 + 12090.8\ \delta^{10} + 4906.12\ x\delta^{10} \\
& - 2224.42\ \delta^{11} - 902.609\ x\delta^{11} + 179.6\ \delta^{12} + 72.88\ x\delta^{12} + 2.23517 \times 10^{-8}\lambda \\
& + 374414\ \delta\lambda - 809069\ \delta^2\lambda + 814976\ \delta^3\lambda - 532678\ \delta^4\lambda + 296877\ \delta^5\lambda \\
& - 151786\ \delta^6\lambda + 61172.8\ \delta^7\lambda - 21286.1\ \delta^8\lambda + 6869.4\ \delta^9\lambda - 1656.78\ \delta^{10}\lambda \\
& + 323.13\ \delta^{11}\lambda - 59.87\ \delta^{12}\lambda
\end{aligned}$$

To get the differential equation form, we employ the usual technique, to get

$$dx(\lambda) = -3.84725 \times 10^{-14} d\lambda - 0.644452 \, dw(\lambda).$$

Generating 2500 trajectories with step size 0.0002, the mean value of the homotopy trajectories can be seen in Fig. 4.13d.

The mean value and the standard deviation of $x(1)$ solution per definition are $m_x = -2.46524$ and $\sigma_x = 0.0649435$. Fitting probability normal distribution to the randomly generated data for normal distribution, we get $N(-2.46342, 0.0875472)$ or Gumbel distribution $G(-2.43146, 0.0653486)$.

The density functions resulting from the algebraic and the stochastic methods are presented in Fig. 4.13e.

4.4.3 Gauss-Jacobi Approach

The Gauss-Jacobi method is extendable to nonlinear systems. Let us consider a simple example $n = 3$ and $m = 2$.

$$f_1(x, y) = 0$$
$$f_2(x, y) = 0$$
$$f_3(x, y) = 0$$

In this case the algorithm is as follows:

- Let us compute the solutions of the nonlinear subsystems via Dixon Resultant or Gröbner Basis or linear homotopy methods, namely $\eta_{12} = (x_{12}, y_{12})$, $\eta_{13} = (x_{13}, y_{13})$ and $\eta_{23} = (x_{23}, y_{23})$,
- Compute the means of these solutions, $\eta = (x_s, y_s)$

$$\eta = \frac{\eta_{12} + \eta_{13} + \eta_{23}}{3}$$

- Let us linearize the overdetermined system at this point.

The linearized form of the j-th nonlinear equation at $\eta = (x_s, y_s)$, $j = 1, 2, 3$.

$$f_j(\eta) + \frac{\partial f_j}{\partial x}(\eta)(x - x_s) + \frac{\partial f_j}{\partial y}(\eta)(y - y_s) = 0$$

or

$$\frac{\partial f_j}{\partial x}(\eta)x + \frac{\partial f_j}{\partial y}(\eta)y + (f_j(\eta) - \frac{\partial f_j}{\partial x}(\eta)x_s - \frac{\partial f_j}{\partial y}(\eta)y_s) = 0$$

Let us recall the computation of the weight of the $\{1, 2\}$ subsystem in linear case ($n = 3$ and $m = 2$)

$$\pi_{12} = \begin{bmatrix} a_{11} & a_{12} \\ a_{21} & a_{22} \end{bmatrix}^2 = (a_{11}a_{22} - a_{21}a_{12})^2$$

Similarly, the weight of the solution of the linearized subsystem $\{1, 2\}$ is

$$\pi_{12}(\eta) = \begin{bmatrix} \frac{\partial f_1}{\partial x}(\eta) & \frac{\partial f_1}{\partial y}(\eta) \\ \frac{\partial f_2}{\partial x}(\eta) & \frac{\partial f_2}{\partial y}(\eta) \end{bmatrix}^2 = \left(\frac{\partial f_1}{\partial x}(\eta) \frac{\partial f_2}{\partial y}(\eta) - \frac{\partial f_2}{\partial x}(\eta) \frac{\partial f_1}{\partial y}(\eta) \right)^2$$

Then computing the other two weights similarly, the approximate Gauss-Jacobi solution of the overdetermined nonlinear system is

$$x_{GJ} = \frac{\pi_{12}x_{12} + \pi_{13}x_{13} + \pi_{23}x_{23}}{\pi_{12} + \pi_{13} + \pi_{23}}$$

and

$$y_{GJ} = \frac{\pi_{12}y_{12} + \pi_{13}y_{13} + \pi_{23}y_{23}}{\pi_{12} + \pi_{13} + \pi_{23}}$$

The precision of the solution can be increased by linearization of the system at (x_{GJ}, y_{GJ}) and using pseudoinverse solution.

Remarks Keep in mind that we did not need initial values and we can get all solutions! After computing the average of the solutions of the subsystems, one can employ standard iteration techniques.

4.4.3.1 Solution of the Subsystems

We have 3 subsystems equations $\{1, 2\}$, $\{1, 3\}$ and $\{2, 3\}$. Let us consider the solution for $\text{pdf}(x_{12}(\delta))$.

To eliminate the variable y we employ reduced Gröbner basis (let $x = x_{12}$),

$$\begin{aligned} \text{grx } 12 &= \text{GroebnerBasis}(\{f_1, f_2\}, \{y\}) = \\ &= 4 + 8x + 9x^2 + 5x^3 + x^4 + 13x\delta + 9x^2\delta - x^4\delta + 4x\delta^2 \\ &\quad + 10x^2\delta^2 + 4x^3\delta^2 + x^4\delta^2 + 6x^2\delta^3 + 2x^3\delta^3 + x^2\delta^4 \end{aligned}$$

Considering the real solution of the error free form ($\delta = 0$),

$$4 + 8x + 9x^2 + 5x^3 + x^4 = 0$$

which is $x = xP = -2$.

The corresponding symbolic solution of $grx12(x, \delta) = 0$ is very complicated, see *Mathematica notebook*, therefore we employ second order Taylor expansion of this solution at $\delta = 0$,

$$x_{12}(\delta) = -2 - \sqrt{2}\sqrt{\delta} - \frac{\delta}{2} - \frac{19\delta^{3/2}}{24\sqrt{2}}$$

Assuming $\sigma = 0.1$ and using distribution transformation for $x(\delta)$ we get the probability distribution $pdf(x_{12}(\delta))$.

The mean and standard deviation can be computed directly as $m_{12} = -2.36659$ and $\sigma_{12} = 0.171549$.

Generating random dataset in order the fit a proper distribution function, the best fitted probability distribution is a Weibull distribution $W_{12}(18.4862, 2.92601, -5.22038)$, see Fig. 4.14a,

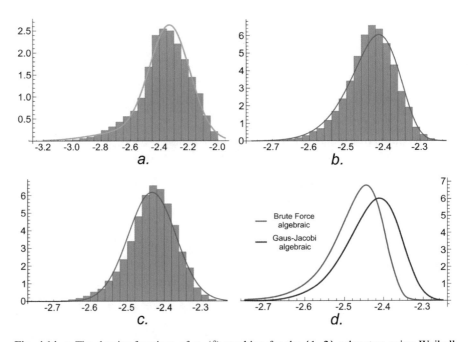

Fig. 4.14 a The density functions of $x_{12}(\delta)$ resulting for the $\{1, 2\}$ subsystem using Weibull distribution as best fit, **b** the density functions of $x(\delta)$ resulting as the average of subsystem solutions using Weibull distribution as best fit, **c** the density functions of $x(\delta)$ resulting as the average of subsystem solutions using the fitted normal distribution, **d** the effect of dimension reduction on the probability distribution

Similarly for the other two subsystems, we get,

$$x_{1,3}(\delta) = -2.43426 - 0.905911 \, \delta + 0.023924 \, \delta^2$$

with normal distribution $N_{13}(-2.43234, 0.0908095)$, and

$$x_{2,3}(\delta) = -2.5 - 0.291667 \, \delta + 0.146991 \, \delta^2$$

with normal distribution $N_{13}(-2.49861, 0.0292661)$.

4.4.3.2 The Averaged Solution of the Subsystems

The distribution transformation in case of simple average is,

$$x(\delta) = (x_{12}(\delta) + x_{13}(\delta) + x_{23}(\delta))/3$$

The best fitted distribution for the randomly generated data is a Weibull distribution $W_x(8.96727, 0.551206, -2.9551)$, with mean value and standard variation, $m_x = -2.43305$ and $\sigma(x) = 0.0645987$, see Fig. 4.14b.

This result is quite close to the solution of the brutal force technique: $m_x = -2.47554$ and $\sigma(x) = 0.06872$. However, this is a normal distribution.

Now the best fitted normal distribution of the Gauss-Jacobi solution is $N(-2.43305, 0.0645971)$, see Fig. 4.14c.

One can employ stochastic homotopy in the case of the Gauss-Jacobi reduction too, as done in the previous sections.

The effect of the methods of the equation reduction are compared in Fig. 4.14d, where the brutal force reduction technique seems to be more simple, requiring less computation efforts. However, in certain cases, the Gröbner basis could be too complicated for symbolic computations even when using the Taylor expansion. Then Gauss-Jacobi technique could provide a feasible solution.

In the following parts the overdetermined case of spatial resection and the GNSS problem are solved.

4.5 Positioning by 3D-Resection

4.5.1 Introduction

In this part a real problem is solved via algebraic as well as stochastic homotopy methods. The resection problem leads to overdetermined multivariate polynomial system that can be reduce via Gröbner basis. However, the degree of the resulting polynomial is too high for a direct symbolic solution necessitating the use of the

Fig. 4.15 The Stuttgart
central geodetic test network

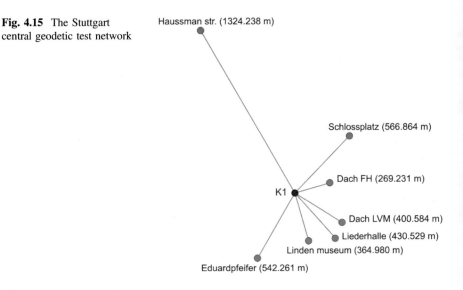

Taylor expansion for the error free solution. The algebraic and stochastic homotopy methods provide practically the same results.

Let us consider a real overdetermined problem exemplified by the Stuttgart Central Test Network presented in Awange and Paláncz (2016), see Fig. 4.15,

We are looking for the error distribution of the coordinates (x_0, y_0, z_0) of the reference point ($K1$).

The GPS coordinates (x_i, y_i, z_i) in the global reference frame, $i = 1, 2, \ldots 7$ are presented in Table 4.1.

We have in the measured distances s_i with errors δ.

$$Y = \{566.864 + \delta, \; 1324.24 + \delta, \; 542.261 + \delta, \; 364.98 + \delta,$$
$$430.529 + \delta, \; 400.584 + \delta, \; 269.231 + \delta\}$$

Table 4.1 The GPS coordinates of the stations in the WGS 84 reference frame

Stations	x(m)	y(m)	z(m)
K1 (reference) point	4,157,066.1116	671,429.6655	4,774,879.3704
1	4,157,246.5346	671,877.0281	4,774,581.6314
2	4,156,749.5977	672,711.4554	4,774,981.5459
3	4,156,748.6829	671,171.9385	4,775,235.5483
4	4,157,066.8851	671,064.9381	4,774,865.8238
5	4,157,266.6181	671,099.1577	4,774,689.8536
6	4,157,307.5147	671,171.7006	4,774,690.5691
7	4,157,244.9515	671,338.5915	4,774,699.9070

The prototype of the equations based on the *implicit* distance definition is,

$$e_i = (x_i - x_0)^2 + (y_i - y_0)^2 + (z_i - z_0)^2 - s_i^2.$$

The objective function to be minimized is

$$obj = \sum_{i=1}^{7} e_i^2.$$

The error free solution via global minimization is

$$\{x_0 = 4157066.111, \ y_0 = 671429.665, \ z_0 = 4774879.370\}.$$

Now, consider the objective function employing the prototype of the equations based on the *explicit* distance definition as,

$$e_i = \sqrt{(x_i - x_0)^2 + (y_i - y_0)^2 + (z_i - z_0)^2} - s_i$$

The solution is quite similar,

$$x_0 = 4157066.111, \ y_0 = 671429.665, \ z_0 = 4774879.370.$$

Consequently, we employ the equations based on the *implicit* definition, since this is a polynomial type of problem predisposed to the Gröbner basis elimination approach. To use Gröbner basis we need rationalize the data. Then the system of the polynomial equations representing the necessary condition of the optimum is,

$$eq_\xi = \frac{\partial \, obj}{\partial \xi} = 0, \quad \xi = x_0, y_0, z_0.$$

4.5.2 *Algebraic Solution*

Let us find the solution, the probability distribution of the coordinate x_0. Eliminating variables y_0 and z_0 employing reduced Gröbner basis, we get

$$gr \, x_0 = \text{GroebnerBasis}[\text{eqs}, x_0, \{y_0, z_0\}, \ \text{MonomialOrder} \rightarrow \text{EliminationOrder}]$$

Since the constant term is very big, we normalize the coefficients of this polynomial so that the constant term is equal to 1.

This polynomial cannot be solved symbolically since the exponent of x_0 and δ are

$$\text{Exponent}[\text{gr}\, x_0 n, \{x_0, \delta\}] \rightarrow \{7, 6\}$$

Therefore, we employ Taylor series expansion at the error free solution of x_0.

$$x_0 P = 4157066.111.$$

Second order expansion is used,

$$grx_0 nS = \text{Series}[\text{gr}\, x_0 n, \{x_0, x_0 P, 2\}].$$

Now, let us solve $grx_0 nS(x_0, \delta) = 0$ polynomial equation symbolically. Considering the first solution and neglecting the higher order terms of δ beyond the third one, we get

$$\beta = 4157066.111 + 0.339286\, \delta + 3.33517 \times 10^{-7}\delta^2 + 4.90642 \times 10^{-13}\delta^3.$$

Considering the standard deviation of the error of the distant measurement as, $\sigma = 0.1$, the variable x_0 as stochastic process is,

$$Dx0 = \text{TransformedDistribution}[\beta, \delta \approx \text{NormalDistribution}[0, \sigma]]$$

The mean and the standard deviation are,

$$\mu_{x0} = 4157066.111 \quad and \quad \sigma_{x0} = 0.0339286.$$

The probability density function is then computed by employing random data set generated by the process. The displayed density function of x_0 is presented in Fig. 4.16a.

4.5.3   Stochastic Homotopy Solution

Now, let us see how to solve the problem via stochastic homotopy. The target system can be the second order system, $grx_0 nS(x_0, \delta)$. Let us employ now x as independent variable,

$$q = grx_0 nS/.(x_0 \rightarrow x).$$

Employing the affine homotopy, the function becomes

$$H(x, \lambda) = (1 - \lambda)q'(x_0)(x - x_0) + \lambda q(x),$$

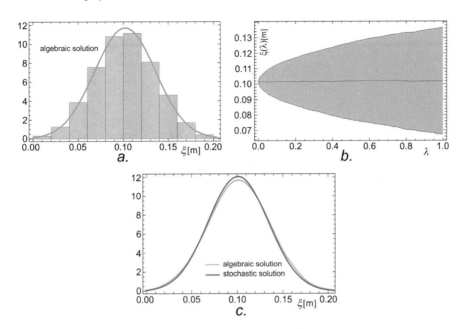

Fig. 4.16 **a** The effect of dimension reduction on the probability distribution, scaling $\rightarrow \xi = \{\mu - 3\delta, \mu + 3\delta\}$, **b** the trajectories of the mean and the $\pm$ standard deviation, **c** the density functions resulting from the two different methods scaling $\rightarrow \xi = \{\mu - 3\delta, \mu + 3\delta\}$

where $x_0 = x_0 P$, see in Section of *Algebraic* solution. The start system is

$$p = x - x_0$$

and derivative at x_0

$$\frac{dq}{dx} = D[q, x]/.(x \rightarrow x_0)$$

$$x_0 = 1.48231 \times 10^{-21} - 7.06819 \times 10^{-28}\delta - 2.70834 \times 10^{-35}\delta^2$$
$$+ 2.93084 \times 10^{-40}\delta^3 + 8.03014 \times 10^{-43}\delta^4 + 2.66134 \times 10^{-46}\delta^5$$
$$- 7.31937 \times 10^{-51}\delta^6$$

Therefore the homotopy function can be written as,

$$H = (1 - \lambda)\frac{dq}{dx}p + \lambda q$$

We can then compute the Ito form as usual,

$$proc = \text{ItoProcess}\,(dx(\lambda) = p\lambda\mathrm{d}\lambda + pw\,dw(\lambda), x(\lambda), \{x, x_0\}, \lambda,$$
$$w \approx \text{WienerProcess}\,(0, \sigma))$$

where

$$pw = (3.0963 \times 10^{-41} + 8.44425 \times 10^{-42}\lambda - 1.45379 \times 10^{-47}x(\lambda)$$
$$- 2.0313 \times 10^{-48}\lambda x(\lambda) + 1.74858 \times 10^{-54}x(\lambda)^2)$$
$$/(1.48231 \times 10^{-21} + 1.67903 \times 10^{-20}\lambda - 4.03897 \times 10^{-27}\lambda x(\lambda))^2$$

and

$$p\lambda = \frac{2.01948 \times 10^{-27}(-415707 + x(\lambda))^2}{1.48231 \times 10^{-21} + \lambda(1.67903 \times 10^{-20} - 4.03897 \times 10^{-27}x(\lambda))}$$

Generating 2000 trajectories with step size 0.001, we get the trajectories of the mean and ± standard deviation, see Fig. 4.16a where the mean value $\mu_{x0} = 4157066.112$ and the standard deviation $\sigma_{x0} = 0.034783$.

The density function of the distribution is then estimated from the generated random values as,

$$\text{estimated } D = \text{NormalDistribution}[4157066.112, \ 0.0351093].$$

This density function together with its comparison resulting from the algebraic solution is presented in Fig. 4.16c.

The techniques can be applied to the other two (y_0, z_0) coordinates too.

The *Mathematica notebook* is available from the "Electronic supplementary material" of Springer relevant web page.

4.6 Ranging by GNSS

4.6.1 Introductory Remarks

In this part we shall compute the error distribution of the position coordinates in case of six satellites. Traditionally, four satellites are required hence having six provides two redundant measurements. Using the *implicit* form of the distances leads to an easy solution, while employing the *explicit* form of the distances results in a significantly more difficult problem. In both cases, we employ the necessary conditions to transform the overdetermined system into a determined one. To reduce the multivariable case to a single variable one Gröbner basis is applied in the first case. In the second case, one can utilize the knowledge of the error free solution.

From the measured distance of the receiver from the i-th GNSS satellite, the pseudo-range observations, d_i is related to the unknown position of the receiver, $\{x_1, x_2, x_3\}$ by

$$d_i = \sqrt{(x_1 - a_i)^2 + (x_2 - b_i)^2 + (x_3 - c_i)^2} + x_4,$$

where $\{a_i, b_i, c_i\}$, $i = 0, 1, 2, 3$ are the coordinates of the i-th satellite.

The distance is influenced also by the satellite and receiver' clock biases. The satellite clock bias can be modeled while the receiver' clock bias has to be determined as an unknown variable, x_4. This means, therefore, that we have four unknowns, consequently needing four satellites to provide a minimum observation. The general (implicit) form of the equation for the i-th satellite is,

$$f_i = (x_1 - a_i)^2 + (x_2 - b_i)^2 + (x_3 - c_i)^2 - (x_4 - d_i)^2.$$

Now, we consider $m = 6 > 4$ satellites. In the first case, we consider the following equations using implicit distance definition, f_i, where $\{a_i, b_i, c_i\}$, $i = 0, 1, 2, 3, 4, 5$ are the coordinates of the i-th satellite, assuming that there are δ errors in the first coordinates of the satellites.

The residual of this type of equations represents the error implicitly. However, in geodesy, the explicit distance error definition is usual, namely,

$$g_i = d_i - \sqrt{(x_1 - a_i)^2 + (x_2 - b_i)^2 + (x_3 - c_i)^2} - x_4$$

In the second case, we consider this type of distance definitions. The two types of definition give only the same results if the residuals are zero, which is not the case in practice.

Then find the probability distributions of the x_j, $j = 1, 2, 3, 4$ as stochastic variables, which minimize the residual,

$$R_f = \sum_{i=0}^{5} f_i^2 \rightarrow \min_{x_j}$$

or

$$R_g = \sum_{i=0}^{5} g_i^2 \rightarrow \min_{x_j}$$

In both cases, we consider the determined equation systems resulting from the necessary conditions of the optimum,

$$\frac{\partial R}{\partial x_j} = 0$$

The main problem lies with the second case where the determined system is not a polynomial and consequently, Gröbner basis cannot be used for equation reduction. Furthermore, we shall employ algebraic solution in both cases.

Let us consider the data for six satellites' case as the following numerical values,

$$
\begin{aligned}
&a_0 = 14177553.47 + \delta, &&a_1 = 15097199.81 + \delta, &&a_2 = 2.460342.33 + \delta, \\
&a_3 = -8206488.95 + \delta, &&a_4 = 1399988.07 + \delta, &&a_5 = 6995655.48 + \delta, \\
&b_0 = -18814768.09, &&b_1 = -4636088.67, &&b_2 = -9433518.58, \\
&b_3 = -18217989.14, &&b_4 = -17563734.90, &&b_5 = -23537808.26, \\
&c_0 = 12243866.38, &&c_1 = 21326706.55, &&c_2 = 8174941.25, \\
&c_3 = 17605231.99, &&c_4 = 19705591.18, &&c_5 = -9927906.48, \\
&d_0 = 21119278.32, &&d_1 = 22527064.18, &&d_2 = 23674159.88, \\
&d_3 = 20951647.38, &&d_4 = 20155401.42, &&d_5 = 24222110.91
\end{aligned}
$$

where δ is $N(0, \delta)$ as usual. It means we assume error in the first coordinates of the data points, in a_i, $i = 0, 1, \ldots 5$.

4.6.2   Solution for Implicit Error Definition

In the first case, the implicit representation of the distance is,

$$e_i = (x_1 - a_i)^2 + (x_2 - b_i)^2 + (x_3 - c_i)^2 - (x_4 - d_i)^2, \quad i = 0, 1, \ldots 5.$$

In order to employ Gröbner basis the rationalization of the data set is necessary, then objective to be minimized is the sum of the residual of the equations, f_i. The global minimum can be found by using the built-in function,

$$\{x_1 = 596928.9, \quad x_2 = -4847849.3, \quad x_3 = 4088224.4, \quad x_4 = 13.4526\}.$$

The equations representing the necessary condition of the optimum of the objective function are

$$\frac{\partial R_f}{\partial x_i} = 0, \quad i = 0, 1, \ldots 5.$$

Then the eliminating the variables x_2, x_3, x_4 reduced Gröbner basis is employed,

$$
\begin{aligned}
grx_1 = {}&\text{GroebnerBasis}[\{eq_1, eq_2, eq_3, eq_4\}, x_1, \{x_2, x_3, x_4\}, \\
&\text{MonomialOrder} \rightarrow \text{EliminationOrder}]
\end{aligned}
$$

The resulted polynomial $grx_1(x_1, \delta)$ has a big constant term, so it is reasonable to normalize the polynomial. Unfortunately this normalized polynomial $grx_1\,n(x_1, \delta) = 0$ cannot be solved symbolically, since

$$\text{Exponent}[grx_1 n, x_1] = 9$$

Therefore, we employ Taylor expansion at the error free solution of x_1.

$$x_1 P = 596928.9.$$

Let us solve $grx_1 nS(x_1, \delta) = 0$ polynomial symbolically. The second solution should be considered and in addition we neglect the higher order term of the δ, too. We get,

$$x_1(\delta) = 596928.9 + 1.0\delta - 3.09899 \times 10^{-21}\delta^2 - 1.93882 \times 10^{-13}\delta^3$$

Let us assume, $\sigma = 0.1$, then using distribution transform of β, the stochastic process of x_1 is,

$$Dx_1 = \text{TransformedDistribution}[x_1(\delta), \delta \approx \text{NormalDistribution}[0, \sigma]]$$

The mean value and the standard deviation are $\mu_{x1} = 596929$ and $\sigma_{x1} = 0.1$. To get the probability distribution function of x_1, we generate random data from Dx_1, and fit a distribution to the data. The result are presented in Fig. 4.17.

Now, let us employ this technique for the last coordinate, x_4.

Then, eliminating the variables x_1, x_2, x_3,

$$gr\,x_4 = \text{GroebnerBasis}[\{eq_1, eq_2, eq_3, eq_4\}, x_4, \{x_1, x_2, x_3\},$$
$$\text{MonomialOrder} \rightarrow \text{EliminationOrder}]$$

Employing the same technique as before in case of x_1, we surprisingly get that

$$x_4(\delta) = 13.4526,$$

Fig. 4.17 The probability density function of x_1 as stochastic variable, scaling $\rightarrow \xi = \{\mu - 3\delta, \mu + 3\delta\}$

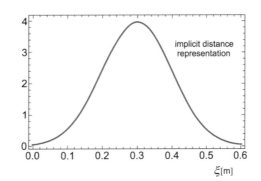

which means the last coordinate (x_4) does not depend on δ, on the error of the first coordinates $(a_i + \delta)$. This could be recognized earlier, since

$$\text{Coefficient}(\text{gr}\,x_4, \delta) = 0$$

Even the Gröbner basis is free from δ!

4.6.3 Solution for Explicit Error Definition

Now, we employ explicit representation of the distance,

$$g_i = d_i - \sqrt{(x_1 - a_i)^2 + (x_2 - b_i)^2 + (x_3 - c_i)^2} - x_4, \quad i = 0, 1, \ldots 5$$

Then the objective function to be minimized is the sum of the residual of the equations,

$$R_g = \sum_{i=0}^{5} g_i^2 \rightarrow \min_{x_i}.$$

The global minimum is found by using the built-in Mathematica function as,

$$\{x_1 = 596929.6, \quad x_2 = -4847851.5, \quad x_3 = 4088226.8, \quad x_4 = 15.5181\}$$

Now, the last coordinate value (x_4) is different. The application of the Gröbner basis is not possible since our equations are not polynomials,

$$\text{Polynomial } Q[g_i, \{x_1, x_2, x_3, x_4\}], \quad i = 0, 1, \ldots 5 \quad \rightarrow \textit{False}$$

However, we know the error free solutions and "just" looking for their error distributions. Let us assume that the mean values of these distributions are equal to the values of the error free solutions. For example one can solve the problem for the first coordinate, considering the $g_1(x_1) = g(x_1, x_2P, x_3P, x_4P)$ objective function, where x_2P, x_3P, x_4P are the optimal solutions in error free case. The error free solution in case of explicit distance definition is,

$$\begin{aligned}\{x_1P, x_2P, x_3P, x_4P\} &= \{x_1, x_2, x_3, x_4\} \\ &= \{596929.6, \quad -4847851.5, \quad 4088226.8, \quad 15.5181\}\end{aligned}$$

The first objective $g_1(x_1, \delta)$ is

$$g_1(x_1, \delta) = g_1(x_2 = x_2P, \quad x_3 = x_3P, \quad x_4 = x_4P)$$

Then the necessary condition of the optimum is

$$eq_1(x_1, \delta) = \frac{\partial g_1}{\partial x_1} = 0$$

which is not a polynomial of neither x_1 nor δ, i.e.

$$\text{Polynomial } Q[eq_1, \{x_1, \delta\}] \rightarrow False.$$

In order to solve $eq_1(x_1, \delta)$ symbolically, the Taylor series expansion is employed. Here, we employed the second order expansion. This is a polynomial already and it can be solved for x_1. We just take into consideration the first three terms,

$$x_4(\delta) = 596930.0 + 1.0\delta - 1.94387 \times 10^{-23}\delta^2 + 1.51819 \times 10^{-16}\delta^3$$

In case of implicit distance definition, obtained practically the same solution,

$$596928.9 + 1.0\delta - 3.09899 \times 10^{-21}\delta^2 - 1.93882 \times 10^{-13}\delta^3$$

Let us check the last coordinate! The first objective $g_4(x_4, \delta)$ is

$$g_4(x_4, \delta) = g_4(x_1 = x_1 P, \quad x_2 = x_2 P, \quad x_3 = x_3 P)$$

Then the necessary condition of the optimum is

$$eq_4(x_4, \delta) = \frac{\partial g_4}{\partial x_4} = 0$$

which is now a first order polynomial of x_4.

$$
\begin{aligned}
eq_4(x_4, \delta) = &-2(\frac{591853997}{25} - x_4 - \sqrt{3.77296 \times 10^{13} + (-2.28634 \times 10^7 - \delta)^2}) \\
&-2(\frac{1126353209}{50} - x_4 - \sqrt{2.9721 \times 10^{14} + (-1.45003 \times 10^7 - \delta)^2}) \\
&-2(\frac{527981958}{25} - x_4 - \sqrt{2.61589 \times 10^{14} + (-1.35806 \times 10^7 - \delta)^2}) \\
&-2(\frac{2422211091}{100} - x_4 - \sqrt{5.45766 \times 10^{14} + (-639873 - \delta)^2}) \\
&-2(\frac{1007770071}{50} - x_4 - \sqrt{4.05596 \times 10^{14} + (-803058.4 - \delta)^2}) \\
&-2(\frac{1047582369}{50} - x_4 - \sqrt{3.6147 \times 10^{14} + (880342 - \delta)^2})
\end{aligned}
$$

then the solution for x_4,

$$x_4(\delta) = \{0.0833333$$

$$(2.65299 \times 10^8 - 2\sqrt{3.77296 \times 10^{13} + (-2.28634 \times 10^7 - \delta)^2}$$

$$- 2\sqrt{2.9721 \times 10^{14} + (-1.45003 \times 10^7 - 1.\delta)^2}$$

$$- 2\sqrt{2.61589 \times 10^{14} + (-1.35806 \times 10^7 - \delta)^2}$$

$$- 2\sqrt{5.45766 \times 10^{14} + (-639873 - \delta)^2}$$

$$- 2\sqrt{4.05596 \times 10^{14} + (-803058.4 - 1.\delta)^2}$$

$$- 2\sqrt{3.6147 \times 10^{14} + (88034 - \delta)^2)}\}$$

and respecting the third order term of δ as the highest one

$$x_4(\delta) = \{15.5181 - 0.356053\delta - 1.53212 \times 10^{-8}\delta^2 + 1.19417 \times 10^{-16}\delta^3\}$$

which is now dependent on δ.

The overdetermined GNSS problem has been solved using two different distance definitions. It could be seen that in case of the explicit distance definition, the errors in the first coordinates $(a_i + \delta)$ can generate error distribution even in the last variable $(x_4 + 0.356053 \delta)$, while in the case of the implicit distance definition, these errors affect only the first variable.

4.7 GPS Meteorology—Bending Angles

4.7.1 Introduction

The bending angles problem of GPS meteorology with errors in the input coefficients of the quartic polynomial has been solved using both techniques, namely algebraic solution using Gröbner basis and the transform of the density function, and employing stochastic homotopy as well. We got practically the same results, which correspond to the results published in the literature.

In this section, we demonstrate how to employ our method for the equations for atmospheric bending angles problem!

Atmospheric bending angles are derived from a system of the two nonlinear trigonometric equations. We refer to the text book for the physical model, see Awange and Paláncz (2016), see Fig. 4.18.

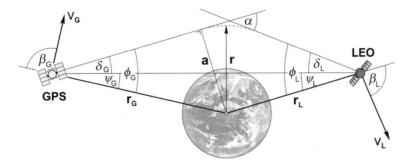

Fig. 4.18 The geometric model of the bending angle problem

The model equations are the followings,

$$f_1 = v_L \cos(\beta_L - \phi_L) - v_G \cos(\phi_G + \beta_G) - v_L \cos(\beta_L - \psi_L)$$
$$+ v_G \cos(\psi_G + \beta_G) - \omega$$
$$f_2 = r_G \sin(\phi_G) - r_L \sin(\phi_L)$$

where

v_L, v_G	the projected LEO and GPS satellite velocities in the occultation plane,
r_L, r_G	the radius of tangent points at LEO and GPS, respectively,
ω	Doppler shift,
ϕ_L, ϕ_G	the bending angles, unknowns,
β_L, β_G, ψ_L, ψ_G	known angles.

Let us introduce new variables,

$$x = \sin(\phi_G), \quad z = \cos(\phi_G), \quad y = \sin(\phi_L), \quad w = \cos(\phi_L),$$
$$\frac{a_1}{v_L} = \cos(\beta_L), \quad \frac{a_2}{v_L} = \sin(\beta_L), \quad -\frac{a_3}{v_G} = \cos(\beta_G), \quad \frac{a_4}{v_G} = \sin(\beta_G),$$
$$a_5 = r_G, \quad -a_6 = r_L$$

Then our equations become,

$$g_1 = -\omega + w a_1 - \cos(\psi_L) a_1 + y a_2 - \sin(\psi_L) a_2 + z a_3 - \cos(\psi_G) a_3$$
$$+ x a_4 - \sin(\psi_G) a_4$$
$$g_2 = x a_5 + y a_6$$

In order to separate known variables in equation g_1, let us introduce the variable α,

$$\alpha = -\omega - \cos(\beta_G - \psi_G)v_G - \cos(\beta_L - \psi_L)v_L$$

Now, the first equation becomes,

$$g_1 = \alpha + wa_1 + ya_2 + za_3 + xa_4$$

Equations g_3 and g_4 represent the trigonometric Pythagorean theorem,

$$g_3 = -1 + x^2 + z^2$$

and

$$g_4 = -1 + w^2 + y^2$$

Application of reduced Gröbner basis to get x variable

$$
\begin{aligned}
\mathrm{gr}\,x &= \mathrm{GroebnerBasis}[\{g_1, g_2, g_3, g_4\}, \{x, y, z, w\}, \{y, z, w\}] \\
&= \Big\{ a_1^4(-x^2 a_5^2 + a_6^2)^2 + 2a_1^2(x^2 a_5^2 - a_6^2) \\
&\quad (x^2 a_2^2 a_5^2 - 2xa_2(\alpha + xa_4)a_5 a_6 + (-((-1 + x^2)a_3^2) + (\alpha + xa_4)^2)a_6^2) \\
&\quad + (x^2 a_2^2 a_5^2 - 2xa_2(\alpha + xa_4)a_5 a_6 + ((-1 + x^2)a_3^2 + (\alpha + xa_4)^2)a_6^2)^2 \Big\}
\end{aligned}
$$

This is a quartic polynomial, which has the following coefficients,

$$
\begin{aligned}
c_0 &= (a_1^4 + (\alpha^2 - a_3^2)^2 - 2a_1^2(\alpha^2 + a_3^2))a_6^4 \\
c_1 &= 4\alpha(\alpha^2 - a_1^2 - a_3^2)a_6^3(-a_2 a_5 + a_4 a_6) \\
c_2 &= 2a_6^2(-a_1^4 a_5^2 + a_2^2(3\alpha^2 - a_3^2)a_5^2 + 2a_2(-3\alpha^2 + a_3^2)a_4 a_5 a_6 \\
&\quad + (-a_3^4 + 3\alpha^2 a_4^2 + a_3^2(\alpha^2 - a_4^2))a_6^2 + a_1^2((\alpha^2 - a_2^2 + a_3^2)a_5^2 \\
&\quad + 2a_2 a_4 a_5 a_6 + (a_3^2 - a_4^2)a_6^2)) \\
c_3 &= -4\alpha a_6(a_2 a_5 - a_4 a_6)(a_1^2 a_5^2 + a_2^2 a_5^2 - 2a_2 a_4 a_5 a_6 + (a_3^2 + a_4^2)a_6^2) \\
c_4 &= a_1^4 a_5^4 + 2a_1^2 a_5^2(a_2^2 a_5^2 - 2a_2 a_4 a_5 a_6 + (-a_3^2 + a_4^2)a_6^2) \\
&\quad + (a_2^2 a_5^2 - 2a_2 a_4 a_5 a_6 + (a_3^2 + a_4^2)a_6^2)^2
\end{aligned}
$$

The coefficients of this normalized polynomial are (see Awange et al. (2004)

$$
\begin{aligned}
c_0 &= (1.7077 + \delta)10^{-7} \\
c_1 &= -(4.09 + \delta)10^{-5} \\
c_2 &= (3.166 + \delta)10^{-3} \\
c_3 &= -(9.70547 + \delta)10^{-2} \\
c_4 &= (1 + \delta)
\end{aligned}
$$

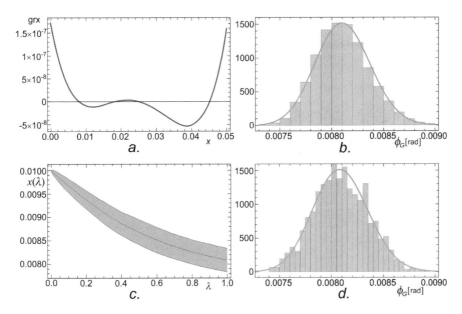

Fig. 4.19 a The reduced Gröbner basis polynomial for the variable x, **b** the density function of the ϕ_G angle, **c** the trajectories of the mean and $\pm$ standard deviation of $x(\lambda)$, **d** the density function of the ϕ_G angle computed via stochastic homotopy

where $\delta = N(0, \sigma)$ assuming error in all coefficients. Then the polynomial is,

$$\operatorname{gr} x = c0 + c1x + c2x^2 + c3x^3 + c4x^4$$

Let us display this polynomial in error free form, see Fig. 4.19a.
The roots of this polynomial are,

$$x_1 = 0.00808895, \quad x_2 = 0.0183432, \quad x_3 = 0.0255148, \quad x_4 = 0.0451077$$

We shall consider the first root of the solution.

4.7.2 Algebraic Solution

To solve this polynomial with error, we employ symbolic solution, hence the use of rationalized data

$$qR = \frac{1}{100}x^3\left(-\frac{970547}{100000} - \delta\right) + \frac{x\left(-\frac{409}{100} - \delta\right)}{100000} + x^4(1 + \delta)$$
$$+ \frac{\frac{17077}{10000} + \delta}{10000000} + \frac{x^2\left(\frac{1583}{500} + \delta\right)}{1000}$$

then the symbolic solution of $qR(x, \delta) = 0$ for $x(\delta)$ can be computed. However, we are looking for the $\phi_G(\delta)$ function, where is $\phi_G(\delta) = \text{Arcsin}(x(\delta))$, which can be computed easily via symbolic computation. This is a quite complicated expression for density transform therefore we employ series expansion in δ, and chopping imaginary tails,

$$\phi\delta S = 0.00808904 + 0.012628\ \delta + 0.0410981\ \delta^2 + 0.23619\ \delta^3.$$

Assuming errors in the coefficients (c_i, $i = 0, 1, 2, \ldots 4$) of the quartic polynomial to be $\sigma = 0.02$, then the density transform,

$$D\phi = \text{TransformedDistribution}(\phi\delta S, \delta \approx \text{NormalDistribution}(0, \sigma))$$

The mean and the standard variation of ϕ_G are

$$\mu\phi = \text{Mean}(D\phi) = 0.00810548$$

and

$$\sigma\phi = \text{StandardDeviation}(D\phi) = 0.000259315.$$

To find the type of the distribution, let us generate random data, and fit the best density function to these data. We get

$$\text{Estimated } D\phi = \text{LogNormalDistribution}(-4.81622, 0.0335365).$$

Let us visualize this function with the histogram of the random data, see Fig. 4.19b.

This seems to be in harmony with the data found in the literature, e.g., Sokolovskiy et al. (2001).

4.7.3   Solution via Stochastic Homotopy

Let us consider the target function as

$$q = qR.$$

Let us take a close value to the error free solution as initial value,

$$x_0 = 0.01.$$

Then our start system can be,

$$p = x - x_0.$$

Now, we employ the affine homotopy whose function is,

$$H(x, \lambda) = (1 - \lambda)q'(x_0)(x - x_0) + \lambda q(x).$$

The derivative of the target system at x_0, is

$$q' = 0.000003\left(-\frac{970547}{100000} - \delta\right) + \frac{-\frac{409}{100} - \delta}{100000} + 0.000004(1 + \delta)$$
$$+ 0.00002\left(\frac{1583}{500} + \delta\right)$$

Therefore the homotopy function can be written as,

$$H = (1 - \lambda)dqdxp + \lambda q$$
$$= (-0.01 + x)0.000003\left(-\frac{970547}{100000} - \delta\right) + \frac{-\frac{409}{100} - \delta}{100000} + 0.000004(1 + \delta)$$
$$+ 0.00002\left(\frac{1583}{500} + \delta\right) + (1 - \lambda) + \frac{1}{100}x^3\left(-\frac{970547}{100000} - \delta\right) + \frac{x\left(-\frac{409}{100} - \delta\right)}{100000}$$
$$+ x^4(1 + \delta) + \frac{\frac{17077}{10000} + \delta}{10000000} + \frac{x^2\left(\frac{1583}{500} + \delta\right)}{1000}\lambda$$

Then

$$proc = \text{ItoProcess}[dx(\lambda) = p\lambda d\lambda + pwdw(\lambda), x(\lambda), \{x, x0\}, \lambda,$$
$$w \approx \text{WienerProcess}(0, \sigma)]$$

where

$$pw = -((0.00544092(-2.46754 \times 10^{-11} - 5.7467 \times 10^{-11}\lambda$$
$$+ 5.47775 \times 10^{-9}x(\lambda) + 1.19707 \times 10^{-8}\lambda x(\lambda) - 4.31021 \times 10^{-7}x(\lambda)^2$$
$$- 9.77692 \times 10^{-7}\lambda x(\lambda)^2 + 0.0000125733x(\lambda)^3 + 0.0000410891x(\lambda)^3$$
$$- 0.000157331x(\lambda)^4 - 0.000158336\lambda x(\lambda)^4 - 0.0497618x(\lambda)^5$$
$$+ 1.\lambda x(\lambda)^6))/(-6.74102 \times 10^{-7} - 0.0000095509\lambda + 0.001583\lambda x(\lambda)$$
$$- 0.072791\lambda x(\lambda)^2 + \lambda x(\lambda)^3)^2)$$

and

$$p\lambda = -((0.25(1.43806 \times 10^{-7} - 0.0000382036x(\lambda) + 0.003166x(\lambda)^2$$
$$- 0.0970547x(\lambda)^3 + x(\lambda)^4))/(-6.74102 \times 10^{-7} - 9.5509 \times 10^{-6}\lambda$$
$$+ 0.001583\lambda x(\lambda) - 0.072791\lambda x(\lambda)^2 + \lambda x(\lambda)^3))$$

Generating 1000 trajectories with step size 0.001, the trajectories of the solution are in Fig. 4.19c.

Since $\sin(\phi_G) = x$, let us compute the density function of the inverse, $\phi_G = \arcsin(x)$. The series expansion of $\arcsin(x)$ at $x = \mu xH$ mean value,

$$\phi_G\delta = -1.03574 \times 10^{-11} + x - 7.92484 \times 10^{-7}x^2 + 0.166716x^3$$

The density function of the stochastics process of ϕ_G is presented in Fig. 4.19d.

The mean and standard deviation are $\mu = 0.00808377$ and $\sigma = 0.000253327$, respectively. The other polynomial for y, where $\sin(\phi_L) = y$ can be solved in the same way.

Appendix—Transforming Density Functions

In case of algebraic solutions, the built-in TransformedDistribution function is used. Here the basic of the method is illustrated.

Definition

Let $f_X(x)$ be the value of the probability density of the continuous random variable X at x. If the function $y = \phi(x)$ is differentiable and either increasing or decreasing (monotonic) for all values within the range of X for which $f_X(x) \neq 0$, then for these values of x, the equation $y = \phi(x)$ can be uniquely solved for x to give $x = \phi^{-1}(y) = w(y)$ where $w(\cdot) = \phi^{-1}(\cdot)$. Then for the corresponding values of y, the probability density of $Y = \phi(X)$ is given by

$$f_Y(y) = f_X(w(y))\left|\frac{dw(y)}{dy}\right|.$$

Examples   Linear Transformation

Considering $Y = \alpha + \beta X$ transform with $f_X(x)$ density function,

$$X = \frac{1}{\beta}(Y - \alpha) \rightarrow w(y) = \phi^{-1}(y) = \frac{1}{\beta}(y - \alpha)$$

then

$$f_Y(y) = f_X(w(y))\left|\frac{dw(y)}{dy}\right| = f_X\left(\frac{1}{\beta}(y-\alpha)\right)\left|\frac{1}{\beta}\right|$$

Nonlinear Transformation

Let us consider our density function as $f_X(x) = e^{-x}$

If $Y = \sqrt{X}$, then $y = \sqrt{x} \rightarrow y^2 = x \rightarrow x = \phi^{-1}(y) = y^2$

then

$$f_Y(y) = f_X\left(\phi^{-1}(y)\right)\left|\frac{d\phi^{-1}(y)}{dy}\right| = e^{-y^2}|2y|$$

Using Mathematica
See Fig. 4.20.

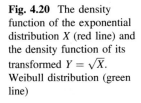

```
⇒ fX = PDF[ExponentialDistribution[1], x];
⇐  ⎧ e^{-x}   x ≥ 0
   ⎨
   ⎩  0       True
⇒ Y = TransformedDistribution[√X, ≈ ExponentialDistribution[1]]
⇐ WeibullDistribution[2, 1]
⇒ fY = PDF[Y, y]
⇐  ⎧ 2e^{-y^2}y   y > 0
   ⎨
   ⎩   0         True
```

Fig. 4.20 The density function of the exponential distribution X (red line) and the density function of its transformed $Y = \sqrt{X}$. Weibull distribution (green line)

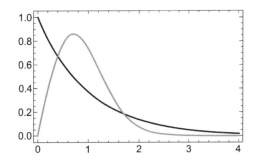

References

Awange JL, Paláncz B (2016) Geospatial algebraic computations: theory and applications, 3rd edn. Springer-Verlag New York, Inc., p 541

Awange JL, Paláncz B, Lewis RH, Völgyesi L (2018) Mathematical geosciences: hybrid symbolic-numeric methods. Springer International Publishing, Cham, p 596

Awange JL, Fukuda Y, Takemoto S, Wickert J, Aoyama Y (2004) Analytic solution of GPS atmospheric sounding refraction angles. Earth Planets Space 56:573–587

Awange J (2012) Environmental monitoring using GNSS: global navigation satellite systems. Berlin Heidelberg

Awange J (2018) GNSS environmental sensing. Springer International Publishers. Transformation (conformal group C7). J Geodesy 76:66–76. https://doi.org/10.1007/s00190-002-0299-9

Grafarend EW and Awange JL (2012) Applications of linear and nonlinear models. Fixed effects, random effects, and total least squares. Springer-Verlag, Berlin, Heidelberge, New York

Grafarend EW, Awange JL (2003) Nonlinear analysis of the three-dimensional datum

Hofmann-Wellenhof B, Lichtenegger H, Wasle E (2008) GNSS–global navigation satellite systems: GPS, GLONASS, galileo, and more. Springer, Vienna, Austria

Jalali F, Seader JD (2000) Use of homotopy-continuation method in stability analysis of multiphase reacting systems. Comput Chem Eng 24(8):1997–2008

Khodabandeh A, Teunissen P (2016) PPP-RTK and inter-system biases. The ISB look-up table as a means to support multi-system PPP-RTK. J Geod 90:837–851

Koch KR (2001) Bermekung zu der Veroeffentlichung "'Zur Bestimmung eindeutiger transformationparameter.'" Z Vermess 126:297

Lenzmann E, Lenzmann L (2001a) Zur Bestimmung Eindeutiger Transformationparameter. Z Vermess 126:138–142

Lenzmann E, Lenzmann L (2001b) Erwiderung auf die Anmerkung von Joerg Reinking und die Bermekungen von Karl-Rudolf Koch zu unserem Meitrag "'Zur Bestimmung eindeutiger transformation parameter.'" Z Vermess 126:298–299

Lewis RH (2021) Computer algebra system Fermat. http://home.bway.net/lewis/

Montenbruck O, Steigenberger P, Prange L, Deng Z, Zhao Q, Perosanz F, Romero I, Noll C, Sturze A, Weber G et al (2017) The multi-GNSS experiment (MGEX) of the international GNSS service (IGS)–achievements, prospects and challenges. Adv Space Res 59:1671–1697

Malissiovas G, Neitzel F, Weisbrich S, Petrovic S (2020) Weighted total least squares (WTLS) solutions for straight line fitting to 3D point data. Stochastic models for geodesy and geoinformation science. In: Neitzel F (ed) Mathematics, pp 11–29

Moritz H, Least-Squares Collocation (1973) Deutsche Geodatische Kommission: Number 75 in Reihe A; Munchen, Germany

Niemeier W, Tengen D (2017) Uncertainty assessment in geodetic network adjustment by combining GUM and Monte-Carlo-simulations. J Appl Geodesy. https://doi.org/10.1515/jag-2016-0017

Paláncz B, Awange JL, Lovas T, Lewis R, Molnár B, Heck B, Fukuda Y (2017) Algebraic method to speed up robust algorithm: example of laser-scanned point clouds. Surv Rev 49:408–418

Paláncz B (2021a) Solution techniques for nonlinear equations with uncertain parameters part 1- A toy example, https://www.researchgate.net/publication/349960205_Solution_Techniques_for_Nonlinear_Equations_with_Uncertain_Parameters_Part_1_-A_toy_example

Paláncz B (2021b) Solution techniques for nonlinear equations with uncertain parameters part 2 - System of Equations, https://www.researchgate.net/publication/349966654_Techniques_for_Nonlinear_Equations_with_Uncertain_Parameters_Part_2_-System_of_equations

Prange L, Orliac E, Dach R, Arnold D, Beutler G, Schaer S, Jaggi A (2017) CODE's five-system orbit and clock solution—the challenges of multi-GNSS data analysis. J Geod 91:345–360

Reinking J (2001) Anmerkung zu "'Zur Bestimmung eindeutiger transformation-parameter.'" Z Vermess 126:295–296

Schubert T, Korte J, Brockmann JM, Schuh WD (2020) A generic approach to covariance function estimation using ARMA-models. Mathematics 8(591):71–89. https://doi.org/10.3390/math8040591

Sokolovskiy SV, Rocken C, Lowry AR (2001) Use of GPS for estimation of bending angles of radio waves at low elevations. Radio Sci 36(3):473–482

Tian Y, Ge M, Neitzel F (2020) Variance reduction of sequential Monte Carlo approach for GNSS phase bias estimation. Stochastic models for geodesy and geoinformation science. In: Neitzel F (ed) Mathematics pp 107–121

Part II
Optimization of Systems

Chapter 5
Simulated Annealing

5.1 Metropolis Algorithm

This algorithm finds the equilibrium state s, belonging to *a given temperature*. Let us consider a state s_i and then perturb it to get another state s_{i+1}. This perturbed state will be accepted as a new state if its free energy is less than that of the original state, namely.

$$\Delta E = E(s_{i+1}) - E(s_i) < 0.$$

However, even in case of $E(s_{i+1}) \geq E(s_i)$, the new state can be accepted with a small

$$\exp\left(-\frac{\Delta E}{kT}\right)$$

probability. This feature can ensure that the optimization process *will not get stuck at a local optimum*.

During the computation, the Boltzmann constant is considered as $k = 1$.

5.2 Realization of the Metropolis Algorithm

Let us consider the following example where someone thinks of a string consisting of 13 characters, let say "tobeornottobe", and the requirement is to find out this word. If we suggest a string as a possible solution, then we will receive information

---

Supplementary Information The online version contains supplementary material available at https://doi.org/10.1007/978-3-030-92495-9_5.

J. L. Awange et al., *Mathematical Geosciences*,
https://doi.org/10.1007/978-3-030-92495-9_5

about how bad our trial is, namely how many wrong characters are in our suggestion.

Remark Without any strategy, we would need a maximum of 2.48115×10^{18} trials to hit the target. Considering the English alphabet, there are 26 characters (small letters). The probability that one character is in the right place is

$$\Rightarrow \frac{1}{26} //\text{N}$$
$$\Leftarrow 0.0384615$$

Then the probability that all of the 13 characters are in the right place is

$$\Rightarrow \left(\frac{1}{26}\right)^{13} //\text{N}$$
$$\Leftarrow 4.03038 \times 10^{-19}$$

So the maximum number of the necessary trials is

$$\Rightarrow 1/\%$$
$$\Leftarrow 2.48115 \times 10^{18}$$

5.2.1   *Representation of a State*

The trial string can be represented by the ASCII code of the characters

$$\Rightarrow \text{s1} = \text{Table[RandomInteger[\{97, 122\}], \{13\}]}$$
$$\Leftarrow \{108, 111, 115, 108, 109, 98, 97, 116, 117, 107, 114, 102, 111\}$$

in text form

$$\Rightarrow \text{FromCharacterCode[s1]}$$
$$\Leftarrow \text{loslmbatukrfo}$$

5.2.2 *The Free Energy of a State*

The free energy of a state can be represented by the number of the wrong characters in the trial string, namely the difference between the target string ("tobeornottobe") and the trial string

$$\Rightarrow \texttt{keyPhrase} = \texttt{ToCharacterCode}[''\texttt{tobeornottobe}'']$$
$$\Leftarrow \{116, 111, 98, 101, 111, 114, 110, 111, 116, 116, 111, 98, 101\}$$

The difference between the two strings can be associated with the free energy of the actual trial string

$$\Rightarrow \texttt{E}[\texttt{s\_}] := \texttt{Count}[\texttt{keyPhrase} - \texttt{s}, \texttt{Except}[0]]$$
$$\Rightarrow \texttt{E}[\texttt{s1}]$$
$$\Leftarrow 12$$

5.2.3 *Perturbation of a State*

In order to perturb a state s, we should consider the string of s, character by character, and change them one by one with a probability μ. To do that, we define a function which produces true or false with a μ probability,

$$\Rightarrow \texttt{flip}[\texttt{x\_}] := \texttt{If}[\texttt{RandomReal}[] \pounds \texttt{x}, \texttt{True}, \texttt{False}].$$

Then we apply it to carry out the perturbation of s with μ probability. In case we get true for a certain character, then we change it with another character (ASCII code) randomly from the English alphabet.

$$\Rightarrow \texttt{Perturb}[\mu\_, \texttt{s\_}]$$
$$:= \texttt{Map}[\texttt{If}[\texttt{flip}[\mu], \texttt{RandomInteger}[\{97, 122\}], \#]\&, \texttt{s}];$$

Let us employ it for s_1 with probability $\mu = 0.3$

$$\Rightarrow \texttt{s2} = \texttt{Perturb}[0.3, \texttt{s1}]$$
$$\Leftarrow \{108, 111, 115, 108, 109, 98, 98, 110, 117, 107, 111, 119, 117\}$$

In text form

$$\Rightarrow \texttt{FromCharacterCode}[\texttt{s2}]$$
$$\Leftarrow \texttt{loslmbbnukowu}$$

The free energy of this state (string) is

$\Rightarrow$ E[s2]

$\Leftarrow$ 11

5.2.4 Accepting a New State

If the free energy of the perturbed state is less than the original one, then we accept it as a new state or reject it with a probability.

$$1 - \exp\left(-\frac{\Delta E}{T}\right).$$

This means that the new state can be accepted with a probability

$$\exp\left(-\frac{\Delta E}{T}\right)$$

even with higher energy level.

$\Rightarrow$ Accept[ΔE_, T_] :

= If[ΔE < 0 URandomReal[] < Exp[ΔE/T], True, False].

For example let $T = 0.5$

$\Rightarrow$ Accept[E[s2] − E[s1], 0.5]

$\Leftarrow$ True

Remark If the energy level does not change then the new state will be accepted.

5.2.5 Implementation of the Algorithm

Collecting these functions, the function of the Metropolis algorithm is the following, see the flow chart Fig. 5.1.

Fig. 5.1 The metropolis
algorithm

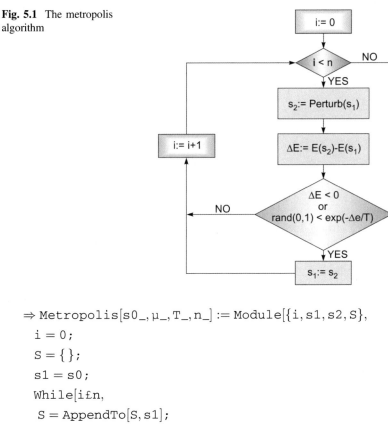

$\Rightarrow$ Metropolis$[\text{s0_}, \mu_, \text{T_}, \text{n_}] := $ Module$[\{\text{i}, \text{s1}, \text{s2}, \text{S}\},$
    i = 0;
    S = {};
    s1 = s0;
    While$[\text{i} \leq \text{n},$
    S = AppendTo$[\text{S}, \text{s1}]$;
    s2 = Perturb$[\mu, \text{s1}]$;
    If$[\text{Accept}[\text{E}[\text{s2}] - \text{E}[\text{s1}], \text{T}], \text{s1} = \text{s2}]$;
    i = i + 1; $]$;
    S$]$;

The flow chart demonstrates the Metropolis algorithm.
The output is the series of strings of the states computed during the iteration steps.
Let us test our function with temperature $T = 1$ and iteration number $n = 100$.
The probability of the perturbation is $\mu = 0.3$.
First, we randomly generate an initial state,

$\Rightarrow$ s0 = Table$[\text{RandomInteger}[\{97, 122\}], \{13\}]$;

Then, employing the algorithm,

$\Rightarrow$ M = Map$[\text{E}[\#]\&, \text{Metropolis}[\text{s0}, 0.3, 1., 100]]$;

The energy level of the series of the subsequent states can be seen in Fig. 5.2.

Fig. 5.2 The energy levels belonging to the different states computed in the iteration steps at $T = 1$

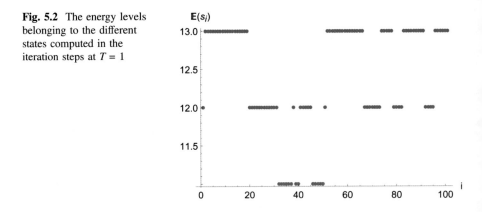

We can see a strong fluctuation of the energy level at this relatively high temperature. Let us repeat this numerical simulation at a lower temperature $T = 0.1$.

$$\Rightarrow \mathtt{M = Map[E[\#]\&, Metropolis[s0, 0.3, 0.1, 100]]};$$

Figure 5.3 shows, that the energy level reaches a stable equilibrium at this low temperature.

5.3   Algorithm of the Simulated Annealing

In order to get the minimum of the energy level, the Metropolis algorithm will be executed at different decreasing temperature levels, step by step. The simplest "cooling" strategy can be,

$$T_{i+1} = \alpha T_i$$

Fig. 5.3 The energy levels belonging to the different states computed in the iteration steps at $T = 0.1$

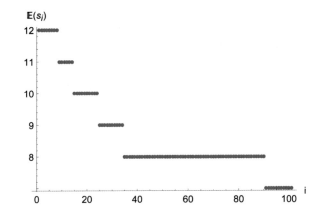

Fig. 5.4 The algorithm of the simulated annealing

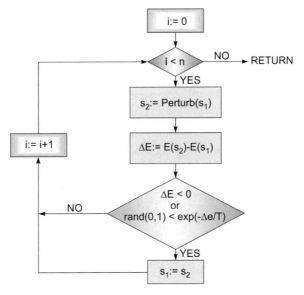

where $0 < \alpha < 1$. The region where the best convergence can be achieved is $\alpha \in [0.8, 0.99]$. Figure 5.4 shows the flow chart of the algorithm.

Where T_{min}, the lowest temperature is input as a simulation parameter.

5.4   Inplementation of the Algorithm

Now, the function of the simulated annealing method can be easily defined as,

```
⇒ SimulatedAnnealing[s0_,T0_,Tf_,μ_,α_,n_] := Module[{T,s,S,TS},
    T = T0;
    s = s0;
    S = {s0};
    TS = {T0};
    While[T > Tf,
    s = Last[Metropolis[s,μ,T,n]];
    S = AppendTo[S,s];
    T = αT; (*annealingschedule*)
    TS = AppendTo[TS,T]; ];
    {S,TS}]
```

The output is the actual cooling temperatures and the corresponding steady states of the energy levels. Let us carry out a simulation with the initial state.

$\Rightarrow$ s0 = Table[RandomInteger[{97, 122}], {13}]; l.

The initial temperature T_0 is,

$$T_0 = 0.5.$$

The termination temperature is,

$$T_{min} = 0.005.$$

The probability of the perturbation is,

$$\mu = 0.3.$$

The cooling factor,

$$\alpha = 0.92.$$

Let us carry out 200 iteration steps in the Metropolis algorithm at every temperature levels,

$\Rightarrow$ MS = SimulatedAnnealing[s0, 0.5, 0.005, 0.3, 0.92, 200];

The cooling temperatures (Fig. 5.5),
The steady state energy levels at these temperatures are shown in Fig. 5.6.
Figure 5.6 shows that at the end of the iteration, the energy level is zero. This means that all of the characters are in the proper place in the string representing the terminal state,

$\Rightarrow$ FromCharacterCode[Last[MS[[1]]]]
$\Leftarrow$ tobeornottobe.

Fig. 5.5 Actual temperatures during the cooling process

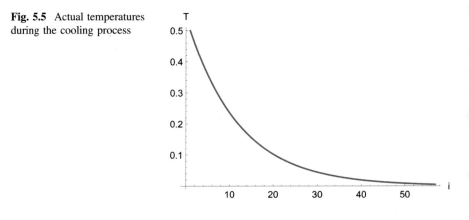

Fig. 5.6 The equilibrium energy levels

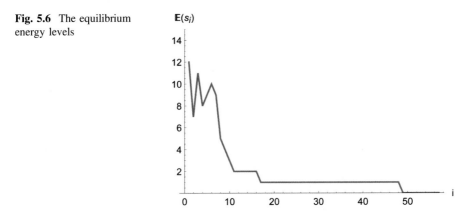

5.5   Application to Computing Minimum of a Real Function

Now, let us employ the method for finding global minimum of a real function using a simple example. Let us find the global minimum of a function $f(x)$ in the region $x \in [0, 50]$ where (Fig. 5.7)

$$f(x) = |2(x - 24) + (x - 24) \ \sin(x - 24)|$$

$$\Rightarrow \ \texttt{f[x\_] := Abs[2(x - 24) + (x - 24)Sin[(x - 24)]];}$$

In this case one point of the region is considered as an i-th state, x_i and the corresponding free energy is $f_i = f(x_i)$. The x_i can be represented in binary form as a string. Using more digits (characters) in the string, we have more precise representation of a real value (higher resolution). Let us consider 15 bits then the biggest number which can be represented is $s_{max} = \{1, 1, 1, 1, 1, 1, 1, 1, 1, 1, 1, 1, 1, 1, 1\}$, or

Fig. 5.7 A multimodal real function with a global minimum

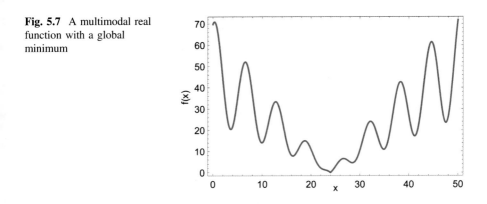

$\Rightarrow \texttt{maxi} = 2^{15}-1$

$\Leftarrow 32767.$

The smallest number $s_{\min} = \{0, 0, 0, 0, 0, 0, 0, 0, 0, 0, 0, 0, 0, 0, 0\}$

$\Rightarrow \texttt{mini} = 2^{0}-1$

$\Leftarrow 0$

This means that the interval [0, 32767] should be projected into the interval [0, 50]. Let us consider the string s_i. The corresponding real value d_i can then be computed with the following function

$\Rightarrow \texttt{Horner}[\texttt{u\_,base\_} : 2] := \texttt{Fold}[\texttt{base \#1} + \texttt{\#2\&}, 0, \texttt{u}]$

For example, the binary form is

$\Rightarrow \texttt{Subscript}[\texttt{s,i}] = \{1, 0, 1, 0, 0, 0, 0, 0, 0, 0, 0, 0, 0, 1, 0\};$

and its decimal representation is,

$\Rightarrow \texttt{d}_{\texttt{i}} = \texttt{Horner}[\texttt{s}_{\texttt{i}}]$

$\Leftarrow 20482$

Indeed

$\Rightarrow 2^{14} + 2^{12} + 2^{1}$

$\Leftarrow 20482$

and the scaled decimal value x_i is

$\Rightarrow \texttt{x}_{\texttt{i}} = \dfrac{50.}{32767} \texttt{d}_{\texttt{i}}$

$\Leftarrow 31.254$

The energy level of s_i can be computed as

$\Rightarrow \texttt{E}[\texttt{s\_}] := \texttt{f}\left[\frac{50.}{32767}\texttt{Horner}[\texttt{s}]\right]$

In our case

$\Rightarrow \texttt{E}[\texttt{s}_{\texttt{i}}]$

$\Leftarrow 20.4951$

It goes without saying, that the function of the perturbation should be changed, since now there are only two characters 0 or 1,

$\Rightarrow$ Perturb[μ_, s_] :
= Map[If[flip[μ], RandomInteger[{0, 1}], #]&, s];

Then we are looking for the string of 15 characters, which provides the minimum of the $E(s)$, which is the same as the minimum of $f(x)$.

Let the initial state be

$\Rightarrow$ s0 = Table[RandomInteger[{0, 1}], {15}]
$\Leftarrow$ {1, 0, 0, 0, 0, 1, 0, 0, 0, 0, 1, 0, 0, 0, 1}

$\Rightarrow$ x0 = $\dfrac{50.}{32767}$ Horner[s0]

$\Leftarrow$ 25.808

Then using the parameters $T_{max} = 0.5$, $T_{min} = 0.005$, $\mu = 0.3$, $\alpha = 0.9$ and $n = 200$,

$\Rightarrow$ MS = SimulatedAnnealing[s0, 0.5, 0.005, 0.3, 0.9, 200];

The values of the free energy levels during the iteration process are (Fig. 5.8)

$\Rightarrow$ Map[E[#]&, MS[[1]]];

The convergence is quite fast, see Figs. 5.9 and Fig. 5.10.

The location of the global minimum is,

$\Rightarrow$ $\dfrac{50.}{32767}$ Horner[Last[MS[[1]]]]

$\Leftarrow$ 23.9998

Fig. 5.8 The free energy levels as function of the number of the iterations

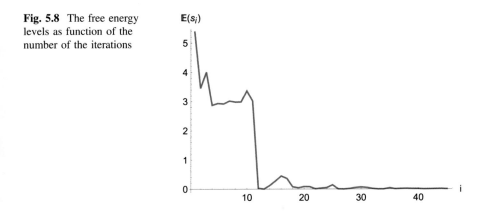

Fig. 5.9 Convergence of the
iteration

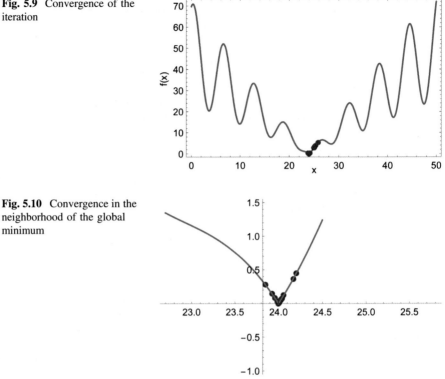

Fig. 5.10 Convergence in the
neighborhood of the global
minimum

The value of the global minimum

⇒ E[Last[MS[[1]]]]
⇐ 0.000488237

The solution of the built-in function is

⇒ NMinimize[{f[x], 0£x£50}, x]
⇐ $1.27898 \times 10^{-13}, \{x \rightarrow 24.\}\}$

Using more digits, we can get better and more precise solutions.

5.6 Generalization of the Algorithm

The simulated annealing method is built into most computing systems like
Mathematica, *Maple* and *Matlab*.

Now, let us employ *Mathematica* to solve the following optimization problem. The function to be maximized is

$$f(x,\ y) = g(x+11,\ y+9) + g(x-11, y-3) + g(x+6, y-9),$$

where (Fig. 5.11)

$$g(x,y) = \frac{50 \sin \sqrt{x^2 + y^2}}{\sqrt{x^2 + y^2}} - \sqrt{x^2 + y^2}.$$

We are looking for the *global maximum* in the region $(x, y) \in [-20, 20]^2$ employing simulated annealing method,

$\Rightarrow$ sol = Reap[NMaximize[{f[x,y], $-20 \leq x \leq 20, -20 \leq y \leq 20$}, {x,y},
   EvaluationMonitor : $\rightarrow$ Sow[{x,y}],
   Method $\rightarrow$ { "SimulatedAnnealing","PerturbationScale" $\rightarrow$ 5}]];

The maximum,

$\Rightarrow$ sol[[1]]
$\Leftarrow$ {10.8432, {x $\rightarrow$ -6.01717, y $\rightarrow$ 9.06022}}

Fig. 5.11 A function of two variables with many local maximums

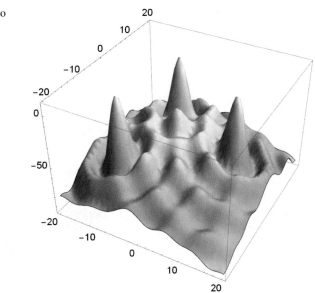

The number of the iteration is

$\Rightarrow \mathrm{ni} = \mathrm{Length[hist]}$

$\Leftarrow 317$

Let us display the states belonging to the actual iteration steps, see Fig. 5.12.

Let us display every 5th point of the function values computed during the iteration, see Fig. 5.13.

Remarks Keep in mind that $\max(f(x)) = \min(-f(x))$.

Fig. 5.12 The states versus number of iterations

Fig. 5.13 The function values computed during the iteration process

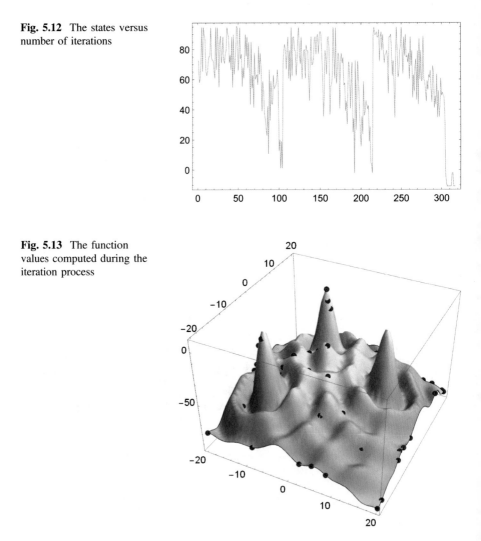

5.7 Applications

5.7.1 A Packing Problem

Given n disks with different radii r_i. Let us arrange them in the smallest square area without overlapping each other! Let us consider 10 disks,

$\Rightarrow n = 10;$

Their radii are chosen randomly between a lower and an upper limit

$\Rightarrow \texttt{maxR} = 0.2;$

$\Rightarrow \texttt{minR} = 0.8;$

$\Rightarrow \texttt{Table}[r_i = \texttt{Random}[\texttt{Real}, \{\texttt{minR}, \texttt{maxR}\}], \{i, 1, n\}]$

$\Leftarrow \{0.466063, 0.396305, 0.706345, 0.471148,$

$0.5559, 0.642377, 0.25306, 0.512608, 0.62259, 0.792981\}$

We assume that the area of the square is $2w \times 2w$, and w has an upper limit $w \leq B$. Now let $B = 10$

$\Rightarrow B = 10;$

The first group of the constrains hinders the overlapping of the disks,

$\Rightarrow \texttt{constraints01} =$

$\quad \texttt{Table}[(x_i - x_j)^2 + (y_i - y_j)^{2^{2^3}} ((r_i + r_j)^2), i, 1, n, j, i+1, n]//\texttt{Flatten};$

The second group of the constrains ensures that $2w \times 2w \leq 2B \times 2B$

$\Rightarrow \texttt{constraints02} = \texttt{Table}[$

$\{r_i + x_i - w \leq 0, -r_i + x_i + w \geq 0, r_i + y_i - w \leq 0, -r_i + y_i + w \geq 0,$

$-B \leq x_i \leq B, -B \leq y_i \leq B\}, \{i, 1, n\}]\}, \{i, 1, n\}]//\texttt{Flatten};$

Combining these constrains with the upper limit of the positive radius,

$\Rightarrow \texttt{constraints} = \texttt{Join}[\texttt{constraints01}, \texttt{constraints02}, \{0 <$
$\qquad\qquad = w < = B\}];$

The unknown variables we are looking for are the center coordinates of the disks $\{x_i, y_i\}$, which minimize w,

$\Rightarrow \texttt{vars} = \texttt{Join}[\{w\}, \texttt{Table}[\{x_i, y_i\}, \{i, 1, n\}]]//\texttt{Flatten};$

Employing simulated annealing method,

$\Rightarrow$ soln = NMinimize[Join[{w}, constraints], vars,
Method $\rightarrow$ {"SimulatedAnnealing", "PerturbationScale" $\rightarrow$ 6,
"SearchPoints" $\rightarrow$ 100}]//Timing//Quiet

$\Leftarrow$ {37.0658, {1.84394, {w $\rightarrow$ 1.84394, x_1 $\rightarrow$ 1.37788, y_1 $\rightarrow$ -0.392879,
$\quad$ x_2 $\rightarrow$ 0.513346, y_2 $\rightarrow$ -1.44764, x_3 $\rightarrow$ 1.13759, y_3 $\rightarrow$ 0.75464,
$\quad$ x_4 $\rightarrow$ 1.3728, y_4 $\rightarrow$ -1.33008, x_5 $\rightarrow$ -0.0064139, y_5 $\rightarrow$ 1.28804,
$\quad$ x_6 $\rightarrow$ -1.20157, y_6 $\rightarrow$ 1.20157, x_7 $\rightarrow$ -0.135096, y_7- $\rightarrow$ 1.58773,
$\quad$ x_8 $\rightarrow$ -0.644569, y_8 $\rightarrow$ 0.189765, x_9 $\rightarrow$ 0.290907, y_9 $\rightarrow$ -0.45332,
$\quad$ x_{10} $\rightarrow$ -1.05096, y_{10} $\rightarrow$ -1.05096 }}}

Figure 5.14 shows the positions of the different disks.

5.7.2 The Traveling Salesman Problem

The problem is to visit each town once and minimize the length of the tour. We have to find an ordering of the visits that minimizes the total distance and visits all the v_i's once, where v_i's are the locations of the towns.

Fig. 5.14 Optimal placement of disks in a minimal square area

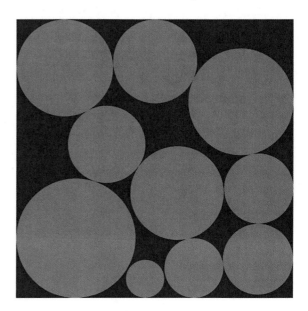

Let us generate 20 town- coordinates in 2D,

$\Rightarrow$ SeedRandom[2]; points = RandomReal[10, {20, 2}];

The coordinates of the town,

$\Rightarrow$ points
$\Leftarrow$ {{7.2224, 1.09449}, {4.70703, 5.35582}, {5.83178, 2.93942},
  {1.65154, 6.01258}, {7.54218, 7.71123}, {7.78574, 0.236104},
  {9.22757, 9.92454}, {3.50409, 0.450047}, {5.01359, 6.33756},
  {6.42208, 3.89875}, {6.64971, 8.43882}, {5.6904, 3.98212},
  {2.38652, 6.73513}, {4.19507, 5.87398}, {0.0833523, 9.42441},
  {7.71263, 1.47503}, {9.64774, 8.98747}, {3.32963, 2.04548},
  {8.39035, 2.50388}, {2.38638, 6.16616}}

Then we employ simulated annealing method

$\Rightarrow$ {length, tour} =
    FindShortestTour[points, Method → "SimulatedAnnealing"];

The total length of the distance of the travel is,

$\Rightarrow$ length
$\Leftarrow$ 41.7378

The order of the optimal visit is

$\Rightarrow$ tour
$\Leftarrow$ {9, 5, 17, 7, 11, 15, 13, 20, 4, 18, 8, 6, 1, 16, 19, 3, 10, 12, 2, 14, 9}

Figure 5.15 displays the route.
Let us solve the problem in space (3D). We are visiting planets in space (Fig. 5.16),

$\Rightarrow$ SeedRandom[2];
$\Rightarrow$ With[{p = RandomReal[20, {9, 3}]},
  Graphics3D[
  {Line[
  p[[Last[FindShortestTour[p,
  Method → "SimulatedAnnealing"]]]]], Sphere/@p},
  ImageSize → 300]]//Quiet

Fig. 5.15 Optimal route of
the tour

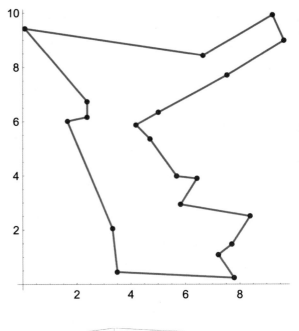

Fig. 5.16 Optimal route of
the flight in space

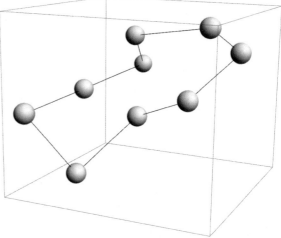

Now, let us visit the capitals of countries of the world.
Function to give a location on the globe (sphere),

⇒ SC[{lat_, lon_}] :=
   r{Cos[lon°]Cos[lat°], Sin[lon°]Cos[lat°], Sin[lat°]};

Locations of the towns

```
⇒ r = 6378.7;
⇒ places = CountryData["Countries"];
⇒ centers = Map[CountryData[#, "CenterCoordinates"]&, places];
⇒ distfun[{lat1_, lon1_}, {lat2_, lon2_}] :=
    VectorAngle[SC[{lat1, lon1}], SC[{lat2, lon2}]]r;
```

The countries to be visited are

```
places
```

{Afghanistan, Albania, Algeria, American Samoa, Andorra, Angola, Anguilla, Antigua and Barbuda, Argentina, Armenia, Aruba, Australia, Austria, Azerbaijan, Bahamas, Bahrain, Bangladesh, Barbados, Belarus, Belgium, Belize, Benin, Bermuda, Bhutan, Bolivia, Bosnia and Herzegovina, Botswana, Brazil, British Virgin Islands, Brunei, Bulgaria, Burkina Faso, Burundi, Cambodia, Cameroon, Canada, Cape Verde, Cayman Islands, Central African Republic, Chad, Chile, China, Christmas Island, Cocos Keeling Islands, Colombia, Comoros, Cook Islands, Costa Rica, Croatia, Cuba, Curacao, Cyprus, Czech Republic, Democratic Republic of the Congo, Denmark, Djibouti, Dominica, Dominican Republic, East Timor, Ecuador, Egypt, El Salvador, Equatorial Guinea, Eritrea, Estonia, Ethiopia, Falkland Islands, Faroe Islands, Fiji, Finland, France, French Guiana, French Polynesia, Gabon, Gambia, Gaza Strip, Georgia, Germany, Ghana, Gibraltar, Greece, Greenland, Grenada, Guadeloupe, Guam, Guatemala, Guernsey, Guinea, Guinea-Bissau, Guyana, Haiti, Honduras, Hong Kong, Hungary, Iceland, India, Indonesia, Iran, Iraq, Ireland, Isle of Man, Israel, Italy, Ivory Coast, Jamaica, Japan, Jersey, Jordan, Kazakhstan, Kenya, Kiribati, Kosovo, Kuwait, Kyrgyzstan, Laos, Latvia, Lebanon, Lesotho, Liberia, Libya, Liechtenstein, Lithuania, Luxembourg, Macau, Macedonia, Madagascar, Malawi, Malaysia, Maldives, Mali, Malta, Marshall Islands, Martinique, Mauritania, Mauritius, Mayotte, Mexico, Micronesia, Moldova, Monaco, Mongolia, Montenegro, Montserrat, Morocco, Mozambique, Myanmar, Namibia, Nauru, Nepal, Netherlands, New Caledonia, New Zealand, Nicaragua, Niger, Nigeria, Niue, Norfolk Island, Northern Mariana Islands, North Korea, Norway, Oman, Pakistan, Palau, Panama, Papua New Guinea, Paraguay, Peru, Philippines, Pitcairn Islands, Poland, Portugal, Puerto Rico, Qatar, Republic of the Congo, Réunion, Romania, Russia, Rwanda, Saint Helena, Ascension and Tristan da Cunha, Saint Kitts and Nevis, Saint Lucia, Saint Pierre and Miquelon, Saint Vincent and the Grenadines, Samoa, San Marino, São Tomé and Príncipe, Saudi Arabia, Senegal, Serbia, Seychelles, Sierra Leone, Singapore, Sint Maarten, Slovakia, Slovenia, Solomon Islands, Somalia, South Africa, South Korea, South Sudan, Spain, Sri Lanka, Sudan, Suriname, Svalbard, Swaziland, Sweden, Switzerland, Syria, Taiwan, Tajikistan, Tanzania, Thailand, Togo, Tokelau, Tonga,

Trinidad and Tobago, Tunisia, Turkey, Turkmenistan, Turks and Caicos Islands, Tuvalu, Uganda, Ukraine, United Arab Emirates, United Kingdom, United States, United States Virgin Islands, Uruguay, Uzbekistan, Vanuatu, Vatican City, Venezuela, Vietnam, Wallis and Futuna Islands, West Bank, Western Sahara, Yemen, Zambia, Zimbabwe}.

The number of the countries to be considered is

$\Rightarrow$ Length[%]

$\Leftarrow$ 240

The traveling salesman problem is a famous example of an NP-complete problem. There is no known algorithm that is guaranteed to solve every-city problem in polynomial time. Brute force is completely impractical. The total number of possible tours when there are cities is. So, for instance, with 30 cities there are 4,420,880,996,869,850,977,271,808,000,000 possible tours. Suffice it to say that calculating the length of each and then identifying the shortest among them is well beyond the capacity of today's computers. Optimal solution is possible in practice up to 20 via dynamic programming. Although it might rise up to 50 if integer linear programming is utilized.

Other approximate methods are available for big problems, but they provide, only suboptimal solutions. Let us carry out the computation some different sub-optimal methods.

Computing the route with Simulated Annealing method provide a quite good result, but takes time,

$\Rightarrow$ AbsoluteTiming[{dist, route} = FindShortestTour[centers,
DistanceFunction $\rightarrow$ distfun, Method $\rightarrow$ "SimulatedAnnealing";
//Quiet]
$\Leftarrow$ {181.848, Null}

The total distance,

$\Rightarrow$ dist
$\Leftarrow$ 191706

The 2-*opt* algorithm is much faster but provide somewhat longer distance

$\Rightarrow$ AbsoluteTiming[{dist, route} = FindShortestTour[centers,
DistanceFunction $\rightarrow$ distfun, Method $\rightarrow$ "TwoOpt"];
//Quiet]
$\Leftarrow$ {5.14926, Null}

The total distance,

> ⇒ dist
> ⇐ 198191

The *greedy* algorithm is faster than the *2-opt* but its solution is quite bad

> ⇒ AbsoluteTiming[{dist, route} = FindShortestTour[centers,
> DistanceFunction → distfun, Method → "Greedy"];
> //Quiet]
> ⇐ {1.0654, Null}

The total distance,

> ⇒ dist
> ⇐ 235646

The integer linear programming can not be employed for so big problem, it takes enormous time, after one hour running we could not get any result.

> ⇒ AbsoluteTiming[{dist, route} = FindShortestTour[centers,
> DistanceFunction → distfun, Method → "IntegerLinearProgramming"];
> //Quiet]
> ⇐ $Aborted

The fastest computation and the shortest tour is provided by the *Mathematica* built-in algorithm

> ⇒ AbsoluteTiming[{dist, route} = FindShortestTour[centers,
> DistanceFunction → distfun, Method → "Automatic",
> PerformanceGoal → "Quality"]; //Quiet]
> ⇐ {0.906354, Null}

The total distance,

> ⇒ dist
> ⇐ 181071.

Figure 5.17 shows the visualization of the globe with the optimal route,

Fig. 5.17 Optimal route of the travel on the Earth surface

```
⇒ surfaceCenters = Map[SC, centers[[route]]];
⇒ GreatCircleArc[{lat1_, lon1_}, {lat2_, lon2_}] :=
    Module[{u = SC[{lat1, lon1}], v = SC[{lat2, lon2}], a},
    a = VectorAn − gle[u, v]; Table[Evaluate[RotationTransform[θ,
    {u, v}][u]], {θ,   0, a, a/Ceiling[10a]}]]]
⇒ tourLine =
    Apply[GreatCircleArc, Partition[centers[[route]], 2, 1], {1}];
⇒ Graphics3D[{Sphere[{0, 0, 0}, 0.99r],
    Map[Line[Map[SC, CountryData[#"SchematicCoordinates", ],
    {−2}]]&, places], {Red, Thick, Line[tourLine]},
    {Yellow, PointSize[Medium],
    Map[Tooltip[Point[SC[CountryData[#"CenterCoordinates", ]]], #]&,
    places]}}, Boxed → False, SphericalRegion → True]
```

5.8   Exercise

Let us solve the following nonlinear parameter estimation problem. Our model is

$$f(x) = \alpha \sin(\beta x)$$

⇒ f[x_] := 2.5Sin[1.5x]

The parameters to be estimated are (α, β). We have noisy measurements $\{x_i, f_i\} + \epsilon$, where ϵ is a random noise in the region $x \in [-2, 8]$ with uniform distribution, see Fig. 5.18.

```
⇒ SeedRandom[12];
⇒ XF = Table[{-2 + i0.5, Random[Real, {0.7, 1.3}]f[-2 + I0.5]},
  {i, 0, 20}];
⇒ p1 = ListPlot[XF, PlotStyle → {RGBColor[1, 0, 0], PointSize[0.02]},
  Frame → True, Axes → False, AspectRatio → 0.7,
  ImageSize → {300, 250}];
⇒ Show[{p1, Plot[f[x], {x, -2, 8}]}]
```

Our task is to compute the parameters α and β on the bases of the *noisy data*. To do that we employ the least squares method, namely we are going to minimize the following objective.

$$G(\alpha, \beta) = \sum_{i=1}^{m} (f_i - \alpha \sin(\beta x_i))^2,$$

where m is the number of the measured data pairs.

Fig. 5.18 Noisy measurements (points) for the parameter estimation problem. The continuous line represents the function with parameters $\alpha = 2.5$ and $\beta = 1.5$.

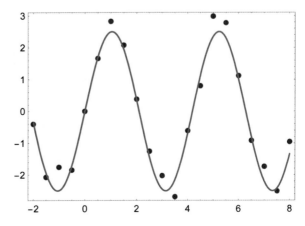

$\Rightarrow \mathtt{m} = \mathtt{Length}[\mathtt{XF}]$

$\Leftarrow \mathtt{21}$

Then our objective function is

$$\Rightarrow \mathtt{G}(\{\alpha\_, \beta\_\}) := \sum_{i=1}^{m} (\mathtt{XF}[[i, 2]] - \alpha \mathtt{Sin}(\beta \mathtt{XF}[[i, 1]]))^2$$

The contour plot of this function in Fig. 5.19 indicates three local minimums,
Computing the local minimums via a local technique (Newton's method), one gets a minimum which is "close" to the initial guess. These local minimums are, see Fig. 5.20. They are local minimums indeed, since

$\Rightarrow \mathtt{min1} = \{\alpha, \beta\}/.\mathtt{FindMinimum}[\mathtt{G}[\{\alpha, \beta\}], \{\{\alpha, 0.75\}, \{\beta, 0.5\}\}][[2]]$

$\Leftarrow \{0.755528, 0.537572\}$

$\Rightarrow \mathtt{G}[\mathtt{min1}]$

$\Leftarrow \mathtt{63.6745}$

$\Rightarrow \mathtt{min2} = \{\alpha, \beta\}/.\mathtt{FindMinimum}[\mathtt{G}[\{\alpha, \beta\}], \{\{\alpha, 0.5\}, \{\beta, 2.5\}\}][[2]]$

$\Leftarrow \{0.357603, 2.44236\}$

$\Rightarrow \mathtt{G}[\mathtt{min2}]$

$\Leftarrow \mathtt{67.6705}$

$\Rightarrow \mathtt{min3} = \{\alpha, \beta\}/.\mathtt{FindMinimum}[\mathtt{G}[\{\alpha, \beta\}], \{\{\alpha, 2.5\}, \{\beta, 1.5\}\}][[2]]$

$\Leftarrow \{2.52079, 1.49648\}$

$\Rightarrow \mathtt{G}[\mathtt{min3}]$

$\Leftarrow \mathtt{2.33031}$

Fig. 5.19 The contour plot of the objective function in the region of $[0, 3] \times [0, 3]$

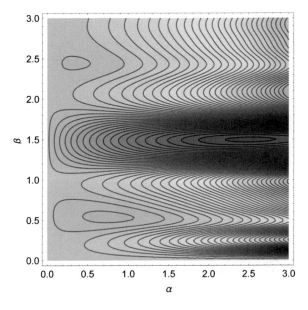

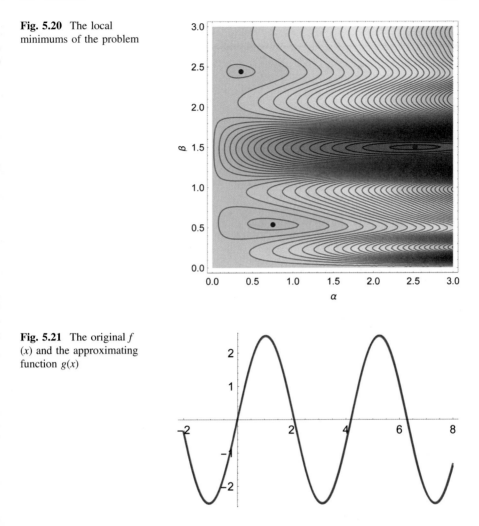

Fig. 5.20 The local minimums of the problem

Fig. 5.21 The original f (x) and the approximating function $g(x)$

Let us employ global minimization via simulated annealing,

$\Rightarrow$ sol = NMinimize[G[α, β}], α, β}, Method $\rightarrow$ "SimulatedAnnealing"]

$\Leftarrow$ {2.33031, {$\alpha \rightarrow$ 2.52079, $\beta \rightarrow$ 1.49648}}

which is the global minimum.

The approximation of the function $f(x)$ is $g(x)$, and Fig. 5.21 shows the approximation of the original function (not the measurement points!). It is important to keep in mind that the original function should be approximated on the basis of the noisy measurements and not noisy points.

$\Rightarrow$ g[x_] := α Sin[β x]/.sol[[2]]

Chapter 6
Genetic Algorithms

6.1 The Genetic Evolution Concept

This stochastic method is based on Darwin's principle of natural selection. It means the fittest individual organisms of a population will survive. An individual can be represented as a string of chromosomes (characters). The string will be the binary form of a value of a real x variable (an individual), and $f(x)$ will be the measure of the fitness of x. Therefore finding the fittest individual means maximizing the $f(x)$ function, see Freeman (1994).

6.2 Mutation of the Best Individual

First, a population of individuals will be generated with random chromosome strings. This will be the first generation. Then the fittest individual will be selected form this population. The chromosomes of this fittest individual will be mutated (changed) randomly in order to generate a new population as the second generation. This strategy leads to the fittest generation in general and probably its fittest individual will have higher fitness value than that of the best individual of the previous generation.

Let us consider again the "tobeornottobe" problem. The ASCII codes of the English alphabet will represent the chromosomes. First, we generate the population of the first generation with 200 individuals.

Supplementary Information The online version contains supplementary material available at https://doi.org/10.1007/978-3-030-92495-9_6.

J. L. Awange et al., *Mathematical Geosciences*,
https://doi.org/10.1007/978-3-030-92495-9_6

⇒ P1 = Table[RandomInteger[{97, 122}], {i, 1, 200}, {j, 1, 13}];
⇒ Short[P1, 5]
⇐ {{119, 113, 117, 102, 100, 99, 110, 100, 110, 106, 117, 106, 115},
  {106, 114, 105, 99, 113, 97, 108, 98, 98, 111, 108, 106, 122}, < <197 > > ,
  {106, 114, 114, 109, 109, 112, 106, 121, 113, 113, 118, 97, 100}}

We are looking for the string

⇒ tobe ="tobeornottobe";

Let us convert the text into a string of ASCII codes

⇒ keyPhrase = ToCharacterCode[tobe]
⇐ {116, 111, 98, 101, 111, 114, 110, 111, 116, 116, 111, 98, 101}

The string of the fittest individual will have the most characters (ASCII code) in the proper place

⇒ diff = Map[(keyPhrase−#)&, P1];
⇒ matches = Map[Count[#, 0]&, diff];
⇒ fitness = Max[matches];
⇒ index = Position[matches, fitness];

in our case,

⇒ pbest = P1[[First[Flatten[index]]]]
⇐ {109, 111, 112, 118, 115, 109, 97, 122, 102, 97, 106, 98, 98}

In character form

⇒ FromCharacterCode[pbest]
⇐ mopvsmazfajbb

Its fitness is

⇒ fitness
⇐ 2

This means that two characters are in the proper place.

Now we are going to create the second generation. Here is a function which gives us the value True with probability μ,

⇒ flip[μ_]:= If[RandomReal[] ≤ μ, True, False].

This function will mutate a string changing its characters one by one with a probability *pmutate*,

```
Mutate[pmute_, letter_]: =
   If[flip[pmute], RandomInteger[{97, 122}], letter];
```

Let us consider the best individual of the first generation,

⇒ pbest

⇐ {109, 111, 112, 118, 115, 109, 97, 122, 102, 97, 106, 98, 98}

and carry out the mutation of all its characters with probability 0.3,

⇒ Map[Mutate[0.3,#]&, pbest]

⇐ {115, 111, 112, 121, 98, 120, 97, 97, 102, 97, 106, 121, 98}

or in text form,

⇒ FromCharacterCode[%]

⇐ sopybxaafajyb

In order to get the population of the second generation, one may have to carry out this mutation 200 times,

⇒ P2 = Table[Map[Mutate[0.3,#]&, pbest], {200}];

Figure 6.1 shows the flow chart of the algorithm.

Fig. 6.1 The flow chart of the algorithm of the *mutation of the best* principle

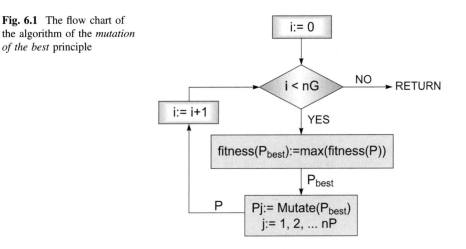

The *Mathematica* function for this algorithm is the following,

```
⇒ Mutation[pmutate_, keyPhrase_, pop_, numGens_] : =
    Module[{i, P, pbest, diff, matches, index, fitness, n, S},
     n = Length[pop];
     P = pop;
     S = {};
     For[i = 1, i ≤ numGens, i + +,
     diff = Map[(keyPhrase−#)&, P];
     matches = Map[Count[#, 0]&, diff];
     fitness = Max[matches];
     index = Position[matches, fitness];
     pbest = P[[First[Flatten[index]]]];
     AppendTo[S, fitness];
     FromCharacterCode[pbest];
     If[fitness == 13, Break[]];
     P = Table[Map[Mutate[pmutate, #]&, pbest],
      {n}];
     ];
     S];
```

The input of the **Mutation**:

pmutate—the probability of the chromosome change (mutation of the string of an individual)

keyPhrase—the string we are looking for, we need it to define the fitness function

pop—the population of the first generation

numGens—the number of the iterations (the maximum number of the generations)

In this example, if the fitness value of the best element of a generation is 13 then the iteration will stop.

The output: the string of the best individuals of the generations, (*S*).

Let us carry out a simulation,

```
⇒ AbsoluteTiming[G1 = Mutation[0.1, keyPhrase, P1, 200]; ]
⇐ {0.189405, Null}
⇒ G1
⇐ {2, 3, 4, 5, 6, 7, 8, 9, 10, 10, 11, 12, 12, 12, 13}
```

The result can be computed in finite steps!

6.3 Solving a Puzzle

Problem Let us consider a puzzle in order to illustrate how one can formulate a problem to be solved by the Mutation of the Best method.

Given a 5×5 grid, see Fig. 6.2, and 9 positive integer numbers $\{1, 2, 3, \ldots, 9\}$. The 9 numbers should be arranged in the grid in such way that the sum shown in Fig. 6.2 should simultaneously add up to 25. The arrangement is as follows:

(a) the sum of the numbers in horizontal direction, $(x_1 + x_2 + x_9 + x_5 + x_6)$,
(b) the sum of the numbers in vertical direction, $(x_4 + x_3 + x_9 + x_8 + x_7)$,
(c) the sum of the numbers staying at the ends of the cross, $(x_1 + x_4 + x_6 + x_7)$.

Solution
One individual can be randomly generated without repetition of the numbers

$\Rightarrow$ s = RandomSample$[\{1, 2, 3, 4, 5, 6, 7, 8, 9\}, 9]$
$\Leftarrow \{3, 8, 5, 7, 2, 9, 6, 1, 4\}$

The function measuring the fitness is

$$1/(1 + |(x_1 + x_2 + x_5 + x_6 + x_9) - 25| + |(x_3 + x_4 + x_7 + x_8 + x_9) - 25|$$
$$+ |(x_1 + x_4 + x_6 + x_7) - 25|)$$

$\Rightarrow$ fit$[x\_] :=$
   $1/(1 + \text{Abs}[x[[1]] + x[[2]] + x[[9]] + x[[5]] + x[[6]] - 25] +$
   $\text{Abs}[x[[4]] + x[[3]] + x[[9]] + x[[8]] + x[[7]] - 25] +$
   $\text{Abs}[x[[4]] + x[[1]] + x[[7]] + x[[6]] - 25])$
$\Rightarrow$ fit$[s]//N$
$\Leftarrow 0.25$

The optimal value is 1 but in other cases, it is less than 1. A function which generates a population of n individuals is

Fig. 6.2 The grid of the puzzle

$\Rightarrow$ generation[n_] : = Table[RandomSample[$\{1,2,3,4,5,6,7,8,9\}$, 9],$\{$n$\}$]

A generation in case of $n = 10$.

$\Rightarrow$ u = generation[10]

$\Leftarrow \{\{9,8,5,6,3,2,7,4,1\},\{6,5,9,7,1,2,8,4,3\},$
$\qquad \{4,3,6,1,9,8,7,2,5\},\{2,5,7,6,4,8,9,3,1\},$
$\qquad \{5,3,2,1,4,7,8,9,6\},\{8,3,1,9,4,7,5,6,2\},$
$\qquad \{6,8,4,1,9,7,5,3,2\},\{6,8,4,9,3,7,2,5,1\},$
$\qquad \{6,7,3,8,4,5,2,1,9\},\{5,2,1,4,8,3,6,9,7\}\}$

The fitness value of the fittest individual is,

$\Rightarrow$ fu = Map[fit[#] &, u];

$\Rightarrow$ Max[fu]//N

$\Leftarrow 0.166667$

The string of the fittest individual itself is

$\Rightarrow$ parent = u[[First[Flatten[Position[fu, Max[fu]]]]]]

$\Leftarrow \{9,8,5,6,3,2,7,4,1\}$

Now, we mutate this individual with probability μ,

$\Rightarrow$ flip[μ_]: = If[RandomReal[] $\leq$ μ, True, False]

After the mutation there should be 9 different numbers. We choose two positions randomly and then change their numbers,

$\Rightarrow$ Mutation[x_, pmutate_] := Module[$\{$x0, i1, i2, s$\}$,
   x0 = x;
   If[flip[pmutate], i1 = RandomInteger[$\{1,9\}$];
    i2 = RandomInteger[$\{1,9\}$];
    s = x0[[i1]];
    x0[[i1]] = x0[[i2]];
    x0[[i2]] = s];
   x0];

The mutation of the best individual is

$\Rightarrow$ Mutation[parent, 0.3]

$\Leftarrow \{9,8,5,3,6,2,7,4,1\}$

$\Rightarrow$ fit[%]//N

$\Leftarrow 0.0909091$

The following function creates the subsequent generations. Its input is a generation g and the number of the generations $ngen$,

```
⇒ Gena[g_,ngen_] : = Module[{fi,newg,fitg,parent,maxfit,n,i},
    fi = {};
    i = 1;
    n = Length[g];
    newg = g;
    For[i = 1,i< = ngen,i++,
    fitg = Map[fit[#]&,newg];
    maxfit = Max[fitg];
    AppendTo[fi,maxfit];
    parent = newg[[First[Flatten[Position[fitg,Max[fitg]]]]]];
  If[maxfit == 1,Print["Hit!"];Break[]];
    newg = Table[Mutation[parent,0.25],{n}];
    ];
  parent];
```

Let us carry out the simulation with a population of 1000 individuals. The initial generation is

```
⇒ fg = generation[1000];
```

The iteration for $ngen = 100$ results in

```
⇒ sc = Gena[fg,100]
⇐ Hit!
⇐ {9,4,7,2,1,6,8,3,5}
```

We have found a proper solution! Let us visualize it in matrix form

```
⇒ M = Table[0,{5},{5}]; M[[3,1]] = sc[[1]]; M[[3,2]] = sc[[2]];
⇒ M[[2,3]] = sc[[3]]; M[[1,3]] = sc[[4]]; M[[3,4]] = sc[[5]];
⇒ M[[3,5]] = sc[[6]]; M[[5,3]] = sc[[7]]; M[[4,3]] = sc[[8]];
⇒ M[[3,3]] = sc[[9]];
```

```
⇒ MatrixForm[M]
```

$$\Leftarrow \begin{pmatrix} 0 & 0 & 2 & 0 & 0 \\ 0 & 0 & 7 & 0 & 0 \\ 9 & 4 & 5 & 1 & 6 \\ 0 & 0 & 3 & 0 & 0 \\ 0 & 0 & 8 & 0 & 0 \end{pmatrix}$$

It is clear that there are many different solutions:

$$\Rightarrow S1 = \begin{pmatrix} 0 & 0 & 6 & 0 & 0 \\ 0 & 0 & 4 & 0 & 0 \\ 8 & 7 & 5 & 3 & 2 \\ 0 & 0 & 1 & 0 & 0 \\ 0 & 0 & 9 & 0 & 0 \end{pmatrix} ;$$

$$\Rightarrow S2 = \begin{pmatrix} 0 & 0 & 9 & 0 & 0 \\ 0 & 0 & 1 & 0 & 0 \\ 2 & 3 & 5 & 7 & 8 \\ 0 & 0 & 4 & 0 & 0 \\ 0 & 0 & 6 & 0 & 0 \end{pmatrix} ;$$

$$\Rightarrow S3 = \begin{pmatrix} 0 & 0 & 9 & 0 & 0 \\ 0 & 0 & 1 & 0 & 0 \\ 3 & 8 & 5 & 2 & 7 \\ 0 & 0 & 4 & 0 & 0 \\ 0 & 0 & 6 & 0 & 0 \end{pmatrix} ;$$

$$\Rightarrow S4 = \begin{pmatrix} 0 & 0 & 4 & 0 & 0 \\ 0 & 0 & 1 & 0 & 0 \\ 8 & 2 & 5 & 3 & 7 \\ 0 & 0 & 9 & 0 & 0 \\ 0 & 0 & 6 & 0 & 0 \end{pmatrix} ;$$

$$\Rightarrow S5 = \begin{pmatrix} 0 & 0 & 8 & 0 & 0 \\ 0 & 0 & 2 & 0 & 0 \\ 9 & 6 & 5 & 4 & 1 \\ 0 & 0 & 3 & 0 & 0 \\ 0 & 0 & 7 & 0 & 0 \end{pmatrix} ;$$

6.4   Application to a Real Function

Let us compute the global maximum of the function $f(x)$, see Fig. 6.3 in the region of $[-40, 40]$,

$$f(x) = 1 + \frac{\cos(x)}{1 + 0.01x^2}$$

Fig. 6.3 The function $f(x)$

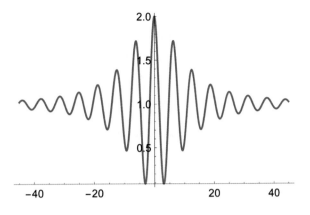

$\Rightarrow f[x\_] := 1 + \mathrm{Cos}[x]/(1 + 0.01x^2)$

Now, we employ 10 bits binary representation for x_i as an individual and its fitness value $f_i = f(x_i)$.

In this case x_i represents an individual with fitness $f_i = f(x_i)$. The maximum number is $\{1, 1, 1, 1, 1, 1, 1, 1, 1, 1\}$, namely

$\Rightarrow \mathrm{maxi} = 2^{10} - 1$

$\Leftarrow 1023$

The smallest one is $\{0, 0, 0, 0, 0, 0, 0, 0, 0, 0\}$ giving

$\Rightarrow \mathrm{mini} = 2^0 - 1$

$\Leftarrow 0$

The conversion can then be carried out using the function

$\Rightarrow \mathrm{Horner}[u\_, base\_ : 2] := \mathrm{Fold}[base \,\#1 + \#2\,\&, 0, u]$

For example,

$\Rightarrow s_i = \{1, 0, 1, 0, 0, 0, 0, 0, 0, 0\};$

$\Rightarrow d_i = \mathrm{Horner}[s_i]$

$\Leftarrow 640$

Indeed,

$\Rightarrow 2^9 + 2^7$

$\Leftarrow 640$

Now, we should project this $d_i \in [0, 1023]$ interval on $x_i \in [-40, 40]$, then the corresponding value of 640 is

$$\Rightarrow x_i = \frac{d_i\,80}{1023} - 40.$$

$\Leftarrow 10.0489$

Considering these two steps together we get,

```
⇒ decodeBGA[chromosome_] : =
    Module[{phenotype, decimal}, decimal = Horner[chromosome];
    phenotype = (decimal 80)/1023 − 40.;
    Return[phenotype]];
⇒ decodeBGA[sᵢ]
⇐ 10.0489
```

The smallest positive value, which can be represented on 10 binary bits is,

```
⇒ decodeBGA[{1, 0, 0, 0, 0, 0, 0, 0, 0, 0}]
⇐ 0.0391007
```

The highest negative value is,

```
⇒ decodeBGA[{0, 1, 1, 1, 1, 1, 1, 1, 1, 1}]
⇐ −0.0391007
```

So these are the best approaches for zero!
The corresponding fitness value is

```
⇒ f[%]
⇐ 1.99922
```

This function will generate the initial generation,

```
⇒ initPop[psize_, csize_] : = RandomInteger[{0, 1}, {psize, csize}];
```

where

psize   the size of the population,
csize   individuals of the population in binary string.

Let us consider 50 individuals in the population,

```
⇒ InitialPop = initPop[15, 10];
⇒ Short[InitialPop, 5]
⇐ {{0, 1, 1, 0, 0, 1, 1, 0, 0, 1},
    {1, 0, 0, 1, 0, 0, 1, 1, 0, 1}, {0, 1, 0, 1, 1, 0, 0, 1, 1, 1},
    {0, 0, 0, 0, 1, 0, 0, 0, 0, 0}, < <8 > > , {0, 1, 0, 0, 1, 0, 1, 1, 1, 1},
    {1, 1, 1, 1, 1, 1, 1, 0, 1, 1}, {1, 1, 1, 1, 0, 0, 0, 1, 0, 0}}
```

The function for the mutation is,

```
⇒ mutateBit[pmute_, Bit_] := If[flip[pmute], RandomInteger[], Bit];
```

The function for the Mutation of the Best algorithm is,

```
⇒ bgaM[pmutate_, popInitial_, fitFunction_, numGens_] :=
    Module[{i, newPop, index, pheno, f, fpheno, parent, fitness, m,
      n, Slide, p1, p2, x},
      newPop = popInitial;
      n = Length[newPop];
      f = fitFunction;
    Slide = {}; p1 = Plot[f[x], {x, −45, 45}, PlotRange → {−0.5, 2.1}];
    For[i = 1, i < = numGens, i + +,
      pheno = Map[decodeBGA, newPop];
      fpheno = f[pheno];
      p2 = ListPlot[Transpose[{pheno, fpheno}],
      PlotStyle → {RGBColor[1, 0, 0], PointSize[0.015]}];
      AppendTo[Slide, Show[{p1, p2}]];
      fitness = Max[fpheno];
      index = Position[fpheno, fitness];
      parent = newPop[[First[Flatten[index]]]];
      Print["Generation", i, ":",
      decodeBGA[parent],
      "      Fitness =", fitness];
      newPop = Table[Map[mutateBit[pmutate, #]&, parent],
    {n}]; ]; (*endofFor*)
    m = Max[Map[f[#]&, pheno]];
    {Select[pheno, f[#] == m&]//First, Slide}];
```

Let us employ it,

$\Rightarrow$ sol = bgaM[0.25, InitialPop, f, 25];

$\Leftarrow$ Generation1 : 6.06061   Fitness = 1.71332

$\Leftarrow$ Generation2 : 6.21701   Fitness = 1.71966

$\Leftarrow$ Generation3 : 6.21701   Fitness = 1.71966

$\Leftarrow$ Generation4 : 6.21701   Fitness = 1.71966

$\Leftarrow$ Generation5 : 6.21701   Fitness = 1.71966

$\Leftarrow$ Generation6 : 6.21701   Fitness = 1.71966

$\Leftarrow$ Generation7 : 6.21701   Fitness = 1.71966

$\Leftarrow$ Generation8 : 6.21701   Fitness = 1.71966

$\Leftarrow$ Generation9 : 6.21701   Fitness = 1.71966

$\Leftarrow$ Generation10 : 0.50831   Fitness = 1.87132

$\Leftarrow$ Generation11 : 0.50831   Fitness = 1.87132

$\Leftarrow$ Generation12 : 0.03911   Fitness = 1.99922

$\Leftarrow$ Generation13 : 0.03911   Fitness = 1.99922

$\Leftarrow$ Generation14 : 0.03911   Fitness = 1.99922

$\Leftarrow$ Generation15 : 0.03911   Fitness = 1.99922

$\Leftarrow$ Generation16 : 0.03911   Fitness = 1.99922

$\Leftarrow$ Generation17 : 0.03911   Fitness = 1.99922

$\Leftarrow$ Generation18 : 0.03911   Fitness = 1.99922

$\Leftarrow$ Generation19 : 0.03911   Fitness = 1.99922

$\Leftarrow$ Generation20 : 0.11730   Fitness = 1.99299

$\Leftarrow$ Generation21 : 0.11730   Fitness = 1.99299

$\Leftarrow$ Generation22 : 0.03911   Fitness = 1.99922

$\Leftarrow$ Generation23 : 0.03911   Fitness = 1.99922

$\Leftarrow$ Generation24 : 0.03911   Fitness = 1.99922

$\Leftarrow$ Generation25 : 0.03911   Fitness = 1.99922

Then, since no further improvement occurs

$\Rightarrow$ sol[[1]]

$\Leftarrow$ 0.0391007

$\Rightarrow$ f[sol[[1]]]

$\Leftarrow$ 1.99922

Fig. 6.4 Simulation of computing the global maximum in case of 10 bits representation at the beginning (top) and at the end of the process (bottom)

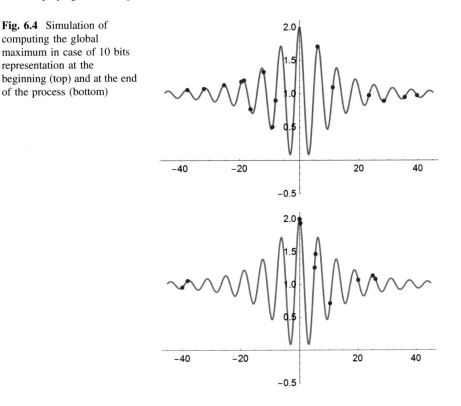

Figure 6.4 shows the positions of the individuals at the beginning and at the end of the process.

The solution with high precision is

$\Rightarrow$ Chop[NMaximize[f[x], x]]

$\Leftarrow \{2., \{x \rightarrow -8.18433 \times 10^{-9}\}\}$

To improve the method, we introduce the concept of sexuality in Sect. 6.5.

6.5 Employing Sexual Reproduction

The effectiveness of the natural selection can be improved when the principle of sexuality is introduced. Namely we employ mating of two individuals as parents and using a crossover method to interchange part of their chromosomes, resulting in two new individuals as children partly having the chromosomes of both parents.

6.5.1 Selection of Parents

The selection of the parents is carried out by the *roulette-wheel method*. In principle, we construct a roulette wheel on which each individual of the population is given a sector, whose size is proportional to the individual's fitness. Then we spin the wheel and whichever individual comes up becomes a parent. In this way the individual having higher fitness will become a parent with higher probability, but not deterministically.

Let us consider the example of a maximization function discussed in the previous section.

We generate a population of ten individuals,

$\Rightarrow$ pop = Table[RandomInteger[{0, 1}], {i, 10}, {j, 10}]
$\Leftarrow$ {{1, 0, 0, 1, 1, 1, 1, 0, 1, 1}, {0, 0, 1, 1, 1, 1, 0, 1, 1, 0},
  {1, 1, 1, 0, 0, 0, 1, 1, 0, 1}, {1, 0, 1, 0, 0, 0, 0, 1, 0, 1},
  {1, 0, 1, 0, 0, 1, 1, 1, 0, 0}, {1, 1, 1, 0, 0, 1, 1, 1, 0, 0},
  {0, 0, 0, 0, 1, 0, 1, 0, 0, 0}, {1, 1, 0, 0, 0, 1, 0, 1, 1, 1},
  {0, 0, 1, 0, 1, 1, 0, 1, 1, 1}, {1, 1, 0, 1, 1, 0, 1, 1, 1, 1}}

Or decoding them as real numbers,

$\Rightarrow$ pheno = Map[decodeBGA, pop]
$\Leftarrow$ {9.65787, −20.7625, 31.085, 10.4399, 12.2385,
  32.2581, −36.8719, 21.8573, −25.6891, 28.739}

Let us compute their fitness,

$\Rightarrow$ fitList = Map[f, pheno]
$\Leftarrow$ {0.496593, 0.936831, 1.0887, 0.747582, 1.37903,
  1.05838, 1.04638, 0.828461, 1.11174, 0.903452}

The cumulative fitness of the population is

$\Rightarrow$ fitListSum = FoldList[Plus, First[fitList], Rest[fitList]]
$\Leftarrow$ {0.496593, 1.43342, 2.52212, 3.2697,
  4.64873, 5.70711, 6.75349, 7.58195, 8.69369, 9.59714}

Then the total fitness of the population is

$\Rightarrow$ fitSum = Last[fitListSum]
$\Leftarrow$ 9.59714

This function simulates the selection of the parents using roulette-wheel operation

```
⇒ selectOne[foldedFitnessList_, fitTotal_] :=
    Module[{randFitness, elem, index}, randFitness =
    RandomReal[] fitTotal;
    elem = Select[foldedFitnessList, #1 randFitness&, 1];
    index = Flatten[Position[foldedFitnessList, First[elem]]];
    Return[First[index]];];
```

The first parent is

```
⇒ parent1Index = selectOne[fitListSum, fitSum]
⇐ 6
```

the second one is,

```
⇒ parent2Index = selectOne[fitListSum, fitSum]
⇐ 3
```

The binary representation of the parents is,

```
⇒ parent1 = pop[[parent1Index]]
⇐ {1, 1, 1, 0, 0, 1, 1, 1, 0, 0}
⇒ parent2 = pop[[parent2Index]]
⇐ {1, 1, 1, 0, 0, 0, 1, 1, 0, 1}
```

While the fitness of the parents is,

```
⇒ fitList[[parent1Index]]
⇐ 0.818018
```

and

```
⇒ fitList[[parent2Index]]
⇐ 0.954068
```

They are not the fittest individuals in the population, since

```
⇒ Max[fitList]
⇐ 1.37903
```

So as we mentioned, the roulette-wheel algorithm provides a bigger chance for individuals having high fitness (bigger size of selection) to be set as parents, but it will not select deterministically the two best individuals.

6.5.2   Sexual Reproduction: Crossover and Mutation

The crossover process will produce two new individuals as children via partly changing the chromosomes characteristics in the strings of the two parents, see Fig. 6.5. The steps of the algorithm are the following:

(1) We decide with a certain probability whether we create new individuals or just copy the parents' chromosomes, and the children become clones.
(2) In case of creating new individuals we carry out the crossover process as in Fig. 6.5 below.
(3) Mutation of the children.

In Fig. 6.5, two parents are selected according to their fitness, and a crossover point illustrated by the vertical line is chosen by a uniform random selection. The children's chromosomes are formed by combining opposite parts of each parents' chromosomes and a crossover point illustrated by the vertical line is chosen by a uniform random selection. The children's chromosomes are formed by combining opposite parts of each parents' chromosomes.

The process described above can be evaluated using the following functions

$\Rightarrow$ myXor[x_,y_] : = If[x == y, 0, 1];
$\Rightarrow$ mutateBGA[pmute_,allel_] : = If[flip[pmute], myXor[allel, 1], allel];

$\Rightarrow$ crossOver[pcross_,pmutate_,parent1_,parent2_] : =
　　Module[{child1, child2, crossAt, lchrom}, lchrom = Length[parent1];
　　If[flip[pcross], crossAt = RandomInteger[{1, lchrom − 1}];
　　child1 = Join[Take[parent1, crossAt], Drop[parent2, crossAt]];
　　child2 = Join[Take[parent2, crossAt], Drop[parent1, crossAt]],
　　child1 = parent1;
　　child2 = parent2;];
　　child1 = (mutateBGA[pmutate,#1]&)/@child1;
　　child2 = (mutateBGA[pmutate,#1]&)/@child2;
　　Return[{child1, child2}];];

where *pcross* is the probability of the crossover operation and *pmute* is the probability of the mutation.

Fig. 6.5 The crossover operation

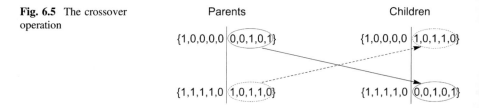

Let us see an example for creating two children via crossover and the subsequent mutation,

```
⇒ pcross = 0.75; pmute = 0.005;
⇒ MatrixForm[children = crossOver[pcross,pmute,parent1,parent2]]
⇐ ( 1  1  1  0  0  1  1  1  0  1 )
    ( 1  1  1  0  0  0  1  1  0  0 )
```

The real number is represented by

```
⇒ decodeList = Map[decodeBGA, children]
⇐ {32.3363,31.0068}
```

The corresponding function values (fitnesses) is

```
⇒ newfitness = Map[f, decodeList]
⇐ {1.05286,1.08644}
```

And the average is given by

```
⇒ Apply[Plus,newfitness]/2
⇐ 1.06965
```

The average fitness of the parents is therefore

```
⇒ (fitList[[parent1Index]] + fitList[[parent2Index]])/2
⇐ 1.07354
```

This means that the average fitness of the children is not necessarily higher than that of their parents. This technique is called the *generational replacement*.

6.6   The Basic Genetic Algorithm (BGA)

These functions can be collected in order to implement the basic genetic algorithm in *Mathematica* see Freeman (1994).

This function will print out the best individuals of the generation during the iteration process

```
⇒ displayBest[fitnessList_,number2Print_] :=
  Module[{i, sortedList}, sortedList = Sort[fitnessList, Greater];
  For[i = 1, i < = number2Print, i + +,
  Print["fitness =", sortedList[[i]]]; ]; (*endofFori*)];
  (*end of Module*)
```

where *number2Print* is for the number of the best individuals to be printed.

The main function is,

```
⇒ bga[pcross_, pmutate_, popInitial_, fitFunction_, numGens_,
   printNum_] :=
  Module[{i, newPop, parent1, parent2, diff, matches,
   oldPop, reproNum, index, fitList, fitListSum, fitSum, pheno,
   pIndex, pIndex2, f, children, m},
   oldPop = popInitial; (*initialize first population*)
   reproNum = Length[oldPop]/2;
   (*calculate number of reproductions*)
   f = fitFunction; (*assign the fitness function*)
   For[i = 1, i < = numGens, i + +, (*perform numGens generations*)
    pheno = Map[decodeBGA, oldPop]; (*decode the chromosomes*)
    fitList = f[pheno]; (*determine the fitness of each phenotype*)
    Print[" "]; (*print out the best individuals*)
    Print["Generation", i, "Best", printNum];
    Print[" "];
    displayBest[fitList, printNum];
    fitListSum = FoldList[Plus, First[fitList], Rest[fitList]];
    fitSum = Last[fitListSum]; (*find the total fitness*)
    newPop = Flatten[Table[(*determine the new population*)
     pIndex1 = selectOne[fitListSum, fitSum];
      (*select parent indices*)
     pIndex2 = selectOne[fitListSum, fitSum];
     parent1 = oldPop[[pIndex1]]; (*identify parents*)
     parent2 = oldPop[[pIndex2]];
     children = crossOver[pcross, pmutate, parent1, parent2];
      (*crossover and mutate*)
     children, {reproNum}], 1        (*add children to list;
       flatten to first level*)]; (*end of Flatten[Table]*)
     oldPop = newPop; (*new becomes old for next gen*)
   ]; (*end of For i*);
   m = Max[Map[f[#1]&, pheno]];
   Select[pheno, f[#] == m&]//First
  ]; (*end of Module*)
```

where

pcross   the probability of the crossover,
pmutate   probability of the mutation of the children,
popInitial   the initial population,
fitFunction   function of fitness,
numGens   the maximum number of the generations,
printNum   the number of the best individuals of the actual generation to be
                printed.

Pay attention to the fact that the function **decodeBGA** should be properly
defined depending on the problem!

Now let us test this implementation to find the maximum of the function
employed in the previous section, but this time with 12 bit representation. The
initial population of 200 individuals is

⇒ initialPopulation = initPop[200, 12];

Using 12 bits, then the maximum number $x_i \in [-40, 40]$ is

⇒ decodeBGA[chromosome_] :=
   Module[{phenotype, decimal}, decimal = Horner[chromosome];
   phenotype = (decimal 80)/4095 − 40.';
   Return[phenotype]];

Now, the approximation of the zero is

⇒ decodeBGA[{1, 0, 0, 0, 0, 0, 0, 0, 0, 0, 0, 0}]
⇐ 0.00976801
⇒ decodeBGA[{0, 1, 1, 1, 1, 1, 1, 1, 1, 1, 1, 1}]
⇐ −0.00976801
⇒ f[%]
⇐ 1.99995

Let us consider 10 generations and print out the best individual of the actual
generation

⇒ bga[0.75, 0.01, initialPopulation, f, 10, 1]

⇐ Generation 1   Best 1

⇐ fitness = 1.99412

⇐ Generation 2   Best 1

⇐ fitness = 1.99606

⇐ Generation 3   Best 1

⇐ fitness = 1.99606

⇐ Generation 4   Best 1

⇐ fitness = 1.99606

⇐ Generation 5   Best 1

⇐ fitness = 1.99878

⇐ Generation 6   Best 1

⇐ fitness = 1.99995

⇐ Generation 7   Best 1

⇐ fitness = 1.99878

⇐ Generation 8   Best 1

⇐ fitness = 1.99878

⇐ Generation 9   Best 1

⇐ fitness = 1.99878

⇐ Generation 10   Best 1

⇐ fitness = 1.99956

⇐ 0.029304

Indeed

⇒ f[%]

⇐ 1.99956

6.7 Applications

First let us solve the nonlinear parameter estimation problem discussed in the previous chapter.

Fig. 6.6 The function and its noisy measured points

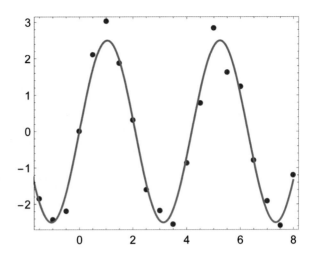

6.7.1 Nonlinear Parameter Estimation

Our function is

$$f(x) = 2.5 \sin(1.5\,x).$$

We have noisy measurements in $x \in [-2, 8]$, see the points in Fig. 6.6.

$\Rightarrow$ f[x_] : = 2.5 Sin[1.5 x]

$\Rightarrow$ XF = Table[{-2 + i 0.5, Random[Real,{0.7, 1.3}] f[-2 + i 0.5]}, {i, 1, 20}];

We are looking for the parameters α and β

$$f(x) = \alpha \sin(\beta\,x)$$

employing least squares method the function to be minimized is then (Fig. 6.7)

$$G(\alpha, \beta) = \sum_{i=1}^{n} (f_i - \alpha \sin(\beta x_i))^2$$

$\Rightarrow$ n = Length[XF];

$\Rightarrow$ G({α_, β_}) := $\displaystyle\sum_{i=1}^{n}$ (XF[[i, 2]] - α Sin(β XF[[i, 1]]))^2

There are more local minimums, see Fig. 6.8.

Let us employ the built-in function (**NMinimize**) with a built-in evalutionary algorithm (Method $\rightarrow$ "DifferentialEvolution").

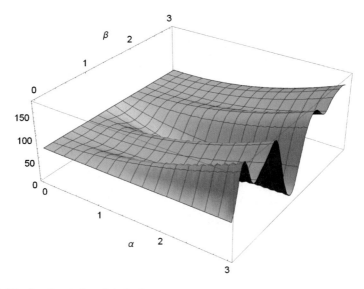

Fig. 6.7 The function to be minimized

Fig. 6.8 More local
minimums

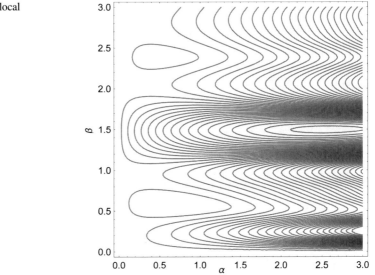

Remark Genetic algorithms is regarded as a subset of evalutionary algorithm.

⇒ sol = Reap[NMinimize[{G[{α, β}], 0 ≤ α ≤ 3, 0 ≤ β ≤ 3}, {α,β},
   EvaluationMonitor : → Sow[{α,β}],
   Method →″ DifferentialEvolution″]];

The global minimum and its place,

⇒ sol[[1]]
⇐ {1.9324,{α → 2.57349,β → 1.49834}}

The number of iterations is (Fig. 6.9),

⇒ {hist} = sol[[2]];
⇒ nh = Length[hist]
⇐ 2029

In general, using the genetic algorithm, the neighborhood of the global minimum has been reached quickly, but to find a precise solution with high accuracy takes a long time with global method. Therefore often we combine the global and the local methods. The global method finds the neighborhood of the global minimum (processing), then the local method—which converges faster—finds the precise location of the minimum (post processing); see the next application.

Fig. 6.9 The trial points of the genetic algorithm during the iteration process, and the global minimum (white point with arrow)

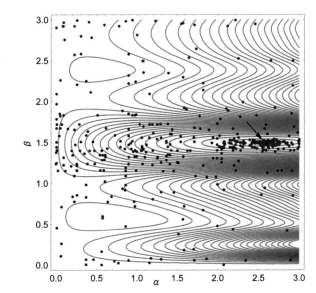

6.7.2 Packing Spheres with Different Sizes

Let us solve the problem discussed in the previous chapter but now in 3D! Consider five spheres with radius ranging between 0.5 and 0.8

$\Rightarrow$ n = 5; minR := 0.5; maxR := 0.8;

$\Rightarrow$ Table[r_i = Random[Real, {minR, maxR}], {i, 1, n}];

$\Rightarrow$ B = 1.5;

$\Rightarrow$ constraints01 =

   Table[$(x_i - x_j)^2 + (y_i - y_j)^2 + (z_i - z_j)^{2\,3}(r_i + r_j)^2$,
   {i, 1, n}, {j, i + 1, n}]//Flatten;

$\Rightarrow$ constraints02 =

   Table[{$x_i + r_i - w \leq 0, x_i - r_i + w \geq 0, y_i + r_i - w \leq 0, y_i - r_i + w \geq 0$,
   $z_i + r_i - w \leq 0, z_i - r_i + w \geq 0, -B \leq x_i \leq B, -B \leq y_i \leq B, -B \leq z_i \leq B$},
   {i, 1, n}]//Flatten;

$\Rightarrow$ constraints = Join[constraints01, constraints02, {$0 \leq w \leq B$}];

$\Rightarrow$ vars = Join[{w}, Table[{x_i, y_i, z_i}, {i, 1, n}]]//Flatten;

First, we try to solve the problem with the *simulated annealing* method,

$\Rightarrow$ soln = NMinimize[Join[{$8w^3 - \sum_{k=1}^{n} \frac{4 r_k^3 \pi}{3}$}, constraints], vars,

   Method $\rightarrow$ {"SimulatedAnnealing", "PerturbationScale" $\rightarrow$ 6,
   "SearchPoints" $\rightarrow$ 100}]//Timing

MessageTemplate[NMinimize , incst ,

   NMinimize was unable to generate any initial points satisfying the inequality constraints
   {0.742592 – w – $x_1 \leq 0$, 0.742592 – w + $x_1 \leq 0$, 0.721401 – w – $x_2 \leq 0$, 0.721401

   – w + $x_2 \leq 0$, «33», 1.18996 – $(x_4 - x_5)^2 - (y_4 - y_5)^2 - (z_4 - z_5)^2 \leq 0$, 0.551895

   – w – $z_5 \leq 0$, 0.551895 – w + $z_5 \leq 0$}. The initial region

   specified may not contain any feasible points. Changing the initial region
   or specifying explicit initial points may provide a better solution.

2 , 217 , 45 , 16816622134118316439 , Local]

Then, let us employ the *genetic algorithm with a post processing Newton's method* as a local method.

$\Rightarrow$ soln = NMinimize[Join[{$8w^3 - \sum_{k=1}^{n} \frac{4 r_k^3 \pi}{3}$}, constraints], vars,

   Method $\rightarrow$ {"DifferentialEvolution", "ScalingFactor" $\rightarrow$ 0.9,
   "CrossProbability" $\rightarrow$ 0.1,
   "PostProcess" $\rightarrow$ {FindMinimum, Method $\rightarrow$ "QuasiNewton"}}]//Timing

Fig. 6.10 The packed
spheres with different radius

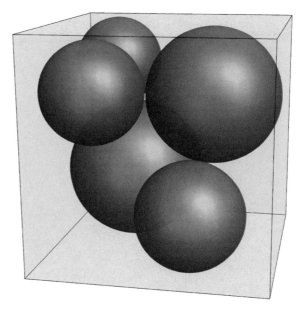

The results show the computation time, the free volume, the size of the box, and
the coordinates of the spheres.

$\Leftarrow \{15.3349, \{10.5258, \{w \rightarrow 1.26476,$
$x_1 \rightarrow -0.714915, y_1 \rightarrow -0.714915, z_1 \rightarrow -0.714915,$
$x_2 \rightarrow -0.481048, y_2 \rightarrow 0.481048, z_2 \rightarrow -0.17328,$
$x_3 \rightarrow 0.679, y_3 \rightarrow 0.692456, z_3 \rightarrow 0.681176,$
$x_4 \rightarrow -0.683168, y_4 \rightarrow -0.683168, z_4 \rightarrow 0.683168,$
$x_5 \rightarrow 0.541146, y_5 \rightarrow -0.541145, z_5 \rightarrow 0.253675\}\}\}$

The smallest results can be visualized in Fig. 6.10.

6.7.3 Finding All the Real Solutions of a Non-Algebraic System

This is a fairly difficult problem, and in general there is no computer system that has
a built-in function for that. Using brute force, however, *Mathematica* can solve this
task. Let us consider the following problem see Wagon (2000),

$\Rightarrow f = -Cos[y] + 2yCos[y^2]Cos[2x];$
$\Rightarrow g = -Sin[x] + 2Sin[y^2]Sin[2x];$

Fig. 6.11 The non-algebraic
system has 67 real zeros in
this region

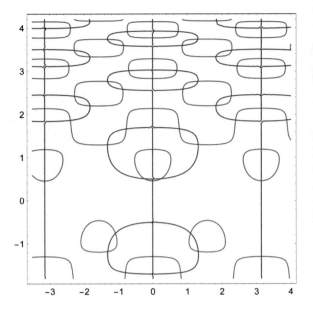

The zero contours of the functions can be seen on Fig. 6.11. The crossing points represent the common zeros, namely the solutions of the system.

The first step is the *discretization* of the region (Fig. 6.12).

$\Rightarrow$ RR = DiscretizeRegion[Rectangle[{−3.5, −1.8}, {4, 4.2}],
    MaxCellMeasure → 0.08]

We compute the specifications of the subregions

$\Rightarrow$ S = MeshCoordinates[RR];
$\Rightarrow$ U = MeshCells[RR, 2];
$\Rightarrow$ V = Map[MeshRegion[S, #] &, U];

The number of the subregions.

$\Rightarrow$ Length[V]
$\Leftarrow$ 872

Let define the objective as

$\Rightarrow$ G = f^2 + g^2
$\Leftarrow$ $(-\text{Cos}[y] + 2y\text{Cos}[2x]\text{Cos}[y^2])^2 + (-\text{Sin}[x] + 2\text{Sin}[2x]\text{Sin}[y^2])^2$

Over every subregion we look for the global minimum of this objective using the Genetic Algorithm. To reduce the computation time, parallel evaluation is employed.

Fig. 6.12 The region is
discretized

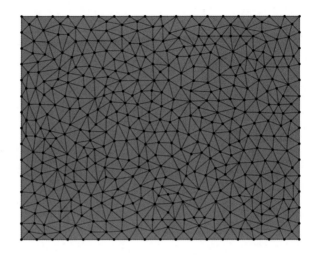

```
⇒ AbsoluteTiming[sol =
   ParallelMap[NMinimize[G,(x,y)Î#]&,V]//Chop;
⇐ {45.4897,Null}
```

We select the solutions where the objective is zero. It turns out that there are 67
real solutions (Fig. 6.13).

Fig. 6.13 The zeros of the
system

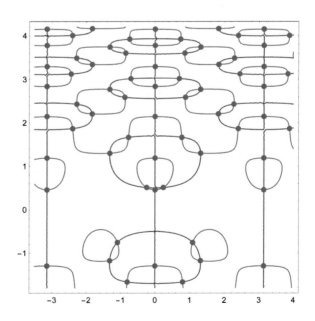

$\Rightarrow$ solv = Select[sol, (#[[1]] == 0)&];

$\Rightarrow$ solvv = Map[({x, y}/.#[[2]])&, solv];

$\Rightarrow$ Length[solvv]

$\Leftarrow$ 67

6.8   Exercise

6.8.1   Foxhole Problem

Let us consider the *Shekels* foxhole problem. We want to find the global minimum of the following function (see Fig. 6.14).

$$\Leftarrow f = 1/(0.002 + \frac{1}{25 + (-32 + x)^6 + (-32 + y)^6} + \frac{1}{24 + (-16 + x)^6 + (-32 + y)^6}$$

$$+ \frac{1}{23 + x^6 + (-32 + y)^6} + \frac{1}{22 + (16 + x)^6 + (-32 + y)^6} + \frac{1}{21 + (32 + x)^6 + (-32 + y)^6}$$

$$+ \frac{1}{20 + (-32 + x)^6 + (-16 + y)^6} + \frac{1}{19 + (-16 + x)^6 + (-16 + y)^6} + \frac{1}{18 + x^6 + (-16 + y)^6}$$

$$+ \frac{1}{17 + (16 + x)^6 + (-16 + y)^6} + \frac{1}{16 + (32 + x)^6 + (-16 + y)^6} + \frac{1}{15 + (-32 + x)^6 + y^6}$$

$$+ \frac{1}{14 + (-16 + x)^6 + y^6} + \frac{1}{13 + x^6 + y^6} + \frac{1}{12 + (16 + x)^6 + y^6} + \frac{1}{11 + (32 + x)^6 + y^6}$$

$$+ \frac{1}{10 + (-32 + x)^6 + (16 + y)^6} + \frac{1}{9 + (-16 + x)^6 + (16 + y)^6} + \frac{1}{8 + x^6 + (16 + y)^6}$$

$$+ \frac{1}{7 + (16 + x)^6 + (16 + y)^6} + \frac{1}{6 + (32 + x)^6 + (16 + y)^6} + \frac{1}{5 + (-32 + x)^6 + (32 + y)^6}$$

$$+ \frac{1}{4 + (-16 + x)^6 + (32 + y)^6} + \frac{1}{3 + x^6 + (32 + y)^6} + \frac{1}{2 + (16 + x)^6 + (32 + y)^6}$$

$$+ \frac{1}{1 + (32 + x)^6 + (32 + y)^6})$$

Let us employ a Quasi-Newton method as a post processing method,

$\Rightarrow$ solf =

NMinimize[f,{{x, -50, 50},{y, -50, 50}},

Method $\rightarrow$ {"DifferentialEvolution","CrossProbability" $\rightarrow$ 0.3,

"PostProcess" $\rightarrow$ {FindMinimum, Method $\rightarrow$" QuasiNewton"}}]//Timing

$\Leftarrow$ {0.187201,{0.998004,{x $\rightarrow$ -31.9786, y $\rightarrow$ -31.9783}}}

The global minimum value of the function value is, see Fig. 6.15.

$\Rightarrow$ {{x, y, solf[[2, 1]]}}/.solf[[2, 2]]

$\Leftarrow$ {{-31.9786, -31.9783, 0.998004}}

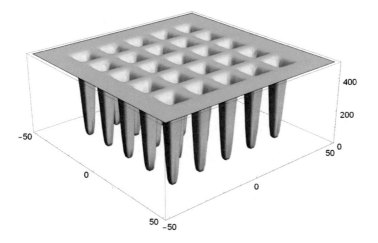

Fig. 6.14 The foxhole region

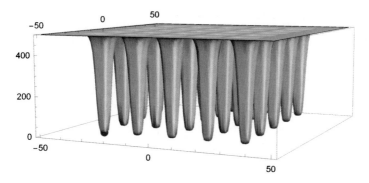

Fig. 6.15 The global minimum

References

Freeman JA (1994) Simulating Neural networks with mathematica. Addison-Wesley, New York
Wagon S (2000) Mathematica in action, 2nd edn. Springer-TELOS, New York

Chapter 7
Nature Inspired Global Optimization

7.1 Particle Swarm Optimization

7.1.1 Global Optimization

In global optimization problems, where an objective function $f(x)$ having many local minima are considered, finding the global minimizer (also known as global solution or global optimum) x^* and the corresponding f^* value is of concern (Awange et al. 2021).

Global optimization problems arise frequently in engineering, decision-making, optimal control, etc. (Awange et al. 2018). There exist two huge but almost completely disjoint communities (i.e., they have different journals, different conferences, different test functions, etc.) solving these problems; (i) a broad community of practitioners using stochastic nature-inspired meta-heuristics, and (ii), academicians studying deterministic mathematical programming methods.

To solve global optimization problems, where evaluation of the objective function is an expensive operation, one needs to construct an algorithm that is able to stop after a fixed number of evaluations M of $f(x)$, from which the lowest obtained value of $f(x)$ is used. For global optimization, it is not the dimension of the problem that is important in local optimization but rather the number of allowed function evaluations (often called budget). In other words, when one has the possibility to evaluate $f(x)$ M times (these evaluations are hereinafter called trials), in the global optimization problem of the dimension 5, 10 or 100, the quality of the found solution after M evaluations is crucial and not the dimensionality of $f(x)$. This happens because it is not possible to adequately explore the multi-dimensional search region D at this limited budget of expensive evaluations of $f(x)$. For instance, if $D \subset R20$ is a hypercube, then it has 220 vertices. This means that one million of

Supplementary Information The online version contains supplementary material available at https://doi.org/10.1007/978-3-030-92495-9_7.

trials are not sufficient not only to explore well the whole region D, but even to evaluate $f(x)$ at all vertices of D. Thus, the global optimization problem is frequently the well know *NP-hard* problem.

Meta-heuristic algorithms widely used to solve real-life global optimization problems have a number of attractive properties that have ensured their success among engineers and practitioners. First, they have limpid nature-inspired interpretations explaining how these algorithms simulate behaviors of populations of individuals. Other reasons that have led to a wide spread of meta-heuristics are the following; they do not require high level of mathematical preparation to understand them, their implementations are usually simple and many codes are freely available. Finally, they do not need a lot of memory as they work at each moment with only a limited population of points in the search domain. On the flip side, meta-heuristics have some drawbacks, which include usually a high number of parameters to tune, and the absence of rigorously proven global convergence conditions ensuring that sequences of trial points generated by these methods always converge to the global solution x^*. In fact, populations used by these methods can degenerate prematurely, returning only a locally optimal solution instead of a global one or even non-locally optimal point if it has been obtained at one of the last evaluations of $f(x)$ and the budget of M evaluations has not allowed it to proceed with an improvement of the obtained solution.

7.1.2   Particle Swarm Optimization

The idea of the Particle Swarm Optimization (PSO; Awange et al. 2021) was inspired by the social behaviors of big groups of animals, like flocking and schooling patterns of birds and fish and suggested in 1995 by *Russell Eberhart* and *James Kennedy*, see in (Yang 2010).

Let us imagine a flock of birds circling over an area where they can smell a hidden source of food. The one who is closest to the food chirps the loudest and the other birds swing around in his/her direction. If any of the other circling birds comes closer to the target than the first, it chirps louder and the others veer over toward him/her. This tightening pattern continues until one of the birds happens to land upon the food. So the motion of the individuals of the group is influenced by the motion of their neighborhood (Fig. 7.1).

Fig. 7.1 Birds are swarming
Source Awange et al (2021)

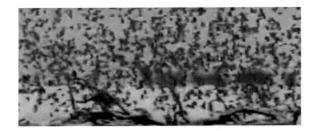

In the searching space, discrete points are defined as individuals (particles), which are characterized by their positional vectors determining their data (fitness) values to be maximized (similar to fitness function) and their velocity vectors indicating how much the data values can change, and a personal best value indicating the closest the particle's data has ever come to the optimal value (target value).

Let us consider a population of N individuals, where their location vectors are $\vec{p}_i$, the objective function to be maximized is $\mathcal{F}$ and the fitness of an individual is $\mathcal{F}(\vec{p}_i)$. These values measure the attraction of the individual, namely the higher its fitness value, the more the followers in its neighborhood following its motion. The local velocity of the individual $\vec{v}_i$, will determine its new location after Δt time step.

The velocity value is calculated according to how far an individual's data is from the target. The further it is, the larger is the velocity value. In the birds' example, the individuals furthest from the food would make an effort to keep up with the others by flying faster towards the best bird (the individual having the highest fitness value in the population). Each individual's personal best value only indicates the closest the particle's data has ever come to the target since the algorithm started.

The best bird's value only changes when any particle's personal best value comes closer to the target than best bird value. Through each iteration of the algorithm, best bird's value gradually moves closer and closer to the target until one of the particles reaches the target.

7.1.3 Optimization: Definitions and Basic Algorithm

We can give the following somewhat simplified basic algorithm of the PSO:

(1) Generate the position and velocity vectors of each particle randomly are

$$P = \left(\vec{p}_0, \ldots, \vec{p}_i, \ldots, \vec{p}_N \right) \text{ and } V = \left(\vec{v}_0, \ldots, \vec{v}_i, \ldots, \vec{v}_N \right).$$

(2) Compute the fitness of the particle:

$$\mathcal{F}(P) = \left(\mathcal{F}\left(\vec{p}_0\right), \ldots, \mathcal{F}\left(\vec{p}_i\right), \ldots, \mathcal{F}\left(\vec{p}_N\right) \right).$$

(3) Store and continuously update the position of each particle where its fitness has the highest value

$$P = \left(\vec{p}_0^{\,best}, \ldots, \vec{p}_i^{\,best}, \ldots, \vec{p}_N^{\,best} \right).$$

(4) Store and continuously update the position of the best particle where its fitness has the highest value (global best):

$$\vec{p}_g^{best} = \max_{\vec{p} \in P} \left(\mathcal{F}\left(\vec{p}\right)\right).$$

(5) Modify the velocity vectors of each particle considering its personal best and the global best

$$\vec{v}_i^{new} = \vec{v}_i + \varphi_1\left(\vec{p}_i^{best} - \vec{p}_i\right) + \varphi_2\left(\vec{p}_g^{best} - \vec{p}_i\right) \quad \text{for} \quad 1 \le i \le N$$

Remark Coefficient φ_1 ensures escaping from local optimum, since the motion of a particle does not follow the boss blindly!

(6) Update the positions with considered stepsize ($\Delta t = 1$):

$$\vec{p}_i^{new} = \vec{p}_i + \vec{v}_i^{new} \quad \text{for} \quad 1 \le i \le N.$$

(7) Stop the algorithm if there is no improvement in the objective function during a certain number of iterations or the number of iterations exceeds the limit.
(8) Go back to (2) and compute new fitness value with the new positions.

7.1.4   Illustrative Example

Let us demonstrate the algorithm with a 2D problem. Let us suppose that the local geoid can be described by the following function, see Fig. 7.2.

$$f(x, y) = -\sqrt{(6+x)^2 + (-9+y)^2} - \sqrt{(-11+x)^2 + (-3+y)^2}$$
$$- \sqrt{(11+x)^2 + (9+y)^2} + \frac{50\sin[\sqrt{(6+x)^2 + (-9+y)^2}]}{\sqrt{(6+x)^2 + (-9+y)^2}}$$
$$+ \frac{50\sin[\sqrt{(-11+x)^2 + (-3+y)^2}]}{\sqrt{(-11+x)^2 + (-3+y)^2}} + \frac{50\sin[\sqrt{(11+x)^2 + (9+y)^2}]}{\sqrt{(11+x)^2 + (9+y)^2}}$$

Fig. 7.2 The 2D objective function

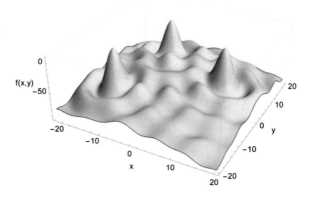

The parameters to be initialized are

N 100; (number of individuals),
λ $\{\{-20, 20\}, \{-20, 20\}\}$; (location ranges),
μ $\{\{-0.1, 0.1\}, \{-0.1, 0.1\}\}$; (velocity ranges),
M 300; (number of iterations),
φ_1 0.2; (local exploitation rate),
φ_2 2.0; (global exploitation rate).

Figure 7.3 shows the distribution of the individuals after 300 iterations, while Table 7.1 shows the results of different global methods, see location of the global optimum and its value.

Fig. 7.3 The population distribution after 300 iterations

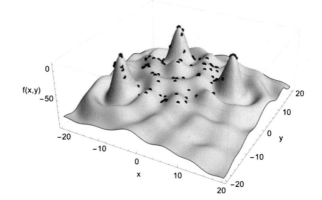

Table 7.1 The results of different global methods

Method	x-opt	y-opt	f-opt
Particle swarm	−6.01717	9.06022	10.8406
Simulated annealing	−6.01717	9.06022	10.8406
Differential evolution	−11.0469	−9.07418	5.3015
Nelder mead	−11.0469	−9.07418	5.3015

7.1.5 Variants of PSO

Although Particle Swarm Optimization (PSO) has demonstrated competitive performance in solving global optimization problems, it exhibits some limitations when dealing with optimization problems with high dimensionality and complex landscape.

Numerous variants of even a basic PSO algorithm are possible in an attempt to improve optimization performance. There are certain research trends; one is to make a hybrid optimization method using PSO combined with other optimizers such as the genetic algorithm. Another is to try and alleviate premature convergence (that is, optimization stagnation), e.g., by reversing or perturbing the movement of the PSO particles. Another approach of dealing with premature convergence is the use of multiple swarms (multi-swarm optimization).

Another school of thought is that PSO should be simplified as much as possible without impairing its performance, leading to the parameters being easier to fine tune such that they perform more consistently across different optimization problems. Initialization of velocities may require extra inputs, however, PSO variant has been proposed that does not use velocity at all.

As the PSO equations given above work on real numbers, a commonly used method to solve discrete problems is to map the discrete search space to a continuous domain, to apply a classical PSO, and then to remap the results, for example by just using rounded values.

In general, an important variant and strategy of global optimization is to employ global method to find the close neighborhood of the global optimum then apply local method to improve the result (Paláncz 2021).

7.1.6 PSO Applications in Geosciences

In this Section, some case studies are introduced to illustrate the applicability of PSO algorithm in geosciences (see also Awange et al. 2021 for more details).

(a) Particle Swarm Optimization for GNSS Network Design

The Global Navigation Satellite Systems is increasingly becoming the official tool for establishing geodetic networks. In order to meet the established aims of a geodetic network, it has to be optimized, depending on some design criteria

(Grafarend and Sansò 1985). Optimization of a GNSS network can be carried out by selecting baseline vectors from all of the probable baseline vectors that can be measured in a GNSS network. Classically, a GNSS network can be optimized using the trial and error method or analytical methods such as linear or nonlinear programming or in some cases by generalized or iterative generalized inverses. Optimization problems may also be solved by intelligent optimization techniques such as Genetic Algorithms (GAs), Simulated Annealing (SA) and Particle Swarm Optimization (PSO) algorithms. The efficiency and the applicability of PSO were demonstrated using a GNSS network, which has been solved previously using a classical method. The result shows that the PSO is effective, improving efficiency by 19.2% over the classical method (Doma 2013).

(b) **GNSS Positioning Using PSO-RBF Estimation Model**

Positioning solutions need to be more accurate, precise and obtainable at minimal effort. The most frequently used method nowadays employs a GNSS receiver, sometimes supported by other sensors. Generally, GNSS suffer from signal perturbations (e.g., from the atmosphere, nearby structures) that produce biases on the measured pseudo-ranges. With a view to optimize the use of the satellites signals received, a positioning algorithm with pseudo range error modeling with the contribution of an appropriate filtering process include, e.g., Extended Kalman Filter (EKF) and Rao Black Wellized Filtering (RBWF), which are among the most widely used algorithms to predict errors and to filter the high frequency noise. A new method of estimating the pseudo-range errors based on the PSO-RBF model is suggested by Jgouta and Nsiri (2017), which achieves an optimal training criterion. The PSO is used to optimize the parameters of neural networks with Radial Basis Function (RBF) in their work. This model offers appropriate method to predict the GNSS corrections for accurate positioning, since it reduces the positioning errors at high velocities by more than 50% compared to the RBWF or EKF methods (Jgouta and Nsiri 2017).

(c) **Using PSO to Establish a Local Geometric Geoid Model**

There exist a number of methods for approximating local geoid surfaces (i.e., equipotential surfaces approximating mean sea level) and studies carried out to determine local geoids. In Kao et al. (2017), the PSO method as a tool for modeling local geoid is presented and analyzed. The ellipsoidal heights (h), derived from GNSS observations, and known orthometric heights (H) from first-order levelling from benchmarks were first used to create local geometric geoid model. The PSO method was then used to convert ellipsoidal heights (h) into orthometric heights (H). The resulting values were compared to those obtained from spirit leveling and GNSS methods. The adopted PSO method improves the fitting of local geometric geoid by quadratic surface fitting method, which agrees with the known orthometric heights within ± 1.02 cm (Kao et al. 2017).

(d) **Application of PSO for Inversion of Residual Gravity Anomalies Over Geological Bodies with Idealized Geometries**

A global particle swarm optimization (GPSO) technique was developed and applied by Singh and Biswas (2016) for the inversion of residual gravity anomalies caused by buried bodies with simple geometry (spheres, horizontal, and vertical cylinders). Inversion parameters, such as density contrast of geometries, radius of body, depth of body, location of anomaly, and shape factor, were optimized. The GPSO algorithm was tested on noise-free synthetic data, synthetic data with 10% Gaussian noise, and five field examples from different parts of the world. Singh and Biswas (2016) showed that the GPSO method is able to determine all the model parameters accurately even when shape factor is allowed to change in the optimization problem. However, the shape was fixed a priori in order to obtain the most consistent appraisal of various model parameters. For synthetic data without noise or with 10% Gaussian noise, estimates of different parameters were very close to the actual model parameters. For the field examples, the inversion results showed excellent agreement with results from previous studies that used other inverse techniques. The computation time for the GPSO procedure was very short (less than 1 s) for a swarm size of less than 50. The advantage of the GPSO method is that it is extremely fast and does not require assumptions about the shape of the source of the residual gravity anomaly (Singh and Biswas 2016).

(e) **Application of a PSO Algorithm for Determining Optimum Well Location and Type**

Determining the optimum type and location of new wells is an essential component in the efficient development of oil and gas fields. The optimization problem is, however, demanding due to the potentially high dimension of the search space and the computational requirements associated with function evaluations, which, in this case, entail full reservoir simulations. In Onwunalu and Durlofsky (2010), the particle swarm optimization (PSO) algorithm is applied to determine optimal well type and location. Four example, cases are considered that involve vertical, deviated, and dual-lateral wells and optimization over single and multiple reservoir realizations. For each case, both the PSO algorithm and the widely used genetic algorithm (GA) are applied to maximize net present value. Multiple runs of both algorithms are performed and the results are averaged in order to achieve meaningful comparisons. It is shown that, on average, PSO outperforms GA in all cases considered, though the relative advantages of PSO vary from case to case (Onwunalu and Durlofsky 2010).

(f) **Introducing PSO to Invert Refraction Seismic Data**

Seismic refraction method is a powerful geophysical technique in near surface study. In order to achieve reliable results, processing of refraction seismic data in particular inversion stage should be done accurately. In Poormirzaee et al. (2014) refraction travel times' inversion are considered using PSO algorithm. This algorithm, being a metaheuristic optimization method, is used in many fields of geophysical data inversion, showing that the algorithm is powerful, fast and easy. For efficiency evaluation, different synthetic models are inverted. Finally PSO inversion code was investigated in a case study of apart of Tabriz city in North-West of Iran for hazard assessment, where the field dataset were inverted using the PSO code. The obtained model was compared to the geological information of the study area. The results emphasize the reliability of the PSO code to invert refraction seismic data with an acceptable misfit and convergence speed (Poormirzaee et al. 2014).

(g) **PSO algorithm to solve geophysical inverse problems: Application to a 1D-DC resistivity case**

The performance of the algorithms was first checked by Fernández-Martínez et al. (2010) using synthetic functions showing a degree of ill-posedness similar to that found in many geophysical inverse problems having their global minimum located on a very narrow flat valley or surrounded by multiple local minima. Finally, they present the application of these PSO algorithms to the analysis and solution of an inverse problem associated with seawater intrusion in a coastal aquifer in southern Spain. PSO family members were successfully compared to other well known global optimization algorithms (binary genetic algorithms and simulated annealing) in terms of their respective convergence curves and the sea water intrusion depth posterior histograms (Fernández-Martínez et al. 2010).

(h) **Anomaly shape inversion via model reduction and PSO**

Most of the inverse problems in geophysical exploration consist of detecting, locating and outlining the shape of geophysical anomalous bodies imbedded into a quasi-homogeneous background by analyzing their effects on the geophysical signature. The inversion algorithm that is currently in use creates very fine mesh in the model space to approximate the shapes and the values of the anomalous bodies and the geophysical structure of the geological background. This approach results in discrete inverse problems with a huge uncertainty space, and the common way of stabilizing the inversion consists of introducing a reference model (through prior information) to define the set of correctness of geophysical models. A different way of dealing with the high underdetermined character of these kinds of problems, consist of solving the inverse problem using a low dimensional parameterization that provides an approximate solution of the anomaly via Particle Swarm Optimization (PSO), e.g., Fernández-Muñiz et al. (2020). This method has been designed for anomaly detection in geological set-ups that correspond with this kind of problem. These authors show its application to synthetic and real cases in gravimetric inversion performing at the same time uncertainty analysis of the solution. The two different parameterizations for the geophysical anomalies (polygons and ellipses) show that similar results are obtained. This method outperforms the common least squares method with regularization (Fernández-Muñiz et al. 2020).

(i) One-dimensional forward modeling in direct current (DC) resistivity

One-dimensional forward modeling in direct current (DC) resistivity is actually computationally inexpensive, allowing the use of global optimization methods (GOMs) to solve 1.5D inverse problems with flexibility in constraint incorporation. GOMs can support computational environments for quantitative interpretation in which the comparison of solutions incorporating different constraints is a way to infer characteristics of the actual subsurface resistivity distribution. To this end, the chosen GOM must be robust to changes in the cost function and also be computationally efficient. The performance of the classic versions of the simulated annealing (SA), genetic algorithm (GA), and particle swarm optimization (PSO) methods for solving the 1.5D DC resistivity inverse problem was compared using synthetic and field data by Barboza et al. (2018). The main results were as follows: (1) All methods reproduced synthetic models quite well, (2) PSO and GA were comparatively more robust to changes in the cost function than SA, (3) PSO first and GA second presented the best computational performances, requiring less forwarding modeling than SA, and (4), GA gave higher performance than PSO and SA with respect to the final attained value of the cost function and its standard deviation. To put them into effective operation, the methods were classified from easy to difficult in the order PSO, GA, and SA as a consequence of robustness to changes in the cost function and of the underlying simplicity of the associated equations. To exemplify a quantitative interpretation using GOMs, these solutions were compared with least-absolute and least-squares norms of the discrepancies derived from the lateral continuity constraints of the log-resistivity and layer depth as a manner of detecting faults. GOMs additionally provided the important benefit of furnishing not only the best solution but also a set of suboptimal quasi solutions from which uncertainty analyses can be performed (Barboza et al. 2018).

7.2 Black Hole Optimization

The *Black Hole (BH) algorithm* is one of the optimization techniques inspired by the behavior of the real black holes. The black hole terminology is used to improve the convergence rate and efficiency of the Particle Swarm Optimization (PSO) algorithm. In recent years, the BH algorithm and its modified versions have been used to solve optimization and engineering problems, see (Hatamlou 2012).

In the eighteenth-century, *John Mitchell* and *Pierre Laplace* pioneered the concept of black holes. Integrating Newton's law, they formulated the theory of a star becoming invisible to the eye, however, during that period it was not known as a black hole and it was only in 1967 that *John Wheeler,* the American physicist coined the phenomenon of mass collapsing as a black hole.

A black hole in space is what forms when a star of massive size collapses. The gravitational power of the black hole is too high that even the light cannot escape from it. The gravity is so strong because matter has been squeezed into a tiny space.

Anything that crosses the boundary of the black hole will be swallowed by it and vanish and nothing can get away from its enormous powers. The sphere-shaped boundary of a black hole in space is known as the *event horizon*. The radius of the event horizon is termed as the *Schwarzschild radius*. At this radius, the escape speed is equal to the speed of light, and once light passes through, even it cannot be spared. Nothing can escape from within the event horizon because nothing can go faster than light. The Schwarzschild radius is calculated by the following equation:

$$R = \frac{GM}{c^2},$$

where G is the gravitational constant, M is the mass of the black hole, and c is the speed of light. If anything moves close to the event horizon or crosses the Schwarzschild radius, it will be absorbed into the black hole and permanently disappear. The existence of black holes can be discerned by its effect over the objects surrounding it.

7.2.1   Black Hole Algorithm

The BH algorithm is a population-based method that has some common features with other population-based methods. As with other population-based algorithms, a population of candidate solutions to a given problem is generated and distributed randomly in the search space. The population-based algorithms evolve the created population towards the optimal solution via certain mechanisms. For example, in Genetic Algorithm (GA), the evolving is done by mutation and crossover operations. In PSO, this is done by moving the candidate solutions around in the search space using the best found locations, which are updated as better locations are found by the candidates. In the proposed BH algorithm, the evolving of the population is done by moving all the candidates towards the best candidate in each iteration, namely, the black hole and replacing those candidates that enter within the range of the black hole by newly generated candidates in the search space.

The details of the BH algorithms are as follows: Like other population—based algorithms, in the proposed black hole algorithm (BH) a randomly generated population of candidate solutions—the stars—are placed in the search space of some problem or function. After initialization, the fitness values of the population are evaluated and the best candidate in the population, which has the best fitness value, is selected to be the black hole and the rest form the normal stars.

The black hole has the ability to absorb the stars that surround it. The absorption of stars by the black hole is formulated as follows:

$$x_i(t+1) = x_i(t) + \text{rand}(x_{BH} - x_i(t)), \quad i = 1, 2, \ldots N,$$

where $x_i(t)$ and $x_i(t+1)$ are the locations of the i-th star at iterations t and $t + 1$, respectively. x_{BH} is the location of the black hole in the search space and rand is a random number in the interval [0, 1]. N is the number of stars (candidate solutions).

While moving towards the black hole, a star may reach a location with a higher fitness value than the black hole. In such a case, the black hole moves to the location of that star and vice versa. Then the BH algorithm will continue with the black hole in the new location and then stars start moving towards this new location.

In addition, there is the probability of crossing the event horizon during moving stars towards the black hole. Every star (candidate solution) that crosses the event horizon of the black hole will be sucked by the black hole. Every time a candidate (star) dies − it is sucked in by the black hole − another candidate solution (star) is born and distributed randomly in the search space and starts a new search. This is done to keep the number of candidate solutions constant. The next iteration takes place after all the stars have been moved.

The radius of the event horizon in the black hole algorithm is calculated using the following equation:

$$R = \frac{f_{BH}}{\sum\limits_{i=1}^{N} f_i},$$

where f_{BH} is the fitness value of the black hole and f_i is the fitness value of the i-th star. When the distance between a candidate solution and the black hole (best candidate) is less than R, that candidate is collapsed and a new candidate is created and distributed randomly in the search space.

The key idea behind this step is the following:

If we would continue the attraction-absorption process then all the stars will be closer and closer to the actual black hole, the actual local fitness maximum, then local maximum can be reached, and we would stacked in this local maximum.

However, we are looking for global maximum and that is why we eliminate this dense population around the local maximum via swallowing them by the local black hole. Newborn stars distributed randomly will replace the swallowed stars. This technique ensures the escape from the local maximum.

Now let us carry out these steps in a simple case:
Number of stars

⇒ ns = 5;

We are looking for the minimum of the function $y(x) = x^2$ in the search space (Fig. 7.4) $x \epsilon [-1, 1]$.

```
⇒ p = Plot[x^2,{x,−1,1},AxesLabel → {"x","y"},ImageSize → 300,
    Frame → True]
```

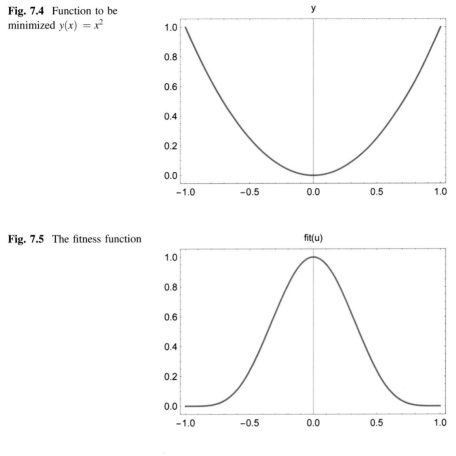

Fig. 7.4 Function to be minimized $y(x) = x^2$

Fig. 7.5 The fitness function

Let define the fitness function (see Fig. 7.5)

$\Rightarrow$ fit[u_] := $(1 - u^2)^5$

Now, let us generate the locations of the stars randomly,

$\Rightarrow$ stars = {0.19451299585105497, -0.19051000373410165,
0.7851016190620821, -0.29885053165836073, -0.9320376959349024}
$\Leftarrow$ {0.194513, -0.19051, 0.785102, -0.298851, -0.932038}

- *First iteration*:

(A) Figure 7.6 shows the function to be minimized, the fitness function, and the actual location of the stars.

Fig. 7.6 The positions of the generated stars (green points)

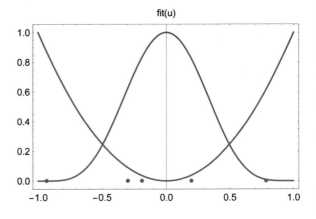

The fitness values are

⇒ starsfit = Map[fit[#]&, stars]
⇐ {0.824607, 0.831233, 0.00830769, 0.626396, 0.0000390318}

The star having the best fitness value is selected as black hole

⇒ blackhole = First[Position[starsfit, Max[starsfit]]//Flatten]
⇐ 2

Its fitness,

⇒ opt = fit[stars[[blackhole]]]
⇐ 0.831233

and the location

⇒ optloc = stars[[blackhole]]
⇐ −0.19051

The radius of the event horizon,

⇒ radius = starsfit[[blackhole]]/Total[starsfit]
⇐ 0.362891

Let us see this situation (see Fig. 7.7),

Fig. 7.7 The stars and the black hole (•)

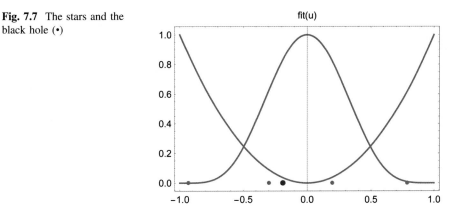

Now, the distances of the stars are computed from the black hole

$\Rightarrow$ popdist $=$ Map$[\sqrt{\text{Norm}[\{\text{optloc}-\#\}]}\&,\text{stars}]$

$\Leftarrow \{0.620502, 0., 0.987731, 0.329151, 0.86112\}$

(the second distance is 0 between the two black holes). The swallowed stars,

$\Rightarrow$ swallowed $=$ Complement$[$Map$[$Position$[$popdist,$\#]\&,$
Select$[$popdist,$\# < = $radius$\&]]//$Flatten,$\{$blackhole$\}]$

$\Leftarrow \{4\}$

(B) If this set is not empty, then the stars inside of the radius of the event horizon will disappear. The stars remained

$\Rightarrow$ remainedstars $=$ Complement$[$stars, Map$[$stars$[[\#]]\&,$ swallowed$]]$

$\Leftarrow \{-0.932038, -0.19051, 0.194513, 0.785102\}$

the swallowed stars will be randomly replaced by new stars

$\Rightarrow$ newstars $=$ Union$[$remainedstars, Table$[$RandomReal$[\{-1,1\}],$
$\{$i, 1, Length$[$swallowed$]\}]]$

$\Leftarrow \{-0.932038, -0.19051, 0.194513, 0.753217, 0.785102\}$

The Fig. 7.8 shows the new situation after one iteration
Then the star population will be refreshed,

$\Rightarrow$ stars $=$ newstars; and back to (A).

- **Second iteration**:

The fitness values are

$\Rightarrow$ starsfit $=$ Map$[$fit$[\#]\&,$ stars$]$

$\Leftarrow \{0.0000390318, 0.831233, 0.824607, 0.0151619, 0.00830769\}$

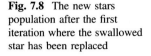

Fig. 7.8 The new stars
population after the first
iteration where the swallowed
star has been replaced

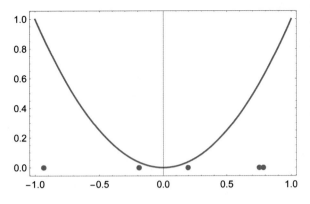

The star having the best fitness value selected as black hole

$\Rightarrow$ blackhole = First[Position[starsfit, Max[starsfit]]//Flatten]
$\Leftarrow$ 2

Its fitness,

$\Rightarrow$ opt = fit[stars[[blackhole]]]
$\Leftarrow$ 0.831233

and the location

poptloc = stars[[blackhole]]
$\Leftarrow$ -0.19051

The radius of the event horizon,

$\Rightarrow$ radius = starsfit[[blackhole]]/Total[starsfit]
$\Leftarrow$ 0.494973

Let us assess this situation, (Fig. 7.9)

Now we compute the distances of the stars from the black hole.

$\Rightarrow$ rpopdist = Map[$\sqrt{\text{Norm}[\{\text{optloc}-\#\}]}$&, stars]
$\Leftarrow$ {0.86112, 0., 0.620502, 0.971456, 0.987731}

Fig. 7.9 The new black hole
of the new stars population

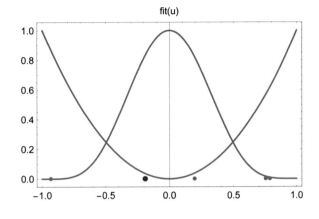

The swallowed stars,

⇒ swallowed = Complement[Map[Position[popdist,#]&,
    Select[popdist,#< = radius&]]//Flatten,{blackhole}]]

⇐ {}

(C) If this set is empty, no stars will be swallowed then the attracted stars move
    towards the black hole. Their new positions are

⇒ newpositions =
    Map[(# + RandomReal[ ](stars[[blackhole]]−#))&, stars]
    ⇐ {−0.544987, −0.19051, 0.0490078, 0.742228, −0.11955}

Then the star population will be refreshed

⇒ stars = newpositions
    ⇐ {−0.544987, −0.19051, 0.0490078, 0.742228, −0.11955}

and back to (B).

The number of the iterations can be given directly or can be controlled in other
ways. In general an iteration process is controlled by the convergency criterium
measured by the differences of two subsequent results and overruled by the number
of iterations, since we do not want infinite loop in case no convergency occured.
This is the basic principle, but any other logical condition can be freely defined.

In the code below it can easily be realized that we used here the number of
iterations as control.

7.2.2   Code for the 1D Algorithm

Here is a simple function for this algorithm,

```
⇒ BlackHole01[fit_,xmin_,xmax_,ns_,ni_] : =
   Module[{stars,starsfit,blackhole,opt,optloc,radius,popdist,
   swallowed,newpositions,remainedstars,newstars,beg1,sol,i,j,n},
   sol ={}; n = 1;
   stars = Table[RandomReal[{xmin,xmax}],{i,1,ns}];
   While[n < ni,
    starsfit = Map[fit[#]&, stars];
    blackhole =
    First[Position[starsfit,Max[starsfit]]//Flatten];
    opt = fit[stars[[blackhole]]];
    optloc = stars[[blackhole]];
    radius = starsfit[[blackhole]]/Total[starsfit];
    popdist = Map[√Norm[{optloc−#}]&, stars];
    swallowed =
    Complement[
    Map[Position[popdist,#]&,
    Select[popdist,# < = radius&]]//Flatten,
    {blackhole}];
    If[swallowed =={},
    newpositions =
     Map[(# + RandomReal[ ](stars[[blackhole]]−#))&, stars];
    stars = newpositions,
    remainedstars = Complement[stars,Map[stars[[#]]&, swallowed]];
    newstars = Union[remainedstars,
     Table[RandomReal[{xmin,xmax}],{i,1,Length[swallowed]}]];
    stars = newstars];
   AppendTo[sol,optloc]; n + +]; sol]
```

where

fit	fitness function should be maximized,
xmin, xmax	boundary of the search space,
ns	number of stars,
ni	number of iterations.

Now, let us see the 2D code.

7.2.3   *Code for the 2D Algorithm*

```
⇒ BlackHole02[fit_,xmin1_,xmin2_,xmax1_,xmax2_,ns_,ni_] :=
    Module[{stars,starsfit,blackhole,opt,optloc,radius,popdist,
     swallowed,newpositions,remainedstars,newstars,beg1,sol,i,j,n},
     sol ={}; n = 1;
     stars =
     stars =
      Partition[
       Flatten[Table[{RandomReal[{xmin1,xmax1}],
        RandomReal[{xmin2,xmax2}]},{i,1,ns}]],2];
     While[n<ni,
      starsfit = Map[fit[#]&,stars];
      blackhole = First[Position[starsfit,Max[starsfit]]//Flatten];
      opt = fit[stars[[blackhole]]];
      optloc = stars[[blackhole]];
      radius = starsfit[[blackhole]]/Total[starsfit];
      popdist = Map[√Norm[{optloc−#}]&,stars];
      swallowed =
       Complement[
        Map[Position[popdist,#]&,Select[popdist,#<= radius&]]//
         Flatten,{blackhole}];
      If[swallowed =={},
       newpositions = Map[(#+RandomReal[](stars[[blackhole]]−#))&,
        stars];
       stars = newpositions,
       remainedstars = Complement[stars,Map[stars[[#]]&,swallowed]];
       newstars = Union[remainedstars,
        Table[{RandomReal[{xmin1,xmax1}],
         RandomReal[{xmin2,xmax2}]},{i,1,Length[swallowed]}]];
       stars = newstars];
      AppendTo[sol,optloc]; n++]; sol]
```

where

fit	fitness function should be maximized,
*xmin*1, *xmin*2	boundaries of the search space,
*xmax*1, *xmax*2	boundaries of the search space,
ns	number of stars,
ni	number of iterations.

7.3 Test Examples

Let us test the 1D BH algorithm on two functions $f_1(x)$ and $f_2(x)$. Four methods will be employed. Three of them, Simulated Annealing, Differential Evolution and the Random Search are built in functions, which are compared to the Code of *BlackHole*01.

7.3.1 Example 1

$\Rightarrow$ f1[x_] : = $-$Abs$[2(x - 24) + (x - 24)$Sin$[(x - 24)]]$;
    See Fig. 7.10.

7.3.1.1 Simulated Annealing

$\Rightarrow$ NMaximize$[\{$f1$[x], -20 \le x \le 60\}, x,$ Method $\rightarrow''$ SimulatedAnnealing$'']$
$\Leftarrow \{-1.93135 \times 10^{-8}, \{x \rightarrow 24.\}\}$

Running time

$\Rightarrow$ AbsoluteTiming$[$
        sol $=$ Reap$[$NMaximize$[\{$f1$[x], -20 \le x \le 60\}, x,$
        EvaluationMonitor $: \rightarrow$ Sow$[\{x\}],$ Method $\rightarrow''$ SimulatedAnnealing$'']];]$
$\Leftarrow \{1.64279, $Null$\}$

Function value and the solution

$\Rightarrow$ sol$[[1]]$
$\Leftarrow \{-1.93135 \times 10^{-8}, \{x \rightarrow 24.\}\}$

Fig. 7.10 The function of $f_1(x\_)$

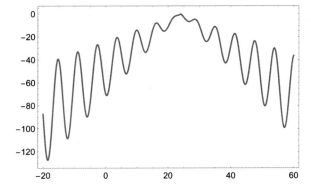

Number of iterations

⇒ {hist} = sol[[2]];

⇒ Length[hist]

⇐ 66

7.3.1.2   Differential Evolution

⇒ NMaximize[{f1[x], −20 ≤ x ≤ 60}, x, Method →″ DifferentialEvolution″]

⇐ {−1.97021 × 10⁻⁸, {x → 24.}}

Running time

⇒ AbsoluteTiming[
   sol = Reap[NMaximize[{f1[x], −20 ≤ x ≤ 60}, x,
    EvaluationMonitor : → Sow[{x}],
    Method →″ DifferentialEvolution″]];]

⇐ {1.62727, Null}

Function value and the solution

⇒ sol[[1]]]]

⇐ {−1.97021 × 10⁻⁸, {x → 24.}}

Number of iterations

⇒ {hist} = sol[[2]];

⇒ Length[hist]

⇐ 615

7.3.1.3   Random Search

⇒ NMaximize[{f1[x], −20 ≤ x ≤ 60}, x, Method →″ Randomsearch″]

⇐ {−2.00897 × 10⁻⁸, {x → 24.}}

Running time

⇒ AbsoluteTiming[
   sol = Reap[NMaximize[{f1[x], −20 ≤ x ≤ 60}, x,
    EvaluationMonitor : → Sow[{x}], Method →″ RandomSearch″]];]

⇐ {1.8846, Null}

Function value and the solution

$\Rightarrow \texttt{sol}[[1]]]$
$\Leftarrow \{-2.00897 \times 10^{-8}, \{x \rightarrow 24.\}\}$

Number of iterations

$\Rightarrow \{\texttt{hist}\} = \texttt{sol}[[2]];$
$\Rightarrow \texttt{Length}[\texttt{hist}]$
$\Leftarrow 115$

7.3.1.4   Black Hole

$\Rightarrow s = \texttt{AbsoluteTiming}[\texttt{BlackHole01}[f1, -20, 60, 100, 10]];$

Running time

$\Rightarrow s[[1]]$
$\Leftarrow 0.0122811$

Function value and the solution (Fig. 7.11)

$\Rightarrow \texttt{Last}[s[[2]]]$
$\Leftarrow 24.$
$\Rightarrow f1[\%]$
$\Leftarrow -0.0000939287$

The Table 7.2 shows the results. The superiority of the BH algorithm can be clearly seen even taking into consideration that the built-in methods are written in compiled code but have big overhead while the BH Code is written in very simple interpretable mode.

Fig. 7.11 The iteration performance of the BH algorithm $f_1(x)$

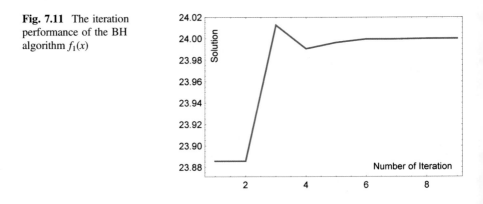

Table 7.2 Results of the different algorithms on $f_1(x)$

Method	Solution	Function value	Number of iteration	Time [s]
Simulated annealing	24	-1.93×10^{-8}	66	1.64
Differential evolution	24	-1.97×10^{-8}	615	1.63
Random search	24	-2.00×10^{-8}	115	1.88
Black hole	24	-9392.87×10^{-8}	10	1.23×10^{-2}

7.3.2   *Example 2*

⇒ f2[x_] :=Sin[N[Pi]x]*Mod[9x, 1];
  See Fig. 7.12.

7.3.2.1   **Simulated Annealing**

⇒ NMaximize[{f2[x], 0 ≤ x ≤ 1}, x, Method →" SimulatedAnnealing"]
⇐ {0.866025,{x → 0.333333}}

Running time

⇒ AbsoluteTiming[
     sol = Reap[NMaximize[{f2[x], 0 ≤ x ≤ 1}, x,
         EvaluationMonitor : → Sow[{x}], Method →" SimulatedAnnealing"]];]
⇐ {0.158268, Null}

Function value and the solution

⇒ sol[[1]]
⇐ {0.866025,{x → 0.333333}}

Fig. 7.12 The function of $f_2(x\_)$

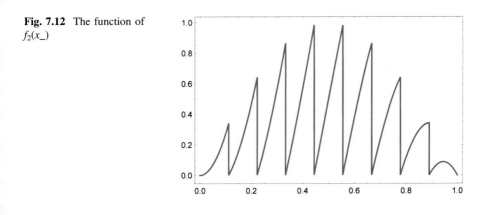

Number of iterations

⇒ {hist}= sol[[2]];
⇒ Length[hist]
⇐ 525

7.3.2.2   Differential Evolution

⇒ NMaximize[{f2[x], 0 ≤ ≤ 1}, x, Method →″DifferentialEvolution″]
⇐ {0.984808, {x → 0.555556}}

Running time

⇒ AbsoluteTiming[
   sol = Reap[NMaximize[{f2[x], 0 ≤ x ≤ 1}, x,
    EvaluationMonitor : → Sow[{x}],
    Method →″DifferentialEvolution″]];]
⇐ {0.237173, Null}

Function value and the solution

⇒ sol[[1]]
⇐ {0.984808, {x → 0.555556}}

Number of iterations

⇒ {hist}= sol[[2]];
⇒ Length[hist]
⇐ 1024

7.3.2.3   Random Search

⇒ NMaximize[{f2[x], 0 ≤ x ≤ 1}, x, Method →″RandomSearch″]
⇐ {0.984808, {x → 0.555556}}

Running time

⇒ AbsoluteTiming[
   sol = Reap[NMaximize[{f2[x], 0 ≤ x ≤ 1}, x,
    EvaluationMonitor : → Sow[{x}], Method →″RandomSearch″]];]
⇐ {0.132212, Null}

Function value and the solution

⇒ sol[[1]]

⇐ {0.984808,{x → 0.555556}}

Number of iterations

⇒ {hist}= sol[[2]];

⇒ Length[hist]

⇐ 52

7.3.2.4   Black Hole

⇒ s = AbsoluteTiming[BlackHole01[f2, 0, 1, 100, 10]];

Running time

⇒ s[[1]]

⇐ 0.0082624

Function value and the solution (Fig. 7.13 and Table 7.3).

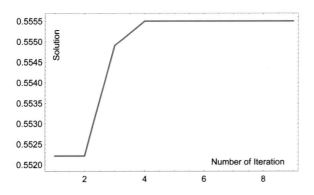

Fig. 7.13 The iteration performance of the BH algorithm on $f_2(x)$

Table 7.3 Results of the different algorithms on $f_2(x)$

Method	Solution	Function value	Number of iteration	Time [s]
Simulated annealing	0.333	0.866	525	0.158
Differential evolution	0.555	0.895	1024	0.237
Random search	0.555	0.895	52	0.132
Black hole	0.555	0.895	10	0.008

⇒ Last[s[[2]]]
⇐ 0.5555
⇒ f1[%]
⇐ 0.984345

7.3.3   *Example 3*

We are looking for the global maximum of our first 2D test function, see Fig. 7.14.

Let

$$\Rightarrow \mathtt{g}[\{\mathtt{u\_},\mathtt{v\_}\}] := \frac{50\mathrm{Sin}[\sqrt{\mathtt{u}\wedge 2 + \mathtt{v}\wedge 2}]}{\sqrt{\mathtt{u}\wedge 2 + \mathtt{v}\wedge 2}} - \sqrt{\mathtt{u}\wedge 2 + \mathtt{v}\wedge 2}$$

Then

$$\Rightarrow \mathtt{ff}[\{\mathtt{u\_},\mathtt{v\_}\}] := \mathtt{g}[\{\mathtt{u}+11,\mathtt{v}+9\}] + \mathtt{g}[\{\mathtt{u}-11,\mathtt{v}-3\}] + \mathtt{g}[\{\mathtt{u}+6,\mathtt{v}-9\}]$$

7.3.3.1   Simulated Annealing

Let us employ built in function

⇒ s = AbsoluteTiming[
    NMaximize[{ff[{u, v}], −20 ≤ u ≤ 20, −20 ≤ v ≤ 20}, {u, v},
    Method →″SimulatedAnnealing″]]
⇐ {0.11415, {10.8432, {u → −6.01717, v → 9.06022}}}

Fig. 7.14 The first 2D test function

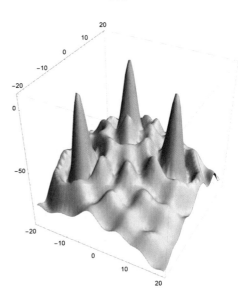

Running time

```
⇒ AbsoluteTiming[
    sol = Reap[NMaximize[{ff[{u,v}], −20 ≤ u ≤ 20, −20 ≤ v ≤ 20}, {u,v},
      EvaluationMonitor : → Sow[{u,v}],
      Method →″ SimulatedAnnealing″]]; ]
⇐ {0.186414, Null}
```

Function value and the solution

```
⇒ sol[[1]]
⇐ {10.8432, {u → −6.01717, v → 9.06022}}
```

Number of iterations

```
⇒ {hist} = sol[[2]];
⇒ Length[hist]
⇐ 305
```

7.3.3.2   Differential Evolution

Let us employ built in function

```
⇒ s = AbsoluteTiming[
    NMaximize[{ff[{u,v}], −20 ≤ u ≤ 20, −20 ≤ v ≤ 20}, {u,v},
      Method →″ DifferentialEvolution″]]
⇒ {0.867097, {5.73015, {u → −11.0469, v → −9.07418}}}
```

Running time

```
⇒ AbsoluteTiming[
    sol = Reap[NMaximize[{ff[{u,v}], −20 ≤ u ≥ 20, −20 ≤ v ≤ 20}, {u,v},
      EvaluationMonitor : → Sow[{u,v}],
      Method →″ DifferentialEvolution″]]; ]
⇐ {0.84417, Null}
```

Function value and the solution

```
⇒ sol[[1]]
⇐ {5.73015, {u → −11.0469, v → −9.07418}}
```

Number of iterations

```
⇒ {hist}= sol[[2]];
⇒ Length[hist]
⇐ 2029
```

7.3.3.3 Random Search

Let us employ built in function

$\Rightarrow$ s = AbsoluteTiming[
    NMaximize[{ff[{u,v}], $-20 \le u \le 20$, $-20 \le v \le 20$}, {u,v},
    Method $\rightarrow$" RandomSearch"]]
$\Leftarrow$ {0.258476, {10.8432, {u $\rightarrow$ -6.01717, v $\rightarrow$ 9.06022}}}

Running time

$\Rightarrow$ AbsoluteTiming[
    sol = Reap[NMaximize[{ff[{u,v}], $-20 \le u \le 20$, $-20 \le v \le 20$}, {u,v},
    EvaluationMonitor : $\rightarrow$ Sow[{u,v}], Method $\rightarrow$" RandomSearch"]];]
$\Leftarrow$ {0.356922, Null}

Function value and the solution

$\Rightarrow$ sol[[1]]
$\Leftarrow$ {10.8432, {u $\rightarrow$ -6.01717, v $\rightarrow$ 9.06022}}

Number of iterations

$\Rightarrow$ {hist}= sol[[2]];
$\Rightarrow$ Length[hist]
$\Leftarrow$ 95

7.3.3.4 Black Hole

$\Rightarrow$ s = AbsoluteTiming[BlackHole02[ff, -20, -20, 20, 20, 150, 30]];
$\Leftarrow$ {0.258476, {10.8432, {u $\rightarrow$ -6.01717, v $\rightarrow$ 9.06022}}}

Running time

$\Rightarrow$ s[[1]]
$\Leftarrow$ 0.1267581
Function value and the solution (Figs. 7.15 and 7.16, Table 7.4)

$\Rightarrow$ Last[s[[2]]]
$\Leftarrow$ {-11.0469, -9.07414}

$\Rightarrow$ ff[%]
$\Leftarrow$ 5.73015

Fig. 7.15 The iteration performance of the BH algorithm on the 2D test function

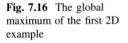

Fig. 7.16 The global maximum of the first 2D example

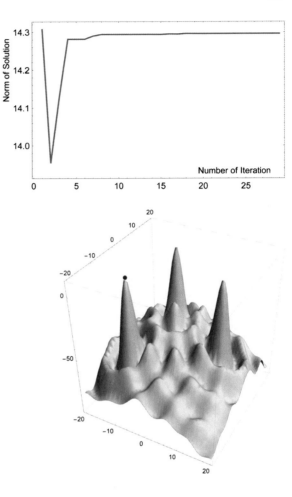

Table 7.4 Results of the different algorithms on the first 2D test function

Method	Solution	Function value	Number of iteration	Time [s]
Simulated annealing	{−6.017, 9.063}	10.8432	305	0.186
Differential evolution	{−11.047, −9.074}	5.7300	2029	0.844
Random search	{−6.017, 9.063}	10.8432	95	0.357
Black hole	{−6.017, 9.063}	10.8432	30	0.127

7.3.4   Example 4

Our second 2D example of the egg-cup function, see Figs. 7.17 and 7.18.

$$\Rightarrow gg[\{x\_, y\_\}] := Exp[-x^2 - y^2] + Sin[5x]Sin[5y];$$

Fig. 7.17 The second 2D
example, egg-cup function

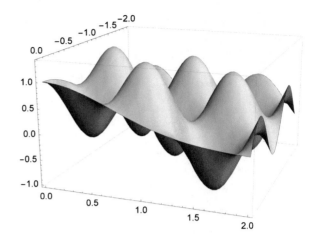

Fig. 7.18 The contour plot of
the egg-cup function

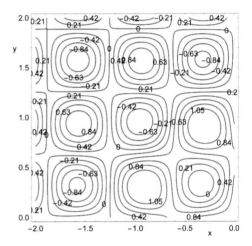

7.3.4.1 Simulated Annealing

Let us employ built in function

$\Rightarrow$ s = AbsoluteTiming[
   NMinimize[{gg[{u, v}]}, $-2 \le u \le 0, 0 \le v \le 2$}, {u, v},
   Method $\rightarrow$ "SimulatedAnnealing"]]

$\Leftarrow$ {0.0185003, {-0.924347, {u $\rightarrow -0.316045$, v $\rightarrow 1.58022$}}}

Running time

```
⇒ AbsoluteTiming[
    sol = Reap[NMinimize[{gg[{u,v}], −2 ≤ u ≤ 0, 0 ≤ v ≤ 2},{u,v},
      EvaluationMonitor : → Sow[{u,v}],
      Method →" SimulatedAnnealing"]]; ]
⇐ {0.0182054, Null}
```

Function value and the solution

```
⇒ sol[[1]]
⇐ {−0.924347,{u → −0.316045, v → 1.58022}}
```

Number of iterations

```
⇒ {hist}= sol[[2]];
⇒ Length[hist]
⇐ 55
```

7.3.4.2   Differential Evolution

Let us employ built in function

```
⇒ s = AbsoluteTiming[NMinimize[{gg[{u,v}], −2 ≤ u ≤ 0, 0 ≤ v ≥ 2},
    {u,v}, Method →" DifferentialEvolution"]]
⇐ {0.206257,{−0.992828,{u → −1.5717, v → 1.5717}}}
```

Running time

```
⇒ AbsoluteTiming[
    sol = Reap[NMinimize[{gg[{u,v}], −2 ≤ u ≤ 0, 0 ≤ v ≤ 2},{u,v},
      EvaluationMonitor : → Sow[{u,v}],
      Method →" DifferentialEvolution"]]; ]
⇐ {0.172884, Null}
```

Function value and the solution

```
⇒ sol[[1]]
⇐ {−0.992828,{u → −1.5717, v → 1.5717}}
```

Number of iterations

```
⇒ {hist}= sol[[2]];
⇒ Length[hist]
⇐ 1629
```

7.3.4.3 Random Search

Let us employ built in function

$\Rightarrow$ s = AbsoluteTiming[NMinimize[{gg[{u, v}], $-2 \leq u \leq 0, 0 \leq v \leq 2$},
    {u, v}, Method $\rightarrow$ " RandomSearch"]]
$\Leftarrow$ {0.196878, {-0.992828, {u $\rightarrow -1.5717$, v $\rightarrow 1.5717$}}}

Running time

$\Rightarrow$ AbsoluteTiming[
    sol = Reap[NMinimize[{gg[{u, v}], $-2 \leq u \leq 0, 0 \leq v \leq 2$}, {u, v},
        EvaluationMonitor : $\rightarrow$ Sow[{u, v}], Method $\rightarrow$ " RandomSearch"]];]
$\Leftarrow$ {0.118955, Null}

Function value and the solution

$\Rightarrow$ sol[[1]]
$\Leftarrow$ {-0.992828, {u $\rightarrow -1.5717$, v $\rightarrow 1.5717$}}

Number of iterations

$\Rightarrow$ {hist}= sol[[2]];
$\Rightarrow$ Length[hist]
$\Leftarrow$ 90

7.3.4.4 Black Hole

$\Rightarrow$ ggB[{u_, v_}] := $-$gg[{u, v}]
$\Rightarrow$ s = AbsoluteTiming[BlackHole02[ggB, $-2, 0, 0, 2, 150, 20$]];

Running time

$\Rightarrow$ s[[1]]
$\Leftarrow$ 0.034576

Function value and the solution (Figs. 7.19, 7.20, and 7.21, Table 7.5).

$\Rightarrow$ Last[s[[2]]]
$\Leftarrow$ {$-1.5717, 1.5717$}

$\Rightarrow$ gg[%]
$\Leftarrow -0.992828$

Fig. 7.19 The iteration performance of the BH algorithm on the second 2D (egg-cup) example

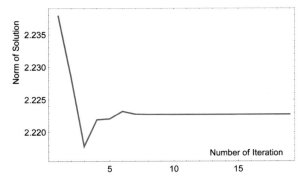

Fig. 7.20 The solutions of the second 2D (egg-cup) test problem

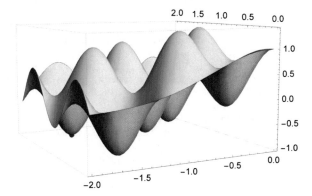

Fig. 7.21 The contour plot of the second 2D (egg-cup) test problem

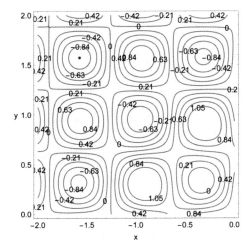

Table 7.5 Results of the different algorithms on the second 2D (egg-cup) example

Method	Solution	Function value	Number of iteration	Time [s]
Simulated annealing	{−0.316, 1.58}	-0.924	55	0.018
Differential evolution	{−1.572, 1.572}	-0.993	1629	0.172
Random search	{−1.572, 1.572}	-0.993	90	0.119
Black hole	{−1.572, 1.572}	-0.993	20	0.034

References

Abdulwahab HA, Noraziah A, Alsewari AA, Salih SQ (2019) An enhanced version of black hole algorithm via levy flight for optimization and data clustering problems. IEEE Access 7:142085–142096

Awange JL, Paláncz B, Lewis RH, Völgyesi L (2018) Mathematical geosciences: hybrid symbolic-numeric methods. Springer International Publishing, Cham, ISBN: 798-3-319-67370-7, p 596

Awange J, Paláncz B, Völgyesi L (2021) Particle swarm optimization in geosciences. In: Daya Sagar B, Cheng Q, McKinley J, Agterberg F (eds) Encyclopedia of mathematical geosciences. Encyclopedia of earth sciences series. Springer, Cham. https://doi.org/10.1007/978-3-030-26050-7_240-1

Barboza FM, Medeiros WE, Santana JM (2018) Customizing constraint incorporation in direct current resistivity inverse problems: a comparison among three global optimization methods. Geophysics 83:E409–E422

Doma MI (2013) Particle swarm optimization in comparison with classical optimization for GPS network design. J Geodetic Sci 3:250–257. https://doi.org/10.2478/jogs-2013-0030

Fernández-Martínez JL, Luis J, Gonzalo E, Fernandez P, Kuzma H, Omar C (2010) PSO: a powerful algorithm to solve geophysical inverse problems: application to a 1D-DC resistivity case. J Appl Geophys 71:13–25. https://doi.org/10.1016/j.jappgeo.2010.02.001

Fernández-Muñiz Z, Pallero JL, Fernández-Martínez JL (2020) Anomaly shape inversion via model reduction and PSO. Comput Geosci 140. https://doi.org/10.1016/j.cageo.2020.104492

Grafarend EW, Sansò F (1985) Optimization and design of geodetic networks. Springer Berlin Heidelberg. https://doi.org/10.1007/978-3-642-70659-2

Hatamlou A (Feb 2012) Black hole: a new heuristic optimization approach for data clustering. Inf Sci 222:175–184

Jgouta M, Nsiri B (2017) GNSS positioning performance analysis using PSO-RBF estimation model. Transp Tele-Commun 18(2):146–154. https://doi.org/10.1515/ttj-2017-0014

Kao S, Ning F, Chen CN, Chen CL (2017) Using particle swarm optimization to establish a local geometric geoid model. Boletim de Ciências Geodésicas 23. https://doi.org/10.1590/s1982-21702017000200021

Onwunalu JE, Durlofsky LJ (2010) Application of a particle swarm optimization algorithm for determining optimum well location and type. Computtational Geosci 14:183–198. https://doi.org/10.1007/s10596-009-9142-1

Paláncz B (2021) A variant of the black hole optimization and its application to nonlinear regression. https://doi.org/10.13140/RG.2.2.28735.43680, https://www.researchgate.net/project/Mathematical-Geosciences-A-hybrid-algebraic-numerical-solution

Poormirzaee R, Moghadam H, Zarean A (Sept 2014) Introducing particle swarm optimization (PSO) to invert refraction seismic data. Conference proceedings, near surface geoscience 2014–20th European meeting of environmental and engineering geophysics, vol 2014, pp. 1–5. https://doi.org/10.3997/2214-4609.20141978

Singh A, Biswas A (2016) Application of global particle swarm optimization for inversion of residual gravity anomalies over geological bodies with idealized geometries. Nat Resour Res 25:297–314. https://doi.org/10.1007/s11053-015-9285-9

Yang X (2010) Natured-inspired metaheuristic algorithms, 2nd edn. Luniver press, Frome, United Kingdom

Chapter 8
Integer Programing

8.1 The Integer Problem

Frequently there exist optimization problems where integer valued solutions are desired, e.g., in the solution of the ambiguity problem of the Global Navigation Satellite System (GNSS). Let us consider a simple production planning example where a furniture factory produces three products (bookcase, desks, and chairs) using different combination of three limited resources (machining, finishing, and labor). Resource requirements and unit products are listed in Table 8.1.

The furniture maker's problem is to utilize the available resources (capacity) to produce the best combination of bookcases, chairs and desks: one that achieves the largest possible total profit.

Let us employ linear programming to maximize the resources using the data in Table 8.1, i.e., net profit versus constraint (finish, labor and machine)

$$\Rightarrow \texttt{profit} = 3x + y + 3z;$$
$$\Rightarrow \texttt{constraints} = \{2x + 2y + z \le 30 \&\&$$
$$x + 2y + 3z \le 25 \&\& 2x + y + z \le 26 \&\& x \ge 0 \&\& y \ge 0 \&\& z \ge 0\};$$

The constraints indicate that one should not be allowed to exceed the capacity limits and the number of the different products cannot be negative.

Looking for the solution on real numbers, we get,

$$\Rightarrow \texttt{sol} = \texttt{Maximize}[\{\texttt{profit}, \texttt{constraints}\}, \{x, y, z\}]$$
$$\Leftarrow \left\{ \frac{231}{5}, \left\{ x \to \frac{53}{5}, y \to 0, z \to \frac{24}{5} \right\} \right\}$$

Supplementary Information The online version contains supplementary material available at https://doi.org/10.1007/978-3-030-92495-9_8.

Table. 8.1 Resource requirements, capacity and profit in a furniture production		Bookcases (*x*)	Chairs (*y*)	Desks (*z*)	Capacity
	Finishing	2	2	1	30
	Labor	1	2	3	25
	Machining	2	1	1	26
	Net profit	3	1	3	–

or

> $\Rightarrow$ sol//N
> $\Leftarrow \{46.2, \{x \rightarrow 10.6, y \rightarrow 0., z \rightarrow 4.8\}\}$

Which are *rounded* off to integer result,

> $\Rightarrow$ solr = MapThread[#2 $\rightarrow$ Round[#1[[2]]]&,{sol[[2]],{x,y,z}}]
> $\Leftarrow \{x \rightarrow 11, y \rightarrow 0, z \rightarrow 5\}$

but now the constraints are not satisfied,

> $\Rightarrow$ constraints/.solr
> $\Leftarrow$ False

An alternative way to get integer solution is to use *floor function*, which converts a real number "x" into the greatest integer less than or equal to "x", we get

> $\Rightarrow$ solf = MapThread[#2 $\rightarrow$ Floor[#1[[2]]]&,{sol[[2]],{x,y,z}}]
> $\Leftarrow \{x \rightarrow 10, y \rightarrow 0, z \rightarrow 4\}$

The constraints are okay,

> $\Rightarrow$ constraints/.solf
> $\Leftarrow$ True

and the profit is

> $\Rightarrow$ profit/.solf
> $\Leftarrow$ 42

Is there a better integer solution in which profit is closer to 46.2. If we employ *integer programming* (IP), we get

⇒ Maximize[
  {profit, 2x + 2y + z <= 30 && x + 2y + 3z <= 25 && 2x + y + z <= 26 &&
  x >= 0 && x >= 0 && y >= 0 && z >= 0 && Element[x|y|z, Integers]},
  {x, y, z}]
  ⇐ {45, {x → 10, y → 0, z → 5}}

This example illustrates that one needs a special method to achieve the best integer result. Neither rounding off nor flooring the real number (i.e., results) will ensure the best profit.

8.2 Discrete Value Problems

A special situation is when the solution can be only a discrete value (which is not necessarily an integer) belonging to a given set. Let us consider the following objective function,

⇒ $p = 0.2x_1 + 0.3x_2 + 0.23x_3$;

The constraint is

⇒ $c_0 = x_1 + x_2 + x_3 \leq 15.5$;

The possible candidates for the solution are in the following sets (not interval),

⇒ $x_1 \in \{1.2, 2.8\}$;
⇒ $x_2 \in \{1.4, 3.8\}$;
⇒ $x_3 \in \{2.5, 3.8, 4.2, 5.8\}$;

Every discrete programming problem can be transformed into a binary (0/1) programming problem!
Let us introduce the following binary variables $\delta_{i,j}$.

⇒ $x_1 = 1.2\, \delta_{1,1} + 2.8\, \delta_{1,2}$;
⇒ $x_2 = 1.4\, \delta_{2,1} + 3.8\, \delta_{2,2}$;
⇒ $x_3 = 2.5\, \delta_{3,1} + 3.8\, \delta_{3,2} + 4.2\, \delta_{3,3} + 5.8\, \delta_{3,4}$;

Only one value from the discrete domains can be assigned to each x_i, therefore we have constraints

⇒ $c_1 = \delta_{1,1} + \delta_{1,2} == 1$;
⇒ $c_2 = \delta_{2,1} + \delta_{2,2} == 1$;
⇒ $c_3 = \delta_{3,1} + \delta_{3,2} + \delta_{3,3} + \delta_{3,4} == 1$;

Now, substituting the new expressions of x_1, x_2 and x_3 into the objective p, the new objective function becomes.

$\Rightarrow \text{P} = \text{p}//\text{Expand}$
$\Leftarrow 0.24\,\delta_{1,1} + 0.56\,\delta_{1,2} + 0.42\,\delta_{2,1} + 1.14\,\delta_{2,2} +$
$\quad 0.575\,\delta_{3,1} + 0.874\,\delta_{3,2} + 0.966\,\delta_{3,3} + 1.334\,\delta_{3,4}$

and similarly the new constraints are

$\Rightarrow \text{C}_0 = \text{c}_0//\text{Expand}$
$\Leftarrow 1.2\,\delta_{1,1} + 2.8\,\delta_{1,2} + 1.4\,\delta_{2,1} + 3.8\,\delta_{2,2} +$
$\quad 2.5\,\delta_{3,1} + 3.8\,\delta_{3,2} + 4.2\,\delta_{3,3} + 5.8\,\delta_{3,4} \leq 15.5.$

The new variables are $\delta_{i,j} \in \{0, 1\}$,

$\Rightarrow \text{var} = \{\delta_{1,1}, \delta_{1,2}, \delta_{2,1}, \delta_{2,2}, \delta_{3,1}, \delta_{3,2}, \delta_{3,3}, \delta_{3,4}\};$

There are two constraints for the variables. They should be integers,

$\Rightarrow \text{c}_4 = \text{Element}[\text{var}, \text{Integers}]$
$\Leftarrow (\delta_{1,1}|\delta_{1,2}|\delta_{2,1}|\delta_{2,2}|\delta_{3,1}|\delta_{3,2}|\delta_{3,3}|\delta_{3,4}) \in \text{Integers}$

and their values should be 0 or 1,

$\Rightarrow \text{c}_5 = \text{Apply}[\text{And}, \text{Map}[0 \leq \# \leq 1\,\&, \text{var}]]$

Since c_4 should be true, therefore c_5 is a proper constraint

$\Leftarrow 0 \leq \delta_{1,1} \leq 1\,\&\&0 \leq \delta_{1,2} \leq 1\,\&\&0 \leq \delta_{2,1} \leq 1\,\&\&0 \leq \delta_{2,2} \leq 1\,\&\&$
$\quad 0 \leq \delta_{3,1} \leq 1\,\&\&0 \leq \delta_{3,2} \leq 1\,\&\&0 \leq \delta_{3,3} \leq 1\,\&\&0 \leq \delta_{3,4} \leq 1$

Then all of the constraints are

$\Rightarrow \text{constraints} = \text{Apply}[\text{And}, \{c_0, c_1, c_2, c_3, c_4, c_5\}]$
$\Leftarrow 1.2\,\delta_{1,1} + 2.8\,\delta_{1,2} + 1.4\,\delta_{2,1} + 3.8\,\delta_{2,2} +$
$\quad 2.5\,\delta_{3,1} + 3.8\,\delta_{3,2} + 4.2\,\delta_{3,3} + 5.8\,\delta_{3,4} \leq 15.5\&\&$
$\quad \delta_{1,1} + \delta_{1,2} = 1\,\&\&\,\delta_{2,1} + \delta_{2,2} = 1\&\&\,\delta_{3,1} + \delta_{3,2} + \delta_{3,3} + \delta_{3,4} = 1\&\&$
$\quad (\delta_{1,1}|\delta_{1,2}|\delta_{2,1}|\delta_{2,1}|\delta_{3,1}|\delta_{3,2}|\delta_{3,3}|\delta_{3,4}) \in \textit{Integers}\&\&$
$\quad 0 \leq \delta_{1,1} \leq 1\,\&\&0 \leq \delta_{1,2} \leq 1\,\&\&0 \leq \delta_{2,1} \leq 1\,\&\&0 \leq \delta_{2,2} \leq 1\&\&$
$\quad 0 \leq \delta_{3,1} \leq 1\,\&\&0 \leq \delta_{3,2} \leq 1\,\&\&0 \leq \delta_{3,3} \leq 1\,\&\&0 \leq \delta_{3,4} \leq 1$

Therefore the solution is

$$\Rightarrow \texttt{sol} = \texttt{Maximize}[\texttt{Rationalize}[\{P, \texttt{constraints}\}], \texttt{var}]$$

$$\Leftarrow \{\frac{1517}{500}, \{\delta_{1,1} \rightarrow 0, \delta_{1,2} \rightarrow 1, \delta_{2,1} \rightarrow 0,$$

$$\delta_{2,2} \rightarrow 1, \delta_{3,1} \rightarrow 0, \delta_{3,2} \rightarrow 0, \delta_{3,3} \rightarrow 0, \delta_{3,4} \rightarrow 1\}\}$$

which means

$$\Rightarrow x_1/.\texttt{sol}[[2]]$$
$$\Leftarrow 2.8$$
$$\Rightarrow x_2/.\texttt{sol}[[2]]$$
$$\Leftarrow 3.8$$
$$\Rightarrow x_3/.\texttt{sol}[[2]]$$
$$\Leftarrow 5.8$$

8.3   Simple Logical Conditions

Let us consider a periodic investment problem. There are 4 different investment possibilities each having 3 periods. The maximum budgets that are at the disposal of each different time periods are 18, 16 and 19 m\$, respectively. The different investments need the following sums of money in the different time periods,

   I. investment: 6, 9, 3
  II. investment: 8, 0, 11
 III. investment: 0, 5, 7
  IV. investment: 4, 5, 6.

The profits of the different investments at the end of the third period are: 16, 22, 12, 8. Which investment can be and should be made in order to maximize the total profit?

Let x_i=1 if the i-th investment takes place, and x_i=0 otherwise. Then the following 0/1 programming problem can be stated:

$$\Rightarrow \texttt{profit} = 16x_1 + 22x_2 + 12x_3 + 8x_4$$
$$\Leftarrow 16x_1 + 22x_2 + 12x_3 + 8x_4$$

The constraints for the budget in the different time periods are

$$\Rightarrow \texttt{constraints1} = \{6x_1 + 8x_2 + 0x_3 + 4x_4 \leq 18, 9x_1 + 0x_2 + 5x_3 + 5x_4 \leq 16, 3x_1 + 11x_2 + 7x_3 + 6x_4 \leq 19\};$$

In addition there are further constraints:

(1) No more than two investments can take place at the same time,

$$\Leftarrow x_1 + x_2 + x_3 + x_4 \leq 2,$$

(2) If the second investment takes place then the fourth one should be also realized,

$$\Leftarrow x_2 - x_4 \leq 0, \text{since if } x_2 = 1 \text{ then } x_4 = 1, \text{or}$$
$$\text{if } x_2 = 0 \text{ then } x_4 = 0, \text{or if } x_4 = 1 \text{ and } x_2 = 0$$

(3) If the first one takes place, then the third one should not be realized

$$\Leftarrow x_1 + x_3 \leq 1, \text{since if } x_1 = 1 \text{then } x_3 = 0, \text{or if } x_1 = 0 \text{then}$$
$$x_3 = 0, \text{or } x_3 = 1 \text{ and } x_1 = 0$$

Let us summarize these additional constrains as,

$$\Rightarrow \texttt{constraints2} = \{x_1 + x_2 + x_3 + x_4 \leq 2, x_2 - x_4 \leq 0, x_1 + x_3 \leq 1\};$$
$$\Rightarrow \texttt{vars} = \texttt{Table}[x_i, \{i, 1, 4\}]$$
$$\Leftarrow \{x_1, x_2, x_3, x_4\}$$

The variables are between 0 and 1,

$$\Rightarrow \texttt{constraints3} = \texttt{Map}[0 < = \# < = 1 \&, \texttt{vars}]$$
$$\Leftarrow \{0 \leq x_1 \leq 1, 0 \leq x_2 \leq 1, 0 \leq x_3 \leq 1, 0 \leq x_4 \leq 1\}$$

We put these constrains together with the restriction that variables should be integers,

$$\Rightarrow \texttt{constraints} = \texttt{Apply}[\texttt{And}, \texttt{Join}[\texttt{constraints1}, \texttt{constraints2},$$
$$\texttt{constraints3}, \{\texttt{Element}[\texttt{vars}, \texttt{Integers}]\}]]$$
$$\Leftarrow 6x_1 + 8x_2 + 4x_4 \leq 18 \&\& 9x_1 + 5x_3 + 5x_4 \leq 16 \&\&$$
$$3x_1 + 11x_2 + 7x_3 + 6x_4 \leq 19 \&\& x_1 + x_2 + x_3 + x_4 \leq 2 \&\&$$
$$x_2 - x_4 \leq 0 \&\& x_1 + x_3 \leq 1 \&\& 0 \leq x_1 \leq 1 \&\& 0 \leq x_2 \leq 1 \&\&$$
$$0 \leq x_3 \leq 1 \&\& 0 \leq x_4 \leq 1 \&\& (x_1 | x_2 | x_3 | x_4) \in \texttt{Integers}$$

Now, let us carry out the maximization,

$$\Rightarrow \texttt{sol} = \texttt{Maximize}[\{\texttt{profit}, \texttt{constraints}\}, \texttt{vars}]$$
$$\Leftarrow \{30, \{x_1 \rightarrow 0, x_2 \rightarrow 1, x_3 \rightarrow 0, x_4 \rightarrow 1\}\}$$

In a more *general* way, the following problem can be handled via introducing binary variables. Let us consider the constraints as:

$$\sum_{j=1}^{n} a_j x_j = b.$$

There are k different possible values for b: $b_1, ..., b_k$

$$\sum_{j=1}^{n} a_j x_j = b_i \quad (i = 1, ..., k),$$

but only one of them can be true. This can be expressed as,

$$\sum_{j=1}^{n} a_j x_j = \sum_{i=1}^{k} b_i y_i$$

and

$$\sum_{i=1}^{k} y_i = 1$$

where

$$y_i = 0 \text{ or } 1, \quad (i = 1, ..., k).$$

8.4 Some Typical Problems of Binary Programming

8.4.1 Knapsack Problem

We have n objects, and the j-th object has a weight a_j which is a positive integer and the value of the object is p_j. We need to select the most valuable subset of the objects having a total weight less than b.

The total value of the objects

$$\max \sum_{j=1}^{n} p_j x_j$$

should be maximized to satisfy the constraint,

$$\sum_{j=1}^{n} a_j x_j \leq b,$$

where x_j is equal to 1 if the object is selected and 0 otherwise,

$$x_j \in \{0, 1\}.$$

Example Given a ferry, which can transport 5t load as maximum. There are three vehicles to be transported: a car weighting 2t, a van 4t, and a motorcycle 1t. If the fares are 300, 400, and 100 \$, respectively, how can we get the highest profit? Our objective function to be minimized is

> ⇒ profit = 300x + 400y + 100 z;

Under the constraint imposed by the capacity of the ferry

> ⇒ constraints ={2x + 4y + 1z ≤ 5&&0 ≤ x ≤ 1&&0 ≤ y ≤ 1&&0 ≤ z ≤ 1&&
> Element[x|y|z, Integers]};

Then

> ⇒ Maximize[{profit, constraints},{x,y,z}]
> ⇐ {500,{x → 0,y → 1,z → 1}}

This therefore means that the ferry will bring van and the motorcycle and his profit will be \$ 500.

8.4.2   Nonlinear Knapsack Problem

The nonlinear form of the profit function is,

$$\max \sum_{j=1}^{n} f_j(x_j).$$

Assuming that the constrains are

$$\sum_{j=1}^{n} g_j(x_j) \leq b,$$

and

$$0 \le x_j \le d_j,$$

where x_j is integer.

Example Now we have a nonlinear objective as

$\Rightarrow \texttt{profit} = x_1^3 + 3x_2 + 0.25x_3^4$

$\Leftarrow x_1^3 + 3x_2 + 0.25x_3^4$

$\Rightarrow \texttt{vars} = \texttt{Table}[x_i, \{i, 1, 4\}]$

$\Leftarrow \{x_1, x_2, x_3, x_4\}$

$\Rightarrow \texttt{constraints} = \texttt{Join}[\{3x_1^2 + 8x_2 + x_3^3 + x_4 == 10\}, \texttt{Map}[0 \le \# \le 5, \texttt{vars}],$

$\quad \texttt{Element}[\texttt{vars}, \texttt{Integers}]]$

$\Leftarrow \{3x_1^2 + 8x_2 + x_3^3 + x_4 = 10, 0 \le x_1 \le 5, 0 \le x_2 \le 5,$

$\quad 0 \le x_3 \le 5, 0 \le x_4 \le 5, (x_1 | x_2 | x_3 | x_4) \in \texttt{Integers}\}$

Then

$\Rightarrow \texttt{sol} = \texttt{Maximize}[\texttt{Rationalize}[\{\texttt{profit}, \texttt{constraints}\}], \texttt{vars}]$

$\Leftarrow \{4, \{x_1 \to 0, x_2 \to 0, x_3 \to 2, x_4 \to 2\}\}$

8.4.3 Set-Covering Problem

Let us consider a deployment problem of fire stations in a town that has the following districts (Fig. 8.1).

It is desired to erect the fewest fire stations that can handle the fires of all districts assuming that a fire station can operate in the neighboring district, too. This is the so called *set covering problem*.

Let us consider a neighborhood matrix **A**, which can be defined as follows

$A_{i,j} = 1$ if the i-th district and the j-th district has common border or $i = j$

$A_{i,j} = 0$ otherwise. In our case A is

$$\Rightarrow A = \begin{pmatrix} 1 & 1 & 1 & 1 & 0 & 0 & 0 & 0 & 0 \\ 1 & 1 & 1 & 0 & 1 & 0 & 0 & 0 & 0 \\ 1 & 1 & 1 & 1 & 1 & 1 & 0 & 0 & 0 \\ 1 & 0 & 1 & 1 & 0 & 1 & 1 & 0 & 0 \\ 0 & 1 & 1 & 0 & 1 & 1 & 0 & 1 & 1 \\ 0 & 0 & 1 & 1 & 1 & 1 & 1 & 1 & 0 \\ 0 & 0 & 0 & 1 & 0 & 1 & 1 & 1 & 0 \\ 0 & 0 & 0 & 0 & 1 & 1 & 1 & 1 & 1 \\ 0 & 0 & 0 & 0 & 1 & 0 & 0 & 1 & 1 \end{pmatrix};$$

Fig. 8.1 District boundaries

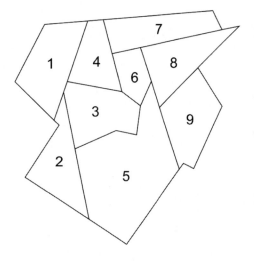

Now, the decision variable x_j can be defined as follows: $x_j = 1$ if in the j-th district a fire station will be built, and $x_j = 0$ otherwise.

The variables are

$$\Rightarrow \texttt{vars} = \texttt{Table}[x_i, \{i, 1, 9\}]$$
$$\Leftarrow \{x_1, x_2, x_3, x_4, x_5, x_6, x_7, x_8, x_9\}$$

The objective function to be minimized is,

$$\Rightarrow \texttt{objective} = \texttt{Apply}[\texttt{Plus}, \texttt{vars}]$$
$$\Leftarrow x_1 + x_2 + x_3 + x_4 + x_5 + x_6 + x_7 + x_8 + x_9$$

In every district or at least in its neighborhood, there should be a fire station,

$$\Rightarrow \texttt{constraints1} = \texttt{Map}[(\# \geq 1)\&, \texttt{A.vars}];$$
$$\Rightarrow \texttt{constraints1}//\texttt{TableForm}$$
$$\Leftarrow x_1 + x_2 + x_3 + x_4 \geq 1$$
$$x_1 + x_2 + x_3 + x_5 \geq 1$$
$$x_1 + x_2 + x_3 + x_4 + x_5 + x_6 \geq 1$$
$$x_1 + x_3 + x_4 + x_6 + x_7 \geq 1$$
$$x_2 + x_3 + x_5 + x_6 + x_8 + x_9 \geq 1$$
$$x_3 + x_4 + x_5 + x_6 + x_7 + x_8 \geq 1$$
$$x_4 + x_6 + x_7 + x_8 \geq 1$$
$$x_5 + x_6 + x_7 + x_8 + x_9 \geq 1$$
$$x_5 + x_8 + x_9 \geq 1$$

The usual constraints for the variables are:

$\Rightarrow$constraints2 = Map$[0\leq\#\leq1\,\&,$vars$]$
$\Leftarrow\{0\leq x_1\leq1,0\leq x_2\leq1,0\leq x_3\leq1,0\leq x_4\leq1,$
   $0\leq x_5\leq1,0\leq x_6\leq1,0\leq x_7\leq1,0\leq x_8\leq1,0\leq x_9\leq1\}$

Let us join the constraints,

$\Rightarrow$constraints = Join$[$constraints1,constraints2,
   $\{$Element$[$vars,Integers$]\}]$
$\Leftarrow\{x_1+x_2+x_3+x_4\geq1, x_1+x_2+x_3+x_5\geq1, x_1+x_2+x_3+x_4+x_5+x_6\geq1,$
   $x_1+x_3+x_4+x_6+x_7\geq1, x_2+x_3+x_5+x_6+x_8+x_9\geq1,$
   $x_3+x_4+x_5+x_6+x_7+x_8\geq1, x_4+x_6+x_7+x_8\geq1,$
   $x_5+x_6+x_7+x_8+x_9\geq1, x_5+x_8+x_9\geq1, 0\leq x_1\leq1, 0\leq x_2\leq1,$
   $0\leq x_3\leq1, 0\leq x_4\leq1, 0\leq x_5\leq1, 0\leq x_6\leq1, 0\leq x_7\leq1, 0\leq x_8\leq1,$
   $0\leq x_9\leq1, (x_1|x_2|x_3|x_4|x_5|x_6|x_7|x_8|x_9)\in$ Integers$\}$
$\Rightarrow$Minimize$[\{$objective,constraints$\}$,vars$]$
$\Leftarrow\{2,\{x_1\rightarrow0, x_2\rightarrow0, x_3\rightarrow1, x_4\rightarrow0, x_5\rightarrow0,$
   $x_6\rightarrow0, x_7\rightarrow0, x_8\rightarrow1, x_9\rightarrow0\}\}$

The 5th and 6th districts are lucky since they have even two neighboring districts with fire stations, see Fig. 8.2.

Fig. 8.2 Solution of the fire-station allocation problem

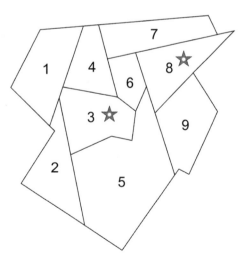

8.5   Solution Methods

8.5.1   Binary Countdown Method

This is a brute force method, which can be useful and effective in case of small sized problems.

Let us consider the following 0/1 programming problem

$\Rightarrow$ profit $= -8x1 - 2x2 - 4x3 - 7x4 - 5x5 + 10$

$\Rightarrow$ constraints $=$ Apply[And,$\{-3x1 - 3x2 + x3 + 2x4 + 3x5 \leq -2,$

$-5x1 - 3x2 - 2x3 - x4 + x5 \leq -4\}$]

$\Leftarrow -3x1 - 3x2 + x3 + 2x4 + 3x5 \leq -2 \&\& -5x1 - 3x2 - 2x3 - x4 + x5 \leq -4$

We generate all the possible combinations

$\Leftarrow$ total $=$ Tuples$[\{0, 1\}, 5]$

$\Leftarrow \{\{0, 0, 0, 0, 0\}, \{0, 0, 0, 0, 1\}, \{0, 0, 0, 1, 0\}, \{0, 0, 0, 1, 1\},$

$\{0, 0, 1, 0, 0\}, \{0, 0, 1, 0, 1\}, \{0, 0, 1, 1, 0\}, \{0, 0, 1, 1, 1\},$

$\{0, 1, 0, 0, 0\}, \{0, 1, 0, 0, 1\}, \{0, 1, 0, 1, 0\}, \{0, 1, 0, 1, 1\},$

$\{0, 1, 1, 0, 0\}, \{0, 1, 1, 0, 1\}, \{0, 1, 1, 1, 0\}, \{0, 1, 1, 1, 1\},$

$\{1, 0, 0, 0, 0\}, \{1, 0, 0, 0, 1\}, \{1, 0, 0, 1, 0\}, \{1, 0, 0, 1, 1\},$

$\{1, 0, 1, 0, 0\}, \{1, 0, 1, 0, 1\}, \{1, 0, 1, 1, 0\}, \{1, 0, 1, 1, 1\},$

$\{1, 1, 0, 0, 0\}, \{1, 1, 0, 0, 1\}, \{1, 1, 0, 1, 0\}, \{1, 1, 0, 1, 1\},$

$\{1, 1, 1, 0, 0\}, \{1, 1, 1, 0, 1\}, \{1, 1, 1, 1, 0\}, \{1, 1, 1, 1, 1\}\}$

$\Rightarrow$ Length$[\%]$

$\Leftarrow 32$

Then select those combinations, which satisfy the constraints

$\Rightarrow$ candid $=$ Select[total, constraints$/.\{x1 \rightarrow \#[[1]],$

$x2 \rightarrow \#[[2]], x3 \rightarrow \#[[3]], x4 \rightarrow \#[[4]], x5 \rightarrow \#[[5]]\}\&$]

$\Leftarrow \{\{0, 1, 1, 0, 0\}, \{1, 0, 0, 0, 0\}, \{1, 0, 1, 0, 0\},$

$\{1, 1, 0, 0, 0\}, \{1, 1, 0, 0, 1\}, \{1, 1, 0, 1, 0\},$

$\{1, 1, 1, 0, 0\}, \{1, 1, 1, 0, 1\}, \{1, 1, 1, 1, 0\}\}$

Finally, choose the eligible solution, which maximizes the objective

$\Rightarrow$ sols $=$ Map[profit$/.\{x1 \rightarrow \#[[1]], x2 \rightarrow \#[[2]],$

$x3 \rightarrow \#[[3]], x4 \rightarrow \#[[4]], x5 \rightarrow \#[[5]]\}\&$, candid]

$\Leftarrow \{4, 2, -2, 0, -5, -7, -4, -9, -11\}$

$\Rightarrow$ candid$[[First[Position[sols, Max[sols]]]]]$

result of the binary countdown:

$$\Leftarrow \{\{0,1,1,0,0\}\}$$

Employing built in function

```
⇒constraints =
    Apply[And,{constraints, 0 ≤ x1 ≤ 1, 0 ≤ x2 ≤ 1, 0 ≤ x3 ≤ 1,
    0 ≤ x4 ≤ 1, 0 ≤ x5 ≤ 1, Element[{x1, x2, x3, x4, x5}, Integers]}]
    ⇐−3x1 − 3x2 + x3 + 2x4 + 3x5 ≤ −2 && −5x1 − 3x2 − 2x3 − x4 + x5 ≤ −4 &&
    0 ≤ x1 ≤ 1 && 0 ≤ x2 ≤ 1 && 0 ≤ x3 ≤ 1 && 0 ≤ x4 ≤ 1 && 0 ≤ x5 < = 1 &&
    (x1|x2|x3|x4|x5) ∈ Integers
⇒Maximize[{profit, constraints}, {x1, x2, x3, x4, x5}]
```

result of the built in Maximize function:

$$\Leftarrow \{4, \{x1 \rightarrow 0, x2 \rightarrow 1, x3 \rightarrow 1, x4 \rightarrow 0, x5 \rightarrow 0\}\}$$

The results are the same.

8.5.2   Branch and Bound Method

Let us consider another technique to solve the task. In this case the value of the original problem can be reduced to a series of solutions of real optimization problems. We demonstrate the algorithm using the following example. Consider the following simple integer-programming problem.

The objective function is

```
⇒ p = 8x₁ + 5x₂;
```

The constraints are

```
⇒cI ={x₁ + x₂ ≤ 6 && 9x₁ + 5x₂ ≤ 45 && x₁³ ≥ && x₂³ ≥ &&
    Element[x₁|x₂, Integers]};
```

The variables

```
⇒ var ={x₁, x₂};
```

Now, we are looking for the solution on the real numbers

```
⇒ cR ={x₁ + x₂ ≤ 6 && 9x₁ + 5x₂ ≤ 45 && x₁ ≥ 0 && x₂ ≥ 0};
```

Step 1.

$$\Rightarrow \texttt{Maximize}[\{p, cR\}, var]$$
$$\Leftarrow \left\{ \frac{165}{4}, \left\{ x_1 \rightarrow \frac{15}{4}, x_2 \rightarrow \frac{9}{4} \right\} \right\}$$

In this case, the objective is $165/4 = 41.25$, which should be smaller in integer case. Considering that $x_1 = 3.75$ let us investigate the cases $x_1 \leq 3$ and $x_1 \geq 4$ (*branching*)

Step 2.

case 2/a:

$$\Rightarrow \texttt{cRL} = \{x_1 + x_2 \leq 6 \,\&\&\, 9x_1 + 5x_2 \leq 45 \,\&\&\, x_1 \geq 0 \,\&\&\, x_2^3 \geq 0 \,\&\&\, x_1 \leq 3\};$$
$$\Rightarrow \texttt{Maximize}[\{p, cRL\}, var]$$
$$\Leftarrow \{39, \{x_1 \rightarrow 3, x_2 \rightarrow 3\}\}$$

Since both solutions are integer this is a candidate *bounding* and the *bound is 39*

case 2/b:

$$\Rightarrow \texttt{cRU} = \{x_1 + x_2 \leq 6 \,\&\&\, 9x_1 + 5x_2 \leq 45 \,\&\&\, x_1 \geq 0 \,\&\&\, x_2 \geq 0 \,\&\&\, x_1 \geq 4\};$$
$$\Rightarrow \texttt{Maximize}[\{p, cRU\}, var]$$
$$\Leftarrow \left\{ 41, \left\{ x_1 \rightarrow 4, x_2 \rightarrow \frac{9}{5} \right\} \right\}$$

Since this is a non-integer solution with bound: $41 > 39$, $\rightarrow$ *branching* $\rightarrow$ step 3 (see Fig. 8.3).

Step 3.

branching: let us consider the cases $x_2 \leq 1$ and $x_2 \geq 2$ since $x_2 = 9/4 = 1.8$.

case 3/a:

Fig. 8.3 First branch and bound (candidate)

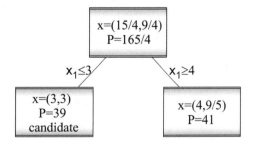

$\Rightarrow \text{cRL} = \{x_1 + x_2 \leq 6 \&\& 9x_1 + 5x_2 \leq 45 \&\& x_1 \geq 0 \&\& x_2 \geq 0 \&\& x_2 \leq 1\};$

$\Rightarrow \text{Maximize}[\{p, \text{cRL}\}, \text{var}]$

$\Leftarrow \left\{ \dfrac{365}{9}, \left\{ x_1 \rightarrow \dfrac{40}{9}, x_2 \rightarrow 1 \right\} \right\}$

Since we have solution and the objective: 365/9 = 40.5556 > 39, but the solution is not integer → *branching* → step 4, see Fig. 8.4..

case 3/b:

$\Rightarrow \text{cRU} = \{x_1 + x_2 \leq 6 \&\& 9x_1 + 5x_2 \leq 45 \&\& x_1 \geq 0 \&\& x_2 \geq 0 \&\& x_2 \geq 2\};$

$\Rightarrow \text{Maximize}[\{p, \text{cRU}\}, \text{var}]$

$\Leftarrow \left\{ \dfrac{165}{4}, \left\{ x_1 \rightarrow \dfrac{15}{4}, x_2 \rightarrow \dfrac{9}{4} \right\} \right\}$

This was the original real solution → infeasible solution → dead end (see Fig. 8.4).

$\Rightarrow 165/4//\text{N}$

$\Leftarrow 41.25$

Step 4.

branching: consider the cases $x_1 \leq 4$ and $x_1 \geq 5$

case 4/a:

$\Rightarrow \text{cRL} = \{x_1 + x_2 \leq 6 \&\& 9x_1 + 5x_2 \leq 45 \&\& x_1 \geq 0 \&\& x_2 \geq 0 \&\& x_1 \leq 4\};$

$\Rightarrow \text{Maximize}[\{p, \text{cRL}\}, \text{var}]$

$\Leftarrow \left\{ \dfrac{165}{4}, \left\{ x_1 \rightarrow \dfrac{15}{4}, x_2 \rightarrow \dfrac{9}{4} \right\} \right\}$

$\Rightarrow 165/4//\text{N}$

$\Leftarrow 41.25$

Fig. 8.4 Second branch and bound (infeasible)

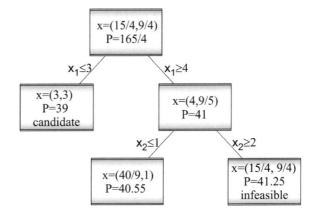

Neither of them is integer → infeasible solution → *dead end*
case 4/b:

$\Rightarrow \text{cRU} = \{x_1 + x_2 \leq 6 \&\& 9x_1 + 5x_2 \leq 45 \&\& x_1 \geq 0 \&\& x_2 \geq 0 \&\& x_2 \geq 5\};$
$\Rightarrow \text{Maximize}[\{p, \text{cRU}\}, \text{var}]$
$\Leftarrow \{40, \{x_1 \rightarrow 5, x_2 \rightarrow 0\}\}$

Both solutions are integer → *candidate with bound: 40* → dead end.
The candidate solution is the solution belonging to the maximum bound $\{x_1 \rightarrow 5, x_2 \rightarrow 0\}$
Indeed, using built-in function,

$\Rightarrow \text{Maximize}[\{p, \text{cI}\}, \text{var}]$
$\Leftarrow \{40, \{x_1 \rightarrow 5, x_2 \rightarrow 0\}\}$

The procedure can be represented by a binary tree, see Fig. 8.5.

8.5.3   Application of Gröbner Basis

Bertsimas et al., (2000) shows that Gröbner basis can be employed to solve 0–1 integer programming. They provided technique to solve problems like

$$c^T x \rightarrow min.$$

Fig. 8.5 Last branch and bound (candidate)

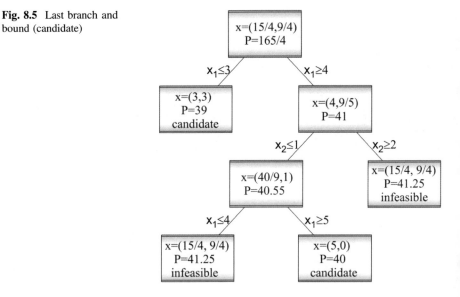

$$\text{subject to } A x = b.$$

with ensuring 0–1 results

$$x_j = x_j^2 \text{ for } \forall j$$

Bertsimas et al., (2000)s idea can be illustrated as follows:

(1) Create the following ideal

$$J = \langle y - c^T x, A x - b, x_j - x_j^2 \rangle$$

(2) Compute the reduced Gröbner basis for y

$$G(y) = p(y)$$

(3) Consider the minimal root of the polynomial

$$p(y) = 0 \rightarrow y_{\min}$$

(4) Compute the full Gröbner basis containing y and all x variables and solve this polynomial system for the x variables. Let us illustrate this algorithm.

We want to minimize the following objective

$$c^T x = x_1 + 2x_2 + 3x_3$$

with the constrain

$$x_1 + 2x_2 + 2x_3 = 3$$

and

$$x_j^2 - x_j = 0 \quad \text{for } j = 1, 2, 3$$

Then our ideal is

$$\Rightarrow J = \{y - x_1 - 2x_2 - 3x_3, x_1 + 2x_2 + 2x_3 - 3, x_1^2 - x_1, x_2^2 - x_2, x_3^2 - x_3\}$$
$$\Leftarrow \{y - x_1 - 2x_2 - 3x_3, -3 + x_1 + 2x_2 + 2x_3, -x_1 + x_1^2, -x_2 + x_2^2, -x_3 + x_3^2\}$$

The reduced Gröbner basis for y

$$\Rightarrow G = \texttt{GroebnerBasis}[J, \{y\}, \{x_1, x_2, x_3\}]$$
$$\Leftarrow \{12 - 7y + y^2\}$$

This polynomial has two roots

$$\Rightarrow \text{soly} = \text{Solve}[G == 0, y]$$
$$\Leftarrow \{\{y \rightarrow 3\}, \{y \rightarrow 4\}\}$$

The full Gröbner basis is

$$\Rightarrow GG = \text{GroebnerBasis}[J, \{y, x_1, x_2, x_3\}]]$$
$$\Leftarrow \{-x_3 + x_3^2, -1 + x_2 + x_3, -1 + x_1, -3 + y - x_3\}$$

This polynomial system can be solved substituting the smaller y solution ($y = 3$)

$$\Rightarrow \text{pp} = \text{Map}[\# == 0\&, GG]/.y - > 3$$
$$\Leftarrow \{-x_3 + x_3^2 == 0, -1 + x_2 + x_3 == 0, -1 + x_1 == 0, -x_3 == 0\}$$

$$\Rightarrow \text{NSolve}[\text{pp}, \{x_1, x_2, x_3\}]$$
$$\Leftarrow \{\{x_1 \rightarrow 1., x_2 \rightarrow 1., x_3 \rightarrow 0.\}\}$$

and the value of the objective is the smaller solution of (y), namely $y = 3$.

Earlier Conti and Traverso (1991) suggested a different algorithm for case of minimizing a cost vector $c.w$ subject to the conditions $A.w = b$, where the components of the optimal solution w are required to be non-negative integers. This algorithm was coded by Kapadia (2002) in *Wolfram Mathematica*. The algorithm is divided into three steps:

1. The coefficients of the matrix A are used to form the exponents in a certain binomial ideal.
2. The Groebner basis of the binomial ideal with respect to a monomial order given by the vector c is found using the kernel function Groebner Basis.
3. The optimal solution is found by reading off the exponents in the answer obtained by reducing a certain monomial formed from the components of b, with respect to the Groebner basis and the same monomial order. The kernel function *PolynomialReduce* is used for this step.

Kapadia (2002)s code is as follows:

```
IntegerProgramming[c_,A_,b_]:=
Module[{m,n,p,q,CostOrder,CostMatrix,e,t},p=Dimensions[A]
[[1]];
    q=Dimensions[A][[2]];
    CostMatrix=Join[
        Table[KroneckerDelta[i,j],{i,1,p+1},{j,1,p+q+1}],
        {Join[Table[0,{i,1,p+1}],If[Apply[And,Table[c[[j]]>=0,
{j,1,q}]],
```

```
        Table[c[[j]],{j,1,q}],
        Table[c[[j]]+Max[Table[-c[[j]]/(Sum[A[[i,j]],{i,1,
p}]]),{j,1,q}]]*
            Sum[A[[i,j]],{i,1,p}],{j,1,q}]]]},
        Table[KroneckerDelta[i,p+q+4-j],{i,3,q+1},{j,1,p+q
+1}]];
    CostOrder=Map[LCM[Denominator[#]]*# &, CostMatrix];
    e=Table[Max[0,Max[Table[-A[[i,j]],{i,1,p}]]],{j,1,q}];
    m=GroebnerBasis[Join[{t*Product[z[i],{i,1,p}]-1},
        Table[t^(e[[j]])*Product[z[i]^(A[[i,j]]+e[[j]]),{i,1,
p}]-w[j],
        {j,1,q}]],Join[{t},Table[z[i],{i,1,p}],Table[w[j],
{j,1,q}]],
    MonomialOrder->CostOrder];
    n=
    PolynomialReduce[t^Max[{0,Max[Table[-b[[i]],{i,1,p}]]}]*
        Product[z[i]^(b[[i]]+(Max[0,Max[Table[-b[[i]],{i,1,
p}]]])),
        {i,1,p}],m,Join[{t},Table[z[i],{i,1,p}],Table[w[j],
{j,1,q}]],
        MonomialOrder->CostOrder][[2]];
    {Sum[c[[If [Length[n[[i]]]==1,n[[i,1]],n[[i,1,1]]]]]*
        (If[Length[n[[i]]]==1,1,n[[i,2]]]),{i,1,Length[n]}],
    Table[If[Length[n[[i]]]==1,n[[i]]->1,n[[i,1]]->n[[i,2]]],
        {i,1,Length[n]}]}}]
```

Here is one of Kapadia (2002)'s examples

This is taken from p. 372 of the book "Using Algebraic Geometry" by Cox et al. (1998), which is the basic reference for our program:

Minimize: $w_1 + 1000w_2 + w_3 + 100w_4$

subject to:

$$3w_1 - 2w_2 + w_3 = -1$$
$$4w_1 + w_2 - w_3 - w_4 = 5$$
$$(w_i \text{ non-negative integers})$$

The problem is translated into the matrix notation and solved using our *IntegerProgramming* function as follows:

$$\Rightarrow \text{A1} = \{\{3, -2, 1, 0\}, \{4, 1, -1, -1\}\};$$
$$\Rightarrow \text{b1} = \{-1, 5\};$$
$$\Rightarrow \text{c1} = \{1, 1000, 1, 100\};$$
$$\Rightarrow \text{IntegerProgramming}[\text{c1}, \text{A1}, \text{b1}]$$
$$\Leftarrow \{2101, \{w[1] \rightarrow 1, w[2] \rightarrow 2, w[4] \rightarrow 1\}\}$$

Let us now employ this code for our problem solved by the proposed by Bertsimas et al. (2000). Now we have 6 variables (w_i, $i = 1, 2, \ldots 6$) since the last three constrains containing slack variables (w_4, w_5, w_6) ensures 0–1 values for the unknowns.

$$\text{Minimize:} \quad w_1 + 2w_2 + 3w_3$$
$$\text{subject to:}$$
$$1w_1 + 2w_2 + 2w_3 = 3$$
$$w_1 + w_4 = 1$$
$$w_2 + w_5 = 1$$
$$w_3 + w_6 = 1$$

Then

$$\Rightarrow \text{cm} = \{1, 2, 3, 0, 0, 0\};$$
$$\Rightarrow \text{Am} = \{\{1, 2, 2, 0, 0, 0\}, \{1, 0, 0, 1, 0, 0\}, \{0, 1, 0, 0, 1, 0\}, \{0, 0, 1, 0, 0, 1\}\};$$
$$\Rightarrow \text{bm} = \{3, 1, 1, 1\};$$
$$\Rightarrow \text{IntegerProgramming}[\text{cm}, \text{Am}, \text{bm}]$$
$$\Leftarrow \{3, \{w[1] \rightarrow 1, w[2] \rightarrow 1, w[6] \rightarrow 1\}\}$$

The $w[6] = 1$ implies that $w[3] = 0$.

8.6   Mixed-Integer Programming

Sometimes the solution of a programming problem can be partly real and partly integer. If our computer system is not prepared for such problems, we can employ the following approximation. Let us consider the following problem

$\Rightarrow 3x_1 + 5x_2 \rightarrow \max$

$\Rightarrow x_1 + 2x_2 \leq 9.75$

$\Rightarrow 3x_1 + 4x_2 \leq 21.75$

$\Rightarrow x_1, x_2 \geq 0$

$\Rightarrow x_2 \in \text{integer}$

Let us introduce a new integer variable ξ_1, for the real variable x_1

$\Rightarrow \xi_1 = 100x_1$

Then our constraints are

$\Rightarrow \text{constraints} = \text{Rationalize}[$
$\quad \text{Apply}[\text{And},\{10^{-2}\,\xi_1 + 2x_2 \leq 9.75, 3 \times 10^{-2}\,\xi_1 + 4x_2 \leq 21.75,$
$\quad 0 \leq \xi_1, 0 \leq x_2, \text{Element}[\{\xi_1, x_2\}, \text{Integers}]\}]]$

$\Leftarrow 2x_2 + \dfrac{\xi_1}{100} \leq \dfrac{39}{4}\ \&\&\,4x_2 + \dfrac{3\xi_1}{100} \leq \dfrac{87}{4}\ \&\&\,0 \leq \xi_1\,\&\&\,0 \leq x_2\,\&\&\,(\xi_1|x_2) \in \text{Integers}$

and the objective function is

$\Rightarrow \text{objective} = 3 \times 10^{-2}\,\xi_1 + 5x_2\,;$

Let us get the solution of this integer programming problem

$\Rightarrow \text{sol} = \text{Maximize}[\{\text{objective}, \text{constraints}\}, \{\xi_1, x_2\}]//N$

$\Leftarrow \{25.25, \{\xi_1 \rightarrow 175, x_2 \rightarrow 4.\}\}$

Then we convert ξ_1 integer value back to x_1 real one.

$\Rightarrow x_1 = \xi_1 10^{-2}/.\text{sol}[[2]]$

$\Leftarrow 1.75$

8.7 Applications

8.7.1 Integer Least Squares

A typical problem is the computation of the integer least-squares estimates of the GNSS cycle ambiguities, see Teunissen (1995) i.e., the integer least-squares problem. Suppose we know the floating point vector Z and the covariance matrix of the errors Q. We seek the integer vector z, which minimize the following objective,

$$(z - Z)^T Q^{-1}(z - Z) \to \min_Z,$$

So we are looking for an integer vector Z, which is closest to Z, but in the meantime, we consider the measure error of every element of the vector Z. The variables are the coordinates of the unknown integer vector.

$\Rightarrow$ var $=$ Table$[z_i, \{i, 1, 3\}]$

$\Leftarrow \{z_1, z_2, z_3\}$

$$\Leftarrow Z = \begin{pmatrix} 2.5602 \\ -2.3256 \\ -4.9076 \end{pmatrix}; Q = \begin{pmatrix} 4.0722 & -1.5606 & 0.1293 \\ -1.5606 & 1.0943 & 0.5632 \\ 0.1293 & 0.5632 & 1.6399 \end{pmatrix};$$

Let us rationalize Z and Q

$\Rightarrow$ ZR $=$ Rationalize$[Z]$; MatrixForm$[ZR]$

$$\Leftarrow \begin{pmatrix} \frac{12801}{5000} \\ -\frac{2907}{1250} \\ -\frac{12269}{2500} \end{pmatrix}$$

$\Rightarrow$ QR $=$ Rationalize$[Q]$; MatrixForm$[QR]$

$$\Leftarrow \begin{pmatrix} \frac{20361}{5000} & -\frac{7803}{5000} & \frac{1293}{10000} \\ -\frac{7803}{5000} & \frac{10943}{10000} & \frac{352}{625} \\ \frac{1293}{10000} & \frac{352}{625} & \frac{16399}{10000} \end{pmatrix}$$

then the objective function in our case is

$\Rightarrow$ objective $=((z_1 \quad z_2 \quad z_3) - $Transpose$[ZR]//$Flatten$).$

$\qquad$ Inverse$[QR].\left(\begin{pmatrix} z_1 \\ z_2 \\ z_3 \end{pmatrix} - ZR\right)//$Simplify$//$First

$\Leftarrow \dfrac{1}{4441351063357500}(3693370825000000z_1^2 +$

$30000z_1(-444839980877 + 438674950000z_2 - 170070485000z_3) +$

$3(26434540496430823 + 5551068575000000z_2^2 + 112108350464500000z_3 +$

$1683946750000000z_3^2 - 20000z_2(291064806319 + 207937385000z_3)))$

The constraints is defined as

$\Rightarrow$constraints $=$ Apply$[$And$, \{-10 < z_1 < 10, -10 < z_2 < 10, -10 < z_3 < 10,$

$\qquad$ Element$[\{z_1, z_2, z_3\},$ Integers$]\}]$

$\Leftarrow -10 < z_1 < 10 \,\&\& -10 < z_2 < 10 \,\&\& -10 < z_3 < 10 \,\&\& (z_1|z_2|z_3) \in$ Integers i.e., a quadratic integer programming problem. The solution is,

$\Rightarrow \texttt{sol} = \texttt{Minimize}[\{\texttt{objective, constraints}\}, \{z_1, z_2, z_3\}]//N$
$\Leftarrow \{0.15277, \{z_1 \rightarrow 2., z_2 \rightarrow -2., z_3 \rightarrow -5.\}\}$

However, employing a floating point solution and simple rounding off leads to

$\Rightarrow \texttt{NMinimize}[\{\texttt{objective, Apply}[\texttt{And},$
$\quad \{-10 < z_1 < 10, -10 < z_2 < 10, -10 < z_3 < 10\}]\}, \{z_1, z_2, z_3\}]$
$\Leftarrow \{-3.60251 \times 10^{-15}, \{z_1 \rightarrow 2.5602, z_2 \rightarrow -2.3256, z_3 \rightarrow -4.9076\}\}$

and blindly rounding it provides

$\Rightarrow \texttt{objective}/.\{z_1 \rightarrow 3., z_2 \rightarrow -2., z_3 \rightarrow -5.\}$
$\Leftarrow 1.12359$

It is integer, but farther from Z.

8.7.2 Optimal Number of Oil Wells

There are five oil-bearing traps, see Shi (2014). The oil revenues of a single well in these different areas are 540, 375 240, 135 and 60, respectively. How many wells should be drilled in the different traps (x_i, i = 1, 2, 3, 4, 5) to maximize the total revenue?

$\Rightarrow \texttt{profit} = 540x_1 + 375x_2 + 240x_3 + 135x_4 + 60x_5$
$\Leftarrow 540x_1 + 375x_2 + 240x_3 + 135x_4 + 60x_5$

We have the following constraints,

(a) 31 wells to be drilled in total

$\Leftarrow x_1 + x_2 + x_3 + x_4 + x_5 = 31$

(b) Minimum one well should be drilled in any trap,

$\Rightarrow 1 \leq x_i, i = 1, 2, 3, 4, 5$

(c) The following relations are among the number of wells drilled in the different traps,

$\Rightarrow x_1 = 2x_2 - 1,$
$\Rightarrow x_2 = 2x_3 - 1,$
$\Rightarrow x_3 = x_4 - 2x_5$

Consequently, the constraints are,

$\Rightarrow$ constraints $=$ Apply[And,

$\{x_1 + x_2 + x_3 + x_4 + x_5 == 31, x_1 == 2x_2 - 1, x_2 == 2x_3 - 1,$

$x_3 == x_4 - 2x_5, 1 \leq x_1, 1 \leq x_2, 1 \leq x_3, 1 \leq x_4, 1 \leq x_5,$

Element[$\{x_1, x_2, x_3, x_4, x_5\}$]]}]

$\Leftarrow x_1 + x_2 + x_3 + x_4 + x_5 == 31 \,\&\&\, x_1 == -1 + 2x_2 \,\&\&$

$x_2 == -1 + 2x_3 \,\&\&\, x_3 == x_4 - 2x_5 \,\&\&\, 1 \leq x_1 \,\&\&\, 1 \leq x_2 \,\&\&$

$1 \leq x_3 \,\&\&\, 1 \leq x_4 \,\&\&\, 1 \leq x_5 \,\&\&\, (x_1 | x_2 | x_3 | x_4 | x_5) \in \textit{Integers}$

The results are

$\Rightarrow$ Maximize[$\{$profit, constraints$\}$, $\{x_1, x_2, x_3, x_4, x_5\}$]

$\Leftarrow \{11475, \{x_1 \rightarrow 13, x_2 \rightarrow 7, x_3 \rightarrow 4, x_x \rightarrow 6, x_5 \rightarrow 1\}\}$

If we drill the wells 13, 7, 4, 6 and 1, in different traps, respectively, then our total revenue will be 11,475.

8.7.3 Solution of GNSS Phase Ambiguity

Solution of the Global Navigation Satellite Systems (GNSS) phase ambiguity is considered as a global quadratic mixed integer programming task, which can be transformed into a pure integer problem with a given digit of accuracy. This section presents three alternative algorithms. Two of them are based on local and global linearization via McCormic Envelopes, respectively. These algorithms can be effective in case of simple configuration and relatively modest number of satellites. The third method is a locally nonlinear, iterative algorithm handling the problem as $\{-1, 0, 1\}$ programming and also lets us compute the next best integer solution easily. However, it should be kept in mind that the algorithm is a heuristic one, which does not always guarantee exactly finding of the global integer optimum. The procedure is very powerful utilizing the ability of the numeric-symbolic abilities of a computer algebraic system, like *Wolfram Mathematica,* and is fast enough for a minimum 4 satellites with normal configuration, which means the Geometric Dilution of Precision (GDOP) should be between 1 and 8. Wolfram Alpha and Wolfram Clouds Apps give possibility to run the suggested code even via cell phones. All of these algorithms are illustrated using numerical examples. The result of the third one was successfully compared with the LAMBDA method, in case of ten satellites sending signals on two carrier frequencies (L1 and L2) with weighting matrix used to weight the GNSS observations and computed as the inverse of the corresponding covariance matrix.

8.7.3.1 Introduction

Highly accurate static Global Navigation Satellite Systems (GNSS) positioning is achieved by the processing of relative phase ranges observed to the GNSS satellites at both the reference and the rover stations (Rózsa 2014; Juni and Rózsa 2019). To eliminate common biases such as the satellite and receiver clock error, the double-differenced phase observations are formed and adjusted using a least squares adjustment. The linearized observation equation of the double-differenced phase observations has the following form, see Leick et al. (2015).

$$\Delta\Delta\,\Phi_{AB}^{jk} = a_1\delta x_B^{jk} + a_2\delta y_B^{jk} + a_3\delta z_B^{jk} + \lambda\delta N_{AB}^{jk}, \tag{8.1}$$

where $\Delta\Delta\,\Phi_{AB}^{jk}$ is the double differences phase observations taken to the j-th and k-th satellite, δx_B, δy_B and δz_B are the relative coordinate differences between the reference (A) and rover (B) stations, λ is the wavelength of the signal, δN_{AB}^{jk} is the double differenced phase ambiguity and j refers to the so-called pivot satellite, that is used as a reference for forming the double differences.

The terms a_i in Eq. (8.1) stands for the coefficients resulting from the partial derivates of the linearized geometrical pseudorange distance equations. Let us assume that five satellites are measured concurrently on both the reference and the rover stations in two consecutive epochs. Since one satellite is used as a pivot satellite, four double differences are formed in each epoch. This means that altogether 8 observation equations are formed, which can be used to evaluate 7 unknowns (3 coordinate differences and 4 double-differenced phase ambiguities).

A usual solution of the problem is to estimate the unknowns using a least-squares adjustment (see e.g., Awange 2012, 2018), where the phase ambiguities are integers, while the coordinates are floating point variables. Consequently, the computation of the integer least-squares estimates of the GNSS cycle ambiguities leads to a mixed integer quadratic problem (see, Grafarend 2003; Teunissen et al. 1997),

$$(y - Ax - Bz)^T Q_y^{-1}(y - Ax - Bz) \to \min_{x,z}, \tag{8.2}$$

where y vector of double differences carrier phase observation in cycles, A design matrix for continuous-valued parameters (baseline components), B design matrix for ambiguities, x is unknown vector of continuous parameters, $x \in \mathcal{R}^3$, z the unknown ambiguity vector in cycles, $z \in Z^m$, where m depends on the number of the satellites and the carrier frequencies. The matrix Q_y^{-1} is the weight matrix (Q_y is the covariance matrix).

Solving the problem (Eq. 8.2) is well known to be *NP hard*. In other words, there exists no algorithm to find the global optimal integer solution to the problem Eq. (8.2) in polynomial time, see, (Xu et al. 2012). Thus, for real time applications such as wireless communication and Global Positioning Systems (GPS) kinematic positioning with many integer ambiguities due to the use of different wavelengths and/or different navigation satellite systems, it may be more realistic to expect some good suboptimal integer solutions than to find the global optimal integer solution. Basically, all the methods to construct suboptimal integer solutions may be classified into two types: simple rounding and sequential rounding. This is the whole point of the development of the LAMBDA approach by Teunissen et al. in the 1990s, i.e. (Teunissen 1990). LAMBDA does not "solve" integer rounding or sequential rounding it is a tool to make Integer Least Square (ILS) more efficient. To solve this task, probably the most popular procedure is the so-called LAMBDA method, see (Teunissen 1995; Jonge et al. 1996).

In this section, three different methods are introduced to solve quadratic integer programming: local linearization, global linearization and sequential nonlinear approach. All these methods can be time-wise effective in case of simple configuration and relatively low number of satellites (less than 8−10). The third method utilizes the numeric-symbolic ability of the computer algebraic system, like *Wolfram Mathematica* and fast enough for normal satellite configuration.

In the first part, the three methods to solve quadratic integer programming are introduced and illustrated via a simple example. Then the third method is demonstrated for different satellite configurations: a simple one with one carrier frequency using synthetic data and with two different carrier frequencies based on real field measured data. The results are compared with those of the latest version of the LAMBDA method.

8.7.3.2 Three Methods to Solve Integer Programming

In this section, three different methods are discussed and illustrated. All of them are based on the global float (floating point) solution of the problem. Let us consider a simple integer quadratic programming example adapted from Li and Sun (2006). We should minimize the following objective function,

$$q = 27x_1^2 - 18x_1x_2 + 4x_2^2 - 3x_2 \tag{8.3}$$

Let us visualize the problem, see Fig. 8.6.

We have to remark that this toy-problem can be solved directly. Excluding trivial solution ($x_1 = 0$, $x_2 = 0$), the integer minimum of $q(x_1, x_2)$ is ($x_1 = 1$, $x_2 = 3$).

However, employing linearization the computation time can be reduced considerably as shown later.

- **Local Linearization**

Let us linearize this q (x_1, x_2) function around the point of $\{x_{10}, x_{20}\}$:

$$
\begin{aligned}
qL = {} & 27x_{10}^2 + (x_1 - x_{10})(54x_{10} - 18x_{20}) - 3x_{20} \\
& - 18x_{10}x_{20} + 4x_{20}^2 + (x_2 - x_{20})(-3 - 18x_{10} + 8x_{20})
\end{aligned}
\tag{8.4}
$$

Fig. 8.6 Contour plot of the objective function $q(x_1, x_2)$

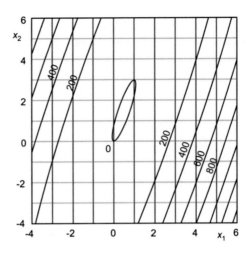

The float minimum of $q(x_1, x_2)$ is ($x_1 = 0.5$, $x_2 = 1.5$). Then the linearization point can be, $x_{10} = 1$, $x_{20} = 2$.

Then linearized model is,

$$qL_0 = -7 + 18x_1 - 5x_2 \qquad (8.5)$$

To minimize (8.5), constraints are required. Here, we use simply a heuristic approach suggested by Champton and Strzebonski (2008), assuming, that $x_{i0} - 1 \leq x_i \leq x_{i0} + 1$, $i = 1, 2$. Therefore, let us introduce new variables $\mu_i = x_i - x_{i0}$ to get a $(-1, 0, 1)$ linear programming problem. Then

$$qL_{0\mu} = 1 + 18\mu_1 - 5\mu_2 \qquad (8.6)$$

The lower and upper bounds for the variables are $\{-1 \leq \mu_1 \leq 1, \ -1 \leq \mu_2 \leq 1\}$.

This linear problem can be solved via linear programming. It can be written in the form of *min* $c\mu$ under the restriction $m\mu \geq b$, where

$$c = (18, \ -5), \quad m = \begin{pmatrix} 1 & 0 \\ 0 & 1 \\ -1 & 0 \\ 0 & -1 \end{pmatrix}, \quad b = \begin{pmatrix} -1 \\ -1 \\ -1 \\ -1 \end{pmatrix} \qquad (8.7)$$

The solution is $\Delta = \{0, 1\}$, then $x_0 = x_0 + \Delta = \{1, 3\}$. Employing this results in to a new linearization point $x_0 = \{1, 3\}$.

Since $\{1, 2\} \rightarrow \{1, 3\}$, the minimum is at $\{1, 3\}$. The running time is considerably smaller than it was in case of the global nonlinear solution.

- **Global Linearization**

In this case linearization is carried out not around a single point but on a restricted domain. The global bound is the domain where the integer solution may exist, and its centre is the float solution.

– Global Bound

The radius of this domain can be computed from the ratio of the maximal and minimal eigenvalues of the following bilinear form Li and Sun (2006). This bilinear form in our case is

$$
(\tilde{x}_1 - 0.5 \quad \tilde{x}_2 - 1.5) \; Q \begin{pmatrix} \tilde{x}_1 - 0.5 \\ \tilde{x}_2 - 1.5 \end{pmatrix} \tag{8.8}
$$

where $\tilde{x}_1$ and $\tilde{x}_2$ are the integer solutions of the optimization problem and the centre of this domain is the float solution $\{x_1 = 0.5, x_2 = 1.5\}$. The matrix Q is computed as

$$
Q = \frac{1}{2} \begin{pmatrix} \frac{\partial^2 q}{\partial x_1^2} & \frac{\partial^2 q}{\partial x_1 \partial x_2} \\ \frac{\partial^2 q}{\partial x_1 \partial x_2} & \frac{\partial^2 q}{\partial x_2^2} \end{pmatrix} \tag{8.9}
$$

Then the matrix of the bilinear form in our case is $Q = \begin{pmatrix} 27 & -9 \\ -9 & 4 \end{pmatrix}$ and the eigenvalues $\{30.1031, 0.896918\}$.

The ratio of the maximum and minimum values of λ's is $\kappa = 33.5628$. Then the radius of a n dimensional hypersphere with the float solution as a centre, $R = 1/2\sqrt{n\kappa}$, now $n = 2$, so $R = 4.09651$.

Using box-type constraint

$$
-3 \le x_1 \le 4, \quad -2 \le x_2 \le 5 \tag{8.10}
$$

Box-bounded region can be seen in Fig. 8.7.

Having global bound for the solution, the problem becomes a constrained nonlinear problem. Further simplification is possible via linearization of the objective function. In order to linearize our function over this region, McCormick Envelopes is employed, which is described in the next paragraph.

– Linearization via McCormick Envelopes

The McCormick envelopes are the convex relaxation of a quadratic problem via introducing new variables for the quadratic terms and employing the additional constraints (Mitsos et al. 2009). In general, we introduce new variables,

Fig. 8.7 Disk and the box-bounded region of the global integer optimum

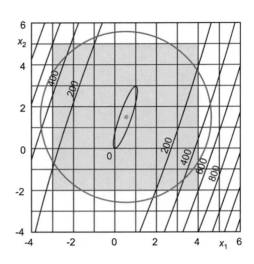

$$w_{ij} = x_i x_j \tag{8.11}$$

with the following constraints,

$$
\begin{aligned}
w_{ij} &\geq x_{iL} x_j + x_i x_{jL} - x_{iL} x_{jL} \\
w_{ij} &\geq x_{iU} x_j + x_i x_{jU} - x_{iU} x_{jU} \\
w_{ij} &\leq x_{iU} x_j + x_{jL} x_i - x_{iU} x_{jL} \\
w_{ij} &\leq x_i x_{jU} + x_{iL} x_j - x_{jU} x_{iL}
\end{aligned}
\tag{8.12}
$$

where $x_L \leq x \leq x_U$.

Let us employ McCormic envelopes' approach to our simple quadratic problem. Employing (8.11), the linear objective function of our example is,

$$qL = 27w_{11} - 18w_{12} + 4w_{22} - 3x_2 \tag{8.13}$$

The box-type bounds are,

$$x_{1L} = -3; \quad x_{1U} = 4; \quad x_{2L} = -2; \quad x_{2U} = 5 \tag{8.14}$$

and the function values at the lower and upper bounds are 157.

The additional inequality constraints, the McCormick envelopes are,

$$w_{11} \geq x_{1L}x_1 + x_1x_{1L} - x_{1L}x_{1L},$$
$$w_{11} \geq x_{1U}x_1 + x_1x_{1U} - x_{1U}x_{1U},$$
$$w_{11} \leq x_{1U}x_1 + x_1x_{1L} - x_{1U}x_{1L},$$
$$w_{11} \leq x_1x_{1U} + x_{1L}x_1 - x_{1U}x_{1L},$$
$$w_{22} \geq x_{2L}x_2 + x_2x_{2L} - x_{2L}x_{2L},$$
$$w_{22} \geq x_{2U}x_2 + x_2x_{2U} - x_{2U}x_{2U},$$
$$w_{22} \leq x_{2U}x_2 + x_2x_{2L} - x_{2U}x_{2L},$$
$$w_{12} \geq x_{1L}x_2 + x_1x_{2L} - x_{1L}x_{2L}, \tag{8.15}$$
$$w_{12} \geq x_{1U}x_2 + x_1x_{2U} - x_{1U}x_{2U},$$
$$w_{12} \leq x_{1U}x_2 + x_1x_{2L} - x_{1U}x_{2L},$$
$$w_{12} \leq x_1x_{2U} + x_{1L}x_2 - x_{2U}x_{1L},$$
$$w_{11} > 0, \quad w_{12} \neq 0, \quad w_{22} > 0,$$
$$x_{1L} \leq x_1 \leq x_{1U}, \quad x_{2L} \leq x_2 \leq x_{2U}, \quad x_1 \neq 0, \quad x_2 \neq 0$$

therefore

$$w_{11} \geq -9 - 6x_1, \quad w_{11} \geq -16 + 8x_1. \quad w_{11} \leq 12 + x_1,$$
$$w_{11} \leq 12 + x_1, \quad w_{22} \geq -4 - 4x_2, \quad w_{22} \geq -25 + 10x_2,$$
$$w_{22} \leq 10 + 3x_2, \quad w_{12} \geq -6 - 2x_1 - 3x_2,$$
$$w_{12} \geq -20 + 5x_1 + 4x_2, \quad w_{12} \leq 8 - 2x_1 + 4x_2, \tag{8.16}$$
$$w_{12} \leq 15 + 5x_1 - 3x_2, \quad w_{11} > 0, \quad w_{12} \neq 0, \quad w_{22} > 0,$$
$$-3 \leq x_1 \leq 4, \quad -2 \leq x_2 \leq 5, \quad x_1 \neq 0, \quad x_2 \neq 0,$$
$$(x_1|x_2|w_{11}|w_{12}|w_{22}) \in \text{Integers}.$$

Now, this is a linear integer programming problem. The price of the linearization is the increase in the number of variables.

Considering the new variables $(x_1, x_2, w_{11}, w_{12}, w_{22})$, in (8.13), the coefficient vector of the objective function is

$$c = \{0, \ -3, \ 27, \ -18, \ 4\} \tag{8.17}$$

We introduce a small positive constant $\varepsilon = 10^{-3}$ in order to exclude the trivial solution $\{0, 0\}$. Then the constraints are,

$$m = \begin{pmatrix} -2x_{1L} & 0 & 1 & 0 & 0 \\ -2x_{1U} & 0 & 1 & 0 & 0 \\ x_{1U}+x_{1L} & 0 & -1 & 0 & 0 \\ 0 & -2x_{2L} & 0 & 0 & 1 \\ 0 & -2x_{2U} & 0 & 0 & 1 \\ 0 & x_{2U}+x_{2L} & 0 & 0 & -1 \\ -x_{2L} & -x_{1L} & 0 & 1 & 0 \\ -x_{2U} & -x_{1U} & 0 & 1 & 0 \\ x_{2L} & x_{1U} & 0 & -1 & 0 \\ x_{2U} & x_{1L} & 0 & -1 & 0 \\ 0 & 0 & 1 & 0 & 0 \\ 0 & 0 & 0 & 1 & 0 \\ 0 & 0 & 0 & 0 & 1 \\ 1 & 0 & 0 & 0 & 0 \\ 1 & 0 & 0 & 0 & 0 \\ 0 & 1 & 0 & 0 & 0 \\ 0 & 1 & 0 & 0 & 0 \\ 1 & 0 & 0 & 0 & 0 \\ 0 & 1 & 0 & 0 & 0 \end{pmatrix} ; \quad b = \begin{pmatrix} -x_{1L}x_{1L} \\ -x_{1U}x_{1U} \\ x_{1U}x_{1L} \\ -x_{2L}x_{2L} \\ -x_{2U}x_{2U} \\ x_{2U}x_{2L} \\ -x_{1L}x_{2L} \\ -x_{1U}x_{2U} \\ x_{1U}x_{2L} \\ x_{2U}x_{1L} \\ \epsilon \\ \epsilon \\ \epsilon \\ x_{1L} \\ -x_{1U} \\ x_{2L} \\ -x_{2U} \\ \epsilon \\ \epsilon \end{pmatrix} \qquad (8.18)$$

Now linear programming can be employed, which results in $\{x_1, x_2\} = \{2, 3\}$. Using this result as a new upper limit, see Fig. 8.8, the second approach can be computed,

$$x_{1L} = -3; \quad x_{1U} = 2; \quad x_{2L} = -2; \quad x_{2U} = 3 \qquad (8.19)$$

Fig. 8.8 Box-regions of the first three iterations of the linear problem. The meaning of the values of the box-regions can be seen in Table 8.1

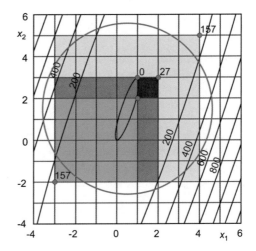

Table. 8.2 Results of the global linearization

Iteration	Bounds for x_1	Bounds for x_2	Solution	Objective at lower bound	Objective at upper bound
0	$\{-3, 4\}$	$\{-2, 5\}$	$\{2, 3\}$	157	157
1	$\{-3, 2\}$	$\{-2, 3\}$	$\{1, 2\}$	157	27
2	$\{1, 2\}$	$\{2, 3\}$	$\{1, 3\}$	1	27
3	$\{1, 1\}$	$\{3, 3\}$	$\{1, 3\}$	0	0

and accordingly, a new McCormick envelopes will be determined. The value of the objective function at the lower bound is 157 and at the upper bound is 27. The results of this iteration process can be seen in Table 8.2.

No more approximation step is necessary since the next iteration will give the same result, so it is a fixed point of the iteration process, see Fig. 8.8.

This method is converging, however now the size of the linear model is 19 × 5. After the linearization techniques, in the next section a non-linear method will be discussed.

- **Successive Nonlinear Method**

Here we employ the heuristic technique suggested by Champton and Strzebonski (2008). We are looking for an improved integer solution x_{i+1} in the neighbourhood $\mathcal{N}$ of the actual one, x_i, assuming that $x_{i+1} \in \mathcal{N}(x_i : L_1(x_i, x_{i+1}) = 1)$ and that x_{i+1} is in the neighbourhood of x_i with L_1 norm equal 1. In this way, we have a $\{-1, 0, 1\}$ quadratic problem.

Starting with $x_0 = \{1, 2\}$, and introducing the new variables $x_1 = x_{01} + \mu_1$ and $x_2 = x_{02} + \mu_2$ we get our objective function

$$q = 1 + 18\mu_1 + 27\mu_1^2 - 5\mu_2 - 18\mu_1\mu_2 + 4\mu_2^2 \qquad (8.20)$$

and the constraints are: $-1 \leq \mu_1 \leq 1, \quad -1 \leq \mu_2 \leq 1$.

Then minimizing q, the solution can be computed as,

$$x_0 = \{1, 2\} + \{0, 1\} = \{1, 3\} \qquad (8.21)$$

Now, no more computation step is necessary.

Until now, we have considered a pure integer problem. However, in the case of mixed integer problem, a part of the variables are continuous variables.

8.7.3.3 Mixed Integer Programming

This type of problem can be transformed into a pure integer one. Let us consider the following illustrative example. Let the function to be maximized be,

$$u = 3x_1 + 5x_2 + y_1 + 2y_2, \tag{8.22}$$

with continuous or "float" variables $(y_1, y_2) \in \mathcal{R}$, and with integer variables $(x_1, x_2) \in \mathcal{Z}$. First, let us solve the continuous version of the problem. The constraints are

$$
\begin{aligned}
2x_1 + x_2 + y_1 + 2y_2 \le 4; & \quad x_1 + 3x_2 + y_1 + y_2 \le 5; \\
x_1 \ge 0; \quad x_2 \ge 0; \quad y_1 \ge 0; & \quad y_2 \ge 0
\end{aligned}
\tag{8.23}
$$

Then the continuous solutions are (employing post rationalization),

$$x_1 = 7/5, \quad x_2 = 6/5. \tag{8.24}$$

Now, we introduce new integer variables as

$$\xi_i = 10^{Accuracy(y_i)} y_i, \quad i = 1, 2 \tag{8.25}$$

In our case let Accuracy $(w_i) = 3$

$$\xi_1 = 1000y_1, \quad \xi_2 = 1000y_2 \tag{8.26}$$

In this way, the continuous variable is considered as an integer one with 3 digit accuracy. Then the objective with the new variables is

$$u = 3x_1 + 5x_2 + \frac{\xi_1}{1000} + \frac{\xi_2}{500}, \tag{8.27}$$

where now all variables are integers,

$$
\begin{aligned}
2x_1 + x_2 + \frac{\xi_1}{1000} + \frac{\xi_2}{500} \le 4; & \quad x_1 + 3x_2 + \frac{\xi_1}{1000} + \frac{\xi_2}{1000} \le 5; \\
x_1 \ge 0; \quad x_2 \ge 0; \quad \xi_1 \ge 0; & \quad \xi_2 \ge 0; \\
(x_1 | x_2 | \xi_1 | \xi_2) \in \text{Integers}
\end{aligned}
\tag{8.28}
$$

The solution is

$$x_1 = 1, \quad x_2 = 1, \quad \xi_1 = 1000, \quad \xi_2 = 0 \tag{8.29}$$

then

$$\{y_1,\ y_2\} = 10^{-3} \quad \{\xi_1,\ \xi_2\} = \{1,\ 0\} \tag{8.30}$$

This technique will be employed in the next section dealing with the GNSS ambiguity solution.

8.7.3.4   Computing the Next Best Integer Solution

With ambiguity resolution, one often also would like to be able to compute the next best integer solution for ambiguity validation purposes using e.g. the ratio test, see (Teunissen and Verhagen 2004). Let us illustrate this computation with the problem Eq. (8.3). The objective function is,

$$q = 27x_1^2 - 18x_1x_2 + 4x_2^2 - 3x_2 \tag{8.31}$$

Then, the first best integer minimum,

$$\{x_1,\ x_2\} = \{1,\ 3\} \tag{8.32}$$

Introducing a new constrain to avoid this minimum,

$$q > q(1,\ 3) \tag{8.33}$$

we solve the problem again,

$$\{x_1,\ x_2\} = \{1,\ 2\} \tag{8.34}$$

That means, after computing the best solution, we can construct a new constrain, and repeat the minimization to avoid best solution and get the next best integer value.

8.7.3.5 Solution of GNSS Phase Ambiguity Problem

In this section, the third algorithm with some modifications will be employed since it has turned out that the first and second algorithms can solve only the simple configuration problem (Paláncz 2018).

Now, let us consider a more serious model configuration. The data are from field measurements, and the theoretical results for the coordinates is a zero vector $\{x, y, z\} \rightarrow \{0, 0, 0\}$, a base-line solution.

The suggested algorithm is a heuristic one that does not ensure obtaining a global integer minimum. However, when this minimum is in the neighborhood of the floating minimum, the method can be very efficient. The flow chart of the algorithm is presented in Fig. 8.9.

In this case of *Successive Nonlinear Solution* for a real satellite configuration, we have 10 satellites. One of them is the reference one, and the other 9 are sending signals on two carrier frequencies (L1 and L2). In total, we have 18 ambiguities and 3 coordinates as unknowns. The actual values of the input arrays are adopted from Khodabandeh (2018). The matrix A and vector y are,

Fig. 8.9 Flow chart of the nonlinear algorithm

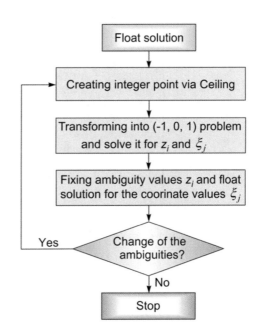

$$
A = \begin{pmatrix}
0.11350 & 0.40225 & 0.50828 \\
-1.09230 & -0.16510 & 1.02080 \\
-0.44717 & -0.17185 & -0.40806 \\
-0.81536 & -0.47265 & -0.13043 \\
-0.37498 & 0.88919 & 1.02100 \\
-0.28402 & 0.51891 & 1.08210 \\
-0.63514 & -0.23136 & 0.96756 \\
-1.59760 & -0.14047 & 0.59082 \\
-1.14130 & -0.32446 & -0.30627 \\
0.11350 & 0.40225 & 0.50828 \\
-1.09230 & -0.16510 & 1.02080 \\
-0.44717 & -0.17185 & -0.40806 \\
-0.81536 & -0.47265 & -0.13043 \\
-0.37498 & 0.88919 & 1.02100 \\
-0.28402 & 0.51891 & 1.08210 \\
-0.63514 & -0.23136 & 0.96756 \\
-1.59760 & -0.14047 & 0.59082 \\
-1.14130 & -0.32446 & -0.30627 \\
0.11350 & 0.40225 & 0.50828 \\
-1.09230 & -0.16510 & 1.02080 \\
-0.44717 & -0.17185 & -0.40806 \\
-0.81536 & -0.47265 & -0.13043 \\
-0.37498 & 0.88919 & 1.02100 \\
-0.28402 & 0.51891 & 1.08210 \\
-0.63514 & -0.23136 & 0.96756 \\
-1.59760 & -0.14047 & 0.59082 \\
-1.14130 & -0.32446 & -0.30627 \\
0.11350 & 0.40225 & 0.50828 \\
-1.09230 & -0.16510 & 1.02080 \\
-0.44717 & -0.17185 & -0.40806 \\
-0.81536 & -0.47265 & -0.13043 \\
-0.37498 & 0.88919 & 1.02100 \\
-0.28402 & 0.51891 & 1.08210 \\
-0.63514 & -0.23136 & 0.96756 \\
-1.59760 & -0.14047 & 0.59082 \\
-1.14130 & -0.32446 & -0.30627
\end{pmatrix}, \;
y = \begin{pmatrix}
-4.75710 \\
2.85700 \\
9.13270 \\
0.19081 \\
1.14110 \\
-4.75720 \\
-4.75910 \\
-4.18520 \\
-12.56000 \\
6.83820 \\
2.68960 \\
4.87710 \\
-0.73147 \\
-1.46970 \\
3.17590 \\
-1.46590 \\
2.20040 \\
1.95520 \\
-0.70882 \\
0.01589 \\
0.27637 \\
-0.08741 \\
0.18193 \\
-0.15592 \\
-0.15185 \\
-0.09746 \\
-0.34652 \\
-0.18282 \\
-0.27811 \\
-0.77963 \\
-0.61641 \\
-0.34707 \\
-0.03891 \\
-0.09485 \\
-0.27346 \\
0.17948
\end{pmatrix}
$$

The structure of matrix B is,

$$
B = \begin{pmatrix}
R_1 & 0 \\
0 & R_2 \\
0 & 0 \\
0 & 0
\end{pmatrix},
$$

where

$$
R_1 = \begin{pmatrix}
0.19 & 0 & 0 & 0 & 0 & 0 & 0 & 0 & 0 \\
0 & 0.19 & 0 & 0 & 0 & 0 & 0 & 0 & 0 \\
0 & 0 & 0.19 & 0 & 0 & 0 & 0 & 0 & 0 \\
0 & 0 & 0 & 0.19 & 0 & 0 & 0 & 0 & 0 \\
0 & 0 & 0 & 0 & 0.19 & 0 & 0 & 0 & 0 \\
0 & 0 & 0 & 0 & 0 & 0.19 & 0 & 0 & 0 \\
0 & 0 & 0 & 0 & 0 & 0 & 0.19 & 0 & 0 \\
0 & 0 & 0 & 0 & 0 & 0 & 0 & 0.19 & 0 \\
0 & 0 & 0 & 0 & 0 & 0 & 0 & 0 & 0.19
\end{pmatrix}
$$

$$
R_2 = \begin{pmatrix}
0.24 & 0 & 0 & 0 & 0 & 0 & 0 & 0 & 0 \\
0 & 0.24 & 0 & 0 & 0 & 0 & 0 & 0 & 0 \\
0 & 0 & 0.24 & 0 & 0 & 0 & 0 & 0 & 0 \\
0 & 0 & 0 & 0.24 & 0 & 0 & 0 & 0 & 0 \\
0 & 0 & 0 & 0 & 0.24 & 0 & 0 & 0 & 0 \\
0 & 0 & 0 & 0 & 0 & 0.24 & 0 & 0 & 0 \\
0 & 0 & 0 & 0 & 0 & 0 & 0.24 & 0 & 0 \\
0 & 0 & 0 & 0 & 0 & 0 & 0 & 0.24 & 0 \\
0 & 0 & 0 & 0 & 0 & 0 & 0 & 0 & 0.24
\end{pmatrix}
$$

The structure of matrix Q_y is,

$$
Q_y = \begin{pmatrix}
R_3 & 0 & 0 & 0 \\
0 & R_3 & 0 & 0 \\
0 & 0 & R_4 & 0 \\
0 & 0 & 0 & R_4
\end{pmatrix} 10^{-4}
$$

where

$$
R_3 = \begin{pmatrix}
0.59566 & 0.30044 & 0.30044 & 0.30044 & 0.30044 & 0.30044 & 0.30044 & 0.30044 & 0.30044 \\
0.30044 & 0.43245 & 0.30044 & 0.30044 & 0.30044 & 0.30044 & 0.30044 & 0.30044 & 0.30044 \\
0.30044 & 0.30044 & 1.43750 & 0.30044 & 0.30044 & 0.30044 & 0.30044 & 0.30044 & 0.30044 \\
0.30044 & 0.30044 & 0.30044 & 0.61351 & 0.30044 & 0.30044 & 0.30044 & 0.30044 & 0.30044 \\
0.30044 & 0.30044 & 0.30044 & 0.30044 & 1.69020 & 0.30044 & 0.30044 & 0.30044 & 0.30044 \\
0.30044 & 0.30044 & 0.30044 & 0.30044 & 0.30044 & 0.53146 & 0.30044 & 0.30044 & 0.30044 \\
0.30044 & 0.30044 & 0.30044 & 0.30044 & 0.30044 & 0.30044 & 0.38545 & 0.30044 & 0.30044 \\
0.30044 & 0.30044 & 0.30044 & 0.30044 & 0.30044 & 0.30044 & 0.30044 & 0.97401 & 0.30044
\end{pmatrix}
$$

$$
R_3 = \begin{pmatrix}
5956.6 & 3004.4 & 3004.4 & 3004.4 & 3004.4 & 3004.4 & 3004.4 & 3004.4 & 3004.4 \\
3004.4 & 4324.5 & 3004.4 & 3004.4 & 3004.4 & 3004.4 & 3004.4 & 3004.4 & 3004.4 \\
3004.4 & 3004.4 & 14375. & 3004.4 & 3004.4 & 3004.4 & 3004.4 & 3004.4 & 3004.4 \\
3004.4 & 3004.4 & 3004.4 & 6135.1 & 3004.4 & 3004.4 & 3004.4 & 3004.4 & 3004.4 \\
3004.4 & 3004.4 & 3004.4 & 3004.4 & 16902. & 3004.4 & 3004.4 & 3004.4 & 3004.4 \\
3004.4 & 3004.4 & 3004.4 & 3004.4 & 3004.4 & 5314.6 & 3004.4 & 3004.4 & 3004.4 \\
3004.4 & 3004.4 & 3004.4 & 3004.4 & 3004.4 & 3004.4 & 3854.5 & 3004.4 & 3004.4 \\
3004.4 & 3004.4 & 3004.4 & 3004.4 & 3004.4 & 3004.4 & 3004.4 & 9740.1 & 3004.4 \\
3004.4 & 3004.4 & 3004.4 & 3004.4 & 3004.4 & 3004.4 & 3004.4 & 3004.4 & 28100.
\end{pmatrix}
$$

The results of the iteration steps are in Table 8.3.

No more iteration is necessary since we get the same result. Applying the LAMBDA method, the same result are achieved (Khodabandeh 2018).

8.7.3.6 Concluding Remarks

The algorithms based on local as well as global linearization were proved to be efficient in cases of one carrier frequency. The third one, a locally nonlinear,

Table. 8.3 Results of the iteration steps

Variable	Float solution	1st iteration	2nd iteration
z_1	−25.160	−25	−25
z_2	14.502	15	15
z_3	48.192	49	48
z_4	0.993	1	1
z_5	5.740	6	6
z_6	−25.403	−25	−25
z_7	−25.546	−25	−25
z_8	−22.234	−22	−22
z_9	−65.855	−65	−66
z_{10}	27.886	28	28
z_{11}	10.614	11	11
z_{12}	20.126	20	20
z_{13}	−3.003	−3	−3
z_{14}	−6.218	−6	−6
z_{15}	12.690	13	13
z_{16}	−6.420	−6	−6
z_{17}	8.823	9	9
z_{18}	8.123	8	8
ξ_1	13.639	7.575	−3.129
ξ_2	−55.421	−7.564	2.466
ξ_3	101.112	20.533	0.484
Residual	2.816	405.690	4.716

iterative algorithm, can be employed successfully when L1 and L2 carrier frequencies are used with weighting matrix having elements of very different magnitudes. For multi-GNSS cases, when more satellite should be tracked simultaneously, one may employ the same strategy. However, at this time, the memory management of CAS does not enable handling of large systems of equations.

8.8 Exercises

8.8.1 Study of Mixed Integer Programming

We have seen that such problems can be handled by introducing new variables of the floating point variables like,

$$\xi = 10^n x,$$

where $x \in R$ (real) and $\xi \in Z$ (integer). The question is, which is the best value for the exponent n? Let us investigate this problem in the following mixed integer quadratic 0/1 programming problem!

Considering the objective to be maximized to be,

$$\Rightarrow f = y_1 + y_2 + y_3 + 5x_1^2$$

The constraints are

$$\Rightarrow g1 = 3x_1 - y_1 - y_2 \leq 0$$
$$\Rightarrow g2 = -x_1 - 0.1y_2 + 0.25y_3 \leq 0$$
$$\Rightarrow g3 = 0.2 - x_1 \leq 0$$

The integer variables y_i, $i = 1, 2, 3$ can be 0 or 1,

$$\Rightarrow \texttt{vars} = \{y_1, y_2, y_3\};$$
$$\Rightarrow \texttt{constraints1} = \texttt{Map}[0 \leq \# \leq 1\&, \texttt{vars}]$$
$$\Leftarrow \{0 \leq y_1 \leq 1, 0 \leq y_2 \leq 1, 0 \leq y_3 \leq 1\}$$

Since x_1 is a floating point variable, first we introduce a new integer variable ξ as,

$$\xi = 10x_1.$$

Then we have further constraints,

$\Rightarrow$ constraints2 =
$\{3\xi10^-1 - y_1 - y_2 \le 0, -\xi10^{-1} - 0.1y_2 + 0.25y_3 \le 0, 0.2 - \xi10^{-1} \le 0\}//$
Rationalize

$$\Leftarrow \left\{\frac{3\xi}{10} - y_1 - y_2 \le 0, -\frac{\xi}{10} - \frac{y_2}{10} + \frac{y_3}{4} \le 0, \frac{1}{5} - \frac{\xi}{10} \le 0\right\}$$

Combining all the constraints, we get

$\Rightarrow$ constraints = Apply[And,
   Join[constraints1, constraints2, {Element[{y_1, y_2, y_3, \xi}, Integers]}]]

$$\Leftarrow 0 \le y_1 \le 1 \&\& 0 \le y_2 \le 1 \&\& 0 \le y_3 \le 1 \&\& \frac{3x}{10} - y_1 - y_2 \le 0 \&\&$$

$$-\frac{x}{10} - \frac{y_2}{10} + \frac{y_3}{4} \le 0 \&\& \frac{1}{5} - \frac{x}{10} \le 0 \&\& (y_1 | y_2 | y_3 | \xi) \in \text{Integers}$$

$\Rightarrow$ objective = $y_1 + y_2 + y_3 + 5(\xi10^{-1})^2$

$$\Leftarrow \frac{\xi^2}{20} + y_1 + y_2 + y_3$$

$\Rightarrow$ Maximize[{objective, constraints}, {y_1, y_2, y_3, \xi}]

$$\Leftarrow \left\{\frac{24}{5}, \{y_1 \to 1, y_1 \to 1, y_3 \to 1, \xi \to 6\}\right\}$$

Increasing n, $(n = 2, 3, \ldots, \infty)$ we get more precise results for x_1, see Table 8.2. Table 8.4 shows how the result can change, i.e., $x_1 = 0.6$, with the increasing exponent.

8.8.2   Mixed Integer Least Square

The mixed integer least squares problem in case of a linear model can be formulated as follows,

$$(y - Ax - Bz)^T Q^{-1}(y - Ax - Bz) \to \min_{x,z},$$

where y, A, B and Q are known real vector and matrices, and x is a real and z is an integer unknown vector,

Table. 8.4 The precision x_1 depending of n

n	objective	x1
1	24/5 = 4.8	0.6
2	2589/500 = 5.178	0.66
...	...	...
∞	47/9 = 5.22...0.2	2/3 = 0.66...0.6

$$x \in R \text{ and } z \in Z.$$

Let us consider the actual values of the input arrays as

$$\Rightarrow y = \begin{pmatrix} 2.5602 \\ -2.3256 \end{pmatrix}; \; Q = \begin{pmatrix} 4.0722 & -1.5606 \\ -1.5606 & 1.0945 \end{pmatrix}; \; A = \begin{pmatrix} 1.2 & 0.97 \\ 0.1 & 1.99 \end{pmatrix}; \; B = \begin{pmatrix} 2.1 & 0.12 \\ 3.12 & 0.5 \end{pmatrix};.$$

Now, two new integer variables for elimination of the real unknown are introduced as,

$$\Leftarrow \xi_1 = 100x_1, \; \xi_2 = 100x_2$$

Then rationalizing the input,

$$\Rightarrow \{yR, QR, AR, BR\} = \texttt{Rationalize}[\{y, Q, A, B\}]$$

$$\Leftarrow \left\{ \left\{ \left\{ \frac{12801}{5000} \right\}, \left\{ -\frac{2907}{1250} \right\} \right\}, \left\{ \left\{ \frac{20361}{5000}, -\frac{7803}{5000} \right\}, \left\{ -\frac{7803}{5000}, \frac{2189}{2000} \right\} \right\}, \right.$$
$$\left. \left\{ \left\{ \frac{6}{5}, \frac{97}{100} \right\}, \left\{ \frac{1}{10}, \frac{199}{100} \right\} \right\}, \left\{ \left\{ \frac{21}{10}, \frac{3}{25} \right\}, \left\{ \frac{78}{25}, \frac{1}{2} \right\} \right\} \right\}$$

the objective function,

$$\Rightarrow \texttt{objective} =$$
$$\left((yR - AR.\{\xi_1 10^{-2}, \xi_2 10^{-2}\} - BR.\{z_1, z_2\})//\texttt{Flatten} \right).\texttt{Inverse}[QR].$$

$$\left(\left(\begin{pmatrix} \frac{12801}{5000} \\ -\frac{2907}{1250} \end{pmatrix} - AR.\begin{pmatrix} \xi_1 10^{-2} \\ \xi_2 10^{-2} \end{pmatrix} - BR.\begin{pmatrix} z_1 \\ z_2 \end{pmatrix} \right) \right)//\texttt{Simplify}//\texttt{First}$$

$$\Leftarrow \frac{1}{2021550540000}$$
$$(10614556745028 + 64917271080000 z_1^2 + 1221082800000 z_2^2 + 30802339440 \, \xi_1 +$$
$$199134600 \, \xi_1^2 + 233946969684 \, \xi_2 + 1192491360 \, \xi_1 \xi_2 + 2318098663 \, \xi_2^2 +$$
$$2400 z_2 (2363910996 + 10969210 \, \xi_1 + 44240009 \, \xi_2) +$$
$$6000 z_1 (6272889330 + 2950454880 z_2 + 33997596 \, \xi_1 + 129192397 \, \xi_2))$$

In order to be able to define the constraints properly it is useful to solve the floating point problem first,

$$\Rightarrow \texttt{NMinimize}[\texttt{objective}, \{z_1, z_2, \xi_1, \xi_2\}]$$
$$\Leftarrow \{-1.2367 \times 10^{-13}, \{z_1 \rightarrow 2.25206, z_2 \rightarrow -17.966, \xi_1 \rightarrow 14.477, \xi_2 \rightarrow -19.2707\}\}$$

Then the constraints can be written concerning the result of floating point problem, namely,

$\Rightarrow$ `constraints` $=$

`Apply[And,`$\{0 < z_1 < 5, -20 < z_2 < 0, 0 < \xi_1 < 20, -25 < \xi_2 < 0,$

`Element[`$\{z_1, z_2, \xi_1, \xi_2\}$`,Integers]}]`

$\Leftarrow 0 < z_1 < 5 \&\& -20 < z_2 < 0 \&\& 0 < \xi_1 < 20 \&\& -25 < \xi_2 < 0 \&\& (z_1 | z_2 | \xi_1 | \xi_2) \in$ *Integers*

This is a quadratic integer programming problem with the solution,

$\Rightarrow$ `Minimize[objective,constraints,`$\{z_1, z_2, \xi_1, \xi_2\}$`]`

$$\Leftarrow \left\{ \frac{16527800527}{2021550540000}, \{z_1 \rightarrow 2, z_2 \rightarrow -17, \xi_1 \rightarrow 19, \xi_2 \rightarrow -1\} \right\}$$

$\Rightarrow$ `%//N`

$\Leftarrow \{0.0081758, \{z_1 \rightarrow 2., z_2 \rightarrow -17., \xi_1 \rightarrow 19., \xi_2 \rightarrow -1.\}\}$

Therefore $x_1 = 0.19$ and $x_2 = -0.01$.

References

Awange JL (2018) GNSS environmental sensing. Springer International Publishing AG, Revolutionizing environmental monitoring

Awange JL (2012) Environmental monitoring using GNSS. Springer, Berlin, Heidelberge, New York, Global navigation satellite systems

Bertsimas D, Perakis G, Tayur S (2000) A new algebraic geometry algorithm for integer programming. Manage Sci 46(7):999–1008

Champton B, Strzebonski A (2008) Constrained optimization. Wolfram Mathematica Tutorial Collection, Wolfram Res., Inc. http://www.johnboccio.com/MathematicaTutorials/08_ConstrainedOptimization

Conti P, Traverso C (1991) Buchberger algorithm and integer programming. In: Proceedings AAECC-9 (New Orleans), Springer, Lecture Notes in Computer Science, vol 539, pp 130–139

Cox DA, Little J, O'Shea D (1998, 2015) Using algebraic geometry, Springer, New York

Grafarend EW (2003) Mixed integer-real valued adjustment (IRA) problems: GPS initial cycle ambiguity resolution by means of the LLL algorithm. In: Grafarend EW, Krumm FW, Schwarze VS (eds) Book geodesy—the challenge of the 3rd millennium, Ed: Springer, ISBN-13:978-3540431602

Jonge P, Tiberius C (1996) The LAMBDA method for integer ambiguity estimation implementation aspects, Publications of the Delft Geodetic Computing Centre, LGR-Series, No 12

Juni I, Sz R (2019) Validation of a New Model for the Estimation of Residual Tropospheric Delay Error Under Extreme Weather Conditions. Periodica Polytechnica Civil Engineering 63 (1):121–129. https://doi.org/10.3311/PPci.12132

Kapadia D (2002) "Integer Programming by the Groebner Basis Method" from the Notebook Archive https://notebookarchive.org/2018-10-10r94ivKhodabandehA (2018): Private Communication

Leick A, Rapoport L, Tatamikov D (2015) GPS satellite surveying. Wiley, New York

Li D, Sun X (2006) Nonlinear Integer Programming. Springer, Berlin, Heidelberg

Mitsos A, Chachuat B, Barton PI (2009) McCormick-based relaxations of algorithms. SIAM J Optim 20(2):573–601. https://doi.org/10.1137/080717341

Paláncz B (2018): Numeric-symbolic solution of GPS phase ambiguity problem with Mathematica, Wolfram Library Archive. http://library.wolfram.com/infocenter/MathSource/9705/

Rózsa S (2014) Modelling tropospheric delays using the global surface meteorological parameter model GPT2. Periodica Polytechnica Civil Eng 58(4):301–308. https://doi.org/10.3311/PPci. 7267

Teunissen PJG, De Jonge PJ, Tiberius CCJM (1997) performance of the LAMBDA method for fast GPS ambiguity resolution. Navigation 44(3):373–400. https://doi.org/10.1002/j.2161-4296.1997.tb02355.x

Teunissen PJG (1990) Quality control in integrated navigation systems. IEEE Aerosp Electron Syst Mag 5(7):35–41

Teunissen PJG (1995) The least-squares ambiguity decorrelation adjustment: a method for fast GPS integer ambiguity estimation. J Geodesy 70(1–2):65–82

Teunissen PJG, Verhagen S (2004) On the foundation of the popular ratio test for GNSS ambiguity resolution. In: ION GNSS 17th international technical meeting of the satellite division, 21–24 Sept., pp 2529–2540. Long Beach, CA

Xu P, Shi C, Liu J (2012) Integer estimation methods for GPS ambiguity resolution: an applications oriented review and improvement. Surv Rev 44(324):59–71. https://doi.org/10. 1179/1752270611Y.0000000004

Chapter 9
Multiobjective Optimization

9.1 Concept of Multiobjective Problem

9.1.1 Problem Definition

It frequently happens that we have two or more competing objectives in case of an optimization problem. As the independent variables change, the value of one of the objectives increases on the one hand, while the value of the other one decreases on the other hand. In this case, we get a set of optimal solutions instead of only a single one. In general, the user should then select his/her favorable solution from this set.

Let us start with a simple example. During the processing of a certain type of product, some harmful material is released. The aim is two-fold: increase the profit by increasing the amount of the product, but decrease the contamination by decreasing the amount of the product. These two objectives are competing. In our case, we have two different products (x_1 and x_2) Suppose the profit objective happens to be

$$z_1 = 140x_1 + 160x_2 \rightarrow \max$$

This should be maximized, while the amount of cost caused by the release of the harmful material should be minimized, let's say

$$z_2 = 500x_1 + 200x_2 \rightarrow \min$$

Since the problem is linear, some restrictions should be introduced. Suppose they are

Supplementary Information The online version contains supplementary material available at https://doi.org/10.1007/978-3-030-92495-9_9.

$$2x_1 + 4x_2 \leq 28$$
$$5x_1 + 5x_2 \leq 50$$
$$0 \leq x_1 \leq 8$$
$$0 \leq x_2 \leq 6$$

9.1.2   Interpretation of the Solution

The feasible solutions can be visualized on Fig. 9.1. In a linear programming problem like this, we know the answer is always one of the vertices.

Table 9.1 contains the feasible solutions and the corresponding values of the objectives.

In the last column, "yes" means that at least one of the objectives z_1 or z_2 in that row has a good value relative to other rows. The feasible solutions B and C are so called "dominated" or "inefficient" solutions, since there is at least one other feasible solution that is better than these, in the sense that at least one of its objective values is better and the other is not worse, see Table 9.1.

Feasible solutions, which are not dominated solutions are called as "non-inferior" or "efficient "solutions, and all of them are candidates for being an optimal solution, see Fig. 9.2.

Fig. 9.1 Feasible solutions in the (x_1, x_2)—parameter space

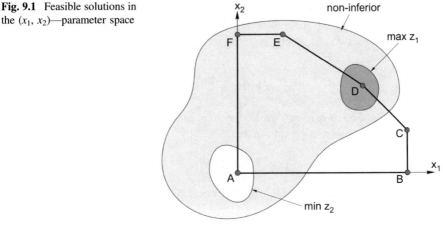

Table. 9.1 Feasible solutions and their objective values

Feasible solution	x_1	x_2	z_1	z_2	Suitability
A	0	0	0	0	Yes
B	8	0	1120	4000	D and E better
C	8	2	1440	4400	D better
D	6	4	1480	3800	Yes
E	2	6	1240	2200	Yes
F	0	6	960	1200	Yes

Fig. 9.2 Objectives (z_1, z_2) —function space

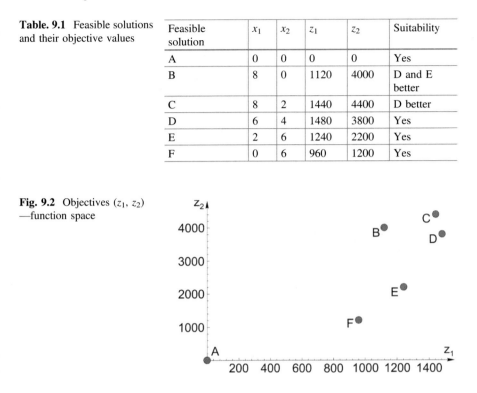

9.2 Pareto Optimum

The (x_1, x_2) values of the feasible, efficient solutions (A, F, E, D) are a *Pareto set*, while the corresponding objective values (z_1, z_2) form a *Pareto front*, see Fig. 9.3.

The region of feasible independent variables is called as *parameter* or *decision* space, and the region of the corresponding function values is called as *function* or *criterion* space.

Conclusions

(a) In case of multiple objectives, there are optimal solutions that are not considered to be solutions of the problem. We call these feasible solutions *inferiority* or *inefficient* solutions.

(b) There are other feasible solutions that can be considered candidates for the solution of the problem. We call these *non-inferiority* or *efficient* solutions.

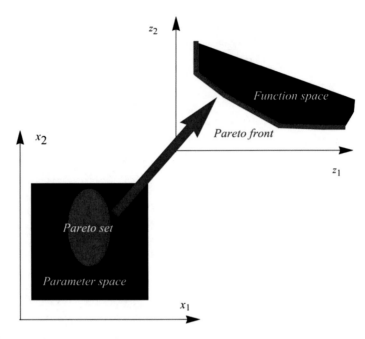

Fig. 9.3 Pareto front (z_1, z_2) and Pareto set

9.2.1 Nonlinear Problems

Let us consider two nonlinear objectives J_1 (x, y) and J_2 (x, y), where

$$J_1 = 3(1-x)^2 e^{-x^2-(1+y)^2} - 10\left(\frac{x}{5} - x^3 - y^5\right)e^{-x^2-y^2} - 3e^{-(x+2)^2-y^2} + 0.5(2x+y)$$
$$J_2 = 3(1+y)^2 e^{-y^2-(1+(-x))^2} - 10\left(-\frac{y}{5} + y^3 - (-x)^5\right)e^{-y^2-(-x)^2} - 3e^{-(-y+2)^2-(-x)^2}$$

See Fig. 9.4.

The problem is to find the maximum of this multi-objective system. As a first naive approach let us compute the maximum of each of the objectives, and as a compromise, consider the maximum of the sum of the objectives in the parameter space, $(x, y) \in [-3, 3]^2$, see Fig. 9.5.

In order to display the feasible function space, let us generate 10^4 random points in the parameter space and compute the corresponding points of the function space. Figure 9.6 shows the parameter space and the corresponding function space with the three optimal points.

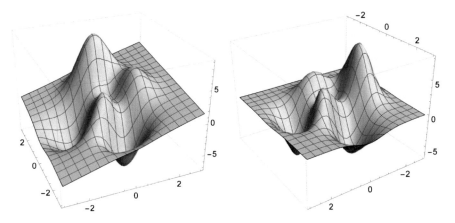

Fig. 9.4 $J_1(x, y)$ and $J_2(x, y)$ nonlinear functions

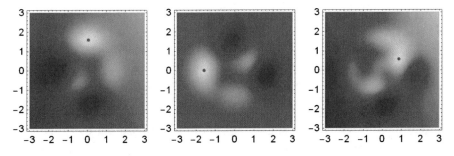

Fig. 9.5 Location of the global maximums of $J_1(x, y)$, $J_2(x, y)$ and their sum $J(x, y)$

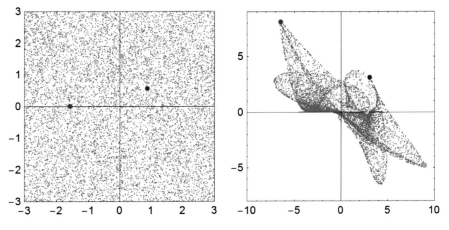

Fig. 9.6 The decision space and the criterion spaces with the three optimum points

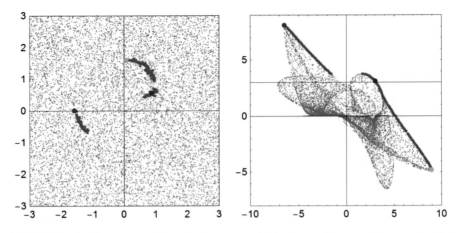

Fig. 9.7 The decision space and the criterion spaces with the three optimum points, as well as the Pareto-set and the Pareto-front

9.2.2 Pareto-Front and Pareto-Set

We call an optimal solution in the decision space *efficient* if by changing the position of this solution a bit, in order to increase the value of one of the optimal solutions, leads to a decrease of the value of the other objective. The corresponding points in the criterion space are called Pareto front; see the upper boundary points of the function set in Fig. 9.7 on the right hand side. However not all of these boundary points belong to an efficient solution, e.g., fixing the value of J_2 at $J_2 = 3$, there are boundary points where the value of J_1 is less than at other boundary points belonging to this fixed J_2 value. Only the boundary point where we have the maximum J_1 value is the element of the Pareto front and has a corresponding efficient solution in the decision space. This efficient solution is dominating the other "unlucky" solutions. The feasible solutions in the decision space are called the Pareto set and the corresponding dominating points in the criterion space are called the Pareto front.

We can see that the Pareto front in the case of the linear problem was convex, but in this nonlinear problem there is a non-convex Pareto front. See the right hand side of Fig. 9.7.

9.3 Computation of Pareto Optimum

9.3.1 Pareto Filter

How can we filter out the Pareto solutions from the feasible solutions? The Pareto filter is a very simple method based on brute force to do that.

Let us consider an optimization problem of an airplane design. We have here four objectives

$$
J = \begin{pmatrix}
J_1 \text{(flying distance, [km])} \rightarrow \max \\
J_2 \text{(specific cost, [\$ /km])} \rightarrow \min \\
J_3 \text{(number of passangers, [person])} \rightarrow \max \\
J_4 \text{(flying velocity, [km/h])} \rightarrow \max
\end{pmatrix}
$$

Let us consider two objective vectors, the i-th and the j-th. Let $i = 1$ and $j = 2$. Suppose their numerical values are,

$$
J_1 = \begin{pmatrix} 7587 \\ 321 \\ 112 \\ 950 \end{pmatrix}, \quad J_2 = \begin{pmatrix} 6695 \\ 211 \\ 345 \\ 820 \end{pmatrix},
$$

Let us compare all the pairs,

$$
J_1 = \begin{pmatrix} 7587 \\ 321 \\ 112 \\ 950 \end{pmatrix} \begin{pmatrix} > \\ > \\ < \\ > \end{pmatrix} \begin{pmatrix} 6695 \\ 211 \\ 345 \\ 820 \end{pmatrix} = J_2
$$

Scoring the relations of the corresponding elements

$$
P_1 = \begin{pmatrix} 1 \\ 0 \\ 0 \\ 1 \end{pmatrix} \quad \begin{pmatrix} 0 \\ 1 \\ 1 \\ 0 \end{pmatrix} = P_2
$$

$$
\sum = 2 \text{ versus } \sum = 2
$$

None of the solution dominates the other one!
Here is another example. Suppose.

$$
J_1 = \begin{pmatrix} 7587 \\ 321 \\ 112 \\ 950 \end{pmatrix}, \quad J_2 = \begin{pmatrix} 6695 \\ 211 \\ 345 \\ 820 \end{pmatrix}.
$$

The objectives are.

$$J_1 = \begin{pmatrix} 7587 \\ 321 \\ 112 \\ 950 \end{pmatrix} \begin{pmatrix} > \\ < \\ > \\ > \end{pmatrix} \begin{pmatrix} 6777 \\ 355 \\ 90 \\ 901 \end{pmatrix} = J_2.$$

Scoring

$$P_1 = \begin{pmatrix} 1 \\ 1 \\ 1 \\ 1 \end{pmatrix} \begin{pmatrix} 0 \\ 0 \\ 0 \\ 0 \end{pmatrix} = P_2$$

$$\sum = 4 \text{ versus } \sum = 0$$

The second solution is dominated by the first one.

If a solution is dominated then the sum of its scores is 0. To represent the result of this analysis, the so called *dominance matrix* can be employed. This matrix has elements 0 or 1.

$$M = \begin{pmatrix} \cdot & \cdot & \cdot & \cdot & \cdot & \cdot & \cdot & \cdot & \cdot & \cdot \\ \cdot & \cdot & \cdot & \cdot & \cdot & \cdot & \cdot & \cdot & \cdot & \cdot \\ \cdot & \cdot & \cdot & \cdot & \cdot & \cdot & \cdot & \cdot & \cdot & \cdot \\ \cdot & \cdot & \cdot & \cdot & \cdot & \cdot & \cdot & \cdot & \cdot & \cdot \\ \cdot & \cdot & \cdot & \cdot & \cdot & 1 & \cdot & 0 & \cdot & \cdot \\ \cdot & \cdot & \cdot & \cdot & \cdot & \cdot & \cdot & \cdot & \cdot & \cdot \\ \cdot & \cdot & \cdot & \cdot & \cdot & \cdot & \cdot & \cdot & \cdot & \cdot \\ \cdot & \cdot & \cdot & \cdot & \cdot & \cdot & \cdot & \cdot & \cdot & \cdot \\ \cdot & \cdot & \cdot & \cdot & \cdot & \cdot & \cdot & \cdot & \cdot & \cdot \\ \cdot & \cdot & \cdot & \cdot & \cdot & \cdot & \cdot & \cdot & \cdot & \cdot \end{pmatrix}$$

if $M_{i,j} = 1$ then the i-th solution dominates the j-th one, if $M_{i,k} = 0$ then the i-th solution does not dominate the j-th one

The sum of the i-th row shows that how many solutions are dominated by the i-th solution. The sum of the j-th column shows how many solutions dominate the j-th solution. Consequently the solutions having zero sum columns are not dominated by any other solution, therefore they are non-inferior, efficient solutions and they belong to the Pareto front.

In the following sections some techniques are introduced to solve multi-objective problems.

Remark It is efficient to organize the computation for the columns independently. It has two advantages,

- if there is a zero in the column, then there is no need to continue the investigation for the further elements of the column, since the solution belonging to this column is a dominated solution,
- the column-wise computation can be evaluated in parallel.

9.3.2 Reducing the Problem to the Case of a Single Objective

This technique tries to solve the multi-objective problem by reducing the number of the objectives to a single one. Let us consider the following two nonlinear objectives, which both to be minimized, see Fig. 9.8.

$$\Rightarrow f_1[x\_] := (2x + 2)^2 - 40$$
$$\Rightarrow f_2[x\_] := (3x - 1)^2 + 5$$

The minimum of each objective,

$$\Rightarrow x_{1\,min} = \text{NMinimize}[f_1[x], x]$$
$$\Leftarrow \{-40., \{x \to -1.\}\}$$

$$\Rightarrow x_{2\,min} = \text{NMinimize}[f_2[x], x]$$
$$\Leftarrow \{5., \{x \to 0.333333\}\}$$

Therefore the parameter space is $x \in \left\{-1, \frac{1}{3}\right\}$.

Let us compute the value of the second objective at the minimum location of the first objective,

$$\Rightarrow f_2[x]/.x_{1\,min}[[2]]$$
$$\Leftarrow 21.$$

and vice versa,

Fig. 9.8 The two objective functions

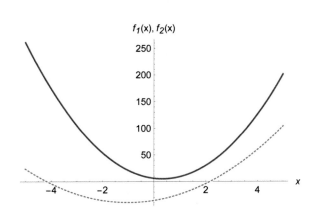

$f_1(x), f_2(x)$

Table. 9.2 Parameter space
and the function space

Optimum	x	f_1	f_2
f_1	-1	-40	21
f_2	$1/3$	32.8889	5

$\Rightarrow \mathtt{f_1[x]/.x_{2\,min}[[2]]}$

$\Leftarrow -32.8889.$

Table 9.2 shows the results.

Therefore the decision space is $x \in \{-1, \ 1/3\}$ and the criterion space is $\{f_1, f_2\} \in \{-40, \ -32.8889\} \times \{5, 21\}$.

Then the *feasible solutions* can be computed as the solution of the following optimization problem,

$$f_1(x) \to \min_x$$

under the restriction,

$$f_2 = \xi, \quad \text{where} \quad 5 \le \xi \le 21.$$

Let us consider $(f_2)_i$ to be the discrete values of $f_2(x_i)$, where

$\Rightarrow \mathtt{xi = Range[5,21,0.5]};$

Then the corresponding minimum locations of $f_i(x)$ are

$\Rightarrow \mathtt{ParetoSet = Map[NMinimize[\{f_1[x], f_2[x] == \#\}, x][[2]]\&, xi]//Flatten};$

The corresponding function values of $f_i(x)$,

$\Rightarrow \mathtt{ParetoFront = Map[\{f_1[x], f_2[x]\}/.\#\&, ParetoSet]};$

Let us display the Pareto front, see Fig. 9.9.

It can be seen that there are dominated solutions among the feasible solutions. We can filter them out by the Pareto filter. Let us construct the dominance matrix. The size of the matrix is $n \times n$, where

$\Rightarrow \mathtt{n = Length[xi]}$

$\Leftarrow 33$

The matrix itself,

$\Rightarrow$ M = Table[If[And[ParetoFront[[i, 1]] < ParetoFront[[j, 1]],
   ParetoFront[[i, 2]] < ParetoFront[[j, 2]]], 1, 0], {i, 1, n}, {j, 1, n}];

$\Rightarrow$ MatrixForm[M]

$\Rightarrow$

$$\begin{pmatrix}
0 0 0 0 1 0 1 0 1 1 0 0 0 0 0 0 0 0 0 0 0 1 1 0 0 0 0 0 0 0 0 0 0 0 \\
0 0 0 0 1 0 1 0 1 1 0 0 0 0 0 0 0 0 0 0 0 1 1 0 0 0 0 0 0 0 0 0 0 0 \\
0 0 0 0 1 0 1 0 1 1 0 0 0 0 0 0 0 0 0 0 0 1 1 0 0 0 0 0 0 0 0 0 0 0 \\
0 0 0 0 1 0 1 0 1 1 0 0 0 0 0 0 0 0 0 0 0 1 1 0 0 0 0 0 0 0 0 0 0 0 \\
0 0 0 0 0 0 1 0 1 1 0 0 0 0 0 0 0 0 0 0 0 1 1 0 0 0 0 0 0 0 0 0 0 0 \\
0 0 0 0 0 0 1 0 1 1 0 0 0 0 0 0 0 0 0 0 0 1 1 0 0 0 0 0 0 0 0 0 0 0 \\
0 0 0 0 0 0 0 0 1 1 0 0 0 0 0 0 0 0 0 0 0 1 1 0 0 0 0 0 0 0 0 0 0 0 \\
0 0 0 0 0 0 0 0 1 1 0 0 0 0 0 0 0 0 0 0 0 1 1 0 0 0 0 0 0 0 0 0 0 0 \\
0 0 0 0 0 0 0 0 1 0 0 0 0 0 0 0 0 0 0 0 0 1 1 0 0 0 0 0 0 0 0 0 0 0 \\
0 0 0 0 0 0 0 0 0 0 0 0 0 0 0 0 0 0 0 0 0 1 1 0 0 0 0 0 0 0 0 0 0 0 \\
0 0 0 0 0 0 0 0 0 0 0 0 0 0 0 0 0 0 0 0 0 1 1 0 0 0 0 0 0 0 0 0 0 0 \\
0 0 0 0 0 0 0 0 0 0 0 0 0 0 0 0 0 0 0 0 0 1 1 0 0 0 0 0 0 0 0 0 0 0 \\
0 0 0 0 0 0 0 0 0 0 0 0 0 0 0 0 0 0 0 0 0 1 1 0 0 0 0 0 0 0 0 0 0 0 \\
0 0 0 0 0 0 0 0 0 0 0 0 0 0 0 0 0 0 0 0 0 1 1 0 0 0 0 0 0 0 0 0 0 0 \\
0 0 0 0 0 0 0 0 0 0 0 0 0 0 0 0 0 0 0 0 0 1 1 0 0 0 0 0 0 0 0 0 0 0 \\
0 0 0 0 0 0 0 0 0 0 0 0 0 0 0 0 0 0 0 0 0 1 1 0 0 0 0 0 0 0 0 0 0 0 \\
0 0 0 0 0 0 0 0 0 0 0 0 0 0 0 0 0 0 0 0 0 1 1 0 0 0 0 0 0 0 0 0 0 0 \\
0 0 0 0 0 0 0 0 0 0 0 0 0 0 0 0 0 0 0 0 0 1 1 0 0 0 0 0 0 0 0 0 0 0 \\
0 0 0 0 0 0 0 0 0 0 0 0 0 0 0 0 0 0 0 0 0 1 1 0 0 0 0 0 0 0 0 0 0 0 \\
0 0 0 0 0 0 0 0 0 0 0 0 0 0 0 0 0 0 0 0 0 1 1 0 0 0 0 0 0 0 0 0 0 0 \\
0 0 0 0 0 0 0 0 0 0 0 0 0 0 0 0 0 0 0 0 0 1 1 0 0 0 0 0 0 0 0 0 0 0 \\
0 0 0 0 0 0 0 0 0 0 0 0 0 0 0 0 0 0 0 0 0 1 1 0 0 0 0 0 0 0 0 0 0 0 \\
0 0 0 0 0 0 0 0 0 0 0 0 0 0 0 0 0 0 0 0 0 0 1 0 0 0 0 0 0 0 0 0 0 0 \\
0 0 0 0 0 0 0 0 0 0 0 0 0 0 0 0 0 0 0 0 0 0 0 0 0 0 0 0 0 0 0 0 0 0 \\
0 0 0 0 0 0 0 0 0 0 0 0 0 0 0 0 0 0 0 0 0 0 0 0 0 0 0 0 0 0 0 0 0 0 \\
0 0 0 0 0 0 0 0 0 0 0 0 0 0 0 0 0 0 0 0 0 0 0 0 0 0 0 0 0 0 0 0 0 0 \\
0 0 0 0 0 0 0 0 0 0 0 0 0 0 0 0 0 0 0 0 0 0 0 0 0 0 0 0 0 0 0 0 0 0 \\
0 0 0 0 0 0 0 0 0 0 0 0 0 0 0 0 0 0 0 0 0 0 0 0 0 0 0 0 0 0 0 0 0 0 \\
0 0 0 0 0 0 0 0 0 0 0 0 0 0 0 0 0 0 0 0 0 0 0 0 0 0 0 0 0 0 0 0 0 0 \\
0 0 0 0 0 0 0 0 0 0 0 0 0 0 0 0 0 0 0 0 0 0 0 0 0 0 0 0 0 0 0 0 0 0 \\
0 0 0 0 0 0 0 0 0 0 0 0 0 0 0 0 0 0 0 0 0 0 0 0 0 0 0 0 0 0 0 0 0 0 \\
0 0 0 0 0 0 0 0 0 0 0 0 0 0 0 0 0 0 0 0 0 0 0 0 0 0 0 0 0 0 0 0 0 0 \\
0 0 0 0 0 0 0 0 0 0 0 0 0 0 0 0 0 0 0 0 0 0 0 0 0 0 0 0 0 0 0 0 0 0 \\
0 0 0 0 0 0 0 0 0 0 0 0 0 0 0 0 0 0 0 0 0 0 0 0 0 0 0 0 0 0 0 0 0 0 \\
0 0 0 0 0 0 0 0 0 0 0 0 0 0 0 0 0 0 0 0 0 0 0 0 0 0 0 0 0 0 0 0 0 0
\end{pmatrix}$$

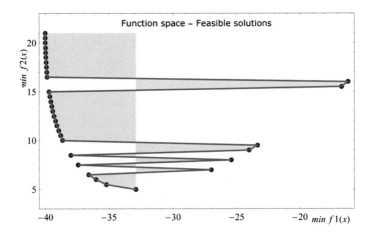

Fig. 9.9 Feasible solutions

The sum of the different columns,

⇒ s = Table[Total[Transpose[M][[i]]], {i, 1, n}]
⇐ {0, 0, 0, 0, 4, 0, 6, 0, 8, 9, 0, 0, 0, 0, 0, 0, 0,
        0, 0, 0, 0, 21, 22, 0, 0, 0, 0, 0, 0, 0, 0, 0, 0}

The dominated feasible solutions are,

⇒ Map[Position[s, #]&, Select[s, #! = 0&]]//Flatten
⇐ {5, 7, 9, 10, 22, 23}

The corresponding points in the function space are

⇒ NonPareto = Map[ParetoFront[[#]]&, %]
⇐ {{−26.9717, 7.}, {−25.3972, 8.}, {−24., 9.},
        {−23.3464, 9.5}, {−16.7009, 15.5}, {−16.2076, 16.}}

Getting rid of them, we have the feasible solutions belonging to the Pareto front,
see Fig. 9.10.

⇒ ParetoFront = Complement[ParetoFront, NonPareto];

This method can be especially efficient if the objective function has more than
one variable, since then one should look for the minimum in a lower dimensional
region than the original one.

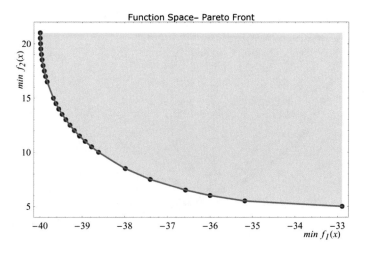

Fig. 9.10 Pareto front

9.3.3   *Weighted Objective Functions*

Similarly to the method introduced above, we can employ the weighted sum of the different objectives as a single objective,

$$F(x) = \lambda f_1(x) + (1 - \lambda)f_2(x) \to \min_x$$

where $0 \le \lambda \le 1$,

$\Rightarrow$ F[x_, λ_] := λf₁[x] + (1 − λ)f₂[x]

Let us consider discrete values for λ

$\Rightarrow$ λᵢ = Range[0, 1, 0.01];

The locations of the minimums,

$\Rightarrow$ ParetoSet = Flatten[Map[NMinimize[F[x, #], x][[2]][[1, 2]] &, λi];

Fig. 9.11 shows the Pareto front

$\Rightarrow$ ParetoFront = Map[f₁[#], f₂[#]} &, ParetoSet];

Now, the parameter space has one dimension $x \in \{-1, 1/3\}$. See Fig. 9.12. This is an easy method; however it works only if the Pareto front is convex.

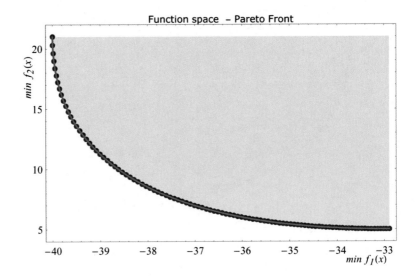

Fig. 9.11 Pareto front

Fig. 9.12 Pareto front

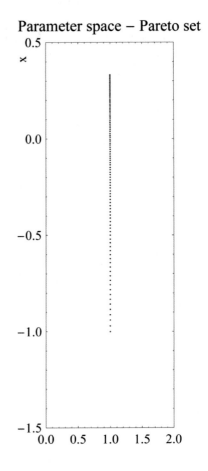

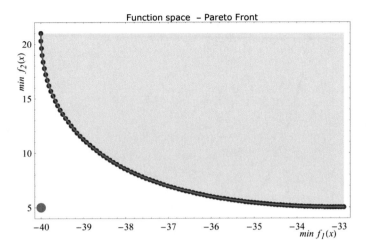

Fig. 9.13 The ideal point in the function space

9.3.4   *Ideal Point in the Function Space*

In Table 9.2 we could see that if the objectives were independent then their minimums would be $f_{1\,min} = -40$ and $f_{2\,min} = 5$.

This is the "ideal point" of the problem, see Fig. 9.13.

9.3.5   *Pareto Balanced Optimum*

The *Pareto balanced optimum* is the point of the Pareto front, which is closest to the ideal point according to some kind of norm. Let us find the Pareto balanced optimum considering Euclidean distance. The distance of the points of the Pareto front are

$\Rightarrow$ ideal $= \{-40, 5\}$;

$\Rightarrow$ distance $=$ Map[Norm[ideal $-$ #] &, ParetoFront];

Now the point which is closest to the ideal point

$\Rightarrow$ ParetoBalanced $=$
   ParetoFront[[Position[distance, Min[distance]]//
   Flatten]]//Flatten

$\Leftarrow \{-37.3996, 7.5\}$

Then the Pareto balanced point can be seen on Fig. 9.14.

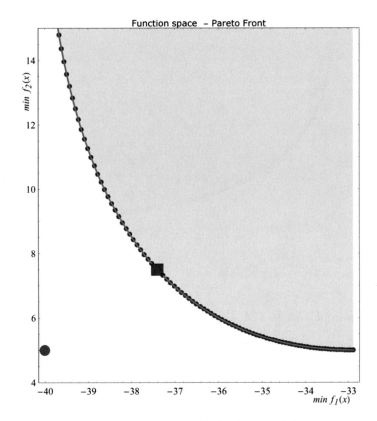

Fig. 9.14 Pareto balanced optimum (square) and ideal point (gray dot)

Sometimes the L_1 norm can be also used

⇒ distance = Map[Norm[ideal − #, 1]&, ParetoFront];

⇒ ParetoBalanced =

  ParetoFront[[Position[distance, Min[distance]]//

  Flatten]]//Flatten

⇐ {−36.5769, 6.5}

This point is belonging to the parameter $\lambda = 0.5$,

⇒ ParetoSet = Flatten[Map[NMinimize[F[x, #], x][[2]][[1, 2]]&, {0.5}]];

⇒ ParetoFront = Map[f₁[#], f₂[#]}&, ParetoSet]

⇐ {{−36.5917, 6.51479}}

9.3.6 Non-Convex Pareto-Front

We mentioned in Sect. 8.3.3 that the method of the weighted objective functions is simple, but can be employed only in case of convex Pareto front. Let us see what happens if we use this method even when the Pareto front is non-convex and how the problem can be cured.

Let us consider the following functions, see Fig. 9.15.

$$f_1(x) = \begin{pmatrix} \text{if } x < 1 \text{ then } -x \\ \text{if } x \leq 3 \text{ then } x - 2 \\ \text{if } x \leq 4 \text{ then } -x \\ \text{else } x - 4 \end{pmatrix}$$

and

$$f_2(x) = (x - 5)^2.$$

$\Rightarrow$ f$_1$[x_] := Which[x $\leq$ 1, -x, x $\leq$ 3, x - 2, x $\leq$ 4, 4 - x, True, x - 4]
$\Rightarrow$ f$_2$[x_] := (x - 5)^2

So the weighted objectives are

$\Rightarrow$ F[x_, λ_] := λf$_1$[x] + (1-λ)f$_2$[x]

The discrete values of the parameter λ are

$\Rightarrow$ λi = Range, [0, 1, 0.01];

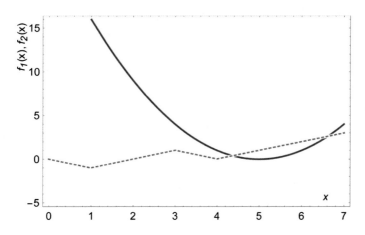

Fig. 9.15 Multi-objective problem leads to non-convex Pareto front

Then the Pareto set

$\Rightarrow$ `Paretoset = Flatten[Map[NMinimize[F[x,#],x][[2]][[1,2]]&,λi];`

Let us visualize the corresponding Pareto front, see Fig. 9.16.

The Pareto set consists of the two disjoint sets, see Fig. 9.17.

Now, the solutions represented by the upper branch in the function space are dominated in the interval of [0, 0.8]. So we should eliminate them. Let us employ the Pareto filter to eliminate dominated solutions; see Fig. 9.18.

The corresponding decision space can be seen on Fig. 9.19.

9.4   Employing Genetic Algorithms

In case of a nonlinear, nonconvex problem, a genetic algorithm can be very effective for finding multi-objective optimum, see Gruna (2009).

$\Rightarrow$ `<<"Pareto`MultiOptimum`"`

$\Rightarrow$ `?MultiOptimum`

```
MultiOptimum[xmin,xmax,F1,F2,isize,psize,mProb,rProb,numgen,
    F1min,F1max,F2min,F2max]
function computes Pareto front (minF2 vs.minF1) and Pareto
    solutions ( x- minimum values).

Input variables:
    xmin - smallest x value ( min(xi)),
    xmax - biggest x value    ( max(xi)),
    F1, F2 - the competing objectives (Fi[x_]:= fi( x[[1]],
        x[[2]],...,x[[i]],..)),
    isize, psize - individual and population size,
    mProb, rProb - mutation and recombination probability,
    numgen - number of generations,
    F1min, F1max, F2min, F2max - PlotRange → {{F1min,F1max},
        {F2min,F2max}} in Pareto plot

Output variable:
    {out1,out2,out3} - out1: plot of Pareto front and dominating
        solutions if they exist in the last generation,
            out2: coordinates of Pareto front {minF1, minF2},
            out3: Pareto solutions in the variables space, {x[[1,j]],
                x[[2,j]],...,x[[i,j]],..}
```

Let us employ this function for the nonconvex problem discussed in the previous section. The variables of the parameter space should be represented as a vector $x = (x_1)$. Then the two objectives are Fig. 9.20

$\Rightarrow$ `f1[x_] := Which[x[[1]] ≤ 1, -x[[1]], x[[1]] ≤ 3, x[[1]] - 2,`
    `x[[1]] ≤ 4, 4 - x[[1]], True, x[[1]] - 4]`

and

$\Rightarrow$ `f2[x_] := (x[[1]] - 5)^2;`

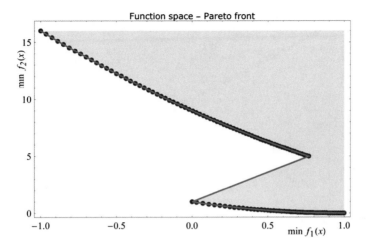

Fig. 9.16 Non-convex Pareto front

Fig. 9.17 Using only
weighted objectives the
Pareto-front without filtering
is disjunct

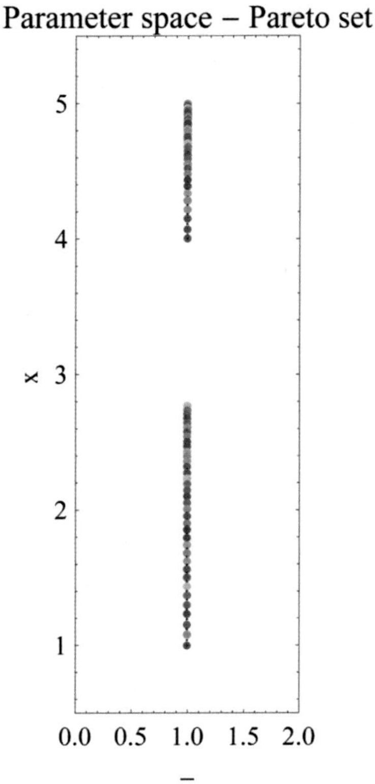

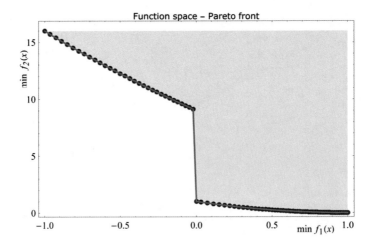

Fig. 9.18 The correct Pareto-front after filtering

Fig. 9.19 The correct
Pareto-set after filtering

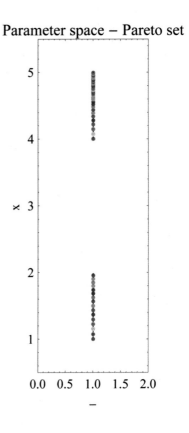

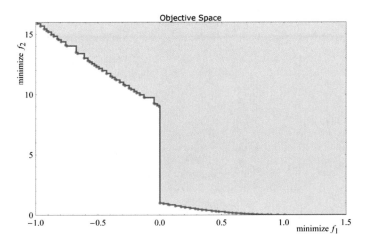

Fig. 9.20 The correct Pareto-front with genetic algorithm

Now, let us employ the genetic optimization,

$\Rightarrow$ R = MultiOptimum[0, 6, f1, f2, 100, 100, 0.02, 0.8, 10, -1., 1.5, 0, 16];
$\Rightarrow$ R[[1]]

As further example, let us consider a two variable problem. We are looking for the minimum of this multi-objective problem

$$f_1(x, y) = x \to \min_{x,y}$$

$$f_2(x, y) = (1 + 10y)\left(1 - \left(\frac{x}{1 + 10y}\right)^2 - \frac{x}{1 + 10y}\sin(8\pi x)\right) \to \min_{x,y}$$

in the $0 \le x, y \le 1$ region. The functions with vector variables

$\Rightarrow$ f1[x_] := x[[1]]
$\Rightarrow$ f2[x_]: = (1 + 10x[[2]]) (1 $-$ (x[[1]]/(1 + 10x[[2]]))^2$-$
    x[[1]]/(1 + 10x[[2]]) Sin[2\[Pi] 4x[[1]]])

First we compute two generations (two iterations), see Pareto-front in Fig. 9.21,

$\Rightarrow$ R = MultiOptimum[0, 1,
    f1, f2, 100, 100, 0.02, 0.8, 2, -0.1, 1.1, -0.3, 11.3];
$\Rightarrow$ R[[1]]

After 10 generations, we get the Pareto-front in Fig. 9.22.
After 100 generations, the front fully develops as in Fig. 9.23.

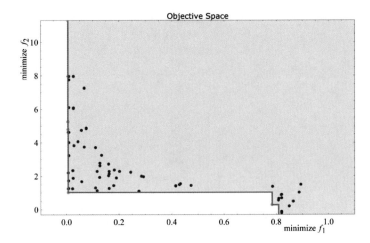

Fig. 9.21 The Pareto-front after two iterations

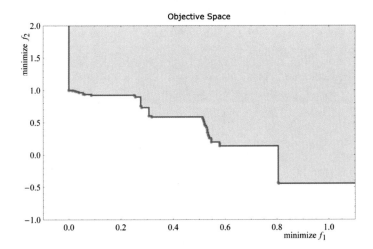

Fig. 9.22 The Pareto-front after 10 iterations

⇒ R = MultiOptimum[0, 1, f1, f2, 100, 100, 0.02, 0.8, 100, −0.1, 1.1, −1, 2];
⇒ R[[1]]

The number of points of the Pareto set is

⇒ R[[3]]//Length
⇐ 100

Figure 9.24 shows the corresponding Pareto set of the problem.

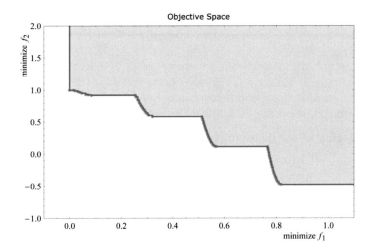

Fig. 9.23 The Pareto-front after 100 iterations

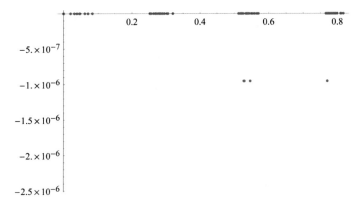

Fig. 9.24 The Pareto set after 100 iterations

This means that the Pareto set consists of 4 disjunct sets of values of x, where in ideal case $y = 0$ for all solutions. The three points having value of 10^{-6} represent numerical error.

9.5 Application

9.5.1 Nonlinear Gauss-Helmert Model

In order to compare the Pareto optimality method with other techniques, let us consider the Gauss-Helmert nonlinear model of weighted 2D similarity transformation,

$$\begin{pmatrix} X \\ Y \end{pmatrix} = F(x, y) = \begin{pmatrix} \cos(\alpha) & -\sin(\alpha) \\ \sin(\alpha) & \cos(\alpha) \end{pmatrix} \begin{pmatrix} \beta & 0 \\ 0 & \beta \end{pmatrix} \begin{pmatrix} x \\ y \end{pmatrix} + \begin{pmatrix} \gamma \\ \delta \end{pmatrix}$$

To determine the 4 unknown parameters α, β, γ, δ from the corresponding measured data pairs $\{(X_i, Y_i), (x_i, y_i)\}$ having the weights $\{(WX_i, WY_i), (wx_i, wy_i)\}$, the following multi-objective problem can be considered in the least squares sense employing the L_2 norm Let us define the two competing objectives

(a) The residual of the transformation $(x, y) \rightarrow (X, Y)$

$$f_1(\alpha, \beta, \gamma, \delta) = \sum_{i=1}^{n} WX_i(X_i - \beta(\cos(\alpha)x_i - \sin(\alpha)y_i) + \gamma)^2$$

$$+ WY_i(Y_i - \beta(\sin(\alpha)x_i + \cos(\alpha)y_i) + \delta)^2$$

Considering the inverse form of the transformation,

$$\begin{pmatrix} x \\ y \end{pmatrix} = F^{-1}(X, Y) = \begin{pmatrix} \frac{1}{\beta} & 0 \\ 0 & \frac{1}{\beta} \end{pmatrix} \begin{pmatrix} \cos(\alpha) & \sin(\alpha) \\ -\sin(\alpha) & \cos(\alpha) \end{pmatrix} \begin{pmatrix} X - \gamma \\ Y - \delta \end{pmatrix}$$

(b) The residual of the transformation $(X, Y) \rightarrow (x, y)$

$$f_2(\alpha, \beta, \gamma, \delta) = \sum_{i=1}^{n} wx_i \left(x_i - \frac{\cos(\alpha)(X_i - \gamma) + \sin(\alpha)(Y_i - \delta)}{\beta} \right)^2$$

$$+ wy_i \left(y_i - \frac{\cos(\alpha)(Y_i - \delta) - \sin(\alpha)(X_i - \gamma)}{\beta} \right)^2 .$$

The multi-objective problem is to find the minimum of the two objective problems in the variable space—in our case they are α, β, γ and δ—in the sense of Pareto optimality. Then select a single optimum from the set of the Pareto optimums. The input data can be found in Tables 9.3 and 9.4.

Table. 9.3 Corresponding coordinates of the two systems

X [m] $\times$ 10^{-6}	Y [m]	x [m] $\times$ 10^{-6}	y [m]
4.5401342780	382,379.89640	4.5401240940	382,385.99800
4.5399373890	382,629.78720	4.5399272250	382,635.86910
4.5399797390	381,951.47850	4.5399695670	381,957.57050
4.5403264610	381,895.00890	4.5403162940	381,901.09320
4.5392163870	382,184.43520	4.5392062110	382,190.52780

Table. 9.4 The weights of the measured coordinates of the two systems

WX	WY	wx	wy
10.0000	14.2857	5.88240	12.5000
0.89290	1.42860	0.90090	1.72410
7.14290	10.0000	7.69230	16.6667
2.22220	3.22590	4.16670	6.66670
7.69230	11.1111	8.33330	16.6667

Table. 9.5 The individual minimum and maximum of the objectives

	f_1	f_2
α[rad]	$-5.80607956729643 \times 10^{-6}$	$-2.44615778001273 \times 10^{-6}$
β	0.9999970185952	0.9999937296126
γ[m]	21.4719793439680	37.7064925325464
δ[m]	21.6518647388158	7.4080798108086
f_{min}[m^2]	0.002113068954524	0.00225632756460
f_{max}[m^2]	0.00241943399890	0.002610007624997

The weights are in Table 9.4.

The first step is to compute the coordinates of the ideal point in the function space.

The individual minimum of the objectives were computed via global minimization by *Mathematica* and the maximum value of the objectives can be computed by simply substituting the minimum of the counterpart objective. The result can be found in Table 9.5.

To compute the Pareto set, the method of the weighted objective functions is used. Therefore the dimensionless single objective has to be defined for the multi-objective problem,

$$F(\alpha, \beta, \gamma, \delta, \eta) = \eta G_1(\alpha, \beta, \gamma, \delta) + (1 - \eta)G_2(\alpha, \beta, \gamma, \delta)$$

where

$$G_i(x) = \frac{f_i(x) - f_{imin}}{f_{imax} - f_{imin}}, \quad 0 \leq G_i(x) \leq 1, \quad i = 1, 2$$

The Perato set is the set of the four-tuples $\{\alpha, \beta, \gamma, \delta\}$, the minimums of $\{G_1, G_2\}$ for $\eta \in [0, 1]$.

The minimization has been carried out for 100 discrete, equidistant $\eta_i \in [0, 1]$ points, using *Mathematica* via parallel computation running on Intel Nehalem processor with 8 threads providing more than 7 times speed-up, see Fig. 9.25.

ID	Name	Host	Process	CPU	RAM	Version	Close	Act	Time	Elapsed: 6.047s, speedup: 7.25
0	master	user–su3b8pwga5	2780	2.751	22M	7.0			0.047	
1	local	user–su3b8pwga5	2384	5.437	13M	7.0	☒	●	5.375	
2	local	user–su3b8pwga5	2584	5.312	12M	7.0	☒	●	5.234	
3	local	user–su3b8pwga5	2608	6.046	13M	7.0	☒	●	5.968	
4	local	user–su3b8pwga5	2492	5.234	13M	7.0	☒	●	5.172	
5	local	user–su3b8pwga5	2560	5.297	13M	7.0	☒	●	5.203	
6	local	user–su3b8pwga5	312	5.687	13M	7.0	☒	●	5.640	
7	local	user–su3b8pwga5	2732	5.922	13M	7.0	☒	●	5.859	
8	local	user–su3b8pwga5	2960	5.453	13M	7.0	☒	●	5.375	

Parallel Kernels

8 kernels running, idle

Close All   Select Columns... | Kernel Configuration...

Fig. 9.25 The status of the kernels at the computation of the Pareto front using parallel evaluation, where the service time of the different threads (Time), the total elapsed time of the computation (Elapsed) as well as the speed-up achieved by parallel evaluation (speedup) can be seen

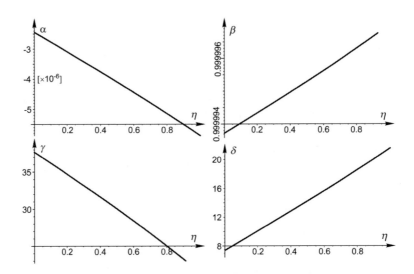

Fig. 9.26 The minimums belonging to the Pareto set as a function the parameter η

The transformation parameters (α, β, γ, δ) as function of η can be seen on Fig. 9.26.

The Pareto front consists of the minimum values of $\{G_1, G_2\}$ with ideal point $\{0, 0\}$ can be seen in Fig. 9.27.

All of the points of the Pareto-front are optimal. In order to select a single optimum solution, one may choose the very point of the front which is the closest to

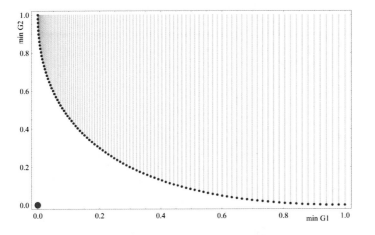

Fig. 9.27 The dimensionless Pareto-front with the ideal point (larger dot)

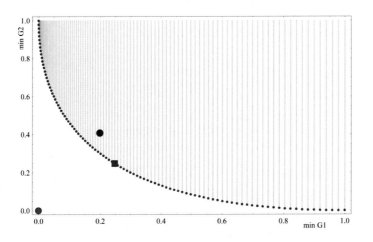

Fig. 9.28 Pareto front with the TLS solution (black dot) and Pareto balanced solution (square)

the ideal point in L_1 norm. This optimum belongs to the parameter value $\eta = 0.5$. Figure 9.28 shows the Pareto balanced solution as well as the Neitzel's TLS (Total Least Square) solution see Neitzel (2010) with the ideal point in the function space.

It can be seen, that although the TLS solution is close to the Pareto front, it is not a Pareto optimal solution and clearly farther from the ideal point than the Pareto balanced solution.

However, the TLS solution gives smaller error for the first objective (G_1) than the Pareto balanced solution. One can find a region on the Pareto front, where the corresponding Pareto solutions in the Pareto set provide smaller errors for both objectives (G_1 and G_2) than the TLS's one. This region belongs to the parameters $\eta \in (0.552846, 0.644113)$, see Fig. 9.29.

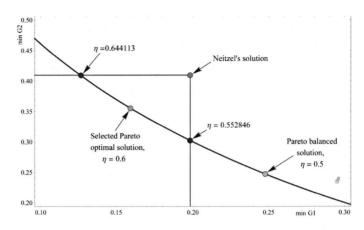

Fig. 9.29 The superiority region of the Pareto optimality $\eta \in (0.552846, 0.644113)$

The selected Pareto optimal solution is roughly in the middle of this region, $\eta = 0.6$. The Pareto optimal solutions belong to this parameter region in the Pareto set, all giving smaller values for G_1 and G_2 than the TLS solution.

9.6 Exercise

The relation between the level of rainfall (h) and the water level in the soil (L) was measured at Cincinnati. They were the following values, see Haneberg (2004)

$$\Rightarrow \text{data} = \{\{1.94, 2.5\}, \{3.33, 1.89\}, \{3.22, 1.67\},$$
$$\{5.67, 1.31\}, \{4.72, 1.02\}, \{3.89, 0.96\}, \{2.78, 1.1\},$$
$$\{10.56, 0.15\}, \{9.44, 3.92\}, \{12.78, 5.23\}, \{14.72, 4.22\},$$
$$\{13.61, 3.63\}, \{20.39, 4.32\}, \{38.89, 5.89\}\};$$

Let us assume a linear model,

$$L = \beta + \alpha h$$

Assuming that the error of the measurement of the rainfall level can be neglected comparing to that of the water in soil, then employing the standard, ordinary least squares method, the parameters of the model (α, β) can be easily computed Fig. 9.30.

$\Rightarrow \{x, y\} = \text{Transpose[data]};$
$\Rightarrow n = \text{Length[data]};$

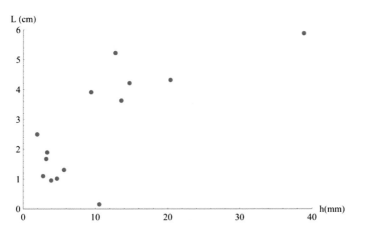

Fig. 9.30 Measured relation between level of rainfall (h) and that of the water in the soil

The function to be minimized is

$$\Rightarrow f_1[\alpha_-, \beta_-] := \sum_{i=1}^{n} (y[[i]] - \beta - \alpha x[[i]])^2$$

$$\Rightarrow f_{1\,\mathrm{min}} = \mathrm{NMinimize}[f_1[\alpha, \beta], \{\alpha, \beta\}]$$

$$\Leftarrow \{18.3111, \{a \rightarrow 0.13763, b \rightarrow 1.26602\}\}$$

Let us display the line Fig. 9.31.

Now, let us suppose the reversed situation, namely the error of the measurement of the water level in soil can be neglected comparing to that of the rainfall. The objective to be minimized is

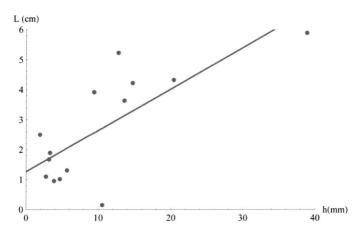

Fig. 9.31 Linear model fitting via least square method

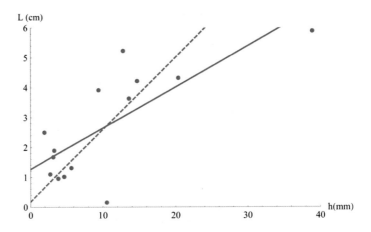

Fig. 9.32 Linear model fitting via least square method in reversed situation

$$\Rightarrow f_2[\alpha_-, \beta_-] := \sum_{i=1}^{n} \left(x[[i]] - \frac{y[[i]] - \beta}{\alpha} \right)^2$$

Therefore

$$\Rightarrow f_{2\min} = \text{NMinimize}[f_2[\alpha, \beta], \{\alpha, \beta\}]$$
$$\Leftarrow \{549.866, \{\alpha \rightarrow 0.24196, \beta \rightarrow 0.178453\}\}$$

Let us visualize the result, see dashed line.

Now, let us suppose that the errors of the measurements are comparable. This is the so called Error-In-all Variables (EIV) model Fig. 9.32.

The problem can be solved using total least squares technique (TLS). Then we employ adjustments for both measured variables (Δh_i and ΔL_i). Then the objective function is

$$\Rightarrow f_3 = \sum_{i=1}^{n} \left(\Delta h_i^2 + \Delta L_i^2 \right);$$

with the restriction

$$\Rightarrow g_3 = y[[i]] - \Delta L_i == \alpha(x[[i]] - \Delta h_i) + \beta, \quad i = 1 \ldots n$$

The restriction is linear, therefore we can transform the problem into a minimization without restriction, namely

$$\Rightarrow f_3 = \sum_{i=1}^{n} \left(\Delta h_i^{\,2} + (y[[i]] - \alpha(x[[x]] - \Delta h_i) - \beta)^2 \right);$$

The variables are

```
⇒ vars = Join[{α, β}, Table[{Δhᵢ}, {i,1,n}]]//Flatten
⇐ {α, β, Δh₁, Δh₂, Δh₃, Δh₄, Δh₅, Δh₆, Δh₇, Δh₈, Δh₉, Δh₁₀, Δh₁₁, Δh₁₂, Δh₁₃, Δh₁₄}
```

Then

```
⇒ f₃ₘᵢₙ=NMinimize[f₃, vars]
⇒ {17.9659, {α → 0.139597, β → 1.24552, Δh₁ → −0.134692, Δh₂ → −0.0245957
    Δh₃ → 0.0034259, Δh₄ → 0.0995512, Δh₅ → 0.121101, Δh₆ → 0.113452
    Δh₇ → 0.0730646, Δhθ → 0.351859, Δh₉ → −0.185769, Δh₁₀ → −0.301302
    Δh₁₁ → −0.125922, Δh₁₂ → −0.0663514, Δh₁₃ → −0.031234, Δh₁₄ → 0.107411}}
```

The result is very close to the result of the ordinary least square, see Fig. 9.33 dotted line,

Now we employ Pareto optimization. Our two competing objectives are $f_1(\alpha, \beta)$ and $f_2(\alpha, \beta)$. Since the dimensions of the two objectives are different, we should normalize the functions. Table 9.6 shows the minimum and maximum of the two objectives. The values resulted from the computations of the two ordinary least squares, see above.

```
⇒ f₁[0.242, 0.179]
⇐ 32.2024
⇒ f₂[0.138, 1.266]
⇐ 961.535
```

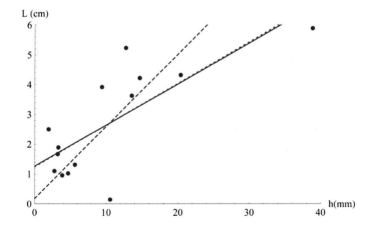

Fig. 9.33 Linear model fitting via total least square (dotted line)

Table. 9.6 The minimum and maximum of the two objectives	Optimum	α	β	f_1	f_2
	f_1	0.138	1.266	18.31	961.54
	f_2	0.242	0.179	32.20	549.87

Then the normalized objectives are,

$$\Rightarrow \text{f1}[\alpha\_, \beta\_] := \frac{\text{f}_1[\alpha, \beta] - 18.31}{32.2 - 18.31}$$

$$\Rightarrow \text{f1}[\alpha\_, \beta\_] := \frac{\text{f}_2[\alpha, \beta] - 549.87}{961.54 - 549.87}$$

In vector form

$$\Rightarrow \text{F1}[\text{X}\_] := \text{f1}[\text{X}[[1]], \text{X}[[2]]]$$
$$\Rightarrow \text{F2}[\text{X}\_] := \text{f2}[\text{X}[[1]], \text{X}[[2]]]$$

Let us employ the genetic algorithm to compute the Pareto-front,

$$\Rightarrow \text{R} = \text{MultiOptimum}[0., 2, \text{F1}, \text{F2}, 100, 100, 0.02, 0.8, 100, 0., 1., 0., 1.];$$
$$\Rightarrow \text{R}[[1]]$$

The ideal point

$$\Rightarrow \text{ideal} = \{0, 0\};$$

To have a single solution, we compute the Pareto balanced solution, Fig. 9.34

$$\Rightarrow \text{ParetoFront} = \text{R}[[2]];$$
$$\Rightarrow \text{ParetoSet} = \text{R}[[3]];$$
$$\Rightarrow \text{ParetoFront}[[1]];$$
$$\Leftarrow \{0.00111459, 100\}$$
$$\Rightarrow \text{distance} = \text{Map}[\text{Norm}[\text{ideal} - \#]\&, \text{ParetoFront}];$$
$$\Rightarrow \text{ParetoBalanced} = \text{ParetoFront}$$
$$[[\text{Position}[\text{distance}, \text{Min}[\text{distance}]]//\text{Flatten}]]//\text{Flatten}$$
$$\Leftarrow \{0.186577, 0.187133, 0.186577, 0.187133\}$$
$$\Rightarrow \text{opt} = \text{Union}[\%]$$
$$\Leftarrow \{0.186577, 0.187133\}$$

Then the parameters Fig. 9.35.

$$\Rightarrow \text{sol}\alpha\beta = \text{FindRoot}[\{\text{F1}[\{\alpha, \beta\}] - \text{opt}[[1]],$$
$$\text{F2}[\{\alpha, \beta\}] - \text{opt}[[2]]\}, \{\{\alpha, 1.5\}, \{\beta, 0.75\}\}]//\text{Flatten}$$
$$\Leftarrow \{\alpha \to 0.182594, \beta \to 0.8268\}$$

The distance of the different solutions from the ideal point:
Ordinary least squares Fig. 9.36

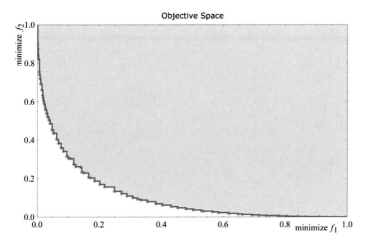

Fig. 9.34 The Pareto front of the linear model fitting problem

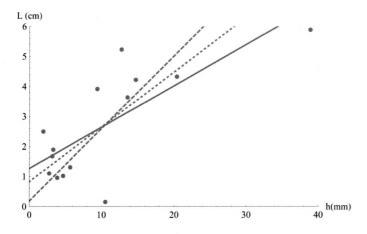

Fig. 9.35 The Pareto balanced solution (dotted line)

$\Rightarrow$ OLSy = Norm[{F1[{0.138, 1.266}], F2[{0.138, 1.266}]}](*Green*)
$\Leftarrow$ 0.999987

$\Rightarrow$ OLSx = Norm[{F1[{0.242, 0.179}], F2[{0.242, 0.179}]}](*Blue*)
$\Leftarrow$ 1.00017

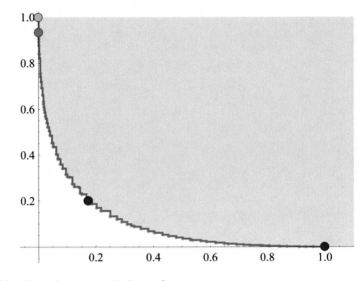

Fig. 9.36 All solutions are on the Pareto front

Total least square

$\Rightarrow$ TLS = Norm[{F1[{0.140, 1.246}], F2[{0.140, 1.246}]}](*Magenta*)

$\Leftarrow$ 0.934608

Pareto balanced solution

$\Rightarrow$ PBO = Norm[{F1[{0.181, 0.850}], F2[{0.181, 0.850}]}](*Red*)

$\Leftarrow$ 0.266064

Although their distances are different, the results of all methods are on the Pareto front!

References

Gruna R (2009) Evolutionary multiobjective optimization, wolfram demonstrations project. http://demonstrations.wolfram.com/EvolutionaryMultiobjectiveOptimization/

Haneberg WC (2004) Computational geosciences with mathematica. Springer, Berlin, Heidelberg

Neitzel F (2010) Generalization of total least-squares on example of unweighted and weighted 2D similarity transformation. J Geodesy 84(12):751–762

Part III
Approximation of Functions and Data

Chapter 10
Approximation with Radial Bases Functions

10.1 Basic Idea of RBF Interpolation

The radial basis method (RBF) is one of the kernel methods, which can be employed for interpolation as well as approximation of data or functions given by values. This is especially true when we do not have a regular grid, but scattered data in many dimensions. The RBF method employs a linear combination of so called *radial basis functions* to carry out a basically local approximation. This method is also applied in machine learning, for example as an activation function of artificial neural networks (ANN).

The wide spread successful application of RBF is based on the theoretical fact that the radial basis functions, like algebraic polynomials and sigmoid activation functions of ANN, are so called *universal approximators*, see Micchelli et al. (2006).

For interpolation, the radial bases functions (RBF) method employs a linear combination of basis functions $\varphi(x)$,

$$f(x) = \sum_{i=1}^{n} c_i \varphi(\|x - x_i\|),$$

where the coefficients c_i can be computed from the known data pairs $f(x_i) = f_i$ via the solution of the following linear system,

Supplementary Information The online version contains supplementary material available at https://doi.org/10.1007/978-3-030-92495-9_10.

$$Mc = \begin{pmatrix} \varphi(\|x_1 - x_1\|) & \varphi(\|x_1 - x_2\|) & \cdots & \varphi(\|x_1 - x_n\|) \\ \varphi(\|x_2 - x_1\|) & \varphi(\|x_2 - x_2\|) & \cdots & \varphi(\|x_2 - x_n\|) \\ \cdots & \cdots & \cdots & \cdots \\ \varphi(\|x_n - x_1\|) & \varphi(\|x_n - x_2\|) & \cdots & \varphi(\|x_n - x_n\|) \end{pmatrix} \begin{pmatrix} c_2 \\ c_2 \\ \cdots \\ c_n \end{pmatrix} = \begin{pmatrix} f_1 \\ f_1 \\ \cdots \\ f_1 \end{pmatrix}$$

$$= f.$$

Let us consider a 1D problem. Suppose the function is,

$$f(x) = x \sin(x).$$

We generate $n = 10$ points in the interval $x_i \in [1/2, 3]$, $i = 1, 2, \ldots, n$, see Fig. 10.1.

$$\Rightarrow \mathtt{data} = \mathtt{Table[\{i0.3, i0.3Sin[i0.3]\}, \{i, 1, 10\}]}$$

$$\Leftarrow \{\{0.3, 0.0886561\}, \{0.6, 0.338785\}, \{0.9, 0.704994\},$$
$$\{1.2, 1.11845\}, \{1.5, 1.49624\}, \{1.8, 1.75293\},$$
$$\{2.1, 1.81274\}, \{2.4, 1.62111\}, \{2.7, 1.15393\}, \{3., 0.42336\}\}$$

To interpolate the function we employ a thin-plate spline as a basis function, which is a type of the radial basis function family,

$$\varphi(r) = r^2 \log(r),$$

where

$$r = \|x - x_i\|.$$

It means this r, is always $r \geq 0$.

Fig. 10.1 The generated discrete points

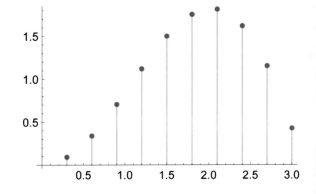

Fig. 10.2 Two basis functions

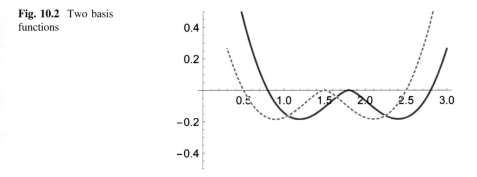

Remark Instead of $r^2 \log(r)$ you use $r^2 \log(\varepsilon + r)$ for suitable ε (1/100, say) then you avoid the singularity at zero and still get reasonable behavior. It makes for a cleaner formula, for instance, and is perhaps a bit faster for large computations.

The symbol r is used often in the following examples. Figure 10.2 shows two basis functions with different locations.

Separating the x_i and y_i coordinates into two arrays,

⇒ np=Length[data]; datax=Transpose[data] [[1]];
⇒ datay=Transpose[data] [[2]]

the matrix of the linear system is

⇒ M = Table[First[TPSpline[datax[[i]],
    datax[[j]]]], {i, 1, np}, {j, 1, np}];

This matrix has a relatively high condition number,

⇒ LinearAlgebra'MatrixConditionNumber[M]
⇐ 505.204

Therefore, to compute the RBF coefficients, we use the pseudoinverse,

⇒ c = PseudoInverse[M].datay

The coefficients can be seen in Fig. 10.3.
The interpolation function will be computed as

⇒ f = First[c.Map[TPSpline[x, #] &, datax]]
⇐ 1.94971 If [x ≠ 0.3, Abs [x − 0.3]² Log [Abs [x − 0.3]], 0]−
  2.26345 If [x ≠ 0.6, Abs [x − 0.6]² Log [Abs [x − 0.6]], 0]−
  0.412961 If [x ≠ 0.9, Abs [x − 0.9]² Log [Abs [x − 0.9]], 0]−
  0.419176 If [x ≠ 1.2, Abs [x − 1.2]² Log [Abs [x − 1.2]], 0]−
  0.282941 If [x ≠ 1.5, Abs [x − 1.5]² Log [Abs [x − 1.5]], 0]−
  0.22277 If [x ≠ 1.8, Abs [x − 1.8]² Log [Abs [x − 1.8]], 0]−
  0.224389 If [x ≠ 2.1, Abs [x − 2.1]² Log [Abs [x − 2.1]], 0]−
  0.178729 If [x ≠ 2.4, Abs [x − 2.4]² Log [Abs [x − 2.4]], 0]−
  1.23521 If [x ≠ 2.7, Abs [x − 2.7]² Log [Abs [x − 2.7]], 0] +
  1.00129 If [x ≠ 3., Abs [x − 3.]² Log [Abs [x − 3.]], 0]

Fig. 10.3 The c_i coefficients

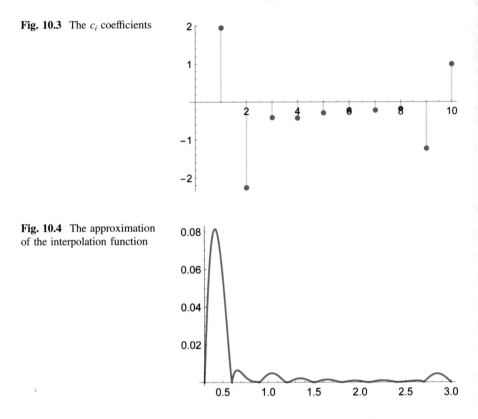

Fig. 10.4 The approximation of the interpolation function

Figure 10.4 shows the relative error of the RBF interpolation function,

In geosciences, frequently we have so called "scattered data interpolation problems", where the measurements do not lie on a uniform or regular grid. Let us consider a 2D problem, where the interpolation points are the discrete values of the following function (Fig. 10.5).

$$f(x, y) = 16x(1 - x)y(1 - y)$$

The interpolation should be carried out on the region $(x, y) \in [0, 1]^2$. We employ $n = 124$ Halton points, see Mangano (2010), see Fig. 10.6.

Now the following basis function is used,

$$\varphi(r) = r^{2\beta} \log r,$$

or

$$\varphi(x, x_i) = \|x - x_i\|^{2\beta} \log \|x - x_i\|.$$

Here we employ $\beta = 1$ and $x = \begin{pmatrix} x_1 \\ x_2 \end{pmatrix}$. Figure 10.7 shows a basis function sitting on the origin.

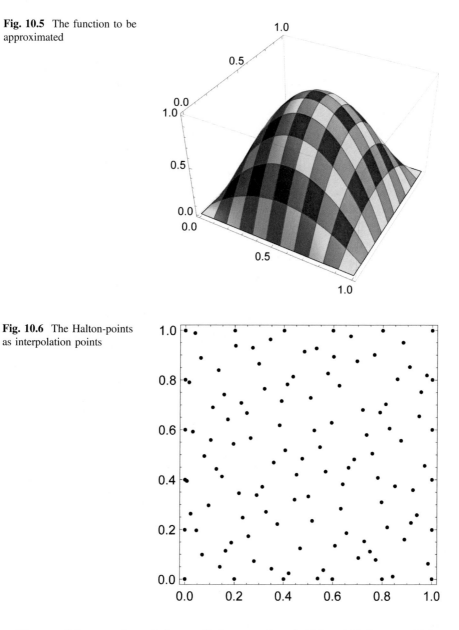

Fig. 10.5 The function to be approximated

Fig. 10.6 The Halton-points as interpolation points

The condition number of the coefficient matrix of 124×124 is very high, therefore our interpolation problem is ill-posed and the pseudoinverse should be used. Figure 10.8 shows the approximation relative errors. As usual, the error is high on the boundary region since the points in this region have neighbors on only one side.

Fig. 10.7 The basis function
of the 2D problem

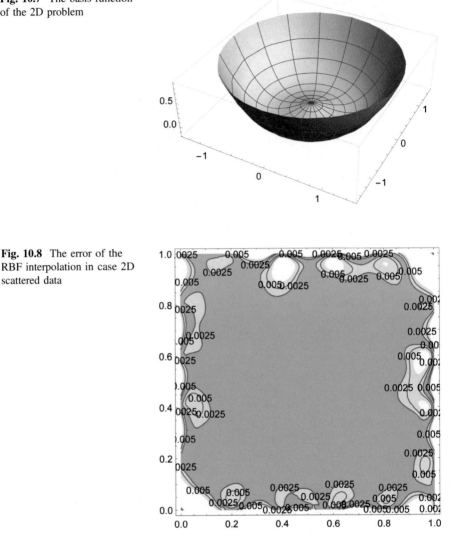

Fig. 10.8 The error of the
RBF interpolation in case 2D
scattered data

10.2 Positive Definite RBF Function

The weakness of the RBF method is the ill-conditioned coefficient matrix, which
leads to an ill-posed interpolation problem. To avoid this problem, we introduce the
definition of a *positive definite* function. A real valued continuous function $\varphi \in R^S \to C$ is called positive definite if

$$\sum_{j=1}^{N}\sum_{k=1}^{N} c_j c_k \varphi\left(x_j - x_k\right) \geq 0,$$

for any pairwise different points $x_1, ..., x_N \in R^s$ and $c = (c_1, ..., c_N)^T \in C^N$. The function φ is called *strictly positive definite* if this quadratic form is zero only for $c \equiv 0$.

In this case, the coefficient matrix will be positive definite, non singular. According to *Bochner's Theorem* Fasshauer (2007) the RBF is positive definite if its Fourier transform is positive. For example, the Gaussian type RBF

$$\varphi(r) = e^{-(\epsilon r)^2}$$

for $\epsilon > 0$ is a positive definite function (Fig. 10.9), since

$$\Rightarrow \varphi = e^{-(\hat{\imath} r)^2}$$

$$\Leftarrow e^{-r^2 \epsilon^2}$$

$$\Rightarrow F = \texttt{FourierTransform}[\varphi, r, \omega]$$

$$\Leftarrow \frac{e^{-\frac{\omega^2}{4\epsilon^2}}}{\sqrt{2}\sqrt{\epsilon^2}}$$

In the special case of $\varepsilon = 1/\sqrt{2}$, the basis function and its Fourier transform are the same

$$F/.\varepsilon \rightarrow \frac{1}{\sqrt{2}}$$

$$e^{-\frac{\omega^2}{2}}$$

Fig. 10.9 The Fourier transform of a Gaussian RBF in case of different ϵ values (0.5, 2 and 5)

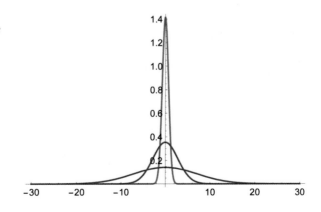

$$\varphi/.\varepsilon \rightarrow \frac{1}{\sqrt{2}}$$

$$e^{-\frac{r^2}{2}}.$$

10.3 Compactly Supported Functions

As we mentioned above, an unwelcome situation appears when the linear systems are solved for large N, since for most radial basis functions the matrix is full and ill-conditioned. An exception is provided by the radial basis functions of compact support.

Compactly supported radial basis functions have been invented for the purpose of getting finite-element type approximations. They give rise to sparse interpolation matrices and can be used to numerically solve partial differential equations, called mesh-less methods. Under suitable conditions on degree and dimension N, they give rise to positive definite interpolation matrices A that are banded, therefore sparse, and then of course also regular.

A compact (closed and bounded) subset $S \subset R$ is a compact support of a function $\varphi(x)$, if $\varphi(x) \neq 0$ when $x \in S$ but it is zero outside of this compact set. If this set S is not bounded then the support is called a global support. Let us consider the following function, see Fig. 10.10.

$$\Rightarrow \varphi= \texttt{Piecewise}[\{\{1-x^2, -1 \leq x < 1\}\}]$$
$$\Leftarrow \begin{cases} 1-x^2 & -1 \leq x < 1 \\ 0 & \text{True} \end{cases}$$

The interval $[-1, 1]$ is obviously the compact support of this function. For the function

Fig. 10.10 A function with interval of $[-1, 1]$ as compact support

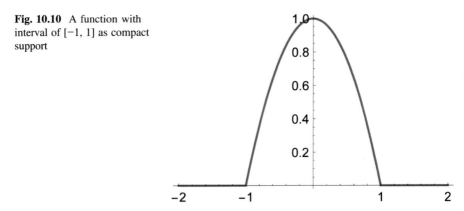

$$\phi = \frac{x}{1+x^2}$$

defined on the domain $x \geq 0$, the interval $x \geq 0$ is the global support (Fig. 10.11) by elementary algebra.

(The interval $[0, \infty]$ is not bounded, therefore it is a global support.)

For another example, the quadratic *Matérn* RBF function Fasshauer (2007), see Fig. 10.12.

$$\Rightarrow \varphi = e^{-r}(3 + 3r + r^2)$$
$$\Leftarrow e^{-r}(3 + 3r + r^2)$$

whose domain is $r \geq 0$, has global support $(r \geq 0)$. This follows from elementary properties of the exponential function.

The *Wendland* $\varphi_{3,2}$ RBF basis function Wendland (2005) has a compact support $([0, 1])$ (Fig. 10.13)

$$\Rightarrow \varphi = (1 - r)^4(1 + 4r);$$

Since, it is well known that the domain of this function is $[0, 1]$.

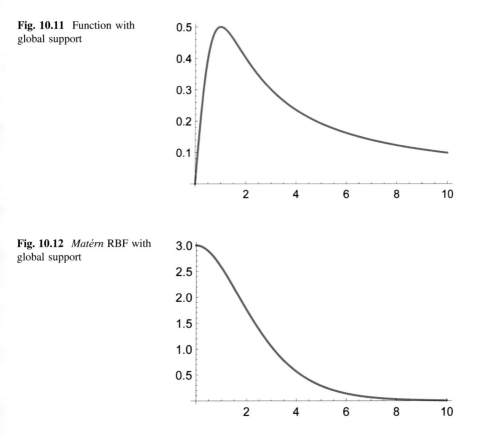

Fig. 10.11 Function with global support

Fig. 10.12 *Matérn* RBF with global support

Fig. 10.13 *Wendland* $\varphi_{3,2}$
RBF with global support

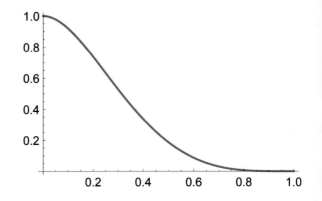

The basis functions with compact support are positive definite functions and their corresponding coefficient matrices are sparse matrices, therefore the corresponding equation system can be solved iteratively.

Remark If we decrease the distance between the basis points (separation distance), the approximation will be better (smaller error), but the condition number of the coefficient matrix will increase, trade-off problem causing high round-off errors.

10.4   Some Positive Definite RBF Function

In this section we list some well-known positive definite RBF functions. The domain is always restricted to $r \geq 0$.

10.4.1   Laguerre–Gauss Function

This is a generalization of the Gauss function (Fig. 10.14). For example,

$$\varphi(r) = e^{-r^2}\left(3 - 3r^2 + \frac{1}{2}r^4\right)$$

$\Rightarrow$ LG = Function[{x, xi}, (3 − 3Norm[x − xi, 2]² + (1/2)Norm[x − xi, 2]⁴)
     exp[−Norm[x − xi, 2]²]];

In two dimensions,

$\Rightarrow$ z[r_,φ_] := LG[{rCos[φ], rSin[φ]}, {0, 0}]

Let us visualize the function, see Fig. 10.15.

Fig. 10.14 Laguerre–Gauss function with compact support

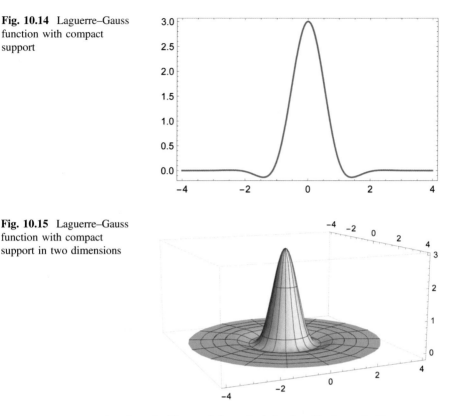

Fig. 10.15 Laguerre–Gauss function with compact support in two dimensions

⇒ ParametricPlot3D[{rCos[φ], rSin[φ], z[r,φ]}, {r, −4, 4},
{φ, 0., 2π}, BoxRatios → {1, 1, 0.5}].

10.4.2   Generalized Multi-quadratic RBF

The RBF basis function,

$$\varphi(r) = \left(1 + r^2\right)^\beta$$

for example with $\beta = -1/2$, we call inverse multi-quadratic RBF, see Fig. 10.16.

⇒ IMQ = Function$\left[\{x, xi\}, \left(1 + Norm[x - xi, 2]^2\right)^{-\frac{1}{2}}\right]$;

In two dimensions, see Fig. 10.17.

⇒ z[r_,φ_] := IMQ[{rCos[φ], rSin[φ]}, {0, 0}]
⇒ ParametricPlot3D[{rCos[φ], rSin[φ], z[r,φ]}, {r, −4, 4},
{φ, 0., 2π}, BoxRatios → {1, 1, 0.5}].

Fig. 10.16 Generalized
multi-quadratic RBF

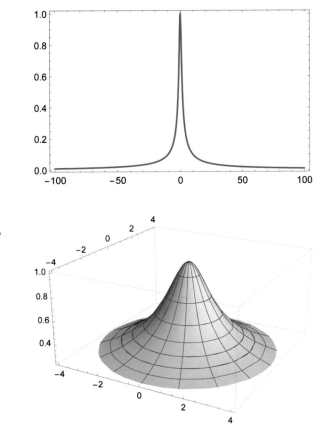

Fig. 10.17 Generalized
multi-quadratic RBF in two
dimensions

In case of $\beta = 1/2$, we get the *Hardy-type multi-quadratic* RBF, see Fig. 10.18.

$$\varphi(r) = \left(1 + r^2\right)^{\beta}$$

$\Rightarrow$ HIMQ = Function$\left[\{x, xi\}, \left(1 + \text{Norm}[x - xi, 2]^2\right)^{\frac{1}{2}}\right]$;
In two dimensions (Fig. 10.19),

$\Rightarrow$ z[r_,φ_] := HIMQ[$\{r\text{Cos}[\varphi], r\text{Sin}[\varphi]\}, \{0, 0\}$]
$\Rightarrow$ ParametricPlot3D[$\{r\text{Cos}[\varphi], r\text{Sin}[\varphi], z[r,\varphi]\}, \{r, -4, 4\}$,
$\{\varphi, 0., 2\pi\}$, BoxRatios $\rightarrow \{1, 1, 0.5\}$].

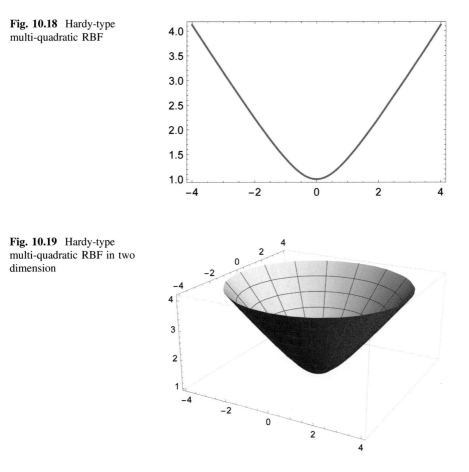

Fig. 10.18 Hardy-type multi-quadratic RBF

Fig. 10.19 Hardy-type multi-quadratic RBF in two dimension

10.4.3   Wendland Function

The Wendland function is a polynomial-type RBF, see Fig. 10.20.

$$\varphi(r) = (1 - r)^6 \left(35r^2 + 18r + 3\right)$$

$\Rightarrow$ WLD = Function[x, xi, $(1 - \text{Norm}[x - xi, 2])^6$
$\left(35\text{Norm}[x - xi, 2]^2 + 18\text{Norm}[x - xi, 2] + 3\right)]$;

In two dimensions, see Fig. 10.21.

$\Rightarrow$ z[r_,φ_] := VLD[{rCos[φ], rSin[φ]}, {0, 0}]
$\Rightarrow$ ParametricPlot3D[{rCos[φ], rSin[φ], z[r,φ]}, {r, -1.0, 1.0},
{φ, 0., 2π}, BoxRatios $\rightarrow$ {1, 1, 0.5}].

Fig. 10.20 Wendland function

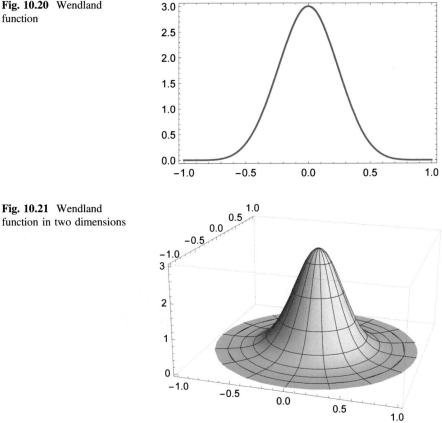

Fig. 10.21 Wendland function in two dimensions

10.4.4   Buchmann-Type RBF

This basis function is a mixed type, see Fig. 10.22.

$$\varphi(r) = 2r^4 \log(r) - \frac{7}{2}r^4 + \frac{16}{3}r^3 - 2r^2 + \frac{1}{6}, \quad 0 \le r \le 1$$

⇒ BH = Function[x, xi, If[x! = xi, 2Norm[x − xi, 2]^4 Log[Norm[x − xi, 2]]−
   7/2 Norm[x − xi, 2]^4 + 16/3 Norm[x − xi, 2]^3 −
   2 Norm[x − xi, 2]^2 + 1/6, 0]];

In two dimensions, see Fig. 10.23.

⇒ z[r_,φ_] := BH[{rCos[φ], rSin[φ]}, {0, 0}]
⇒ ParametricPlot3D[{rCos[φ], rSin[φ], z[r,φ]}, {r, −1.0, 1.0},
   {φ, 0., 2π}, BoxRatios → {1, 1, 0.5}].

Fig. 10.22 Buchmann function

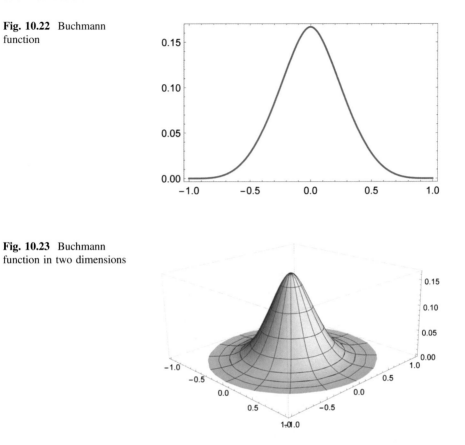

Fig. 10.23 Buchmann function in two dimensions

10.5   Generic Derivatives of RBF Functions

This topic is important for solving partial differential equations.

The formulas for all first and second-order derivatives of RBF of two variables are the followings:

If the basis function is given by

$$\varphi(r) = \varphi(\|u\|) = \varphi\left(\sqrt{x^2 + y^2}\right),$$

then the first order partial derivatives is

$$\frac{\partial \varphi}{\partial x} = \frac{d\varphi}{dr}\frac{\partial r}{\partial x} = \frac{d\varphi}{dr}\frac{x}{\sqrt{x^2 + y^2}} = \frac{x}{r}\frac{d\varphi}{dr},$$

similarly

$$\frac{\partial \varphi}{\partial y} = \frac{y}{r} \frac{d\varphi}{dr}.$$

The second order derivatives,

$$\frac{\partial^2 \varphi}{\partial x^2} = \frac{x^2}{r^2} \frac{d^2 \varphi}{dr^2} + \frac{y^2}{r^3} \frac{d\varphi}{dr},$$

similarly

$$\frac{\partial^2 \varphi}{\partial y^2} = \frac{y^2}{r^2} \frac{d^2 \varphi}{dr^2} + \frac{x^2}{r^3} \frac{d\varphi}{dr},$$

as well as

$$\frac{\partial^2 \varphi}{\partial x \partial y} = \frac{xy}{r^2} \frac{d^2 \varphi}{dr^2} - \frac{xy}{r^3} \frac{d\varphi}{dr}.$$

The Laplacian is

$$\left(\frac{\partial^2}{\partial x^2} + \frac{\partial^2}{\partial y^2} \right) \varphi(r) = \frac{d^2 \varphi}{dr^2} + \frac{1}{r} \frac{d\varphi}{dr}.$$

The generic fourth-order biharmonic (or double Laplacian) is,

$$\left(\frac{\partial^4}{\partial x^4} + 2 \frac{\partial^4}{\partial x^2 \partial y^2} + \frac{\partial^4}{\partial y^4} \right) \varphi(r) = \frac{d^4 \varphi}{dr^4} + \frac{2}{r} \frac{d^3 \varphi}{dr^3} - \frac{1}{r^2} \frac{d^2 \varphi}{dr^2} + \frac{1}{r^3} \frac{d\varphi}{dr}.$$

10.6   Least Squares Approximation with RBF

Up to now we have looked only at interpolation. However, many times it makes more sense to approximate the given data by a least squares fit. This is especially true if the data are contaminated with noise, or if there are so many data points that efficiency considerations force us to approximate from a space spanned by fewer basis functions than data points. We will use n_R as the number of the basis functions. In case of interpolation, this equals the number of data points. Here the user chooses n_R in the beginning. It should be less than the number of data points (otherwise it is the same as interpolation).

If $x \in R^m$, so x is a vector of m dimensions, then the number of the unknowns is $n_R + m\, n_R = (1 + m)n_R$.

We speak about regression if the system is overdetermined $n > (1 + m)n_R$, where n is the number of the data.

Let us consider n "noisy" data points, but n_R RBF basis functions ($n_R < n$), then we have an overdetermined system for the coefficients c_i, $i = 1, 2, ..., n_R$.

$$f_j = \sum_{i=1}^{n_R} c_i \varphi(\|x - x_i\|), \quad j = 1, 2, ..., n > n_R,$$

For example, let us consider the following function, see Fig. 10.24.

$$f(x, y) = 0.3\left(1.35 + e^x \sin\left(13(x - 0.6)^2\right)e^{-y}\sin(7y)\right),$$

To approximate this function on the interval of $[0, 1] \times [0, 1]$ with $n = 100$ "noisy" points

⇒ f[{x_,y_}] := 0.3(1.35 + e^x sin[13(x − 0.6)²]e^{−y} sin[7y]);

We add a white noise $N(0, \sigma)$, with $\sigma = 0.05$ to the functions values (Fig. 10.25),

⇒ fn[{x_,y_}] := f[{x,y}] + Random[NormalDistribution[0., 0.05]]

Then the data triples $\{\{x_i, y_i, f(x_i, y_i)\}$ are

⇒ dataxyfn = Flatten[Table[N[{i, j,
  fn[{i, j}]}], {i, 0, 1, 1/9}, {j, 0, 1, 1/9}], 1];
⇒ Dimensions[dataxyfn]
⇐ {100, 3}

In our case $m = 2$, therefore the maximum number of the basis is $n_R \sim 33$. Let us employ 6 thin plate spline basic functions,

Fig. 10.24 The function to be approximated

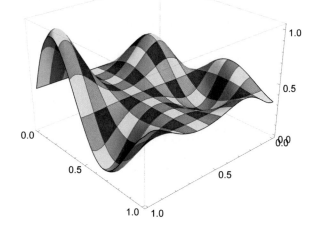

Fig. 10.25 The function with the "noisy" data

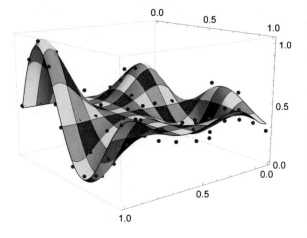

⇒ xym = Map[{#[[1]], #[[2]]}&, dataxyfn]; fm = Map[#[[3]]&, dataxyfn];

We organize the basis function in an RBF neural network,

⇒ < <NeuralNetworks'

⇒ rbf = InitializeRBFNet[xym, fm, 10,

   Neuron → Function[x, If[x¹0, x²*Log[−x], 0]]]//Quiet

⇐ RBFNet[w1, λ, w2}, χ},

   Neuron → Function[x, If[x ≠ 0, x² Log[−x], 0]],

   FixedParameters → None, AccumulatedIterations → 0,

   CreationDate → 2016, 11, 12, 18, 27, 10.9401869,

   OutputNonlinearity → None, NumberOfInputs → 2]

Let us train the network on the noisy data, see Fig. 10.26.

⇒ {rbf2, fitrecord} = NeuralFit[rbf, xym, fm, 500]//Quiet;

The approximation function is

⇒ Short[rbf2[{u, v}], 20]

⇐ {−20.3697 + <<16 > > +

   86084.4 If[−4.01953(−0.709155 + u)² − 4.01953(−0.0616565 + v)²! = 0,

   (−4.01953(−0.709155 + u)² − 4.01953(−0.0616565 + v)²)²

   Log[−(−4.01953(−0.709155 + u)² − 4.01953(−0.0616565 + v)²)], 0]−

   86079.9 If[−4.01961(−0.709153 + u)² − 4.01961(−0.0616512 + v)²! = 0,

   (−4.01961(−0.709153 + u)² − 4.01961(−0.0616512 + v)²)²

   Log[−(−4.01961(−0.709153 + u)² − 4.01961(−0.0616512 + v)²)], 0]}

Fig. 10.26 The error of the network during the training process

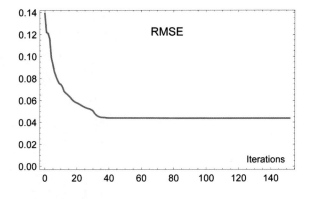

The error of the approximation can be seen in Fig. 10.27.

As an alternative solution let us employ Hardy-type multi-quadratic RBF basis functions,

$\Rightarrow$ rbf = InitializeRBFNet[xym, fm, 10, Neuron $\rightarrow$ Function[x, $\sqrt{1+x^2}$]]

$\Leftarrow$ RBFNet[w1, λ, w2, χ, Neuron $\rightarrow$ Function[x, $\sqrt{1+x^2}$],

    FixedParameters $\rightarrow$ None, AccumulatedIterations $\rightarrow$ 0,

    CreationDate $\rightarrow$ 2016, 11, 12, 18, 27, 30.2686209,

    OutputNonlinearity $\rightarrow$ None, NumberOfInputs $\rightarrow$ 2]

Training the network (Fig. 10.28),

$\Rightarrow$ {rbf2, fitrecord} = NeuralFit[rbf, xym, fm, 500]//Quiet;

The approximation function is

$\Leftarrow$ {30.4649 + 74.3045u −
86.2706 Sqrt[1 + (−1.48345(−0.23777 + u)^2 − 1.48345(−1.01835 + v)^2)^2] +
247.353 Sqrt[1 + (−1.10768(−0.191987 + u)^2 − 1.10768(−0.696926 + v)^2)^2] −
8979.46 Sqrt[1 + (−1.60279(−0.0245169 + u)^2 − 1.60279(−0.47947 + v)^2)^2] +
9063.27 Sqrt[1 + (−1.60339(−0.0285279 + u)^2 − 1.60339(−0.475971 + v)^2)^2] −
7.20718 Sqrt[1 + (−7.58294(−0.459186 + u)^2 − 7.58294(−0.430761 + v)^2)^2] +
0.754657 Sqrt[1 + (−16.7615(−0.912564 + u)^2 − 16.7615(−0.262636 + v)^2)^2] −
1517.98 Sqrt[1 + (−1.34759(−0.437715 + u)^2 − 1.34759(−0.24286 + v)^2)^2] +
1228.5 Sqrt[1 + (−1.42576(−0.487658 + u)^2 − 1.42576(−0.225803 + v)^2)^2] −
4.40349 ∗ 10^6 Sqrt[1 + (−2.85803(−0.272968 + u)^2 − 2.85803(−0.058178 + v)^2)^2] +
4.40355 ∗ 10^6 Sqrt[1 + (−2.85801(−0.272968 + u)^2 − 2.85801(−0.0581755 + v)^2)^2] −
148.471v}

The error of the approximation can be seen in Fig. 10.29.

The error of the approximation is very similar, but the formula is not so complex as was the case of the thin plate spline basic functions.

Fig. 10.27 The error of the approximation

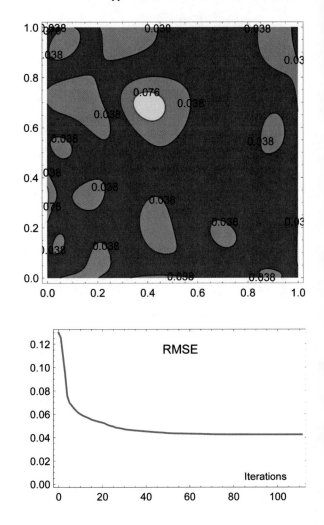

Fig. 10.28 The error of the network during the training process

10.7   Applications

10.7.1   Image Compression

Image compression methods try to reduce the necessary data of an image without loosing the basic information for the proper image reconstruction. Let us consider a well-known image, see Fig. 10.30.

⇒ MM = ImageData[image]

⇒ Image[MM]

Fig. 10.29 The error of the approximation

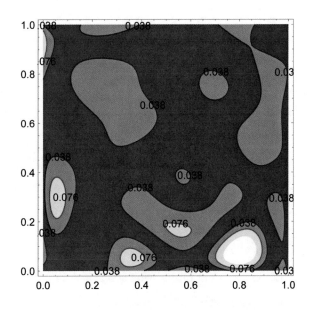

Fig. 10.30 The original image

The intensity values of the image pixels are between zero and one, see Fig. 10.31.

The size of the image is 180×180 pixel giving 32,400 pixels. In order to a create "noisy" image, the non-white pixels will be randomly whitened,

⇒ MMR = MM;

⇒ Do[MMR[[RandomInteger[{1, 180}], RandomInteger[{1, 180}]]] = 1, {i, 1, 10000}];

Fig. 10.31 The intensity
value of the image pixels

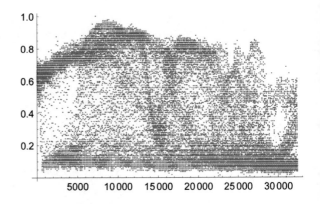

Figure 10.32 shows the image after 10^4 times carried out whitening,

⇒ Image[MMR]

Let us carry out further whitening randomly, see Fig. 10.33.

⇒ MMR = MM;
⇒ Do[MMR[[RandomInteger[{1, 180}], RandomInteger[{1, 180}]]] =
    1, {i, 1, 80000}];
⇒ Image[MMR]

Now, the coordinates of the non black pixels will be normalized, see Fig. 10.34

⇒ dataz = {}; dataxy = {};
⇒ Do[If[MMR[[i, j]]¹1, AppendTo[dataxy, {i/180., j/180.}];
    AppendTo[dataz, MMR[[i, j]]]], {i, 1, 180}, {j, 1, 180}]

Fig. 10.32 The noisy image
after 8×10^4 random
whitening

Fig. 10.33 Strongly blurred
image after 80,000 additional
random whitening

Fig. 10.34 The non-black
pixels with normalized
coordinates

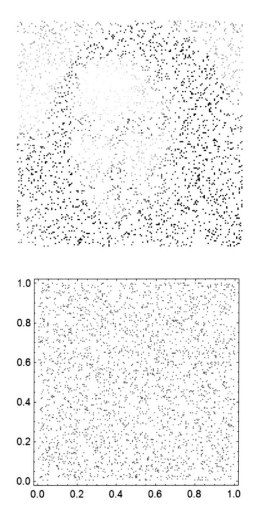

These pixels will be the basic points and their intensity represents the function
value to be approximated. Their number is

$\Rightarrow$ nj = Length[dataxy]

$\Leftarrow$ 2771

which is only

$\Rightarrow$ Length[dataxy]/(nm)100.

$\Leftarrow$ 8.55247

percent of the number of the original 32,400 pixels. Let us employ a second order
TPS approximation

⇒ TPSpline2 =

Function[x, xi, If[x¹xi, Norm[x − xi, 2]⁴Log[Norm[x − xi, 2]], 0]]

⇐ Function[x, xi, If[x ≠ xi, Norm[x − xi, 2]⁴Log[Norm[x − xi, 2]], 0]]

We can employ parallel computation for computing the elements of the coefficient matrix,

⇒ LaunchKernels[8]

⇐ {KernelObject[1, local], KernelObject[2, local], KernelObject[3, local],
    KernelObject[4, local], KernelObject[5, local], KernelObject[6, local],
    KernelObject[7, local], KernelObject[8, local]}

⇒ DistributeDefinitions[TPSpline2]

⇐ {TPSpline2}

⇒ M = ParallelTable[TPSpline2[dataxy[[i]],
    dataxy[[j]]], {i, 1, nj}, {j, 1, nj}];

Using pseudoinverse,

⇒ PM = PseudoInverse[M];

we get the 2771 coefficients, see Fig. 10.35.

⇒ c = PM.dataz;

⇒ ListPlot[c, Filling → Axis]

With these, one can define the reconstruction (approximation) function as,

⇒ g[x_, y_] := c.Map[TPSpline2[{x, y}, #]&, dataxy]

Now, generating 10^4 pixel values in parallel computation, we get the reconstructed image, see Fig. 10.36.

⇒ DistributeDefinitions[g]

⇐ {g, c, TPSpline}

⇒ RImage = ParallelTable[g[x, y], {x, 0, 1, 0.01}, {y, 0, 1, 0.01}];

⇒ Image[RImage]

Fig. 10.35 The coefficients of the image reduction problem

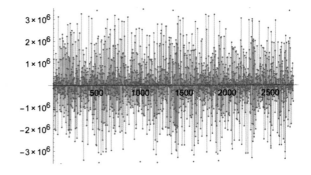

Fig. 10.36 The reconstructed image

It goes without saying that the reconstructed image is less attractive than the original one, but in case of the original image one should store 32,400 intensity values plus their (x, y) coordinates, while in case of the reconstructed one, we need to store only the 2771 coefficients.

10.7.2   RBF Collocation Solution of Partial Differential Equation

To avoid finite element meshing problems, we often try meshless methods.

Let us consider a linear elliptic PDE problem,

$$\mathcal{L}u(x, y) = f(x, y), \quad x, y \in \Omega,$$

where $\mathcal{L}$ is a linear operator. The boundary condition is,

$$u(x, y) = g(x, y) \quad x, y \in \Gamma.$$

The number of the collocation points are $N = N_\Omega + N_\Gamma$. We are looking for the solution in a form

$$u(x, y) = \sum_{\varphi=1}^{N_\Omega} c_j \varphi \left(\| \{x, y\} - \{x_j, y_j\} \| \right) + \sum_{\varphi=N_\Omega+1}^{N} c_j \varphi \left(\| \{x, y\} - \{x_j, y_j\} \| \right).$$

The first term is representing the solution inside the region Ω and the second one is for the boundary. In the region we carry out the substitution for every j index, we get

$$\mathcal{L}u(x,y) = \sum_{\varphi=1}^{N_\Omega} c_j \mathcal{L}\varphi\left(\left\|\{x,y\} - \{x_j, y_j\}\right\|\right).$$

In matrix form,

$$\mathcal{L}\,\mathbf{u} = \mathbf{A}_\mathcal{L}\mathbf{c},$$

where

$$(\mathbf{A}_\mathcal{L})_{i,j} = \mathcal{L}\varphi\left(\left\|\{x,y\} - \{x_j, y_j\}\right\|\right)_{\{x=x_i, y=y_i\}}, \quad i,j = 1,\ldots, N_\Omega.$$

For the boundary

$$u(x_i, y_i) = g(x_i, y_i) \quad x_i, y_i \in \Gamma$$

we can develop our equation as follows,

$$u(x,y) = \sum_{\varphi=N_\Omega+1}^{N} c_j \varphi\left(\left\|\{x,y\} - \{x_j, y_j\}\right\|\right) \quad \rightarrow \mathbf{u} = \mathbf{A}\mathbf{c},$$

where

$$A_{i,j} = \varphi\left(\left\|\{x_i, y_i\} - \{x_j, y_j\}\right\|\right), \quad i,j = N_\Omega + 1,\ldots, N,$$

in matrix form

$$\mathbf{A}\mathbf{c} = \mathbf{g}.$$

Collecting our equations for the unknown coefficients, we get

$$\begin{pmatrix} \mathbf{A}_\mathcal{L} \\ \mathbf{A} \end{pmatrix} \mathbf{c} = \begin{pmatrix} \mathbf{f} \\ \mathbf{g} \end{pmatrix}.$$

where $f_i = f(x_i, y_i)$, $i \in [1,\ldots, N_\Omega]$ and $g_i = g(x_i, y_i)$, $i \in [N_\Omega + 1,\ldots, N]$.
Here, we consider the following problem:
In the region

$$\Delta u(x,y) = -\frac{5\pi^2}{4} \sin(\pi x) \cos\left(\frac{\pi y}{2}\right), \quad x, y \in \Omega,$$

on the boundary

$$\Delta u(x, y) = \sin(\pi x) \cos\left(\frac{\pi y}{2}\right), \quad x, y \in \Gamma,$$

where

$$\Omega = [0, 1]^2 - [0.5, 1]^2$$

and Γ is the boundary of the region Ω, see Fig. 10.37.

The solution will be approximated via the following RBF functions,

$$u(x, y) \approx \sum_{i=1}^{ni} c_i \varphi(\{x, y\}, \{x_i, y_i\}) + \sum_{i=ni+1}^{ni+nb} c_i \varphi(\{x, y\}, \{x_i, y_i\}).$$

The first term is for the region and the second one is for the boundary. Now we use a thin plate spline of second order

$$\varphi(\{x, y\}, \{x_i, y_i\}) = \|\{x, y\}, \{x_i, y_i\}\|^4 \log\|\{x, y\}, \{x_i, y_i\}\| \quad \text{if} \{x, y\} \neq \{x_i, y_i\}$$

and

$$\varphi(\{x, y\}, \{x_i, y_i\}) = 0, \quad \text{otherwise.}$$

The Laplace operator in case of a $\varphi(r)$ basis function can be expressed as (see Sect. 9.5)

Fig. 10.37 The region of problem

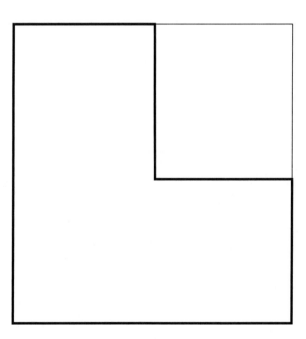

$$\Delta u(x, y) = \frac{\partial^2 u}{\partial x^2} + \frac{\partial^2 u}{\partial y^2},$$

$$\Delta u(x, y)\varphi(r) = \frac{d^2}{dr^2}\varphi(r) + \frac{1}{r}\frac{d}{dr}\varphi(r).$$

For second order TPS, it can be written as

$$\varphi(r) = r^4 \log(r),$$

$$\frac{d}{dr}\varphi(r) = r^3(4\log(r) + 1),$$

$$\frac{d^2}{dr^2}\varphi(r) = r^2(12\log(r) + 7).$$

Then

$$\Delta u(x, y)\varphi(r) = \frac{d^2}{dr^2}\varphi(r) + \frac{1}{r}\frac{d}{dr}\varphi(r)$$

$$= r^2(12\log(r) + 7) + \frac{1}{r}r^3(4\log(r) + 1)$$

and the Laplace operator is,

$$\Delta u(x, y)\varphi(r) = r^2(16\log(r) + 8).$$

The implementation of the problem in *Mathematica* is easy. First we generate Halton points as collocation points inside the region,

```
⇒ corput[n_, b_] :=
    IntegerDigits[n, b].(bRange[−Floor[log[b, n] + 1], −1]);
⇒ SetAttributes[corput, Listable]
⇒ halton[n_, s_] := corput[n, Prime[Range[s]]]
```

Then we define the Laplace function as

```
⇒ Δφ = Function[{x, xi},
       If[x'xi, Norm[x − xi, 2]² (16Log[Norm[x − xi, 2]] + 8), 0]]
  ⇐ Function[{x, xi}, If[x'xi, Norm[x − xi, 2]² (16 Log[Norm[x − xi, 2]] + 8), 0]]
```

Let us visualize this function in 2D, see Fig. 10.38

```
⇒ z[r_, φ_] := Delta φ[x, {0, 0}]/.x → {rCos[φ], rSin[φ]}
```

The basis function is,

```
⇒ TPSpline2 =
  Function[x, xi, If[x! = xi, Norm[x − xi, 2]⁴ Log[Norm[x − xi, 2]], 0]]
  ⇐ Function[{x, xi}, If[x ≠ xi, Norm[x − xi, 2]⁴ Log[Norm[x − xi, 2]], 0]]
```

Fig. 10.38 The second order
Laplace operator in our case

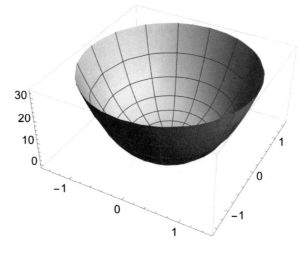

Let the number of the collocation points in the region of $[0, 1] \times [0, 0.5]$ be
ni1, while in the region of $[0, 0.5] \times [0, 0.5]$ is **ni2,**

⇒ ni2 = 50; ni1 = 100;

⇒ dataxy2 =
    Map[{#[[1]]0.5, #[[2]]0.5 + 0.5}&, Table[halton[n, 2], {n, ni2}]];

⇒ dataxy1 = Map[{#[[1]], 0.5#[[2]]}&, Table[halton[n, 2], {n, ni1}]];

See Fig. 10.39.
The boundary points are

⇒ dataA = Map[{#, 0}&, Range[0, 1, 0.05]];

⇒ dataB = Map[{1, #}&, Range[0.1, 0.5, 0.05]]

⇒ dataC = Map[{#, 0.5}&, Range[0.5, 0.95, 0.05]]

⇒ dataD = Map[{0.5, #}&, Range[0.55, 0.95, 0.05]]

⇒ dataE = Map[{#, 1}&, Range[0, 0.5, 0.05]]

⇒ dataF = Map[{0, #}&, Range[0.1, 0.95, 0.05]]

⇒ dataBoundary = Join[dataA, dataB, dataC, dataD, dataE, dataF]//Union;

We have 150 points inside the region and 78 points on the boundary, see Fig. 10.40.

⇒ nb = Length[dataBoundary]

⇐ 78

Let us put these points together

⇒ dataxy12 = Join[dataxy1, dataxy2];

⇒ ni = ni1 + ni2; nj = ni + nb;

⇒ dataTotal = Join[dataxy12, dataBoundary];

Fig. 10.39 The Halton points as collocation points in the region

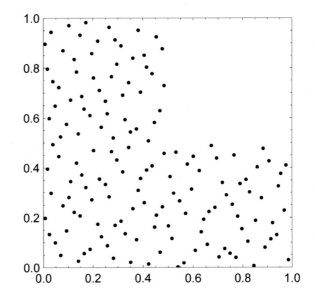

The coefficient matrix is

$$\mathbf{MF} = \begin{pmatrix} \mathbf{A}_{\mathcal{L}} \\ \mathbf{A} \end{pmatrix}.$$

$\Rightarrow$ MF = Table[If[i $\leq$ ni,$\Delta\varphi$[dataTotal[[i]], dataTotal[[j]]],
   TPSpline2[dataTotal[[i]], dataTotal[[j]]]], {i, 1, nj}, {j, 1, nj}];

Its size is

$\Rightarrow$ Dimensions[MF]
$\Leftarrow$ {228, 228}

The right side vector can be generated as $\begin{pmatrix} \mathbf{f} \\ \mathbf{g} \end{pmatrix}$.

$\Rightarrow$ f[x_] := $-\dfrac{5\pi^2}{4}$ sin[πx[[1]]]cos[$\dfrac{\pi}{2}$x[[2]]]

$\Rightarrow$ g[x_] := Sin[πx[[1]]]Cos[$\dfrac{\pi}{2}$x[[2]]]

$\Rightarrow$ fg =
   Table[If[i $\leq$ ni, f[dataTotal[[i]]], g[dataTotal[[i]]]], {i, 1, nj}];
$\Rightarrow$ Dimensions[fg]
$\Leftarrow$ {228}

The linear equation system for the coefficients can be written as

Fig. 10.40 The collocation points in the region and on the boundary

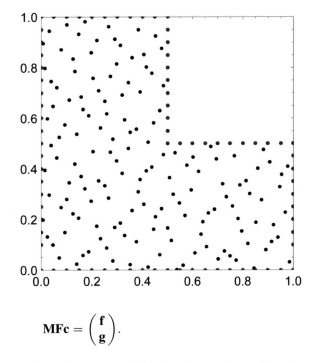

$$\mathbf{MFc} = \begin{pmatrix} \mathbf{f} \\ \mathbf{g} \end{pmatrix}.$$

The solution of this linear system gives the coefficient vector, **c**, whose elements can be seen in Fig. 10.41.

⇒ cc = PseudoInverse[MF].fg;

Therefore the approximation function is

⇒ g[{x_,y_}] := cc.Map[TPSpline2[{x,y}, #]&, dataTotal]

We can compare our RBF approximation with the exact solution, which is

$$u(x, y) = sin(\pi x)cos\left(\pi\frac{y}{2}\right).$$

⇒ G[{x_,y_}] := Sin[πx]Cos[πy/2]

The exact values in the collocation points are

⇒ dataz = Map[G[#]&, dataTotal]//N;
⇒ dataxyz = MapThread[Flatten[{#1, #2}]&, {dataTotal, dataz}];

Figure 10.42 shows the exact values in the collocation points with the surface of the approximation function,

The error function can be expressed as (Fig. 10.43)

Fig. 10.41 The coefficients of our problem

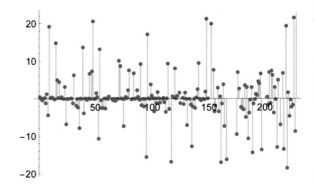

Fig. 10.42 The exact collocation values and the surface of the RBF approximation solution

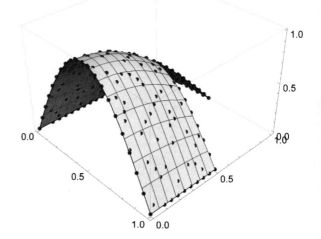

Fig. 10.43 The error function of the RBF approximation solution

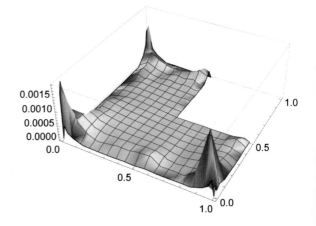

$\Rightarrow \Delta[\{x\_, y\_\}] := \text{Abs}[g[\{x, y\}] - G[\{x, y\}]]$

As expected, the approximation error is highest in the corners of the region, but even here the error is less than 0.002.

10.8 Exercise

10.8.1 Nonlinear Heat Transfer

Let us consider again our nonlinear steady state heat transfer problem in 1D,

$$\frac{d}{dx}\left(\lambda(\theta)\frac{d\theta}{dx}\right) = 0.$$

The boundary conditions are,

$$\theta(0) = 0 \quad \text{and} \quad \theta(1) = 1.$$

The heat transfer coefficient depending on the temperature is

$$\lambda(\theta) = 1 + k\,\theta.$$

Now we should employ RBF collocation with the Hardy-type multi-quadratic RBF.

Solution
The differential equation with $k = 1$ is

$\Rightarrow \text{D}[(1 + k\ \theta[x])\text{D}[\theta[x], x], x]/.k \rightarrow 1$
$\Leftarrow \theta'[x]^2 + (1 + \theta[x])\theta''[x]$

The RBF function can be written as

$\Rightarrow \text{Hardy} = \text{Function}\left[\{x, xi\}, \sqrt{\left(1 + \text{Abs}[x - xi]^2\right)}\right]$

$\Leftarrow \text{Function}\left[\{x, xi\}, \sqrt{1 + \text{Abs}[x - xi]^2}\right]$

Its first derivative is

$\Rightarrow \text{D}\left[\sqrt{(1 + r\hat{2})}, r\right] //\text{Simplify}$

$\Leftarrow \dfrac{r}{\sqrt{1 + r^2}}$

Then substituting our RBF into it, we get the first derivative

$$\Leftarrow d\theta = \texttt{Function}\left[\{x,xi\}, \texttt{Abs}[x-xi]\left(1+\texttt{Abs}[x-xi]^2\right)^{-1/2}\right]$$

$$\Leftarrow \texttt{Function}\left[\{x,xi\}, \texttt{Abs}[x-xi]\,\dfrac{1}{\sqrt{1+\texttt{Abs}[x-xi]^2}}\right]$$

The second derivate can similarly be obtained since

$$\Rightarrow D\left[\sqrt{(1+r\hat{2})}, r, r\right]//\texttt{Simplify}$$

$$\Leftarrow \dfrac{1}{(1+r^2)^{3/2}}$$

$$\Rightarrow dd\theta = \texttt{Function}\left[\{x,xi\}, \left(1+\texttt{Abs}[x-xi]^2\right)^{-3/2}\right]$$

$$\Leftarrow \texttt{Function}\left[\{x-xi\}, \left(1+[\texttt{Abs}[x-xi]^2\right)^{-3/2}\right]$$

Using these derivatives, our differential equation can be written as

$$\Rightarrow d = \texttt{Function}\left[\{x,xi\}, \left(\texttt{Abs}[x-xi]\left(1+\texttt{Abs}[x-xi]^2\right)^{-1/2}\right)^2 + \right.$$

$$\left.\left(1+\sqrt{\left(1+\texttt{Abs}[x-xi]^2\right)}\right)\left(1+\texttt{Abs}[x-xi]^2\right)^{-3/2}\right]$$

Let us chose 4 collocation points

$$\Rightarrow \texttt{dataTotal} = \{\tfrac{1}{3},\tfrac{2}{3},0,1\}$$
$$\Leftarrow \{\tfrac{1}{3},\tfrac{2}{3},0,1\}$$

Then the coefficient matrix is

$$\Rightarrow \texttt{MF} = \texttt{Table}[\texttt{If}[i \leq 2, d[\texttt{dataTotal}[[i]], \texttt{dataTotal}[[j]]],$$
$$\texttt{Hardy}[\texttt{dataTotal}[[i]], \texttt{dataTotal}[[j]]]], \{i,1,4\}, \{j,1,4\}]//\texttt{N};$$
$$\Rightarrow \texttt{MF}//\texttt{MatrixForm}$$

$$\Leftarrow \begin{pmatrix} 2. & 1.85381 & 1.85381 & 1.57603 \\ 1.85381 & 2. & 1.57603 & 1.85381 \\ 1.05409 & 1.20185 & 1. & 1.41421 \\ 1.20185 & 1.05409 & 1.41421 & 1. \end{pmatrix}$$

The boundary conditions in the collocation points are

$$\Rightarrow \texttt{fg} = \{0,0,0,1\}$$
$$\Leftarrow \{0,0,0,1\}$$

The solution of the linear system for the unknown coefficients of the Hardy-Function is

Fig. 10.44 The temperature profile employing the Hardy-type multi-quadratic RBF

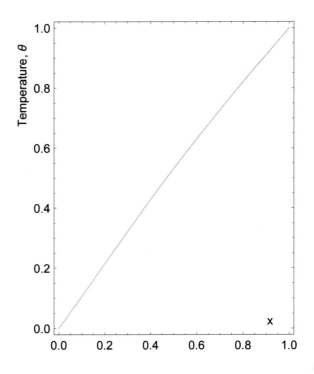

⇒ cc = PseudoInverse[MF].fg//N

⇐ {−8.21461, 6.02496, 4.97703, −2.51671}

Employing these coefficient values, the approximation function for the temperature distribution in analytical form can be written as

⇒ cc.Map[Hardy[x, #]&, dataTotal]

$$\Leftarrow -2.51671\sqrt{1+\text{Abs}[-1+x]^2} + 6.02496\sqrt{1+\text{Abs}\left[-\frac{2}{3}+x\right]^2} -$$

$$8.21461\sqrt{1+\text{Abs}\left[-\frac{1}{3}+x\right]^2} + 4.97703\sqrt{1+\text{Abs}[x]^2}$$

The compact form of this function is

⇒ g[x_] := cc.Map[Hardy[x, #]&, dataTotal]

Fig. 10.44 shows the temperature profile.

References

Fasshauer GE (2007) Meshfree approximation methods with MATLAB, interdisciplinary mathematical sciences, Vol 6. World Scientific, New Jersey, London, Singapore

Mangano S (2010) Mathematica cookbook. O'REILLY, Beijing, Cambridge, Farnham, Köln

Micchelli CA, Xu Y, Zhang H (2006) Universal Kernels. J Mach Learn Res 7:2651–2667

Wendland H (2005) Scattered data approximation. Cambridge University Press, New York, Cambridge Monographs on Applied and Computational Mathematics

Chapter 11
Support Vector Machines (SVM)

11.1 Concept of Machine Learning

In statistical learning theory (regression, classification, etc.) there are many regression models, such as algebraic polynomials, Fourier series, radial basis functions (RBF), neural networks, fuzzy logic models, and time series. In all of these the most important feature is the model's ability to generalize to other, future data.

The Learning machine is the model of the real System. The learning machine estimates the output of the system (y) (see Fig. 11.1) by adjusting the parameters (w) during its training phase. After training, in the generalization or validation phase, the output from the machine is expected to be a good estimate of the system's true response, y. At that point the learning machine will rarely try to interpolate training data pairs, but would rather seek an approximating function that can generalize well. In other words, during the training phase, the data (x) are from a training set (H) but a good approximation is expected on a target set (Z) where $Z \supset H$. There is a trade-off between the approximation error on H and the generalization error on $Z - H$. So we are looking for a *type of approximation function f_B* that has good performance on Z. In practice we can find only an f that has less than optimal behavior. Seeking the best function that can generalize is achieved through *structural risk minimization*. There are different techniques of constructing good estimators:

(a) *cross validation* divides the data into two parts; a training set and a validation set. The validation set is used *after* the training phase.

Supplementary Information The online version contains supplementary material available at https://doi.org/10.1007/978-3-030-92495-9_11.

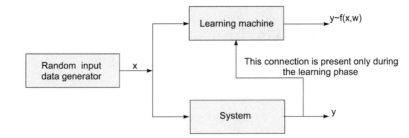

Fig. 11.1 Operation of a learning machine in case of supervised learning

The cross validation can be used for comparing several learning machines, then choosing the model which provides the best trade-off between the error on the training and validation set.

Another technique is when the validation set is *monitored* during the training phase. When the error starts to increase on the validation set, the training will be stopped (early stopping).

(b) *regularization*, when the error function to be minimized on H is extended with an additional term, which is responsible for the smoothness of the approximation function. For example this term can be the square of the gradient of the approximating function.

Remark In general, we use the training set and validation set during developing our model. So the data of the validation set are also involved in the training phase. After that test data can be used as a last check for our model generalization ability.

The *support vector machine* provides a further method to construct an optimal estimator via structural risk minimization.

11.2   Optimal Hyperplane Classifier

The Support Vector Machine (SVM) is a very effective so called "kernel method" for classification as well as regression, even in cases of strongly nonlinear problems. First, we start with classification. We shall show how we can extend a linear maximal margin classifier for use as a nonlinear classifier employing kernel techniques. The SVM technique can be especially competitive when the vectors of the elements to be classified are very long.

11.2.1 Linear Separability

In many problems we wish to separate clusters of data with a line, plane, or hyper-plane. This is called *separable linear classification*. The plane is called a *decision plane*. Obviously, there are almost always infinitely many such planes, but some are better than others. We would like to select a decision plane that is "in the middle", where the distances of closest points of the two different clusters are the maximum; see Fig. 11.2.

Let us consider a hyperplane as a decision plane. Here is a decision line separating the two clusters

$$y = wx + b.$$

For the boundary elements of the cluster dark blue squares (x_1), let

$$wx_1 + b = 1$$

and for the boundary elements of the cluster light yellow circle (x_2), let

$$wx_2 + b = -1.$$

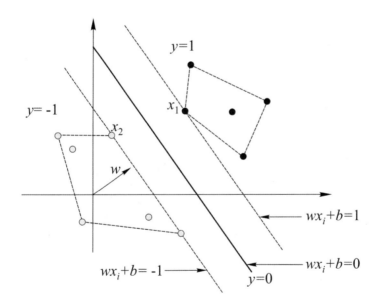

Fig. 11.2 A separable classification problem

Then

$$w\,(x_1 - x_2) = 2.$$

The distance of the boundary points is

$$x_1 - x_2 = \frac{2}{w}.$$

Consequently, w is minimized in order to decrease the structural risk of the linear separation. In general, the scalar product

$$\frac{1}{2}\langle w, w \rangle$$

should be minimized. Figure 11.3 shows two solutions for the clustering problem.

The solution on the left-hand side was computed by a neural network's "perceptron". It provides higher structural risk than the solution on the right-hand side computed via an SVM linear classifier, because it just barely works. Should the data have small errors and the points move slightly, that decision line will no longer be correct.

Since for the light yellow circles,

$$wx_i + b \leq -1,$$

and for the dark blue squares,

$$wx_i + b \geq 1,$$

let $y_i = -1$ be valid for the points belonging to the gray cluster and $y_i = 1$ for the cluster of the black points. Then, the constraints can be written in general form as

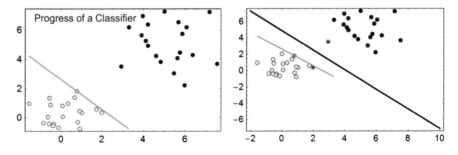

Fig. 11.3 Two different solutions for classification problem. Left: perceptron neural network, Right: support vector machine

$$y_i(wx_i + b) \geq 1, \quad i = 1, 2, ..., n, \quad x_i \in S.$$

We are looking for the weight vector in the following form

$$w = \sum_{i=1}^{n} \alpha_i x_i, \quad \text{where} \quad x_i \in S.$$

Therefore, the decision function can be written as

$$f(x) = \text{sign}\left(\sum_{i=1}^{n} \alpha_i \langle x, x_i \rangle + b \right).$$

where $f(x_i) = 1$ or -1 depending on which is the proper class for the element x_i.

11.2.2   Computation of the Optimal Parameters

The optimal α_i and b parameters can be computed via minimization of the following objective function

$$L(\alpha_i, b) = \frac{1}{2} w^2 + \sum_{i=1}^{n} \alpha_i (y_i(wx_i + b) - 1) \rightarrow \min .$$

This is the Lagrange form of the constrained optimization problem. In practice the optimal parameters will be computed from its dual form, see Berthold and Hand (2003)

$$G(\alpha) = \sum_{i=1}^{n} \alpha_i - \frac{1}{2} \sum_{i,j=1}^{n} y_i y_j \alpha_i \alpha_j \left(\langle x_i, x_j \rangle + \frac{1}{c} \delta_{i,j} \right) \rightarrow \max$$

under the constraints,

$$\sum_{i=1}^{n} y_i \alpha_i = 0, \quad \alpha_i \geq 0, \quad i = 1, ..., n.$$

Let us suppose that the optimal solutions are $\alpha_i, i = 1, ..., n$. Then our decision function is,

$$f(x) = \text{sign}\left(\sum_{i=1}^{n} \alpha_i \langle x, x_i \rangle + b \right) = \sum_{i=1}^{n} y_i \alpha_i \langle x_i, x \rangle + b, .$$

In the primary problem w is the sum of the $\alpha_i x_i$. Here, c is a constant parameter given by the user. The elements for $\alpha_i \neq 0$, i $= 1,2,..., m$ are true and called *support vectors*. Compare this phrase with the support of a function as in the previous chapter.

The value of the parameter b should be considered, that for all i index where $\alpha_i \neq 0$, will be true.

$$y_j f(x_j) = 1 - \frac{\alpha_j}{c}, \quad j = 1, 2, ...n.$$

Considering

$$f(x_j) = \sum_{i=1}^{n} y_i \alpha_i \langle x_i, x_j \rangle + b, \quad j = 1, 2, ...n$$

we get

$$b = \frac{1 - \frac{\alpha_j}{c}}{y_j} - \sum_{i=1}^{n} y_i \alpha_i \langle x_i, x_j \rangle, \quad j = 1, 2, ...n.$$

The reliability of the solution can be tested via controlling that for all j one gets the same b value.

11.2.3   Dual Optimization Problem

Since the dual form of the optimization problem in case of the determination of the parameters of the SVM approach plays a very important role, let us recall this technique.

First, we consider a linear problem,

$$\min \langle c^T, x \rangle, \quad Ax \geq b, \quad x \geq 0,$$

then its dual form is

$$\max \langle b^T, y \rangle, \quad A^T y \leq c, \quad y \geq 0.$$

In a nonlinear case the situation is more complicated. Let us consider the primary problem as

$$f(x) \rightarrow \min,$$
$$g_i(x) \leq 0, \quad i = 1, 2,..., m,$$
$$h_i(x) = 0, \quad i = 1, 2,..., p.$$

The Lagrange form of the problem is,

$$\mathcal{L}(x, \alpha, \beta) = f(x) + \sum_{i=1}^{m} \alpha_i g_i(x) + \sum_{i=1}^{p} \beta_i h_i(x),$$

where the variable x is called the *primary* variable while the variables α and β are called as *dual* variables.

Then, we consider the following primary problem

$$\min_x \left(\max_{a,b} \mathcal{L}(x, a, \beta) \right) = \min_x W_P(x)$$

where

$$W_P(x) = \max_{\alpha, \beta} \mathcal{L}(x, \alpha, \beta).$$

The dual form of this problem is

$$\max_{\alpha, \beta} \left(\min_x \mathcal{L}(x, \alpha, \beta) \right) = \max_{\alpha, \beta} W_P(\alpha, \beta),$$

where

$$W_P(\alpha, \beta) = \min_x \mathcal{L}(x, \alpha, \beta).$$

Let us consider a numerical illustration. Consider a simple problem as

$$x_1^2 + x_2^2 \rightarrow \min,$$
$$4 - 2x_1 - x_2 \leq 0.$$

The primary direct solution is

$$\Rightarrow \texttt{Minimize}[\{x_1^2 + x_2^2, 4 - 2x_1 - x_2 \leq 0\}, \{x_1, x_2\}]$$
$$\Leftarrow \left\{ \frac{16}{5}, \left\{ x_1 \rightarrow \frac{8}{5}, x_2 \rightarrow \frac{4}{5} \right\} \right\}$$

The Lagrange form of the problem is

$$\mathcal{L}(x, \alpha) = x_1^2 + x_2^2 + \alpha(4 - 2x_1 - x_2).$$

Then, the dual form to be maximized is

$$W_D(\alpha) = \min_x \mathcal{L}(x, \alpha).$$

Considering the necessary conditions

$$\frac{\partial \mathcal{L}(x, \alpha)}{\partial x_1} = 2x_1 - 2\alpha = 0 \rightarrow x_1 = \alpha,$$

$$\frac{\partial \mathcal{L}(x, \alpha)}{\partial x_2} = 2x_2 - \alpha = 0 \rightarrow x_2 = \frac{\alpha}{2},$$

then, for the dual form, we get

$$W_D(\alpha) = \min_x \mathcal{L}(x, \alpha) = \mathcal{L}(x, \alpha)|_{x_1 = \alpha, x_2 = \frac{\alpha}{2}} = 4\alpha - \frac{5\alpha^2}{4},$$

hence the dual problem is

$$\max_\alpha \left(4\alpha - \frac{5\alpha^2}{4} \right), \quad \alpha \geq 0,$$

whose solution is

$$\Rightarrow \texttt{Maximize} \left[\left\{ 4\alpha - \frac{5\alpha^2}{4}, \alpha^3 0 \right\}, \alpha \right]$$

$$\Leftarrow \left\{ \frac{16}{5}, \left\{ \alpha \rightarrow \frac{8}{5} \right\} \right\}$$

Consequently

$$x_1 = \alpha = \frac{8}{5} \quad \text{and} \quad x_2 = \frac{\alpha}{2} = \frac{4}{5}.$$

11.3   Nonlinear Separability

It may not be possible to separate clusters with linear hyper-planes, so we have the concept of nonlinear separable problems. However, we can transform such problems into a higher dimensional space, where they can be considered as a linear separable problem. To illustrate such a situation, we consider here the XOR problem.

The input–output logical table can be seen in Table 11.1.

Table 11.1 Input (x_1, x_2)– output (y) logical table of the XOR problem

x_1	x_2	y
0	0	0
0	1	1
1	0	1
1	1	0

Table 11.2 Input (x_1, x_2, x_3)– output (y) logical table of the XOR problem

x_1	x_2	x_3	y
0	0	0	0
0	1	0	1
1	0	0	1
1	1	1	0

The geometrical interpretation of the XOR problem can be seen in Fig. 11.4. It is clear that one linear line cannot separate the points of the class of true $(y = 1)$ from the points of class of false $(y = 0)$.

However, introducing a new input variable,

$$x_3 = x_1 x_2.$$

The input–output logical table is, the XOR problem can be transformed into a linearly separable problem in 3D; see Fig. 11.5. A feasible decision plane can be defined with the following three points $P_1(0.7, 0., 0.)$, $P_2(0., 0.7, 0.)$ and $P_3(1., 1., 0.8)$.

Fig. 11.4 Geometrical representation of the XOR problem

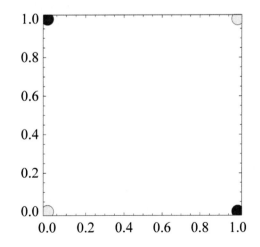

Fig. 11.5 Geometrical
representation of the XOR
problem with three inputs

11.4   Feature Spaces and Kernels

The objective function of the dual form of the linear separable problem contains the
scalar product (linear kernel) of the coordinates of the points to be classified,
$\langle x_i, x_j \rangle$. A nonlinear separable problem can be transformed into a linear one via
changing this linear kernel into a nonlinear one, i.e., $\langle x_i, x_j \rangle \rightarrow \langle \Phi(x_i), \Phi(x_j) \rangle$, see
Fig. 11.6.

This transformation maps the data into some other dot space, called the feature
space F. Recalling the dual form $G(\alpha)$ in Sect. 11.2.2, this only requires the
evaluation of dot products, $K(x_i, x_j)$.

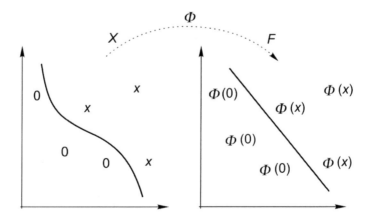

Fig. 11.6 The idea of nonlinear SVM

$$G(\alpha) = \sum_{i=1}^{n} \alpha_i - \frac{1}{2} \sum_{i,j=1}^{n} y_i y_j \alpha_i \alpha_j \left(K(x_i, x_j) + \frac{1}{c} \delta_{i,j} \right) \rightarrow \max .$$

Clearly, if F is a high dimensional feature space, the dot product on the right hand side will be very expensive to compute. In some cases, however there is a simple *kernel* that can be evaluated efficiently. For instance, the polynomial kernel.

$$K(u, v) = \langle u, v \rangle^d,$$

can be shown to correspond to a map Φ into the space spanned by all products of exactly d dimensions of R^n. For $d = 2$ and $u, v \in R^2$, for example, we have

$$\langle u, v \rangle^2 = \left(\left\langle \begin{pmatrix} u_1 \\ u_2 \end{pmatrix}, \begin{pmatrix} v_1 \\ v_2 \end{pmatrix} \right\rangle \right)^2 = \langle \Phi(u), \Phi(v) \rangle = \left\langle \begin{pmatrix} u_1^2 \\ \sqrt{2}u_1 u_2 \\ u_2^2 \end{pmatrix}, \begin{pmatrix} v_1^2 \\ \sqrt{2}v_1 v_2 \\ v_2^2 \end{pmatrix} \right\rangle.$$

defining $\Phi(x) = (x_1^2, \sqrt{2}x_1 x_2, x_2^2)$.

More generally, we can prove that for every kernel that gives rise to a positive matrix (kernel matrix) $M_{i,j} = K(x_i, x_j)$, we can construct a map such that $K(u, v) = \langle \Phi(u), \Phi(v) \rangle$ holds.

11.5   Application of the Algorithm

11.5.1   Computation Step by Step

The dual optimization problem can be solved conveniently using *Mathematica*. In this section, the steps of the implementation of the support vector machine classification (SVMC) algorithm are shown by solving XOR problem. The truth table of XOR, using bipolar values for the output, see Table 11.3.

The input and output data lists are

$\Rightarrow$ xym = {{0,0}, {0,1}, {1,0}, {1,1}};
$\Rightarrow$ zm = {-1,1,1,-1};

Table 11.3 Logical table of the XOR problem

x_1	x_2	y
0	0	-1
0	1	1
1	0	1
1	1	-1

Let us employ Gaussian kernel with β gain

$\Rightarrow \beta = 10.;$

$\Rightarrow \text{K}[\text{u\_},\text{v\_}] := \text{Exp}[-\beta(\text{u}-\text{v}).(\text{u}-\text{v})]$

The number of the data pairs in the training set, m is

$\Rightarrow \text{m} = \text{Length}[\text{zm}]$

$\Leftarrow 4$

Create the objective function $W(\alpha)$ to be maximized, with regularization or penalty parameter, $c = 5$

$\Rightarrow \text{c} = 5.;$

First, we prepare a matrix M, which is an extended form of the kernel matrix,

$\Rightarrow \text{M} = (\text{Table}[\text{N}[\text{K}[\text{xym}[[\text{i}]], \text{xym}[[\text{j}]]]], \{\text{i}, 1, \text{m}\}, \{\text{j}, 1, \text{m}\}] + (1/\text{c})\,\text{IdentityMatrix}[\text{m}]);$

then the objective function can be expressed as,

$\Rightarrow \text{W} = \sum_{i=1}^{m} \alpha_i - \frac{1}{2}\sum_{i=1}^{m}\sum_{j=1}^{m}(\text{zm}[[\text{i}]]\,\text{zm}[[\text{j}]]\alpha_i\alpha_j\text{M}[[\text{i},\text{j}]]);$

The constrains for the unknown variables are

$\Rightarrow \text{g} = \text{Apply}\left[\text{And}, \text{Join}\left[\text{Table}[\alpha_i^3 0, \{\text{i}, 1, \text{m}\}], \left\{\sum_{i=1}^{m} \text{zm}[[\text{i}]]\alpha_i == 0\right\}\right]\right].$

$\Leftarrow \alpha_1 \geq 0 \;\&\&\alpha_2 \geq 0\;\&\&\alpha_3 \geq 0\;\&\&\alpha_4 \geq 0\;\&\& -\alpha_1+\alpha_2+\alpha_3-\alpha_4 == 0$

The solution of this constrained optimization problem will results in the unknown variables α_i. However, this is a quadratic optimization problem for α, we use global method with local post processing,

$\Rightarrow \text{sol} = \text{NMaximize}[\{\text{W}, \text{g}\}, \text{vars}, \text{Method} \rightarrow \{"\text{DifferentialEvolution}",$
$"\text{CrossProbability}" \rightarrow 0.3, "\text{PostProcess}" \rightarrow \text{Automatic}\}]$
$\Leftarrow \{1.66679, \{\alpha_1 \rightarrow 0.833396, \alpha_2 \rightarrow 0.833396, \alpha_3 \rightarrow 0.833396, \alpha_4 \rightarrow 0.833396\}$

The consistency of this solution can be checked by computing values of parameter b for every data point. Theoretically, these values should be the same for any data point, however, in general, this is only approximately true.

$$\Rightarrow \text{bdata} = \text{Table}\left[\left(\left(1 - \frac{\alpha_j}{c}\right) / \text{zm}[[j]] - \sum_{i=1}^{m} \text{zm}[[i]] \alpha_i K[\text{xym}[[i]], \text{xym}[[j]]] \right) \right.$$

$$\left. /.\text{sol}[[2]], \{j, 1, m\} \right]$$

$$\Leftarrow \{-2.3063 \times 10^{-10}, -4.92456 \times 10^{-10}, 1.07811 \times 10^{-9}, -3.55027 \times 10^{-10}\}$$

The value of b can be chosen as the average of these values

$$\Rightarrow b = \text{Apply}[\text{Plus}, \text{bdata}]/m$$

$$\Leftarrow -1.38439 \times 10^{-17}$$

Then the classifier function is,

$$\Rightarrow f[\text{w}_-] := \left(\left(\sum_{i=1}^{m} \text{zm}[[i]] \alpha_i K[\text{w}, \text{xym}[[i]]] \right) + b \right) /.\text{sol}[[2]]$$

In analytic form

$$\Rightarrow f[\{x, y\}]//\text{Chop}$$

$$\Leftarrow -0.833396 e^{-10.\left((-1+x)^2 + (-1+y)^2\right)} + 0.8333962 e^{-10.\left(x^2 + (-1+y)^2\right)} +$$

$$0.833396 e^{-10.\left((-1+x)^2 + y^2\right)} - 0.833396 e^{-10.\left(x^2 + y^2\right)}$$

Let us display the contour lines of the continuous classification function, see Fig. 11.7.

For the discrete classifier, the decision rule using signum function is, see Fig. 11.8.

Let us check the classifier by substituting the input pairs. We get

$$\Rightarrow \text{Map}[\text{Sign}[f[\#]]\&, \text{xym}]$$

$$\Leftarrow \{-1, 1, 1, -1\}$$

The optimal separation boundary lines $x = 1/2$ and $y = 1/2$ crossing at $\{0.5, 0.5\}$, since

$$\Rightarrow \text{Sign}[f[\{0.75, 0.25\}]]$$

$$\Leftarrow 1$$

$$\Rightarrow \text{Sign}[f[\{0.75, 0.75\}]]$$

$$\Leftarrow -1$$

Fig. 11.7 The contour lines
of the continuous
classification function
$f(x, y)$ for XOR problem

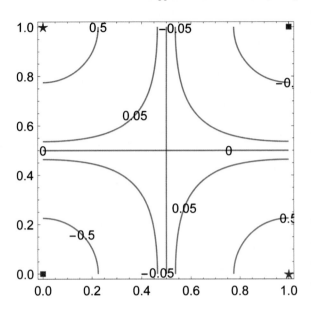

Fig. 11.8 The decision rule,
sign($f(x, y)$)) for XOR problem

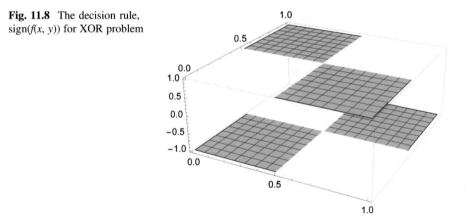

11.5.2   *Implementation of the Algorithm*

The SVM classification algorithm (SVMC) can be implemented in the following
Mathematica function,

⇒ SupportVectorClassifier[xm_, ym_, K_, c_] := Module[
  {m, n, M, i, j, W, g, vars, sol, bdata, b},
  m = Length[ym]; n = Length[xm[[1]]];
  M = Table[K[xm[[i]], xm[[j]]], {i, 1, m}, {j, 1, m}] +

  $\frac{1}{c}$ IdentityMatrix[m];

  $$W = \sum_{i=1}^{m} \alpha_i - \frac{1}{2} \sum_{i=1}^{m} \sum_{j=1}^{m} (ym[[i]] ym[[j]] \alpha_i \alpha_j M[[i, j]]);$$

⇒ g = Apply$\left[$And, Join$\left[$Table$[\alpha_i \geq 0, \{i, 1, m\}], \left\{\sum_{i=1}^{m} ym[[i]] \alpha_i == 0\right\}\right]\right]$;

⇒ vars = Table[α_i, {i, 1, m}];

⇒ sol = FindMaximum[{W, g}, vars, Method → "InteriorPoint"][[2]];

⇒ bdata = Table$\left[\left(\dfrac{1 - \frac{\alpha_j}{c}}{ym[[j]]} - \sum_{i=1}^{m} ym[[i]] \alpha_i K[xm[[i]], xm[[j]]]\right)\right/.$

  sol[[2]], {j, 1, m}];

⇒ b = Apply[Plus, bdata]/m;

⇒ $\left\{\left(\left(\sum_{i=1}^{m} ym[[i]] \alpha_i K[Table[x_j, \{j, 1, n\}], xm[[i]]]\right) + b\right)/.sol, vars/.sol\right\}$;

The output of this function is the analytical form of the SVMC function.
Let us employ this function for a linear problem, see Fig. 11.9.
We use linear kernel,

⇒ K[u_, v_] := u.v

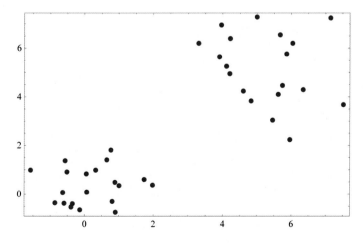

Fig. 11.9 A linear separable problem

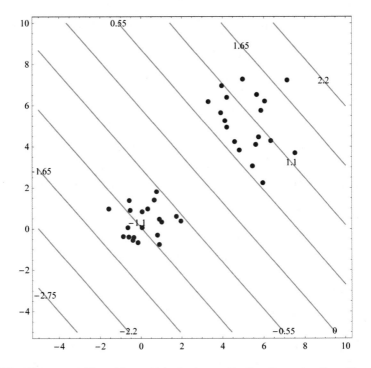

Fig. 11.10 A linear separable problem, with optimal separation line, the contour line with zero value

Let the regularization parameter

$\Rightarrow$ c = 0.025;

Then employing our function, we get

$\Rightarrow$ F = SupportVectorClassifier[xn, yn, K, c];

Figure 11.10 shows the optimal separation line, the contour line with zero value (0).

11.6   Two Nonlinear Test Problems

11.6.1   Learning a Chess Board

Let us consider a 2×2 size chess board in Fig. 11.11. The training points are generated by uniformly distributed random numbers from the interval $[-1, 1] \times [-1, 1]$, see Fig. 11.11.

We create the training set using 100 Halton-points as random sample data points, see Fig. 11.12.

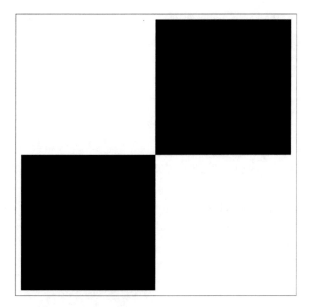

Fig. 11.11 The 2 × 2 chess board problem

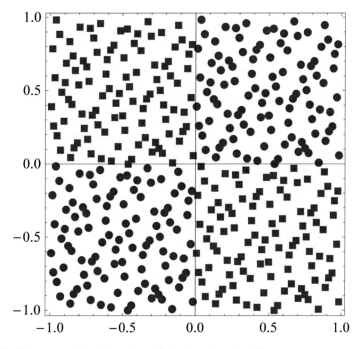

Fig. 11.12 The generated random points of the chess board problem

Let us employ the same Gaussian kernel, but now with gain $\beta = 20$.

$\Rightarrow$ K[u_, v_] := Exp$[-\beta(u - v).(u - v)]$

$\Rightarrow \beta = 20$;

using parameter $c = 100$,

$\Rightarrow$ c = 100;

Then the solution,

$\Rightarrow$ F = SupportVectorClassifier[xn, yn, K, c];

The contour lines can be seen on Fig. 11.13.

The zero contour lines represent the boundary of the clusters, see Fig. 11.14. One can get better results by increasing the number of points.

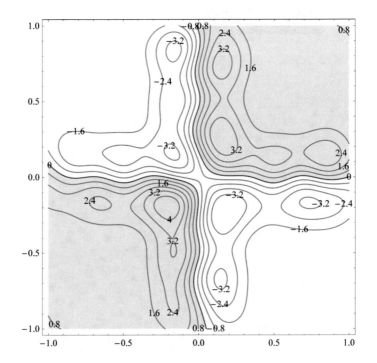

Fig. 11.13 The contour lines of the solution function of the chess board problem

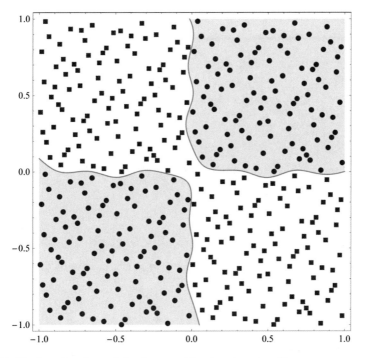

Fig. 11.14 The boundary lines of the clusters of the chess board problem

11.6.2 *Two Intertwined Spirals*

The *two intertwined spirals* is a challenging classification benchmark adopted from the field of neural networks.

Parametric equations of the spirals are

$$\Rightarrow x1[t\_] := 2\,Cos[t]e^{t/10}$$

$$\Rightarrow y1[t\_] := 1.5\,Sin[t]e^{t/10}$$

$$\Rightarrow x2[t\_] := 2.7\,Cos[t]e^{t/10}$$

$$\Rightarrow y2[t\_] := 2.025\,Sin[t]e^{t/10}$$

Generating 26–26 discrete points for each spiral,

$$\Rightarrow s1 = Table\left[\{x1[t], y1[t]\}, \left\{t, \pi, 3.5\,\pi, \frac{2.5\,\pi}{25}\right\}\right];$$

$$\Rightarrow s2 = Table\left[\{x2[t], y2[t]\}, \left\{t, -\frac{\pi}{2}, 2.5\,\pi, \frac{3\,\pi}{25}\right\}\right];$$

and displaying these points, one gets the data points for the two classes, see Fig. 11.15

$\Rightarrow$ S1 = List − Plot[s1, PlotStyle → {RGBCol − or[1, 0, 0], PointSize[0.02]},
    AspectRatio → 1];
$\Rightarrow$ S2 = ListPlot[s2, PlotStyle → {RGBCol − or[0, 0, 1], PointSize[0.015]},
    AspectRatio → 1];
$\Rightarrow$ pspiral = Show[{S1, S2}]

Creating the teaching set, putting these points into one list

$\Rightarrow$ xym = Join[s1, s2];

Generating the labels of the samples,

$\Rightarrow$ zm = Join[Table[1., {26}], Table[−1., {26}]];

Applying a wavelet kernel with parameter $a = 1.8$, in case of dimension $n = 2$

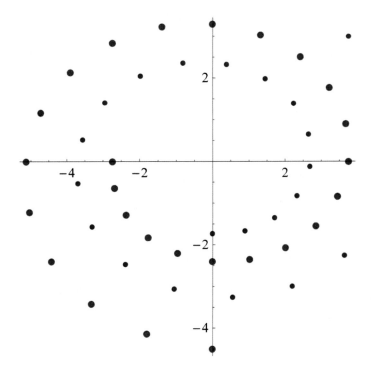

Fig. 11.15 Two intertwined spirals represented by 26 points each cluster

$\Rightarrow$ n = 2; a = 1.8;

$$\Rightarrow \mathrm{K}[\mathrm{u\_,v\_}] := \prod_{i=1}^{n} \left(\mathrm{Cos}\left[1.75\frac{\mathrm{u}[[i]] - \mathrm{v}[[i]]}{\mathrm{a}}\right] \mathrm{Exp}\left[-\frac{(\mathrm{u}[[i]] - \mathrm{v}[[i]])^2}{2\mathrm{a}^2}\right]\right)$$

and with parameter $c = 100$

$\Rightarrow$ c = 100;

the solution is

$\Rightarrow$ F = SupportVectorClassifier[xym, zm, K, c];

Figure 11.16 shows the result of the classification.

The continuous classification function and the decision rule can be displayed in 3D, too (Fig. 11.17).

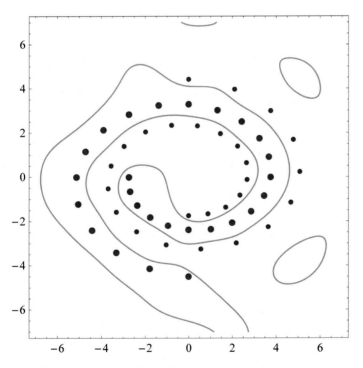

Fig. 11.16 Classification results of SVMC with the nonlinear decision boundary

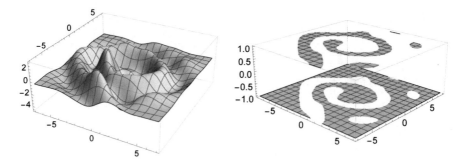

Fig. 11.17 Classification function and the corresponding decision rule using Signum function

11.7   Concept of SVM Regression

11.7.1   ε-Insensitive Loss Function

The problem of regression is that of finding a function that approximates a mapping from an input domain to the real numbers based on a training sample. We refer to the difference between the hypothesis output and its training value as the residual of the output, an indication of the accuracy of the fit at this point. We must decide how to measure the importance of this accuracy, as small residuals may be inevitable while we wish to avoid large ones. The loss function determines this measure. Each choice of loss function will result in a different overall strategy for performing regression. For example, least square regression uses the sum of the squares of the residuals.

Although several different approaches are possible, we will provide an analysis for generalization of regression by introducing a threshold test accuracy ε, beyond which we consider a mistake to have been made. We therefore aim to provide a bound on the probability that a randomly drawn test point will have accuracy less than ε.

The linear ε-insensitive loss function $L^\varepsilon(x, y, f)$ is defined by

$$L^\varepsilon(x, y, f) = (|y - f(x)|)_\varepsilon = \max(0, |y - f(x)| - \varepsilon),$$

where f is a real-valued function on a domain $X, x \in X$ and $y \in \mathbb{R}$, see Cristianini and Taylor (2000). Similarly the quadratic ε-insensitive loss is given by

$$L^\varepsilon(x, y, f) = (|y - f(x)|)_\varepsilon^2.$$

This loss function determines how much a deviation from the true $f(x)$ is penalized; for deviations less than ε, no penalty is incurred. Here is what the loss function looks like, see Fig. 11.18.

Fig. 11.18 ε-insensitivity
loss function

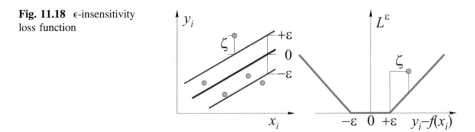

11.7.2 Concept of the Support Vector Machine Regression (SVMR)

Let us suppose that there are data pairs, $\{(x_1, y_1), \ldots, (x_k, y_k) \subset X \times \mathbb{R}$, where $x \in X \equiv \mathbb{R}^n$, namely the input are n dimensional vectors, while the output are scalar values. Vapnik (1995) introduced the definition of the ε-SV regression: we are looking for the $f(x)$ function, which has a maximum ε actual deviation from a y_i value, namely $|y_i - f(x_i)| \le$ ε, for every i-index and at the same time it is as flat as possible. This technique tries to minimize the error, but at the same time to avoids over learning. First, let us consider a linear function

$$f(x) = \langle w, x \rangle + b,$$

where $w \in X$ and $b \in \mathbb{R}$. The gradient of this function should be minimized to avoid over learning

$$\frac{1}{2} w^2 \to \text{min.}$$

Assuming in a tube with radius of the function is $f(x_i) = 0$,

$$y_i - \langle w, x_i \rangle - b \le \varepsilon,$$

$$b + \langle w, x_i \rangle - y_i \le \varepsilon,$$

the error ζ (see Fig. 11.18) also should be minimized. Therefore our objective is

$$c \sum_{i=1}^{n} \zeta_i + \frac{1}{2} w^2 \to \text{min}$$

where c is a regularization constant.

Since

$$y_i = \zeta_i + \varepsilon + f(x_i),$$

similarly

$$y_i + \zeta_i + \varepsilon = f(x_i).$$

Therefore

$$y_i - f(x_i) = \zeta_i + \varepsilon,$$

and

$$\zeta_i + \varepsilon = f(x_i) - y_i,$$

or

$$|y_i - f(x_i)| = \zeta_i + \varepsilon.$$

Considering the ε-insensitivity function,

$$L^{\varepsilon}(x, y, f) = \max(0, |y - f(x)| - \varepsilon).$$

The objective to be minimized is

$$G(\alpha) = c \sum_{i=1}^{n} L^{\varepsilon}(x_j, y_j, f(x_j)) + \frac{1}{2}w^2 \to \min.$$

As we did in case of SVMC, the weight vector will be represented as

$$w = \sum_{i=1}^{n} \alpha_i x_i.$$

Introducing a nonlinear kernel, our function can be generalized for nonlinear case, too

$$f(x_j) = \sum_{i=1}^{n} \alpha_i \langle x_j, x_i \rangle + b \to f(x_j) = \sum_{i=1}^{n} \alpha_i K(x_j, x_i) + b.$$

The dual form of the minimization problem, which should be maximized is

$$W(\alpha) = \sum_{i=1}^{n} y_i \alpha_i - \varepsilon \sum_{i=1}^{n} |\alpha_i| - \frac{1}{2} \sum_{i,j=1}^{n} \alpha_i \alpha_j \left(K(x_i, x_j) + \frac{1}{c} \delta_{ij} \right) \to \max,$$

$$\sum_{i=1}^{n} \alpha_i = 0 \quad \text{and} \quad -c < \alpha_i \leq c \quad i = 1, 2, ...n,$$

Using the α_i as the solution, the approximating function is,

$$f(x) = \sum_{i=1}^{n} \alpha_i K(x_i, x) + b.$$

Similarly to the classification problem, we call the input vectors x_i support vectors if the corresponding $\alpha_i \neq 0$.

The parameter b should be chosen such that for $\alpha_i \neq 0$, we have

$$f(x_i) = y_i - \varepsilon - \frac{\alpha_i}{c}, \quad i = 1, 2, ...n.$$

Considering that,

$$f(x_j) = \sum_{i=1}^{n} \alpha_i K(x_i, x_j) + b,$$

then substituting and expressing b, we get

$$b = y_j - \varepsilon - \frac{\alpha_j}{c} - \sum_{i=1}^{n} \alpha_i K(x_i, x_j), \quad j = 1, 2, ...n.$$

In practice, one may compute b as a mean,

$$b = \frac{1}{n} \sum_{j=1}^{n} \left(y_j - \varepsilon - \frac{\alpha_j}{c} - \sum_{i=1}^{n} \alpha_i K(x_i, x_j) \right).$$

11.7.3 The Algorithm of the SVMR

The algorithm of the support vector regression will be illustrated via the following simple example. Let us consider the following data pairs, see Fig. 11.19.

$\Rightarrow$ xn $= \{1., 3., 4., 5.6, 7.8, 10.2, 11., 11.5, 12.7\}$;

$\Rightarrow$ yn $= \{-1.6, -1.8, -1., 1.2, 2.2, 6.8, 10., 10., 10.\}$;

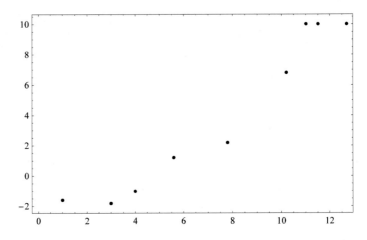

Fig. 11.19 Data pairs of the regression problem

First, we employ a linear kernel,

$\Rightarrow K[u\_, v\_] := u.v$

The parameters are

$\Rightarrow \varepsilon = 0.25; \ c = 0.1;$

The input and output vectors can be given as

$\Rightarrow xm = \texttt{Partition}[xn, 1];$

$\Rightarrow ym = yn;$

The number of the measurements

$\Rightarrow m = \texttt{Length}[xm]$

$\Leftarrow 9$

The kernel matrix is

$\Rightarrow M = \texttt{Table}[K[xm[[i]], xm[[j]]], \{i, 1, m\}, \{j, 1, m\}] +$
$1/c \, \texttt{IdentityMatrix}[m];$

Then the objective function to be maximized is

$\Rightarrow W = \sum_{i=1}^{m} \alpha_i ym[[i]] - \varepsilon \sum_{i=1}^{m} \texttt{Abs}[\alpha_i] - \frac{1}{2} \sum_{i=1}^{m} \sum_{j=1}^{m} (\alpha_i \alpha_j M[[i, j]]);$

and the constrains

$$\Rightarrow g = \text{Apply}\left[\text{And}, \text{Join}\left[\text{Table}[-c < \alpha_i \le c, \{i, 1, m\}], \left\{\left(\sum_{i=1}^{m} \alpha_i == 0\right)\right\}\right]\right];$$

The variables of the optimization problem

$$\Rightarrow \text{vars} = \text{Table}[\alpha_i], \{i, 1, m\}];$$

Let us compute the optimal α's using global maximization method

$$\Rightarrow \text{sol} = \text{NMaximize}[\{W, g\}, \text{vars}, \text{Method} \rightarrow \text{"DifferentialEvolution"}][[2]];$$

The parameter b can be computed as a mean,

$$\Rightarrow b = \frac{1}{m}\text{Apply}\left[\text{Plus}, \text{Table}\right.$$

$$\left[\left(\text{ym}[[j]] - \sum_{i=1}^{m} \alpha_i K[\text{xm}[[i]], \text{xm}[[j]]] - \varepsilon - \frac{\alpha_j}{c}\right) / .\text{sol}, \{j, 1, m\}\right]\right]$$

Then the analytical form of the regression line is

$$\Rightarrow F = \left(\sum_{i=1}^{m} \alpha_i K[\text{xm}[[i]], \text{Table}[x_j, \{j, 1, n\}]] + b\right) / .\text{sol}$$

$$\Leftarrow -4.05732 + 1.04889 \, x_1$$

Figure 11.20 shows the linear SVMR approximation of the regression problem. The SVMR algorithm can be summarize and implemented as the following *Mathematica* function,

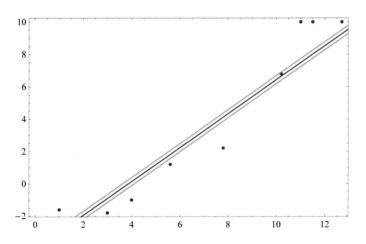

Fig. 11.20 The linear SVMR approximation of the regression problem

$\Rightarrow$ SupportVectorRegression$[\{xm_,ym_\},K_,\epsilon_,c_] := $ Module$[$
$\{m,n,M,i,j,W,g,vars,sol,b\},$
$m = $ Length$[ym]; n = $ Length$[xm[[1]]];$
$M = $ Table$[K[xm[[i]],xm[[j]]],\{i,1,m\},\{j,1,m\}] +$
$1/$cIdentityMa $-$ trix$[m];$

$$W = \sum_{i=1}^{m} \alpha_i ym[[i]] - \epsilon \sum_{i=1}^{m} \text{Abs}[\alpha_i] - \frac{1}{2} \sum_{i=1}^{m} \sum_{j=1}^{m} (\alpha_i \alpha_j M[[i,j]]); n$$

$$g = \text{Apply}\left[\text{And}, \text{Join}\left[\text{Table}[-c < \alpha_i \le c, \{i,1,m\}], \left\{\left(\sum_{i=1}^{m} \alpha_i == 0\right)\right\}\right]\right];$$

vars $= $ Table$[\alpha_i],\{i,1,m\}];$
sol $= $ NMaximize$[\{W,g\},vars,$ Method $\rightarrow$ DifferentialEvolution$][[2]];$
$b = 1/$mApply$[$Plus,

$$\text{Table}\left[\left(ym[[j]] - \sum_{i=1}^{m} \alpha_i K[xm[[i]],xm[[j]]] - \epsilon - \frac{\alpha_j}{c}\right)/.sol,\{j,1,m\}\right]];$$

$$\left\{\left(\sum_{i=1}^{m} \alpha_i K[xm[[i]], \text{Table}[x_j,\{j,1,n\}]] + b\right)/.sol, vars/.sol\right\};$$

where the kernel function *K*, and the parameters ε and *c* should be given by the user! The data points are $\{xm, ym\}$. The main advantage of this *Mathematica* modul is that it provides the solution function in analytic form.

11.8   Employing Different Kernels

Now we employ nonlinear kernels for the same data set considered above.

11.8.1   *Gaussian Kernel*

Employing the Gaussian kernel

$\Rightarrow$ K$[u_,v_] := $ Exp$[-\beta$ Norm$[(u - v)]];$

with the parameters,

$\Rightarrow \varepsilon = 0.3; c = 200; \beta = 0.05;$

we get the following approximation function,

$\Rightarrow$xmp = Partition[xm, 1];

$\quad\Rightarrow$ F = SupportVectorRegression[{xmp, ym}, K, ε, c];

$\quad\Rightarrow$ F[[1]]

$\quad\Leftarrow 3.58965 - 1.908315e^{-0.05\text{Abs}[1.-x_1]} -$

$\quad 8.70777e^{-0.05\text{Abs}[3.-x_1]} - 2.322437e^{-0.05\text{Abs}[4.-x_1]} +$

$\quad 2.13623e^{-0.05\text{Abs}[5.6-x_1]} - 11.61337e^{-0.05\text{Abs}[7.8-x_1]} -$

$\quad 10.8503e^{-0.05\text{Abs}[10.2-x_1]} + 27.6728e^{-0.05\text{Abs}[11.-x_1]} +$

$\quad 2.63874e^{-0.05\text{Abs}[11.5-x_1]} + 2.95439e^{-0.05\text{Abs}[12.7-x_1]}$

Figure 11.21 shows the result,

11.8.2 Polynomial Kernel

The general form of the polynomial kernel is

$$K(\text{u, v}) = (c + \langle u, v \rangle)^d,$$

where $d > 0$, is the order of the kernel. In our case let $d = 3$,

$\quad\Rightarrow$ d = 3;

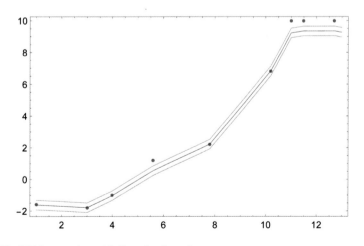

Fig. 11.21 SVM regression with Gaussian kernel

and the kernel

$$\Rightarrow K[u\_, v\_] := (a + u.v)^3$$

with the parameters

```
⇒F = SupportVectorRegression[{xm, ym}, K, ε, c];
⇒F[[1]]
```
$$\Leftarrow -1.09704 - 45.3558(0.01 + 1.0x_1)^3 -$$
$$63.9509(0.01 + 3.0x_1)^3 + 144.114(0.01 + 4.0x_1)^3 + 88.8897(0.01 + 5.6x_1)^3 -$$
$$199.513(0.01 + 7.8x_1)^3 - 50.0423(0.01 + 10.2x_1)^3 + 123.831(0.01 + 11.0x_1)^3 +$$
$$83.8837(0.01 + 11.5x_1)^3 - 81.856(0.01 + 12.7x_1)^3$$

Fig. 11.22 shows the third order SVMR function,

11.8.3   Wavelet Kernel

Let us employ the wavelet kernel, which is more flexible

$$\Rightarrow K[u\_, v\_] := \prod_{i=1}^{n} \left(\cos\left[1.75 \frac{u[[i]] - v[[i]]}{a}\right] \exp\left[-\frac{(u[[i]] - v[[i]])^2}{2a^2}\right] \right)$$

With the wavelet parameters

$$\Rightarrow n = 1; \ a = 4.;$$

The parameter of the SVMR are

$$\Rightarrow \varepsilon = 0.3; \ c = 200.;$$

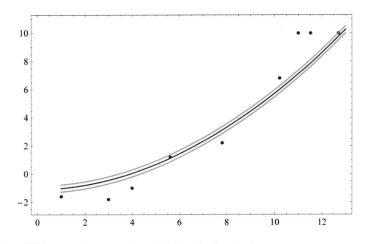

Fig. 11.22 SVM regression with polynomial kernel of order three

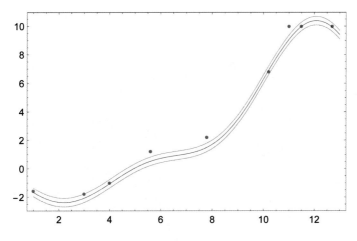

Fig. 11.23 SVM regression with Wavelet kernel, $\varepsilon = 0.3$

Then the analytical solution is

$\Rightarrow$ F = SupportVectorRegression[$\{$xmp, ym$\}$, K, ε, c];

$\Rightarrow$ F[[1]]

$\Leftarrow$ $2.53089 - 3.07833 e^{-0.03125(1.-x_1)^2} \cos[0.4375(1.-x_1)] -$

$9.08084 \times 10^{-9} e^{-0.03125(3.-x_1)^2} \cos[0.4375(3.-x_1)] -$

$3.79962 e^{-0.03125(4.-x_1)^2} \cos[0.4375(4.-x_1)] +$

$0.429218 e^{-0.03125(5.6-x_1)^2} \cos[0.4375(5.6-x_1)] +$

$4.78458 e^{-0.03125(7.8-x_1)^2} \cos[0.4375(7.8-x_1)] -$

$- 33.4234 e^{-0.03125(10.2-x_1)^2} \cos[0.4375(10.2-x_1)] +$

$\Leftarrow$ $2.53089 - 3.07833 e^{-0.03125(1.-x_1)^2} \cos[0.4375(1.-x_1)] -$

$9.08084 \times 10^{-9} e^{-0.03125(3.-x_1)^2} \cos[0.4375(3.-x_1)] -$

$3.79962 e^{-0.03125(4.-x_1)^2} \cos[0.4375(4.-x_1)] +$

Now let us consider a 2D function approximation ($n = 2$) with a wavelet kernel.
The function to be approximated employing 5×5 data points is, see Fig. 11.24,

$\Rightarrow$ 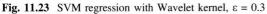 z[$\{$x_, y_$\}$] := $(x^2 - y^2)$Sin[0.5x]

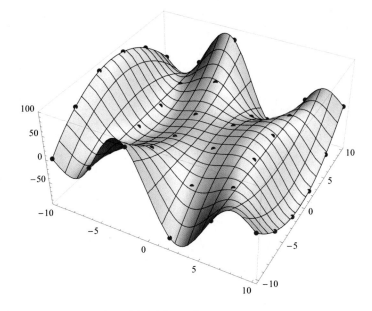

Fig. 11.24 The 2D test function with 25 data points

The selected data triplets are $(x_i, y_i, f(x_i, y_i))$,

$\Rightarrow$ data = Flatten[Table[N[{i − 10, j − 10, z[{i − 10, j − 10}]}],
  {i, 0, 20, 4}, {j, 0, 20, 4}], 1];
$\Rightarrow$ dataxy = Transpose[Take[Transpose[data], {1, 2}]];
$\Rightarrow$ dataz = Transpose[Take[Transpose[data], {3, 3}]]//Flatten;

The wavelet and the SVMR parameters are

$\Rightarrow$ n = 2; a = 4.0; ε = 0.15; c = 200;

The kernel itself is

$$\Rightarrow \text{K}[u\_, v\_] := \prod_{i=1}^{n} \left(\text{Cos}\left[1.75 \frac{u[[i]] - v[[i]]}{a}\right] \text{Exp}\left[-\frac{(u[[i]] - v[[i]])^2}{2a^2}\right] \right)$$

Then the function in short form can be written as

$\Rightarrow$ F = SupportVectorRegression[{dataxy, dataz}, K, ε, c];
$\Rightarrow$ Short[F[[1]], 10]
$\Leftarrow -0.15 + 37.7231 e^{-0.03125(-10.-x_1)^2 - 0.03125(-10.-x_2)^2}$
  $\text{Cos}[0.4375(-10. - x_1)]\text{Cos}[0.4375(-10. - x_2)] + 10.883 \ll 3 \gg + \ll 52 \gg$

Figure 11.25 shows the SVMR function with the data points,
Comparing the two figures, the approximation is quite reasonable.

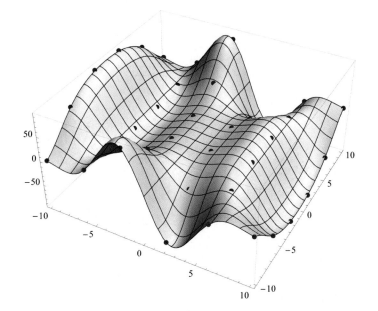

Fig. 11.25 The approximating surface of the SVMR function with the data points

11.8.4 Universal Fourier Kernel

This kernel is valid on the region of $[0, 2\pi]$ or in our case x, y $\in [0, 2\pi]^2$,

$$\Rightarrow K[u\_, v\_] := \prod_{i=1}^{n} \frac{1-q^2}{2(1-2q\,Cos[u[[i]]-v[[i]]]+q^2)}$$

Let us employ the following parameters

$$\Rightarrow q = 0.25; \; n = 2; \; \varepsilon = 0.0025; \; c = 200;$$

Our task is to approximate the following function

$$\Rightarrow z[\{x\_, y\_\}] := \frac{Sin\left[\sqrt{x^2+y^2}\right]}{\sqrt{x^2+y^2}}$$

in the region of $[-5, 5] \times [-5, 5]$. In order to employ the kernel, the function should be rescaled,

$$\Rightarrow z[\{x\_, y\_\}] := \frac{Sin\left[\sqrt{\left(\frac{5x}{\pi}-5\right)^2 + \left(\frac{5y}{\pi}-5\right)^2}\right]}{\sqrt{\left(\frac{5x}{\pi}-5\right)^2 + \left(\frac{5y}{\pi}-5\right)^2}}$$

Our test function seems to be a simple one, however it is not. Therefore, we employ 10×10 data pairs, see Fig. 11.26.

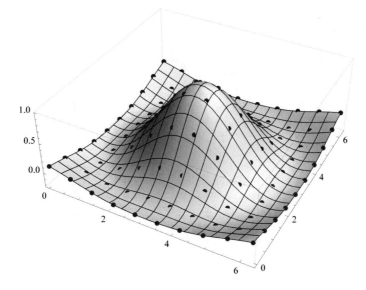

Fig. 11.26 The surface of the test function with the data points

⇒data = Flatten[Table[N[{u, v, z[{u, v}]}], {u, 0, 2π, (2π) 9},
  {v, 0, 2π, (2π)/9}], 1];

⇒ dataxy = Transpose[Take[Transpose[data], {1, 2}]];
⇒ dataz = Transpose[Take[Transpose[data], {3, 3}]]//Flatten;

Employing the Fourier kernel, the SVMR approximation function in short form is

⇒ F = Sup − portVectorRegression[{dataxy, dataz}, K, ε, c];
⇒ *Short*[F[[1]], 15]

$$\Leftarrow 0.0253295 - \frac{0.105729}{(1.0625 - 0.5\cos[0.0 - x_1])(1.0625 - 0.5\cos[0.0 - x_2])} +$$
$$\frac{0.151706}{(1.0625 - 0.5\cos[0.698132 - x_1])(1.0625 - 0.5\cos[0.0 - x_2])} + \ll 150 \gg +$$
$$\frac{0.14956}{(1.0625 - 0.5\cos[5.58505 - x_1])(1.0625 - 0.5\cos[6.28319 - x_2])} -$$
$$\frac{0.107952}{(1.0625 - 0.5\cos[6.28319 - x_1])(1.0625 - 0.5\cos[6.28319 - x_2])}$$

Figure 11.27 shows the approximating function with the data points.
The comparison of the two figures indicates a fairly good approximation.

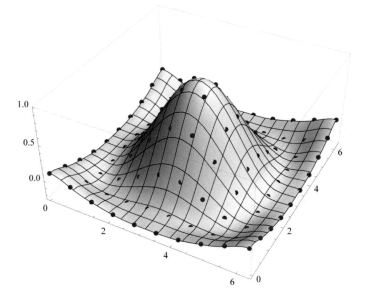

Fig. 11.27 The surface of approximating SVMR function with the data points

11.9 Applications

11.9.1 Image Classification

Let us classify images into two different categories, snowman and dice, see Figs. 11.28 and 11.29.

The feature extraction of these pictures represented by matrices of 128×128 pixels was carried out by a wavelet transform using the second order Daubechies filter, then employing averaging technique for seven non-overlapping bands of the spectrum. Consequently the dimension of these feature vectors is seven, $n = 7$.

Fig. 11.28 Representatives of snowman category

Fig. 11.29 Representatives of dice category

The feature vectors of snowmen,

$\Rightarrow$ SnowMan = Import$["$G : \\Snowman.dat$"]$;

$\Rightarrow$ Dimensions[SnowMan]

$\Leftarrow \{10, 7\}$

and those of the dices,

$\Rightarrow$ DiCes = Import$["$G : \\Dice.dat$"]$;

$\Rightarrow$ Dimensions[DiCes]

$\Leftarrow \{10, 7\}$

From the 10–10 pictures, the first 7–7 ones are considered as elements of the training set, and the last 3–3 represent the testing set. Then the elements of the training set as input,

$\Rightarrow$ xym = Join[Take[SnowMan, $\{1, 7\}$], Take[DiCes, $\{1, 7\}$]];

$\Rightarrow$ Dimensions[xym]

$\Leftarrow \{14, 7\}$

The out value for the snowman class is +1 and for the dice class is −1.

$\Rightarrow$ zm = $\{1, 1, 1, 1, 1, 1, 1, -1, -1, -1, -1, -1, -1, -1\}$;

First, wavelet kernel is employed with parameters

$\Rightarrow$ n = 7; a = 2.;

the kernel,

$$\Rightarrow K[u\_, v\_] := \prod_{i=1}^{n} \left(\mathrm{Cos}\left[1.75 \frac{u[[i]] - v[[i]]}{a}\right] \mathrm{Exp}\left[-\frac{(u[[i]] - v[[i]])^2}{2a^2}\right] \right)$$

and the penalty constant is

$\Rightarrow$ c = 150;

Then the SVMC function is in short form,

$\Rightarrow$ F = SupportVectorClassifier[xym, zm, K, c];

$\Rightarrow$ Short[F[[1]], 25]

$\Leftarrow -0.0893017 + 0.00002233654$

$-0.125(-17.7665 + x_1)^2 - 0.125(-15.3532 + x_2)^2 - 0.125(\ll 1 \gg)^2 - 0.125 \ll 1 \gg -$

e $\quad\quad 0.12 \ll 1 \gg^2 -0.125(-7.32602 + \ll 1 \gg)^2 - 0.125(-5.1615 + x_7)^2$

$\cos[0.875(-17.7665 + x_1)]\cos[0.875(-15.3532 + x_2)] \ll 1 \gg$

$\cos[0.875(- \ll 19 \gg + \ll 1 \gg)]\cos[0.875(-9.49168 + x_5)]$

$\cos[0.875(-7.32602 + x_6)]\cos[0.875(-5.1615 + x_7)] + \ll 20 \gg$

Let us define a function for the picture classification as

$\Rightarrow$ PictureClassifier[u_] := Sign[F[[1]]]/.$\{x_1 \rightarrow u[[1]], x_2 \rightarrow u[[2]],$
$x_3 \rightarrow u[[3]], x_4 \rightarrow u[[4]], x_5 \rightarrow u[[5]], x_6 \rightarrow u[[6]], x_7 \rightarrow u[[7]]\}$

and do employ it for the pictures of the testing set,

$\Rightarrow$ Test = Join[Take[SnowMan, $\{8, 10\}$], Take[DiCes, $\{8, 10\}$]];
$\Rightarrow$ Map[PictureClassifier[#]&, Test]
$\Leftarrow \{1, 1, 1, -1, -1, -1\}$

The result is correct.

In addition, we try another kernel, *Kernel with Moderate Decreasing* (KMOD), which especially effective in case of classification problem,

$\Rightarrow$ K[u_, v_] := Exp[γ/(Norm[u $-$ v]^2 + σ^2)] $-$ 1

Now the parameters are

$\Rightarrow \gamma = 0.5;$ $\sigma = 3;$

Then the SVMC function is,

$\Rightarrow$ F = SupportVectorClassifier[xym, zm, K, c];
$\Rightarrow$ Short[F[[1]], 10]

$\Leftarrow 0.0191626 + 13.127 \left(-1 + e^{\frac{0.5}{9 + Abs[\ll 1 \gg]^2 + \ll 1 \gg^2 + \ll 2 \gg + \ll 1 \gg + Abs[\ll 1 \gg]^2 + Abs[-\ll 18 \gg + x_7]^2}}\right) +$

$39.3118 \left(-1 + e^{\frac{0.5}{9 + \ll 6 \gg + \ll 1 \gg^2}}\right) - 14.9837(-1 + \ll 1 \gg) +$

$\ll 12 \gg + 17.038 \left(-1 + e^{\frac{0.5}{\ll 1 \gg}}\right) -$

$2.30283 \times 10^{-8} \left(-1 + e^{\frac{0.5}{9 + \ll 1 \gg^2 + \ll 4 \gg + \ll 1 \gg^2 + Abs[\ll 1 \gg]^2}}\right) - 37.6952$

$\left(-1 + e^{\frac{0.5}{9 + Abs[-\ll 18 \gg + x_1]^2 + Abs \ll 1 \gg \ll 1 \gg]^2 + \ll 1 \gg \ll 1 \gg \ll 1 \gg + \ll 1 \gg^2 + Abs[\ll 1 \gg]^2 + Abs[-4.3202 + x_7]^2}}\right)$

which works perfectly on the testing set, too

$\Rightarrow$ Map[PictureClassifier[#]&, Test]
$\Leftarrow \{1, 1, 1, -1, -1, -1\}$

11.9.2 *Maximum Flooding Level*

Let us forecast the maximum flooding level of the river Danube (z) at Budapest, Hungary, on the bases of the earlier statistical values, namely the level of the Danube at Budapest just before the flooding wave (x) and the average of the rainfall levels measured at the meteorological stations in the water collection region of the Danube in Austria (y), see Table 11.4

Let us employ the following kernel

$$\Rightarrow \text{K}[\text{u\_}, \text{v\_}] := \text{Exp}[-\beta \text{Norm}[(\text{u} - \text{v})]];$$

with parameters

$$\Rightarrow \varepsilon = 10.; \; c = 500; \; \beta = 0.05;$$

Then the approximating function is

Table 11.4 Data for the flooding level problem

x (mm)	y (cm)	z (cm)
58	405	590
52	450	660
133	350	780
179	285	770
98	330	710
72	400	640
72	550	670
43	480	520
62	450	660
67	610	690
64	380	500
33	460	460
57	425	610
62	560	710
54	420	620
48	620	660
86	390	620
74	350	590
95	570	740
44	710	730
77	580	720
46	700	640
123	560	805
62	430	673

$\Rightarrow \texttt{F} = \texttt{Sup} - \texttt{portVectorClassifier}[\texttt{xymT}, \texttt{zmT}, \texttt{K}, \varepsilon, \texttt{c}];$

$\Leftarrow 658.525 + 90.5948\, e^{-0.05\sqrt{\texttt{Abs}[179-x_1]^2 + \texttt{Abs}[285-x_2]^2}} -$

$\quad 106.158\, e^{-0.05\sqrt{\texttt{Abs}[133-x_1]^2 + \texttt{Abs}[350-x_2]^2}} - 155.401\, e^{-0.05\sqrt{\texttt{Abs}[64-x_1]^2 + \texttt{Abs}[380-x_2]^2}} +$

$\quad 20.4979\, e^{-0.05\sqrt{\texttt{Abs}[72-x_1]^2 + \texttt{Abs}[400-x_2]^2}} - 27.0761\, e^{-0.05\sqrt{\texttt{Abs}[54-x_1]^2 + \texttt{Abs}[420-x_2]^2}} +$

$\quad 26.483\, e^{-0.05\sqrt{\texttt{Abs}[62-x_1]^2 + \texttt{Abs}[430-x_2]^2}} + 66.2043\, e^{-0.05\sqrt{Abs[52-x_1]^2 + \texttt{Abs}[450-x_2]^2}} -$

$\quad 189.417\, e^{-0.05\sqrt{\texttt{Abs}[33-x_1]^2 + \texttt{Abs}[460-x_2]^2}} - 89.7719\, e^{-0.05\sqrt{\texttt{Abs}[43-x_1]^2 + \texttt{Abs}[480-x_2]^2}} -$

$\quad 15.9142\, e^{-0.05\sqrt{\texttt{Abs}[72-x_1]^2 + \texttt{Abs}[550-x_2]^2}} + 30.5367\, e^{-0.05\sqrt{\texttt{Abs}[62-x_1]^2 + \texttt{Abs}[560-x_2]^2}} +$

$\quad 119.777\, e^{-0.05\sqrt{\texttt{Abs}[123-x_1]^2 + \texttt{Abs}[560-x_2]^2}} + 33.7208\, e^{-0.05\sqrt{\texttt{Abs}[95-x_1]^2 + \texttt{Abs}[570-x_2]^2}} +$

$\quad 6.24451\, e^{-0.05\sqrt{\texttt{Abs}[67-x_1]^2 + \texttt{Abs}[610-x_2]^2}} - 2.18785\, e^{-0.05\sqrt{\texttt{Abs}[48-x_1]^2 + \texttt{Abs}[620-x_2]^2}} -$

$\quad 76.3213\, e^{-0.05\sqrt{\texttt{Abs}[46-x_1]^2 + \texttt{Abs}[700-x_2]^2}} + 98.1576\, e^{-0.05\sqrt{\texttt{Abs}[44-x_1]^2 + \texttt{Abs}[710-x_2]^2}}$

Figure 11.30 shows the SVMR function with the measured data points.

The error of the approximation is quite reasonable; see Fig. 11.31 which also shows the region of the validity of the approximation.

11.10   Exercise

11.10.1   Noise Filtration

Let us generate synthetic noisy data of the *sinc(x)* function on the interval $[-10, 10]$. The noise has a normal distribution $N(0, 0.1)$, see Fig. 11.32

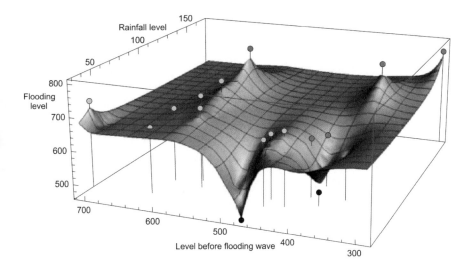

Fig. 11.30 The SVMR approximation function and the measured values

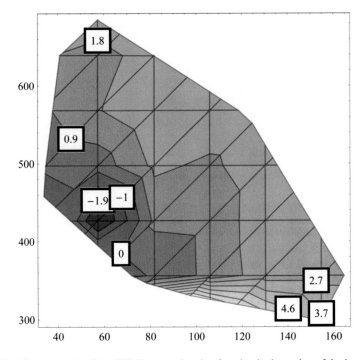

Fig. 11.31 The error in % of the SVMR approximation function in the region of the input, rainfall level versus level of the Danube before the flood wave

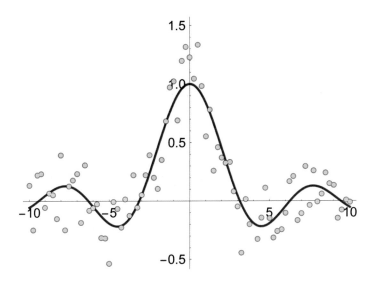

Fig. 11.32 Noisy data of the sinc(x) function and the function itself

$\Rightarrow$ SeedRandom[2];
$\Rightarrow$ data = Table[{x, Sinc[x] + 2Random[NormalDistribution[0, 0.1]]},
{x, −10, 10, 0.25}];

Our task is to represent the *sinc(x)* function employing the noisy data points. Let us employ wavelet kernel with the parameters

$\Rightarrow$ n = 1; a = 4.;

$$\Rightarrow K[u\_, v\_] := \prod_{i=1}^{n} \left(\text{Cos}\left[1.75 \frac{u[[i]] - v[[i]]}{a}\right] \text{Exp}\left[-\frac{(u[[i]] - v[[i]])^2}{2a^2}\right] \right)$$

The parameters for the SVMR approximation are

$\Rightarrow \varepsilon = 0.02$; c = 200.;

Then the solution is

$\Rightarrow$ F = SupportVectorRegression[{xmp, ym}, K, ε, c]

$\Leftarrow 0.124549 + 18.4605\, e^{-0.03125(-10.-x_1)^2} \cos[0.4375(-10.0 - x_1)] -$

$51.5907\, e^{-0.03125(-9.75-x_1)^2} \cos[0.4375(-9.75 - x_1)] +$

$32.0491\, e^{-0.03125(-9.5-x_1)^2} \cos[0.4375(-9.5 - x_1)] + \ll 107 \gg +$

$9.87568\, e^{-0.03125(8.75-x_1)^2} \cos[0.4375(8.75 - x_1)] +$

$11.7166\, e^{-0.03125(9.-x_1)^2} \cos[0.4375(9. - x_1)] -$

$31.539\, e^{-0.03125(9.25-x_1)^2} \cos[0.4375(9.25 - x_1)] -$

$11.2079\, e^{-0.03125(9.5-x_1)^2} \cos[0.4375(9.5 - x_1)] +$

$4.40217\, e^{-0.03125(9.75-x_1)^2} \cos[0.4375(9.75 - x_1)] +$

$7.84421\, e^{-0.03125(10.-x_1)^2} \cos[0.4375(10. - x_1)]$

Figure 11.33 shows the SVMR approximation with the original *sinc(x)* function.

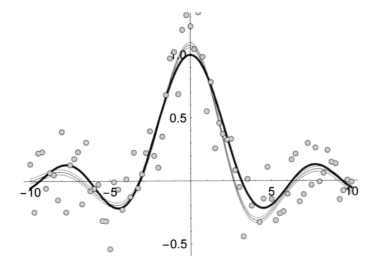

Fig. 11.33 SVM regression with wavelet kernel, $\varepsilon = 0.02$

References

Berthold M, Hand DJ (2003) Intelligent data analysis, an introduction, Springer, Heidelberg, Berlin
Cristianini N, Taylor JS (2000) An introduction to support vector machines and other kernel-based learning methods. Cambridge University Press, Cambridge
Vapnik V (1995) The nature of statistical learning theory. Springer, Berlin, Heidelberg

Chapter 12
Symbolic Regression

12.1 Concept of Symbolic Regression

The symbolic regression SR can be considered as a broad generalization of the class of Generalized Linear Models (GLM), which is a linear combination of basic functions β_i, $i = 1, 2, ..., n$ with a dependent variable y, and an independent variable vector $\mathbf{x}$.

$$y(\mathbf{x}) = c_0 + \sum_{i=1}^{n} c_i \beta_i(\mathbf{x}) + \varepsilon,$$

where c_i are the coefficients and ε is the error term.

SR will search for a set of basic functions (building blocks) and coefficients (weights) in order to minimize the error ε when given y and $\mathbf{x}$. The standard basic functions of the coordinates of $\mathbf{x}$ are: constant, addition, subtraction, multiplication, division, sine, cosine tangent, exponential, power, square root, etc. To select the optimal set of basic functions, Koza (1992) suggested using genetic programming (GP). GP is a biologically inspired machine learning method that evolves computer programs to perform a task. In order to carry out genetic programming, the individuals (competing functions) should be represented by a binary tree. In standard GP, the leaves of the binary tree are called terminal nodes represented by variables and constants, while the other nodes, the so called non-terminal nodes are represented by functions. Let us see a simple example. Suppose for a certain i

$$\beta_i(\mathbf{x}) = x_1 x_2 + \frac{1}{2} x_3.$$

Its binary tree representation can be seen in Fig. 12.1.

Supplementary Information The online version contains supplementary material available at https://doi.org/10.1007/978-3-030-92495-9_12.

In this example, there are three variables (x_1, x_2, x_3), two constants $(1, 2)$, and three elementary functions (*plus, times, rational*). The binary tree of $y(\mathbf{x})$ can be built up from such trees as sub-trees. Mathematically,

$$y(\mathbf{x}) = c_0 + c_1 \, tree_1 + c_2 \, tree_2 + \ldots.$$

GP randomly generates a population of individuals $y_k(\mathbf{x})$, $k = 1, 2, \ldots, n$ represented by tree structures to find the best performing trees.

There are two important features of the function represented by a binary tree: complexity and fitness. We define complexity as the number of nodes in a binary tree needed to represent the function. The fitness qualifies how good a model $(y = y(\mathbf{x}))$ is. Basically, there are two types of measures used in SR; the root mean squared error (RMSE) and the R-square. The latter returns the square of the Pearson product moment correlation coefficient (R) describing the correlation between the predicted values and the target values. Then the goodness of the model, the fitness function, can be defined as,

$$f = \frac{1}{1 + \text{RMSE}} \quad \text{or} \quad f = \text{R}^2,$$

where $0 \leq f \leq 1$.

GP tries to minimize this error to improve the fitness of the population consisting of individuals (competing functions) from generation to generation by mutation and cross-over procedure. Mutation is an eligible random change in the structure of the binary tree, which is applied to a randomly chosen sub-tree in the individual. This sub-tree is removed from the individual and replaced by a new randomly created sub-tree. This operation leads to a slightly (or even substantially) different basic function. Let us consider the binary tree on Fig. 12.2a, where the sub-tree of y^2 is replaced by $y + x^2$. Then the mutated binary tree can be seen in Fig. 12.2b.

Fig. 12.1 The binary tree representation of a basic function β_i

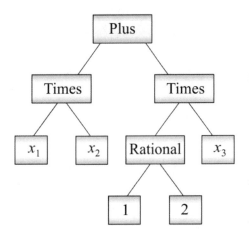

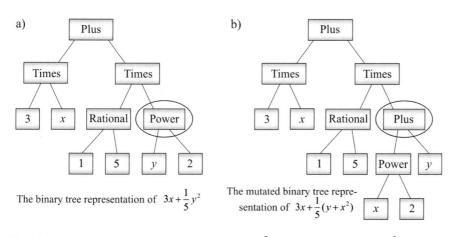

Fig. 12.2 Binary tree representations of the mutation, y^2 in (**a**) is replaced by $y + x^2$ in (**b**)

The operation "cross-over" representing sexuality can accelerate the improvement of the fitness of a function more effectively than mutation alone can do. It is a random combination of two different basic functions (parents), based on their fitness, in order to create a new generation of functions, more fit than the original functions. To carry out cross-over, crossing points (non-terminal nodes) in the trees of both parents should be randomly selected, as can be seen in Fig. 12.3. Then, subtrees belonging to these nodes will be exchanged creating offspring. Let us consider the parents before the cross-over. The first parent $(x - y)/3$, with its crossing point (x) is shown in Fig. 12.3a, the second parent $3x + y^2/5$ with its crossing point (see the sub-tree of y^2) is presented in Fig. 12.3b.

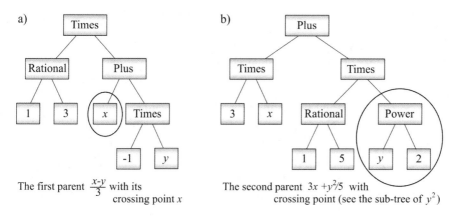

Fig. 12.3 The parents before the crossover

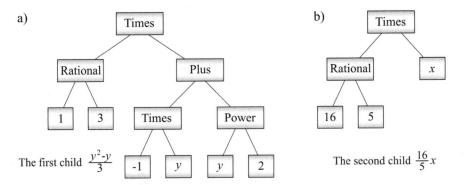

Fig. 12.4 The children produced by the crossover

The children produced by the cross-over are given on Fig. 12.4. The first child $(y^2 - y)/3$ is shown in Fig. 12.4a while the second child $(16/5)x$ is shown in Fig. 12.4b.

The generalization of GP was invented by Cramer (1985) and further developed by Koza (1992). GP is a class of evolutionary algorithms working on executable tree structures (parse trees). Koza (1992) showed that GP is capable of doing symbolic regression (or function identification) by generating mathematical expressions approximating a given sample set very closely or in some cases even perfectly. Therefore, GP finds the entire approximation model and its (numerical) parameters simultaneously. An important goal in symbolic regression is to get a solution, which is numerically robust and does not require high levels of complexity to give accurate output values for given input parameters. Small mean errors may lead to wrong assumptions about the real quality of the expressions found. To be on the safe side, a worst case absolute error should be determined. Sometimes, alternating the worst case absolute error and RMSE or R^2 as the target to be minimized during GP process is the best strategy.

Complexity and fitness are conflicting features leading to a multi-objective problem. A useful expression is both predictive and parsimonious. Some expressions may be more accurate but over-fit the data, whereas others may be more parsimonious but oversimplify. The prediction error versus complexity or 1-fitness versus complexity of the Pareto front represent the optimal solutions as they vary over expression complexity and maximum prediction error. As Fig. 12.5 shows, functions representing the Pareto front have the following features:

- In case of fixed complexity, there is no such solution (function), which could provide less error than the Pareto solution,
- Conversely, in case of fixed error, there is no such solution (function), which would have smaller complexity than the Pareto solution.

The Pareto front tends to contain a cliff where predictive ability jumps rapidly at some minimum complexity. Predictive ability then improves only marginally with

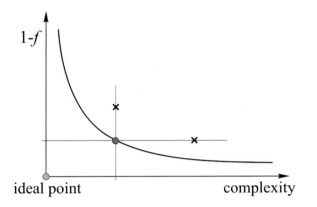

Fig. 12.5 The Pareto front

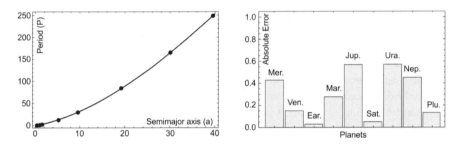

Fig. 12.6 Approximation via algebraic polynomial regression

more complex expressions. Since the Pareto front provides the set of optimal solutions, the user should decide which one is preferable. However, one may select blindly the very solution on the Pareto front, which is closest to the ideal point (zero error, zero complexity).

To carry out an SR computation requires considerable computational power. Fortunately, the algorithm is suited perfectly for parallel computation. There are many software implementations of this method, both commercial and non-commercial. Commercial symbolic regression packages, such as DTREG (http://www.dtreg.com/) are designed for predictive modeling and forecasting. Besides symbolic regression, it can also perform other data-mining tasks (e.g., in neural networks, support vectors machines, etc.). DataModeler is another well documented commercial symbolic regression package available for Mathematica (http://www.evolved-analytics.com). Non-commercial open source symbolic regression packages such as *Eureqa* are available for free download (http://ccsl. mae.cornell.edu/eureqa) as well as *GPLab* and *GPTIPS* toolboxes for *Matlab* (http://gplab.sourceforge.net/) and (http://gptips.sourceforge.net), respectively. In this study, *DataModeler* is used, which has parallel implementation in

Mathematica. We should mention that in the meantime *DataRobot* has integrated the *Nutonian Inc.* and in this way their product *Eureqa*, too, see, https://www.datarobot.com/.

12.2 Problem of Kepler

The third law of *Kepler* states: "The square of the orbital period of a planet is directly proportional to the cube of the semi-major axis of its orbit (average distance from the Sun)."

$$P^2 \propto a^3,$$

where P is the orbital period of the planet and a is the semi-major axis of the orbit.

For example, suppose planet A is 4 times as far from the Sun as planet B. Then planet A must traverse 4 times the distance of planet B each orbit, and moreover it turns out that planet A travels at half the speed of planet B, in order to maintain equilibrium with the reduced gravitational centripetal force due to being 4 times further from the Sun. In total it takes $4 \times 2 = 8$ times as long for planet A to travel an orbit, in agreement with the law ($8^2 = 4^3$).

The third law currently receives additional attention as it can be used to estimate the distance from an exoplanet to its central star, and help to decide if this distance is inside the habitable zone of that star.

The exact relation, which is the same for both elliptical and circular orbits, is given by the equation.

This third law used to be known as the harmonic law, because Kepler enunciated it in a laborious attempt to determine what he viewed as the "music of the spheres" according to precise laws, and express it in terms of musical notation. His result was based on the *Rudolphine* table containing the observations of Tycho Brache 1605, see Table 12.1.

Table 12.1 Normalized observation planetary data

Planet	Period P (year)	Semi-major axis a
Mercury	0.24	0.39
Venus	0.61	0.72
Earth	1.00	1.00
Mars	1.88	1.52
Jupiter	11.86	5.20
Saturn	29.46	9.54
Uranus	84.01	19.19
Neptune	164.79	30.06
Pluto	284.54	39.53

Where a is give in units of Earth's semi-major axis.

Let us assume that Kepler could have employed one of the function approximation techniques like polynomial regression, artificial neural networks, support vector machine, thin plate spline. Could he find this simple relation with these sophisticated methods?

Let us try different type of methods to solve the problem. The data pairs are,

$\Rightarrow$ samplePts $=$
    SetPrecision$\{0.39, 0.72, 1., 1.52, 5.2, 9.54, 19.19, 30.06, 39.53\}, 10$;

$\Rightarrow$ observedResponse $=$ SetPrecision
    $[\{0.24, 0.61, 1., 1.88, 11.86, 29.46, 84.01, 164.79, 248.54\}, 10]$;

$\Rightarrow$ data $=$ Transpose$[\{$samplePts, observedResponse$\}]$;

12.2.1 Polynomial Regression

We compute a third order algebraic polynomial,

$\Rightarrow$ P $=$ Fit$[$data,$\{1, a, a^2, a^3\}, a]$

$\Leftarrow -0.903031 + 1.777511a + 0.15698459a^2 - 0.001072739a^3$

12.2.2 Neural Network

Let us employ an RBF network with a single neuron,

$\Rightarrow$ $<$ $<$NeuralNetworks$'$

$\Rightarrow$ rbf $=$ InitializeRBFNet$[$samplePts, observedResponse, $1]$

$\Leftarrow$ RBFNet$[\{\{$w1,λ, w2$\}$χNeuron $\rightarrow$ Exp, FixedParameters $\rightarrow$ None,
    AccumulatedIterations $\rightarrow$ 0, CreationDate $\rightarrow$ 2017, 1, 14, 17, 9, 9.89298

$\Rightarrow$ $\{$rbf2, fitrecord$\}=$ NeuralFit$[$rbf, samplePts, observedResponse, $100]$

Figure 12.7 shows the error of the Network during the iteration.
The neural network has the following form

$\Rightarrow$ $\{$P$\}=$ rbf2$[\{$a$\}]$

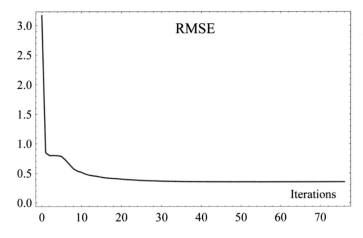

Fig. 12.7 Training process of RBF neural network with a single neuron

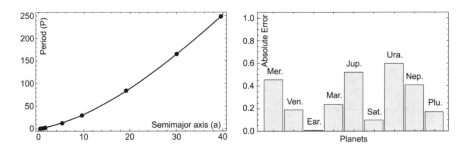

Fig. 12.8 Approximation via RBF neural network with a single neuron

$$\Leftarrow P = 50302.7 - 163.097\,a - 68962.5\,\mathrm{e}^{-0.00000851191(192.52+a)^2}$$

Figure 12.8 shows the error of the different planet.

12.2.3 Support Vector Machine Regression

Our kernel function is a wavelet kernel within the following parameters,

$$\Rightarrow n = 1;\, b = 25.;$$

$$\Rightarrow K[u\_, v\_] := \prod_{i=1}^{n} \left(\mathrm{Cos}\left[1.75\,\frac{u[[i]] - v[[i]]}{b}\right] \mathrm{Exp}\left[-\frac{(u[[i]] - v[[i]])^2}{2b^2}\right] \right)$$

Parameter values for the SVMR are,

$\Rightarrow \varepsilon = 0.025; \; c = 500.; \; tp = \text{Partition}[\text{samplePts}, 1];$

$\Rightarrow sol = \text{SupportVectorRegression}[\{tp, \text{observedResponse}\}, K, \varepsilon, c];$

The analytic form of SVMR is

$\Rightarrow P = sol[[1]]/.x1 \rightarrow a$

$$\Leftarrow 126.361 - 61.1375\, e^{-0.0008(0.39-a)^2} \cos[0.07(0.39 - a)] -$$
$$60.4577\, e^{-0.0008(0.72-a)^2} \cos[0.07(0.72 - a)] -$$
$$47.5422\, e^{-0.0008(1.0-a)^2} \cos[0.07(1.0 - a)] -$$
$$6.7781\, e^{-0.0008(1.52-a)^2} \cos[0.07(1.52 - a)] +$$
$$113.053\, e^{-0.0008(5.2-a)^2} \cos[0.07(5.2 - a)] -$$
$$167.045\, e^{-0.0008(9.54-a)^2} \cos[0.07(9.54 - a)] +$$
$$317.417\, e^{-0.0008(19.19-a)^2} \cos[0.07(19.19 - a)] -$$
$$446.980\, e^{-0.0008(30.06-a)^2} \cos[0.07(30.06 - a)] +$$
$$359.47\, e^{-0.0008(39.53-a)^2} \cos[0.07(39.53 - a)]$$

Figure 12.9 shows the SVMR approach and its error in case of the different planets.

12.2.4 RBF Interpolation

Now our interpolation function is $r^2 \log(r)$ which is the *Thin-Plate Spline* interpolation function

$\Rightarrow \text{Needs}[\text{Imtek`Interpolation`}]$

$\Rightarrow \text{polyharmonic} =$
$\quad \text{Function}[x, xi, \text{If}[x^1 xi, \text{Norm}[x - xi, 2]^2 \text{Log}[\text{Norm}[x - xi, 2]], 0]]$

$\Leftarrow \text{Function}[x, xi, \text{If}[x \neq xi, \text{Norm}[x - xi, 2]^2 \text{Log}[\text{Norm}[x - xi, 2]], 0]]$

$\Rightarrow \text{interpolationFunction} =$
$\quad \text{imsUnstructuredInterpolation}[\text{data}, \text{polyharmonic}];$

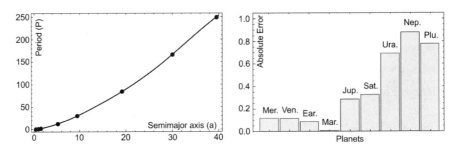

Fig. 12.9 Approximation via SVMR employing wavelet kernel

The Thin-Plate Spline (TPS) function is

$\Rightarrow \text{P} = \text{interpolationFunction}[\text{a}]$
$\Leftarrow -31.9484529 +$

$\quad 0.82232083\text{If}[\text{a} \neq 0.39, \text{Norm}[\text{a} - 0.39, 2]^2 \text{Log}[\text{Norm}[\text{a} - 0.39, 2]], 0] -$
$\quad 0.59163242\text{If}[\text{a} \neq 0.72, \text{Norm}[\text{a} - 0.72, 2]^2 \text{Log}[\text{Norm}[\text{a} - 0.72, 2]], 0] -$
$\quad 0.006\text{If}[\text{a} \neq 1.0, \text{Norm}[\text{a} - 1.0, 2]^2 \text{Log}[\text{Norm}[\text{a} - 1.0, 2]], 0] -$
$\quad 0.149\text{If}[\text{a} \neq 1.52, \text{Norm}[\text{a} - 1.52, 2]^2 \text{Log}[\text{Norm}[\text{a} - 1.52, 2]], 0] -$
$\quad 0.04578\text{If}[\text{a} \neq 5.2, \text{Norm}[\text{a} - 5.2, 2]^2 \text{Log}[\text{Norm}[\text{a} - 5.2, 2]], 0] -$
$\quad 0.0178870114\text{If}[\text{a} \neq 9.54, \text{Norm}[\text{a} - 9.54, 2]^2 \text{Log}[\text{Norm}[\text{a} - 9.54, 2]], 0] -$
$\quad 0.0114\text{If}[\text{a} \neq 19.19, \text{Norm}[\text{a} - 19.19, 2]^2 \text{Log}[\text{Norm}[\text{a} - 19.19, 2]], 0] -$
$\quad 0.02\text{If}[\text{a} \neq 30.06, \text{Norm}[\text{a} - 30.06, 2]^2 \text{Log}[\text{Norm}[\text{a} - 30.06, 2]], 0] +$
$\quad 0.0194\text{If}[\text{a} \neq 39.53, \text{Norm}[\text{a} - 39.53, 2]^2 \text{Log}[\text{Norm}[\text{a} - 39.53, 2]], 0].$

The approximation can be seen in Fig. 12.10.

There are different techniques with different models, but all of them provide a reasonably good approximation; however the complexity of the formulas are quite different. Can one provide a simpler model without loosing accuracy?

12.2.5 Random Models

The simplest form of symbolic regression is when the models are generated just randomly.

$\Rightarrow$ `<<DataModeler`

Now we generate models representing algebraic expressions randomly. Let us consider 60 models as population size.

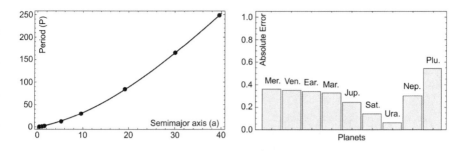

Fig. 12.10 Interpolation via TPS

$\Rightarrow$ randomModels = UpdateModelQuality[RandomModels[DataVariables $\rightarrow$
{a}, PopulationSize $\rightarrow$ 60], samplePts, observedResponse];

$\Rightarrow$ ModelSelectionReport[FitModels $\rightarrow$ randomModels,
SelectionStrategy $\rightarrow$ AllModels]

The result can be seen in Table 12.2.

The best model (31) is $P = \sqrt{a^3}$, which has a very small error, $1 - R^2 = 4.284 \times 10^{-8}$ and low complexity: 29. The random model generation technique is very fast, but no guarantee for the proper result, see later.

Table 12.2 Randomly generated models

Page 1	Page 2	Page 3	Page 4
Model selection report			
	Complexity	$1 - R^2$	Function
31	29	4.284×10^{-8}	$a^{3/2}$
32	29	0.797	$-(2.42 \times 10^{-3}/a^{3/2})$
33	29	0.824	$1/(-8.29 + 3a)$
34	31	1.682×10^{-4}	(4.12×10^{-2}) $(10 + 1/a + a + 6.24$ $a^2)$
35	31	0.013	$9.88 + (1/7 - a)^2$
36	31	0.029	$-\sqrt{a} + (6.88 \times 10^{-2})$ a^6
37	34	0.703	$(1/a^4)^{1/4}$
38	36	0.250	164.42 a^6 $(67.28 + a)^2$
39	41	0.287	$(18.50 + 4$ $a)^9$
40	41	0.749	$a^{1/6}/(0.17 - a)$
41	42	0.705	$(-5.47 + 0.10$ $\sqrt{a})/a$
42	44	0.053	$167^{1/3}$ a^2 $\sqrt{a^{3/2}}$
43	49	0.226	$-(7.37 + a - 677.16$ $a^3)^2$
44	50	0.927	$(1.05$ $(-514.82 + 3/a))/a^{14/3}$
45	53	0.113	$(93.07 + a^{1/3} - a + a^2)^2$

12.2.6 Symbolic Regression

The most effective technique is using multi-objective optimization. The competing objectives are the complexity and accuracy. Employing symbolic regression and generating 537 models, we get nearly the same result, see Table 12.3, model (4).

```
⇒ TimingParetoFrontLogPlot[
    quickModels1 = SymbolicRegression[
    samplePts, observedResponse, DataVariables → {a},
    AlignModel → True,
    RobustModels → True, TimeConstraint → 150, FitnessPrecision → 10
    ]
  ]
```

The Pareto front of the solution can be seen in Fig. 12.11.

Table 12.3 Symbolic regression models

Page 1			Page 2
Model selection table			
	Complexity	$1 - R^2$	Function
1	11	0.022	$-12.55 + 6.12a$
2	15	0.012	$7.17 + 0.16\,a^2$
3	19	1.948×10^{-4}	$-0.98 + 1.21a^{1.45}$
4	20	4.284×10^{-8}	$-(5.59 \times 10^{-3}) + 1.00\,\sqrt{a^3}$
5	28	1.927×10^{-8}	$8.85 \times 10^{-3} - (6.36 \times 10^{-3})\,a + 1.00\,\sqrt{a^3}$
6	32	1.214×10^{-8}	$2.69 \times 10^{-3} + (2.85 \times 10^{-6})\,a^3 + 1.00\,\sqrt{a^3}$
7	36	1.185×10^{-8}	$3.92 \times 10^{-3} + (5.36 \times 10^{-5})\,a^{2.32} + 1.00\,\sqrt{a^3}$
8	39	9.800×10^{-9}	$-(4.86 \times 10^{-3}) + 0.99\,\sqrt{a^3} + (5.96 \times 10^{-4})\,a\,(17 + a)$
9	40	7.278×10^{-9}	$-(4.03 \times 10^{-2}) + (7.08 \times 10^{-2})\,\sqrt{a^3} - (3.01 \times 10^{-2})\,a + 1.00\,\sqrt{a^3}$
10	44	6.624×10^{-9}	$0.14 - (2.00 \times 10^{-2})\,a + 1.00\,\sqrt{a^3} - 0.62/(4 + a)$
11	49	6.351×10^{-9}	$7.14 \times 10^{-2} - (1.56 \times 10^{-2})\,a + 1.00\,\sqrt{a^3} - 0.33/(4.69 + a^2)$
12	50	5.328×10^{-9}	$71.32 + 0.52a + 0.98\,\sqrt{a^3} - 19.50\,(49 + a)^{1/3}$
13	54	4.474×10^{-9}	$546.45 + 0.58a + 0.97\,\sqrt{a^3} - 451.40\,(49.93 + a)^{0.0489}$
14	85	1.688×10^{-9}	$-0.54 + 1.00a^{3/2} - (6.35 \times 10^{-6})\,a^3 - 340.42/(-629 - (a^3)^{3/4})$
15	124	1.305×10^{-9}	$-0.59 + 1.00\,a^{3/2} - (3.99 \times 10^{-5})\sqrt{(10.14 + a^{\sqrt{3}})^3} - 372.19/(-629 - (a^3)^{3/4})$

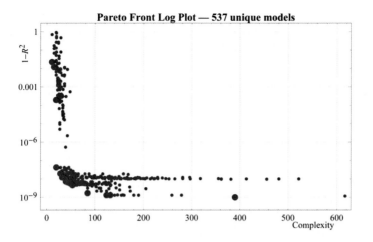

Fig. 12.11 The Pareto front of the Kepler problem solutions

⇒ ModelSelectionTable@quickModels1

The generated functions belonging to the Pareto front are summarized in Table 12.3.

Figure 12.12 shows the comparison of the errors of the two models.

Here, the chosen model is the 4th, which has a similar accuracy to the best random model. However, $\sqrt{a^3}$ represents lower complexity than $a^{3/2}$.

In Table 12.4 we can see the statistics of relative errors of the different models in percents.

It is inevitable that statistically the best model is provided by the symbolic regression even if its mean error is higher than that of the Kepler solution, but the latter one is simple in practice.

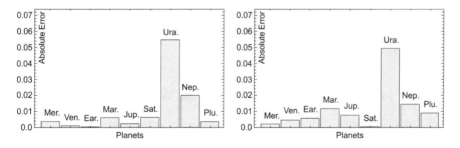

Fig. 12.12 The absolute errors of the randomly generated (to left) and the symbolic regression model (to right)

Table 12.4 Statistics of the relative error (%) of the different approximation methods

Method	Mean error	Max error	Standard deviation
Polynomial regression	25.14	177.49	59.96
Neural network	26.80	191.34	64.22
Support vector machine	8.89	47.47	16.14
Thin plate spline	29.10	149.87	49.59
Kepler solution	0.23	1.48	0.51
Symbolic regression	0.32	0.84	0.37

12.3   Applications

12.3.1   Correcting Gravimetric Geoid Using GPS Ellipsoidal Heights

The accuracy of the gravimetric geoid can be significantly improved using GPS (ellipsoidal) height measurements. The new, adjusted geoid can be constructed as a gravimetric surface plus the so called corrector surface. The difference between the gravimetric and the GPS/leveling geoid is $\Delta N(\varphi, \lambda) = N_{GPS} - N_{grav}$, where ΔN is the discrepancies between the GPS-derived ellipsoidal height and the height from the gravimetric geoid. The variables φ latitude and λ longitude are the ellipsoidal coordinates.

In practice, the various wavelength errors in the gravity solution are usually approximated by different kinds of functions in order to fit the geoid to a set of GPS/leveling points through an integrated least squares adjustment. Several models can be used ranging from simple linear regression to a more complicated seven parameter similarity transformation model. However, as the achievable accuracy of GPS and geoid heights improve, the use of such simplified models may not be sufficient. The problem is further complicated by the fact that selecting the proper model type depends on data distribution, density and quality, which varies for each case.

In order to quantify the effectiveness of the SR method, only global methods are considered here. We use three different parametric models and ANN with one hidden layer and sigmoid activation function, as well as with radial basis activation function (RBF).

12.3.1.1   Dataset for the Numerical Computations

For modeling the corrector surface, 302 GPS ellipsoidal height data points of the Hungarian geoid available from Kenyeres and Virág (1998) were used. As Fig. 12.13 shows, for the parameter estimations, 194 data points for training, and 108 data points for the validation were considered. The results for the different models are presented as follows.

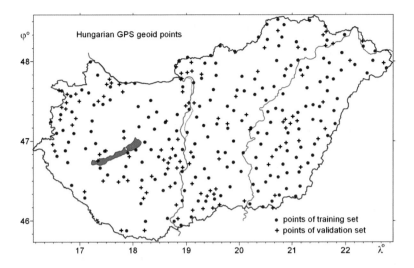

Fig. 12.13 Hungarian GPS geoidal height data: 194 training sets (*marked by circles*) and the 108 validation

12.3.1.2 Parametric Models

The family of these models based on the general 7-parameter similarity datum transformation, with the simplified classic 4-parameter model of Heiskanen and Moritz (1967, p. 213) is given by

$$\Delta N_4(\varphi, \lambda) = p_1 + p_2 \cos \varphi \cos \lambda + p_3 \cos \varphi \sin \lambda + p_4 \sin \varphi$$

An extended version of this model is given with the inclusion of a fifth parameter (Duquenne et al, 1995) as follows,

$$\Delta N_5(\varphi, \lambda) = p_1 + p_2 \cos \varphi \cos \lambda + p_3 \cos \varphi \sin \lambda + p_4 \sin \varphi + p_5 \sin^2 \varphi$$

and a more complicated form of the differential similarity transformation model developed by Kotsakis et al. (2001) is given as,

$$\Delta N_7(\kappa, \lambda) = p_1 \cos \varphi \cos \lambda + p_2 \cos \varphi \sin \lambda + p_3 \sin \varphi + p_4 W \sin \varphi \cos \varphi \sin \lambda +$$
$$p_5 W \sin \varphi \cos \varphi \cos \lambda + p_6 W \sin^2 \varphi + p_7 W (1 - f^2 \sin^2 \varphi)$$

where $W = 1/\sqrt{1 - e^2 \sin^2 \varphi}$, $e^2 = 0.006694379990$ is the numerical eccentricity, and $f = 1/298.257223563$ is the flattening of the reference (WGS84) ellipsoid. The parametric models can be easily achieved via linear parameter estimation

$$\Delta N_4(\varphi, \lambda) = 0.065666 - 0.0465406 \cos \varphi \cos \lambda - 0.00943184 \cos \varphi \sin \lambda - 0.0185872 \sin \varphi,$$

$$\Delta N_5(\kappa, \lambda) = 0.0514108 - 0.0508839 \cos \varphi \cos \lambda - 0.00115442 \cos \varphi \sin \lambda - 0.0195404 \sin \varphi + 0.0247211 \sin^2 \varphi,$$

$$\begin{aligned} \Delta N_7(\varphi, \lambda) = {}& -0.0530829 \cos \varphi \cos \lambda - 0,00618307 \cos \kappa \sin \lambda \\ & - 0.0206765 \sin \varphi + 0.0241447 W \sin \varphi \cos \varphi \sin \lambda \\ & + 0.0161115 \; W \sin \varphi \cos \varphi \cos \lambda + 0.0240435 \; W \sin^2 \varphi \\ & + 0.0518829 \; W(1 - 0.0000112413 \sin^2 \varphi), \end{aligned}$$

where $W = 1/\sqrt{1 - 0.00669438 \sin^2 \varphi}$.

12.3.1.3   Artificial Neural Network (ANN) Models

Two types of ANN models have been studied: ANN with sigmoidal activation function and ANN with radial bases function (RBF). Both of them have universal approximation properties like polynomials, (e.g., Kecman (2001)). Therefore, ANN with such activation functions can be a good candidate for approximating the corrector surface. The general structure of an ANN network with sigmoidal activation function in one hidden layer and with a linear tail is,

$$\Delta N(\varphi, \lambda) = \sum_{i=1}^{n} \frac{p_{1,i}}{1 + \exp(p_{2,i} + p_{3,i}\varphi + p_{4,i}\lambda)} + p_5 \kappa + p_6 \lambda + p_7$$

where n is the number of the neurons in the hidden layer (Fig. 12.14). There are 3 input nodes for φ, λ and for the bias 1. The linear combinations of these inputs are transformed in the hidden layer nodes according to the sigmoid activation function. Then the linear combination of these transformed signals with the linear tail having weights p_5, p_6 and p_7 represent the output of the network. There is no further signal transformation in the output node. Similarly, the ANN with RBF (Fig. 12.15) is

$$\Delta N(\varphi, \lambda) = \sum_{i=1}^{n} p_{1,i} \exp\left[p_{2,i}(p_{3,i} + \varphi)^2 + p_{4,i}(p_{5,i} + \lambda)^2 \right] + p_6 \varphi + p_7 \lambda + p_8.$$

The structure of the RBF network is quite similar to the network with sigmoid activation function. The main differences are that the weights between the input and hidden layer nodes are 1, and the activation functions in the hidden layer nodes are RBF functions. The RBF neural network has 4 nonlinear parameters ($p_{2,i}$, $p_{3,i}$, $p_{4,i}$, ..., $p_{5,i}$) while the feedforward network has only 3, namely $p_{2,i}$, $p_{3,i}$ and $p_{4,i}$. Here 25 ANN models for both activation functions were created with $n = 1, 2, \ldots, 25$ nodes in the hidden layer, respectively.

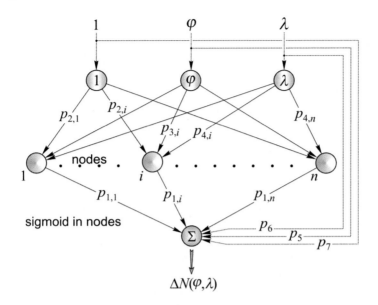

Fig. 12.14 ANN with sigmoid activation function

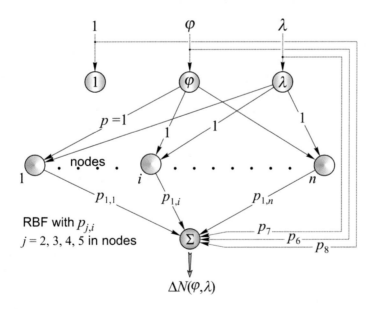

Fig. 12.15 ANN with RBF activation function

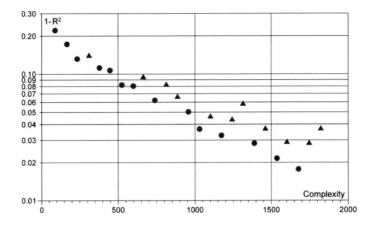

Fig. 12.16 The Pareto front (*circle points*) in case of the ANN models with sigmoidal activation function. The *triangular points* indicate optimal, but not Pareto optimal solutions

Figure 12.16 represents the result for networks with sigmoid activation function. It can be clearly seen that the 23rd model with 23 nodes—represented by the last point of the Pareto Front (circle points)—is the best, since the last two models with 24 and 25 nodes (triangular points) have higher complexity and greater error.

In order to compare the results of the different regression techniques, RMSE is used as an error measure, which is usual in engineering society. However, to illustrate the Pareto front of a model family belonging to a given technique, the fitness measure R^2 is used since its range is restricted to the interval [0, 1]. Therefore in the figures, the model error is characterized by $1 - R^2$. Figure 12.17 shows the Pareto Front of the 25 ANN models with RBF activation functions.

Figure 12.17 shows the Pareto Front of the 25 ANN models with RBF activation functions. Now, the best is the last but one model, since the last model has higher

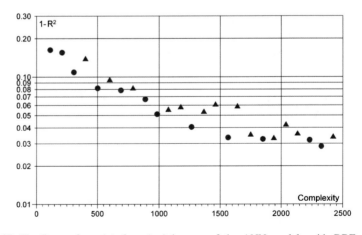

Fig. 12.17 The Pareto front (*circle points*) in case of the ANN models with RBF activation function. The *triangular points* indicate optimal, but not Pareto optimal solutions

complexity as well as greater error than the 24th model. As the figure shows, this model has higher complexity and somewhat greater error than the best ANN model with sigmoid activation function. Both types of the ANN models were evaluated via the Neural Network application package of *Mathematica*. We should mention here that the latest version *Mathematica* 11.1.1 has already built-in functions for ANN.

12.3.1.4 Application of Symbolic Regression

There are many parameters to be adjusted properly in order to make the symbolic regression algorithm work successfully. But, probably, the two very important ones are the type of the functions in the initial population and the type of the functions employed as basic functions in the regression process. Functions included in the basic function set are: plus, times, divide and subtract as basic arithmetic; square, sqrt, inverse as Extended Mathematics; power, exp, log as Power Mathematics and sigmoid, radial basis (RBF), see *DataModeler*.

During this study, the initial population was selected in three different ways. The functions in the initial population can be selected randomly; this is the default, automatic option. Another more intelligent way is to apply Keijzer-expansion, (Keijzer 2003). The name of the method originates from Maarten Keijzer who suggested randomly synthesizing a selection of models, selecting the one having the best fit to the targeted response, removing the model contribution from the response, and iteratively repeating the process on the residual until the desired accuracy is achieved. The resulting model may provide a good starting point for SR. The third method applied here simply employs the ANN models as initial population for SR. The reason for this choice is that SR may improve the robustness as well as the generalization ability—the quality of the fitting on the validation set-of the ANN models. Different models have been applied with the following measured data. As Fig. 12.18 shows, employing SR with randomly generated initial population also results in a function with high complexity.

Applying Keijzer expansion to create the initial population, we obtained the following models (Fig. 12.19). As already mentioned, Keijzer expansion provides a good initial population for symbolic regression. In our case, all of the Keijzer models—all models, not only models which are the elements of the Pareto Front—have relatively low complexity (Fig. 12.19). The last 5 models of the Pareto Front of the 200 Keijzer models can be seen in Table 12.5.

It can be seen that this model has considerable lower complexity than the ANN ones however their errors are rather high. The statistics of the last 5 models of the Pareto front of the Keijzer models can be seen in Table 12.5.

As Fig. 12.20 shows, employing these models as initial population, the result of the SR can be improved considerably.

Comparing this result with those of ANN models (Figs. 12.16 and 12.17), it can be seen that the models generated using Keijzer models as initial population have relative higher errors but considerably lower complexities than those of the ANN

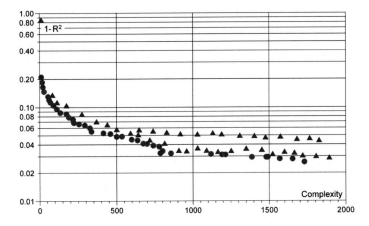

Fig. 12.18 The Pareto front (*circle points*) in case of the SR model with random initial population. The *triangular points* indicate optimal, but not Pareto optimal solutions

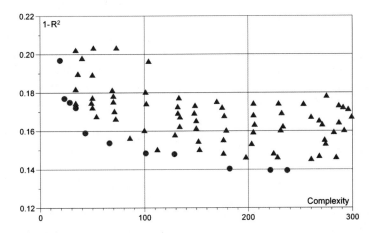

Fig. 12.19 The Pareto front (*circle points*) of the Keijzer models. The *triangular points* indicate optimal, but not Pareto optimal solutions

models. In Table 12.6, the statistics of the models investigated in study are summarized.

Here, η is the ratio of the RMSE of the validation and that of the training set. The generalization ability of the model reliable is ideal if η is close to 1. The parametric models are linear models and they cannot be considerably improved even with increasing the number of their parameters. The ANN models are nonlinear models and provide much better results than the parametric ones; however, their complexities are very high and the teaching process requires a large number of iterations leading to considerable time in both cases. In addition, the value of $\eta > 1$ indicates overlearning on the training set.

Table 12.5 The last five models of the Keijzer expansion

No	Complexity	$1 - R^2$	Functions
1	101	0.149	$0.513 - \frac{6.738 \times 10^{-6}\lambda^4}{\varphi^2} - 5.412 \times 10^{-7}\left(-6.547 + \sqrt{\lambda}\right)\lambda^3\varphi - \frac{4.996 \times 10^{-4}}{5.961 - \lambda + 2\varphi}$
2	128	0.148	$0.513 - \frac{6.738 \times 10^{-6}\lambda^4}{\varphi^2} - 5.412 \times 10^{-7}\left(-6.547 + \sqrt{\lambda}\right)\lambda^3\varphi -$ $-4.541 \times 10^{-27}\lambda^8\varphi^8 - \frac{4.996 \times 10^{-4}}{5.961 - \lambda + 2\varphi}$
3	182	0.140	$1.236 - 0.01\left(5 + 2\lambda - \frac{37.575\,(-0.589 + \lambda)}{\varphi}\right)^{-1} - \frac{0.024}{-3.314 + \lambda - 2\varphi} -$ $-0.002\left(20 - \frac{2.484}{\lambda} - \varphi\right)^2 - 2.044 \times 10^{-9}\lambda^4(-0.197 + 2\lambda + \varphi)$
4	221	0.140	$1.236 - 0.01\left(5 + 2\lambda - \frac{37.575\,(-0.589 + \lambda)}{\varphi}\right)^{-1} - \frac{0.024}{-3.314 + \lambda - 2\varphi} - 0.002 \times$ $\times\left(20 - \frac{2.484}{\lambda} - \varphi\right)^2 - 2.044 \times 10^{-9}\lambda^4(-0.197 + 2\lambda + \varphi) - \frac{220.174}{0.125 - \lambda + \lambda^2 + \varphi}$
5	237	0.140	$1.236 - 0.01\left(5 + 2\lambda - \frac{37.575\,(-0.589 + \lambda)}{\varphi}\right)^{-1} - \frac{0.024}{-3.314 + \lambda - 2\varphi} -$ $-0.002\left(20 - \frac{2.484}{\lambda} - \varphi\right)^2 - 2.044 \times 10^{-9}\lambda^4(-0.197 + 2\lambda + \varphi) -$ $-\frac{220.174}{0.125 - \lambda + \lambda^2 + \varphi} + \frac{0.002}{-16.291 + \varphi}$

Table 12.6 Statistics of errors of different corrector models at the points of the validation set

Model	Standard dev. (cm)	Min (cm)	Max (cm)	RMSE (cm)	$\eta = \frac{RMSE_V}{RMSE_T}$	Complexity
4-Parameter	8.23	−27.79	14.78	6.17	1.01	−
5-Parameter	8.03	−27.18	14.42	8.06	0.99	−
7-Parameter	8.06	−25.87	14.26	8.06	1.00	−
ANN Sigmoid	1.87	−6.01	4.21	1.90	1.64	1675
ANN RBF	2.03	−6.98	4.75	2.08	1.42	2323
SR Automatic	1.43	−3.09	3.52	1.40	0.45	1727
SR Keijzer	1.76	−4.66	4.20	1.73	0.52	873
SR Sigmoid	1.23	−3.61	3.14	1.65	1.34	2329
SR RBF	1.62	−3.98	3.25	1.62	0.45	1382

The SR model with randomly generated initial population (automatic) is quite good and $\eta < 1$ indicate under-learning. Unfortunately, the required running time is about 10 times larger than that of the ANN models. This model also has high complexity. However, employing Keijzer expansion, its complexity can be reduced considerably without significantly worsening the model quality. SR models with initial population with ANN models provide also quite good results. Considering these results, one may select the SR with ANN models as initial population since they provide quite good results; however the complexity of the resulting models are considerably higher. Therefore, on the one hand, one may select moderate complexity but on the other hand, if the model quality is the most important feature,

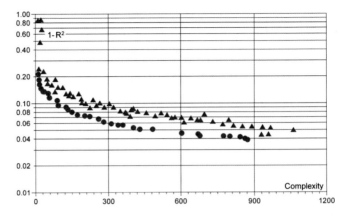

Fig. 12.20 The Pareto front (*circle points*) in case of the SR models based on the Keijzer models as initial population. The *triangular points* indicate optimal, but not Pareto optimal solutions

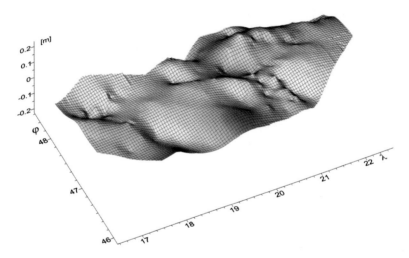

Fig. 12.21 The corrector surface computed using the SR model employing Keijzer expansions

then the SR model with initial population of the sigmoid ANN model is the best choice. The visualization of the corrector surface can be seen in Fig. 12.21. It is not surprising that one needs so high complexity function to approximate it.

12.3.2 Geometric Transformation

Transformation of coordinates is important in computer vision, photogrammetry as well as in geodesy. In this example we consider some standard 2D transformations as

Table 12.7 The coordinates of the corresponding observation points on the comparator and reseau planes

Points	x (mm)	y (mm)	X (mm)	Y (mm)
1	−113.767	−107.4	−110	−110
2	−43.717	−108.204	−40	−110
3	36.361	−109.132	40	−110
4	106.408	−109.923	110	−110
5	107.189	−39.874	110	−40
6	37.137	−39.07	40	−40
7	−42.919	−38.158	−40	−40
8	−102.968	−37.446	−100	−40
9	−112.052	42.714	−110	40
10	−42.005	41.903	−40	40
11	38.051	40.985	40	40
12	108.089	40.189	110	40
13	108.884	110.221	110	110
14	38.846	111.029	40	110
15	−41.208	111.961	−40	110
16	−111.249	112.759	−110	110

similarity, affine and projective transformations between the coordinates of the fiducial marks on the comparator plate and that of the corresponding points on the reseau plate, see Gosh (2005). The 16 observed (x, y) and calibrated (X, Y) coordinates are given in Table 12.7.

12.3.2.1 Similarity Transformation

The advantage of this model is that it is linear in the coordinates as well as in the parameters. Consequently an iterative solution is not required and the inverse transformation is easy. This transformation can be parametrized in the following form,

$$\begin{pmatrix} x \\ y \end{pmatrix} = \begin{pmatrix} a & b \\ -b & a \end{pmatrix} \begin{pmatrix} X \\ Y \end{pmatrix} + \begin{pmatrix} c \\ d \end{pmatrix}$$

For each observed i-th point the following pair of observations equation can be written for the residuals (r_{x_i}, r_{y_i}),

$$\begin{pmatrix} X_i & Y_i & 1 & 0 \\ Y_i & -X_i & 0 & 1 \end{pmatrix} \begin{pmatrix} a \\ b \\ c \\ d \end{pmatrix} + \begin{pmatrix} -x_i \\ -y_i \end{pmatrix} = \begin{pmatrix} r_{x_i} \\ r_{y_i} \end{pmatrix}.$$

Table 12.8 The estimated parameters of the similarity transformation model

Parameter	Value
a	0.999162
b	−0.011416
c	2.4471
d	−1.3878

A minimum of two fiducial marks or reseau crosses are required for a unique solution. Having more observation points, linear least square can be directly applied. In our example the result can be seen in Table 12.8.

12.3.2.2 Affine Transformation

This is also a linear model although it needs six parameters, therefore at least three fiducial marks or reseau crosses are required for a unique solution. Iterative solution is not required and the inverse transformation is easy. This model can be parametrized in the following form.

$$\begin{pmatrix} x \\ y \end{pmatrix} = \begin{pmatrix} a & b \\ d & e \end{pmatrix} \begin{pmatrix} X \\ Y \end{pmatrix} + \begin{pmatrix} c \\ f \end{pmatrix}$$

For each observed ith point the following pair of observations equation can be written for the residuals (r_{x_i}, r_{y_i}),

$$\begin{pmatrix} X_i & Y_i & 1 & 0 & 0 & 0 \\ 0 & 0 & 0 & X_i & Y_i & 1 \end{pmatrix} \begin{pmatrix} a \\ b \\ c \\ d \\ e \\ f \end{pmatrix} + \begin{pmatrix} -x_i \\ -y_i \end{pmatrix} = \begin{pmatrix} r_{x_i} \\ r_{y_i} \end{pmatrix}$$

Having more than 3 observation points, linear least square can be directly applied. In our example the result can be seen in Table 12.9.

Table 12.9 The estimated parameters of the affine transformation model

Parameter	Value
a	0.999167
b	−0.011339
c	2.4470
d	0.011494
e	0.999157
f	−1.3877

12.3.2.3 Projective Transformation

The projective transformation can be expressed by the following two equations,

$$x = \frac{a_1 X + a_2 Y + a_3}{c_1 X + c_2 Y + 1}$$

and

$$y = \frac{b_1 X + b_2 Y + a_3}{c_1 X + c_2 Y + 1}$$

These equations are a special case of the co-linearity condition for mapping of 2D points from one plane onto another. There are 8 unknown parameters, so if only four fiducial marks are available, the solution is not unique. This is perhaps the main reason why it is not frequently used. However, it is relevant for systems that have been retro-fitted with a reseau plate, such as the Hasselblad camera used to acquire the assignment imagery.

The least square solution is nonlinear due to the rational nature of the functions. However, an approximate linear solution can be implemented if both sides of the equations are multiplied by the denominator and partial derivatives with respect to the observables (as in the combined adjustment model) are ignored.

The inverse of the transformation can be also computed by solving the equation system in symbolic form for X and Y, we obtain

$$X = \frac{a_2(-y + a_3) + (x - a_3)b_2 + (-x + y)a_3 c_2}{-x b_2 c_1 + a_2(-b_1 + y c_1) + x b_1 c_2 + a_1(b_2 - y c_2)}$$

and

$$Y = \frac{a_1(y - a_3) + (-x + a_3)b_1 + (x - y)a_3 c_1}{-x b_2 c_1 + a_2(-b_1 + y c_1) + x b_1 c_2 + a_1(b_2 - y c_2)}$$

Having more than four observation points allows application of nonlinear least squares. In our example the result can be seen in Table 12.10.

Table 12.10 The estimated parameters of the projective transformation model

Parameter	Value
a_1	0.999165
a_2	−0.01134
a_3	2.445675
b_1	0.011494
b_2	0.999156
b_3	−1.393161
c_1	0
c_2	−0.000001

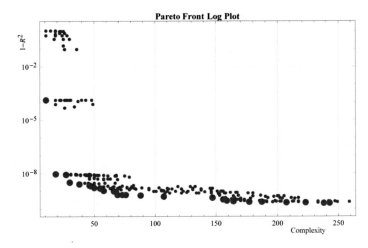

Fig. 12.22 The Pareto front in case of $x = x\,(X,\,Y)$

12.3.2.4   Symbolic Regression

By definition, neither the model nor the parameters of the model are known. However in order to have a chance for computing the inverse of the transformation trigonometric functions are excluded from the function-set. The Pareto front for the $x = x\,(X,\,Y)$ relation can be seen in Fig. 12.22.

Table 12.11 shows the Model selection report. We selected the best linear model, the third one, with complexity 30 and with error $1 - R^2 = 3.029 \times 10^{-9}$.

Remark   In Table 12.11 x_1 and x_2 stand for X and Y.

So our model is

$$x = -2.43296 + 1.0007X + 0.0113571\,Y - 9.51358 \times 10^{-7}X\,Y.$$

Similarly for the relation $y = y\,(X,\,Y)$ we get

$$y = 1.42065 - 0.0115116X + 1.00071\,Y - 1.78795 \times 10^{-7}X\,Y - 5.60054 \\ \times 10^{-7}Y^2.$$

The Fig. 12.23 illustrates how this nonlinear transformation warps (deforms) the original $(X,\,Y)$ plane into the $(x,\,y)$ plane.

In Table 12.12 the performance of the different transformation methods can be seen.

The SR model is better than the other models, and *its inverse transformation can be computed* via *Groebner basis*.

Our model is $(x,\,y) = F(X,\,Y)$, now we should compute F^{-1}, where $(X,Y) = F^{-1}\,(x,\,y)$. Let us consider $F(X,\,Y)$ in general form.

Table 12.11 Model selection report for $x = x(X, Y)$

Page 1			Page 2	
Model selection report				
	Complexity	$1 - R^2$	Function	
1	19	9.312×10^{-9}	$-2.43+1.00x_1+(1.14\times 10^{-2})x_2$	
2	27	8.794×10^{-9}	$-2.43+0.13/x_1+1.00 x_1+(1.14\times10^{-2})x_2$	
3	30	3.029×10^{-9}	$-2.43+1.00x_1+(1.14 \times 10^{-2})x_2- (9.51\times10^{-7})x_1x_2$	
4	38	2.540×10^{-9}	$-2.43+0.13/x_1 +1.00x_1+ (1.14\times10^{-2})x_2- (9.49\times10^{-7})x_1x_2$	
5	46	2.165×10^{-9}	$-2.43+ 0.12/x_1+1.00x_1+ 0.11/x_2+(1.13 \times10^{-2})x_2- (9.50\times10^{-7})x_1x_2$	
6	47	1.934×10^{-9}	$-2.43+1.00x_1-)x_2- (9.53\times10^{-7})x_1x_2$	
7	50	1.547×10^{-9}	$-2.44+1.00x_1 +0.11/x_2+ (1.13\times10^{-2})x_2- (9.54\times10^{-7})x_1x_2+(5.17\times 10^{-7})(x_2)^2$	
8	55	1.460×10^{-9}	$-2.43+0.12/x_1 +1.00 x_1-9.94/(x_2)^2 +(1.14\times 10^{-2})x_2- (9.51\times10^{-7})x_1x_2$	
9	58	1.085×10^{-9}	$-2.44+0.12/x_1 +1.00x_1 +0.11/x_2 +(1.13\times10^{-2})x_2-(9.52\times10^{-7})x_1x_2 +(5.13\times10^{-7})(x_2)^2$	

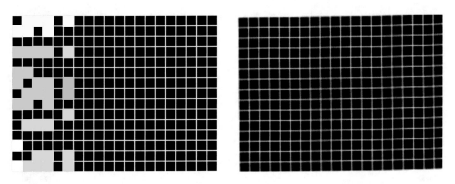

Fig. 12.23 The original plane and below the plane after warping via the nonlinear transformation

Table 12.12 Some statistical values of the different transformation models

Transformation model	Max of absolute errors (mm)	Standard deviation of the absolute errors (mm)	RMSE of the residual errors (mm)
Similarity	0.0268663	0.00882640	0.0106
Affine	0.0180314	0.00518175	0.0089
Projective	0.0128281	0.00342605	0.0066
Symbolic regression	0.0078720	0.00197883	0.0028
Similarity	0.0268663	0.00882640	0.0106
Affine	0.0180314	0.00518175	0.0089

$$F_1 = a_1 + b_1 X + c_1 Y + d_1 XY.$$

$$F_2 = a_2 + b_2 X + c_2 Y + d_2 XY + e_2 Y^2.$$

Assuming that the coordinates (x, y) are given, we have to solve the following multivariate polynomial system.

$$F_1(X, Y) - x = 0.$$

$$F_2(X, Y) - y = 0.$$

Let us compute the Groebner basis for X, we get

$$
\begin{aligned}
gbX(X, x, y) = & -a_1^2 e_2 - c_1 c_2 x - e_2 x^2 - b_2 c_1^2 X + b_1 c_1 c_2 X - c_2 d_1 xX - \\
& c_1 d_2 xX + 2b_1 e_2 xX - 2b_2 c_1 d_1 X^2 + b_1 c_2 d_1 X^2 + b_1 c_1 d_2 X^2 - b_1^2 e_2 X^2 - \\
& d_1 d_2 xX^2 - b_2 d_1^2 X^3 + b_1 d_1 d_2 X^3 - a_2(c_1 + d_1 X)^2 + \\
& a_1(2e_2(x - b_1 X) + c_1(c_2 + d_2 X) + d_1 X(c_2 + d_2 X)) + c_1^2 y + 2c_1 d_1 Xy + d_1^2 X^2 y
\end{aligned}
$$

which is a polynomial for X of third order

$$\delta(x, y)X^3 + \gamma(x, y)X^2 + \beta(x, y)X + \alpha(x, y) = 0$$

So to get X for given (x, y) we have to find the roots of this polynomial.

12.4   Exercise

12.4.1   Bremerton Data

Figures 12.24 and 12.25 represent the digital elevation model of Bremerton West, Washington. We should like to approximate this surface via symbolic regression model,

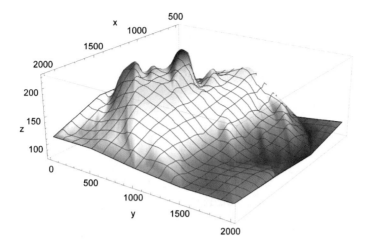

Fig. 12.24 The digital elevation model of the Bremerton West, Washington

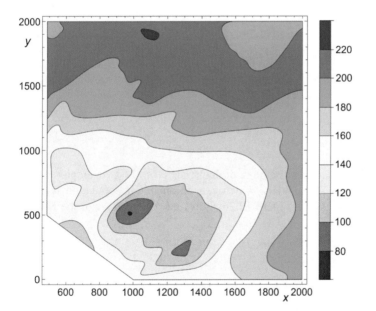

Fig. 12.25 Contour plot of the digital elevation model of the Bremerton West, Washington

$\Rightarrow$ dataFew $=$ Import$[$D $:\backslash\backslash$bremertonM.dat$]$;

$\Rightarrow$ DataModeler`

The data pairs of the digital elevation model,

⇒ samplePts = Map[{#[[1]],#[[2]]}&, dataFew]; (*{x,y}*)

⇒ observedResponse = Map[#[[3]]&, dataFew]; (*{z}*)

First let us try the random model generation technique.

12.4.1.1 Random Models

Now the population size is 200 models.

⇒ randomModels = UpdateModelQuality[
RandomModels[DataVariables → {x,y},
FunctionPatterns → {1, BuildFunctionPatterns[PowerMath, RBF,
Tanh, Sinusoids, SigmoidDM]}, PopulationSize → 200],
samplePts, observedResponse];

The result can be seen in Table 12.13.

⇒ ModelSelectionReport[FitModelsrandomModels,
SelectionStrategy → AllModels]

The best model is the number 50 with 57 complexity and having relatively high
error $(1 - R = 0.443)$.

12.4.1.2 Keijzer Models

The problem seems to be quite difficult; therefore first we employ a Keijzer model
population of 302 models, which can be an initial population for the traditional
symbolic regression. Figure 12.26 shows the Pareto front of the Keijzer models.

Table 12.13 The result of the random model generation

Page 1	Page 2	Page 3	Page 4
Model selection report			
	Complexity	$1 - R^2$	Function
46	48	0.988	$\text{Log}[\text{Log}[x/8]]^2$
47	55	0.999	$\text{Cos}[9169.39\ x^5\ \text{Tan}[y]^2]$
48	55	1.000	$1/(-9 + (\text{Cos}[x] - \text{Sin}[x])^x)$
49	56	1.000	$\text{Tan}[12 - 2.17/(-3.81 - x) + x]$
50	57	0.443	$\text{Sqrt}[-4 + 1/\text{Log}[7]^{10} + x] - y$
51	60	0.377	$-10.05 + \text{Sin}[x] + \sqrt{y} + y^{0.98} + \text{Tan}[\text{Cos}[y]]$

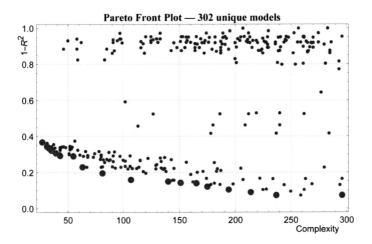

Fig. 12.26 The Pareto front of the generated Keijzer models

```
⇒ ParetoFrontPlot[
  keijzerModels = KeijzerExpansion[samplePts, observedResponse,
  DataVariables → {x, y}, TimeConstraint → 300,
  FunctionPatterns → {1, BuildFunctionPatterns[PowerMath]},
  KeijzerPopulationSize → 300, MaximumKeijzerModelComplexity → 0.3,
  MemoryLimit → 5 GB//]
]
```

```
⇒ ModelSelectionReport[keijzerModels, QualityBox → {All, 0.15}]
```
The Table 12.14 shows statistics of the Keijzer models.

12.4.1.3   Symbolic Regression

Employing these Keijzer models as initial model population, 451 further models
were generated, see Fig. 12.27

```
⇒ ParetoFrontLogPlot[
  quickModels1 = SymbolicRegression[
  samplePts, observedResponse, DataVariables → {x, y},
  TimeConstraint → 10000,
  FunctionPatterns → {1, BuildFunctionPatterns[
   "PowerMath", Sinusoids", "SigmoidDM", "RBF", "Tanh"]},
  InitialPopulation → keijzerModels, MemoryLimit → 20"GB"
  ]
]
```

Table 12.14 The generated Keijzer models

Model selection report

	Complexity	$1 - R^2$	Function
1	140	0.149	$-303.65 + 44.02\, x^{1/3} - 0.15\, x + 3.15\, y^{1/3} - (4.05 \times 10^{-2})\, y + (28.08\, y)/x + (2.76 \times 10^{-5})\, x\, y - (1.29 \times 10^{-4})\, y^2 + (4.31 \times 10^{-8})\, y^3 + 88.72\,(3.96 + y)^{1/9} + 471.85/(7.51 + y)$
2	151	0.143	$194.22 - 19{,}634.99/x - 1.18\,\sqrt{x} + (1.42 \times 10^{-2})\, x - (6.24 \times 10^{-9})\, x^3 - 10.01\, y^{1/3} + 6.11\,\sqrt{y} - (4.18 \times 10^{-2})\, y + (10.74\, y)/x + (1.28 \times 10^{-5})\, x\, y - (9.68 \times 10^{-5})\, y^2 + (6.49 \times 10^{-12})\, y^4 + (1.49 \times 10^{-8})\,(23.10 + y)^3$
3	165	0.140	$165.30 - 43{,}940.05/x + 0.52\,\sqrt{x} - (3.62 \times 10^{-2})\, x - (7.21 \times 10^{-9})\, x^3 + 4.79\,\sqrt{y} - (4.65 \times 10^{-2})\, y - (1.38 \times 10^{-4})\, y^2 + (4.76 \times 10^{-8})\, y^3 + (15.65\,(13.11 + x + y))/x + (6.37 \times 10^{-11})\, x^{5/2}\,(17.98 + x + 2\,y)$
4	175	0.120	$136.76 + 0.14\, x - (4.30 \times 10^{-5})\, x^2 + 18.69\, y^{1/3} - 0.54\, y + 0.11\, y^{4/3} - (6.95 \times 10^{-4})\, y^2 + (1.35 \times 10^{-7})\, y^3 - 1.09\,(x\, y)^{1/3} - 3.34\,\sqrt{-5.65 + x + y} + (1.03 \times 10^{-9})\,(\sqrt{x} + x + y)^3$
5	194	0.104	$-313.81 + 56.52\, x^{1/3} - 0.18\, x - (5.52 \times 10^{-9})\,(x - y)^3 + 7.78\, y^{1/3} - 0.79\,\sqrt{y} - 0.14\, y + (10.52\, y)/x + (7.26 \times 10^{-6})\, x\, y + (2.28 \times 10^{-9})\, x\, y^2 - (2.70 \times 10^{-8})\, y^3 + (1.36 \times 10^{-11})\, y^4 + 15.32\,(1.94 + y)^{1/3} + 1024.05/(12 + 2\,y)$
6	214	0.090	$-7.82 + 16.87\, x^{1/3} + 0.14\, x - (5.96 \times 10^{-5})\, x^2 + 18.69\, y^{1/3} - 0.52\,\sqrt{y} - 0.54\, y + (10.71\, y)/x + 0.11\, y^{4/3} - (6.95 \times 10^{-4})\, y^2 + (1.35 \times 10^{-7})\, y^3 - 1.09\,(x\, y)^{1/3} - 3.34\,\sqrt{-5.65 + x + y} + (1.03 \times 10^{-9})\,(\sqrt{x} + x + y)^3$
7	237	0.072	$140.48 - 8679.84/x + (1.03 \times 10^{-2})\, x - (4.59 \times 10^{-6})\, x^2 - (1.21 \times 10^{-8})\, x^3 + (1.66 \times 10^{-2})\,\sqrt{x^2} + 7.18\,\sqrt{y} - 0.21\, y + (1{,}939{,}078.70\, y)/x^3 + (3.48 \times 10^{-6})\, x\, y + (5.31 \times 10^{-9})\, x^2\, y - (3.12 \times 10^{-6})\, y^2 + (8.01 \times 10^{-9})\, y^3 + 821.90/(-5.38 + y)^2 - ((7.32 \times 10^{-6})\, y^3)/(3 + x + y) - 43{,}233.23/(x^{1/3} + x + y)$
8	296	0.071	$-1074.43 + 3465.78/x^{1/3} - 0.31\, x - (6.73 \times 10^{-5})\, x^2 + 3.77\,(x^2)^{1/3} - (2.84 \times 10^{-9})\,(x - y)^3 - 11.32\, y^{1/3} + 15.35\,\sqrt{y} - 4.45\, y^{2/3} - (6.33 \times 10^{-2})\, y - (4.29 \times 10^{-5})\, x\, y + (2.66 \times 10^{-10})\, x^2\, y^{3/2} - (2.42 \times 10^{-5})\, y^2 + (1.72 \times 10^{-11})\, x^2\, y^2 + (2.29 \times 10^{-8})\, y^3 - (5.39 \times 10^{-3})\,(16.00 + y)^{3/2} + 27.17\,\sqrt{1 + x + y}$

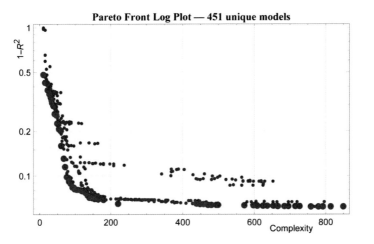

Fig. 12.27 The Pareto front and models via symbolic regression

Table 12.15 The statistics of the symbolic models

Page 1	Page 2		
Model selection table			
	Complexity	$1 - R^2$	Function
1	493	0.063	$425.89 - 87.87 \, \text{Sin}[4527.20 + \text{Sin}[(4514.99 + y)^{1/3}]^4] - 37.87$ $\text{Sin}[\text{Sin}[(20.53 + x + y)^{1/3}]^3]^3 - 6194.88 \, (-0.19 + \text{Sin}[\text{Sin}[\text{Sin} [\text{Sin}[\text{Sin}[\text{Sin}[4 + \text{Sin}[\text{Sin}[(64.53 + x + y)^{1/3}]] + \text{Sin}[x^{1/3}]]]]]]]]] + \text{Sin}[(x + 3 \, y)^{1/9}] + (4527.20 + y)^{1/4})^{-1.34}$
2	499	0.063	$418.07 + 94.42 \, \text{Sin}[1/4 + \text{Sin}[(4514.99 + y)^{1/3}]^4] - 37.72 \, \text{Sin} [\text{Sin}[(20.53 + x + y)^{1/3}]^3]^3 - 6228.04 \, (-0.19 + \text{Sin}[\text{Sin}[\text{Sin}[\text{Sin} [\text{Sin}[\text{Sin}[4 + \text{Sin}[\text{Sin}[(64.53 + x + y)^{1/3}]] + \text{Sin}[x^{1/3}]]]]]]]]] + \text{Sin}[(x + 3 \, y)^{1/9}] + (4527.20 + y)^{1/4})^{-1.34}$
3	572	0.063	$429.71 - 116.02 \, \text{Sin}[\text{Sin}[\text{Sin}[(x + y)^{1/3}]]^3]^3 + 90.23$ $\text{Sin}[1/4 + \text{Sin}[(4527.20 + y)^{1/3}]^4] - 6379.36 \, (-0.19 + \text{Sin}[\text{Sin}[\text{Sin}[\text{Sin}[\text{Sin}[\text{Sin}[\text{Sin}[\text{Sin}[\text{Sin}[4 + \text{Sin}[\text{Sin}[\text{Sin} [(20.83 + x + y)^{1/3}]]] + \text{Sin}[x^{1/3}]]]]]]]]]]] + \text{Sin}[(x + 4 \, y)^{1/9}] + (4514.99 + y)^{1/4})^{-1.34}$
4	597	0.063	$450.00 + 91.34 \, \text{Sin}[1/4 + \text{Sin}[(4527.20 + y)^{1/3}]^4] - 39.53 \, \text{Sin} [\text{Sin}[(20.53 + x + y)^{1/3}]^3]^3 - 6759.17 \, (-0.19 + \text{Sin}[\text{Sin}[\text{Sin}[\text{Sin}[\text{Sin}[\text{Sin}[\text{Sin}[\text{Sin}[4 + \text{Sin}[\text{Sin}[\text{Sin}[(20.83 + x + y)^{1/3}]]] + \text{Sin}[x^{1/3}]]]]]]]]]]]] + \text{Sin}[(x + 4 \, y)^{1/9}] + (4514.99 + y)^{1/4})^{-1.34}$
5	607	0.063	$1203.22 + 90.92 \, \text{Sin}[1/4 + \text{Sin}[(4527.20 + y)^{1/3}]^4] - 38.06 \, \text{Sin} [\text{Sin}[(x^{1/3} + x + y)^{1/3}]^3]^3 - 2906.45 - 0.19 + \text{Sin}[\text{Sin}[\text{Sin}[\text{Sin} [\text{Sin}[\text{Sin}[\text{Sin}[\text{Sin}[4 + \text{Sin}[\text{Sin}[\text{Sin}[(20.83 + x + y)^{1/3}]]] + \text{Sin}[x^{1/3}]]]]]]]]]]]] + \text{Sin}[(x + 4 \, y)^{1/9}] + (4514.99 + y)^{1/4})^{-0.44}$
6	609	0.063	$1190.92 - 121.16 \, \text{Sin}[\text{Sin}[\text{Sin}[(7 + x + y)^{1/3}]]^3]^3 + 92.49$ $\text{Sin}[1/4 + \text{Sin}[(4527.20 + y)^{1/3}]^4] - 2883.90 - 0.19 + \text{Sin}[\text{Sin}[\text{Sin}[\text{Sin}[\text{Sin}[\text{Sin}[\text{Sin}[\text{Sin}[\text{Sin}[4 + \text{Sin}[\text{Sin}[\text{Sin} [(20.83 + x + y)^{1/3}]]] + \text{Sin}[x^{1/3}]]]]]]]]]]]] + \text{Sin}[(x + 3 \, y)^{1/9}] + (4514.99 + y)^{1/4})^{-0.44}$

(continued)

Table 12.15 (continued)

Page 1		Page 2
7	625	0.063	$1243.53 + 91.58 \sin[1/4 + \sin[(4527.20 + y)^{1/3}]^4] - 38.97$ $\sin[\sin[(16.53 + x + y)^{1/3}]^3]^3 - 3012.35\ (-0.19 + \sin[\sin[\sin$ $[\sin[\sin[\sin[\sin[\sin[\sin[\sin[4 + \sin[\sin[\sin$ $[(20.83 + x + y)^{1/3}]]] + \sin[x^{1/3}]]]]]]]]]]]] + \sin[(x + 4\,y)^{1/9}] +$ $(4514.99 + y)^{1/4})^{-0.44}$
8	637	0.063	$1223.44 - 118.31 \sin[\sin[\sin[(7 + x + y)^{1/3}]]^3]^3 + 91.26$ $\sin[1/4 + \sin[(4527.20 + y)^{1/3}]^4] - 2960.64\ (-0.19 + \sin[\sin$ $[\sin[\sin[\sin[\sin[\sin[\sin[\sin[4 + \sin[\sin[\sin$ $[(20.83 + x + y)^{1/3}]]] + \sin[x^{1/3}]]]]]]]]]]] +$ $\sin[(x + 4\,y)^{1/9}] + (4514.99 + y)^{1/4})^{-0.44}$
9	654	0.062	$1279.84 + 92.16 \sin[1/4 + \sin[(4527.20 + y)^{1/3}]^4] - 39.10 \sin$ $[\sin[(16.53 + x + y)^{1/3}]^3]^3 - 3108.27\ (-0.19 + \sin[\sin[\sin[\sin$ $[\sin[\sin[\sin[\sin[\sin[4 + \sin[\sin[\sin$ $[(20.83 + x + y)^{1/3}]]] + \sin[x^{1/3}]]]]]]]]]]]]$ $+ \sin[(x + 4\,y)^{1/9}] + (4514.99 + y)^{1/4})^{-0.44}$
10	666	0.062	$1259.10 - 118.68 \sin[\sin[\sin[(7 + x + y)^{1/3}]]^3]^3 + 91.83$ $\sin[1/4 + \sin[(4527.20 + y)^{1/3}]^4] - 3054.86\ (-0.19 + \sin$ $[\sin[\sin[\sin[\sin[\sin[\sin[\sin[\sin[4 + \sin[\sin[\sin$ $[(20.83 + x + y)^{1/3}]]] + \sin[x^{1/3}]]]]]]]]]]]] +$ $\sin[(x + 4\,y)^{1/9}] + (4514.99 + y)^{1/4})^{-0.44}$
11	684	0.062	$1315.91 + 92.79 \sin[1/4 + \sin[(4527.20 + y)^{1/3}]^4] - 39.79 \sin$ $[\sin[(20.53 + x + y)^{1/3}]^3]^3 - 3203.28\ (-0.19 + \sin[\sin[\sin$ $[\sin[\sin[\sin[\sin[\sin[\sin[\sin[4 + \sin[\sin$ $[(20.83 + x + y)^{1/3}]]] + \sin[x^{1/3}]]]]]]]]]]]]]$ $+ \sin[(x + 4\,y)^{1/9}] + (4514.99 + y)^{1/4})^{-0.44}$
12	696	0.062	$1292.07 - 118.14 \sin[\sin[\sin[(4 + x + y)^{1/3}]^3]^3 + 92.34$ $\sin[1/4 + \sin[(4527.20 + y)^{1/3}]^4] - 3142.20\ (-0.19 + \sin[\sin$ $[\sin[\sin[\sin[\sin[\sin[\sin[\sin[\sin[\sin[4 + \sin[\sin[\sin$ $[(20.83 + x + y)^{1/3}]]] + \sin[x^{1/3}]]]]]]]]]]]]] +$ $\sin[(x + 4\,y)^{1/9}] + (4514.99 + y)^{1/4})^{-0.44}$
13	715	0.062	$1348.79 + 93.32 \sin[1/4 + \sin[(4527.20 + y)^{1/3}]^4] - 39.95 \sin$ $[\sin[(20.53 + x + y)^{1/3}]^3]^3 - 3290.18\ (-0.19 + \sin[\sin[\sin[\sin$ $[\sin[\sin[\sin[\sin[\sin[\sin[\sin[4 + \sin[\sin$ $[(20.83 + x + y)^{1/3}]]] + \sin[x^{1/3}]]]]]]]]]]]]] +$ $\sin[(x + 4\,y)^{1/9}] + (4514.99 + y)^{1/4})^{-0.44}$
14	727	0.062	$1325.14 - 119.54 \sin[\sin[\sin[(7 + x + y)^{1/3}]^3]^3 + 92.90$ $\sin[1/4 + \sin[(4527.20 + y)^{1/3}]^4] -$ $3229.41\ (-0.19 + \sin[\sin[\sin[\sin[\sin[\sin[\sin[\sin[\sin[\sin[\sin$ $[\sin[\sin[4 + \sin[\sin[\sin[(20.83 + x + y)^{1/3}]]] +$ $\sin[x^{1/3}]]]]]]]]]]]]]]]] + \sin[(x + 4\,y)^{1/9}] + (4514.99 + y)^{1/4})^{-0.44}$
15	759	0.062	$1355.85 - 120.00 \sin[\sin[\sin[(7 + x + y)^{1/3}]^3]^3 + 93.41$ $\sin[1/4 + \sin[(4527.20 + y)^{1/3}]^4] - 3310.62(-0.19 +$ $\sin[\sin[\sin[\sin[\sin[\sin[\sin[\sin[\sin[\sin[\sin[\sin[\sin[\sin$ $[4 + \sin[\sin[\sin[(20.83 + x + y)^{1/3}]]] +$ $\sin[x^{1/3}]]]]]]]]]]]]]]]]]] + \sin[(x + 4\,y)^{1/9}] + (4514.99 + y)^{1/4})^{-0.44}$

Table 12.16 The selected model

	Complexity	$1 - R^2$	Function
1	654	0.062	$1279.84 + 92.16\,Sin[1/4 + Sin[(4527.20 + y)^{1/3}]^4] - 39.10$ $Sin[Sin[(16.53 + x + y)^{1/3}]^3]^3 - 3108.27\,(-0.19 + Sin[Sin$ $[Sin[Sin[Sin[Sin[Sin[Sin[Sin[Sin[Sin[4 + Sin[Sin[Sin$ $[(20.83 + x + y)^{1/3}]]]] + Sin[x^{1/3}]]]]]]]]]]] + Sin[(x + 4\,y)^{1/9}]$ $+ (4514.99 + y)^{1/4})^{-0.44}$

$\Rightarrow$ ModelSelectionReport[quickModels1, QualityBox $\rightarrow$ {All, 0.0635}]

The Table 12.15 shows the statistic of the models.
The Pareto front consists of 69 models,

$\Rightarrow$ Length[ParetoFront[quickModels1]]
$\qquad\qquad \Leftarrow 69$

We select the model having number 9, see Table 12.16

$\Rightarrow$ ModelSelectionReport[ParetoFront[quickModels1][[9]]]

Our model in analytic form is

$\Rightarrow$ ModelPhenotype[ParetoFront[quickModels1]][[59]]
$\quad \Leftarrow 1279.84 + 92.1596\,Sin[1/4 + Sin[(4527.2 + y)^{(1/3)}]^4]$
$\quad - 39.1007\,Sin[Sin[(16.5283 + x + y)^{(1/3)}]^3]^3$
$\quad - 3108.27/(-0.188723 + (4514.99 + y)^{(1/4)} + Sin[(x + 4y)^{(1/9)}]$
$\quad + Sin[Sin[Sin[Sin[Sin[Sin[Sin[Sin[Sin[Sin[4 + Sin[x^{(1/3)}]$
$\quad + Sin[Sin[Sin[(20.8302 + x + y)^{(1/3)}]]]]]]]]]]]]])^{0.438417}$

The histogram of the relative model error in percent, see Fig. 12.28.

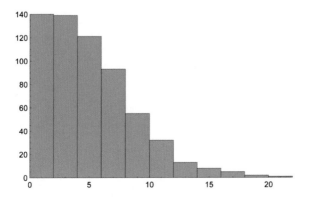

Fig. 12.28 The histogram of the model error of the selected model in %

The histogram indicates that the $\sim 90\%$ of the data points have less than 10 percent model error.

Remark One may choose a different model having somewhat higher error but considerably lower complexity, than using this model structure, can carry out a parameter estimation, see Fig. 12.27.

References

Cramer NL (1985) A representation for the adaptive generation of simple sequential programs. In: Grefenstette JJ (ed) Proceedings of the 1st international conference on genetic algorithm and their applications, Erlbaum, pp 183–187

Duquenne H, Jiang Z, Lemarie C (1995) Geoid determination and levelling by GPS: some experiments on the test network. In: IAG symposia gravity and geoid, vol 113. Springer, pp 559–568

Ghosh (2005) Fundamentals of computational photogrammetry. Concept Publishing Company, New Delhi, pp 40–45

Heiskanen W, Moritz H (1967) Physical geodesy. W H Freeman and Co., San Francisco

Kecman V (2001) Learning and soft computing: support vector machines, neural networks, and fuzzy logic models (complex adaptive systems). The MIT Press, Cambridge, MA, USA

Keijzer M (2003) Regression with interval arithmetic and linear scaling. In: Genetic programming, 6th European conference, EuroGP 2003, vol 2610. Springer, pp 70–82

Kenyeres A, Virág G (1998) Testing recent geoid models with GPS/levelling in Hungary. Reports of the Finnish Geodetic Institute, Masala 98(4):217–223

Kotsakis C, Fotopulos G, Sideris M.G (2001) Optimal fitting of gravimetric geoid undulations to GPS/levelling data using an extended similarity transformation model. In: The 27th annual meeting of the canadian geophysical union, Ottawa, Canada

Koza JR (1992) Genetic programming. MIT Press, Cambridge (Massachusetts) USA, On the programming of computers by means of natural selection

Chapter 13
Quantile Regression

13.1 Problems with the Ordinary Least Squares

In order to introduce this type of regression method, we shall consider some real world examples, where the application of traditional regression leads to misleading results. Then the definition of the quantile will be given and illustrated. With this definition, the method of quantile regression can be constructed.

13.1.1 Correlation Height and Age

Let us suppose that we want to find the correlation between height and age in a human population see Logan and Petscher (2013). It is clear that this correlation is different for teenagers than for adults; see Fig. 13.1.

The correlation between height and age is stronger in the case of teenagers than in that of adults. If we use Ordinary Least Square (OLS) we force the same correlation for the total population!

13.1.2 Engel's Problem

Engel were interested in the relationship between income and expenditures on food for a sample of working class Belgian households in 1857. The collected data are in Fig. 13.2.

Supplementary Information The online version contains supplementary material available at https://doi.org/10.1007/978-3-030-92495-9_13.

Fig. 13.1   Correlation between height and age in the human population

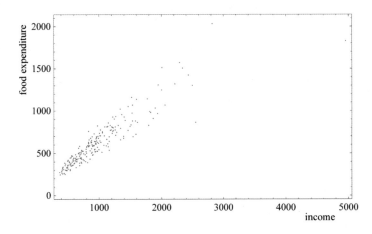

Fig. 13.2   Engel's household data

⇒ engels = Import["D : \\Book_Of_Four\\Engels.dat"]; eng = Drop[engels, −1]

Let us employ linear regression using least squares with the Euclidean metric, Ordinary Least Square (OLS). The regression line is

⇒ lm = LinearModelFit[eng, x, x]; OLS = Normal[LinearModelFit[eng, x, x]]
⇐ 147.475 + 0.485178x

Let us visualize the fitted function with the data, see Fig. 13.3.

In order to qualify this prediction, we also visualize the *prediction error of the food expenditure*. It can be seen that this error increases with the income as well as the food expenditure. This phenomenon is called *heteroscedasticity*, see Fig. 13.4.

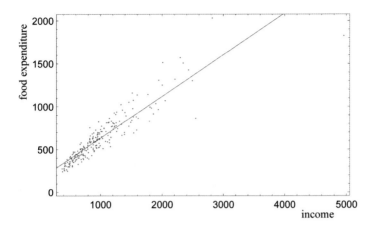

Fig. 13.3 Linear prediction via OLS

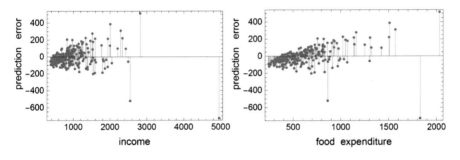

Fig. 13.4 The error of OLS as function of the income as well as that of the food expenditure

From studying these figures the following remarks can be made:

(1) Food expenditure increases with income,
(2) The dispersion of food expenditure increases with income (food expenditure): heteroscedasticity.
(3) The least squares estimates fit low income observations quite poorly, the OLS line passes over most low income households; see Fig. 13.3.

This latest remarks can be more clearly illustrated if we compare the predicted and the observed food expenditures; see Fig. 13.5. In case of low region of food expenditure, there is overestimation, while in high region mainly underestimation occurs.

Considering the Probability Density Function (PDF) of the food expenditure we also can see that the data is skewed to the left; see Fig. 13.6. Since OLS represents the correlation of the mean values, it indicates that this correlation is not representative for the households spending less or more money for food.

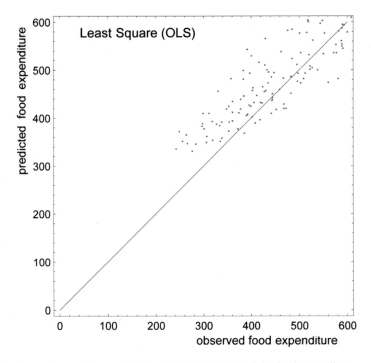

Fig. 13.5 Comparison of the predicted and observed values of the food expenditure

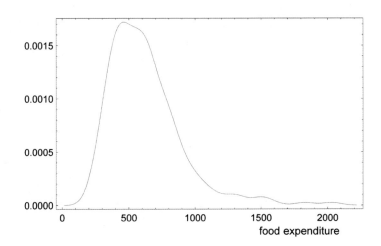

Fig. 13.6 PDF of the food expenditure is skewed to the left

We should like to analyze the correlation between food expenditure and income in the region of the distribution at *lower* and at *higher probability* density *instead of the mean value region*! We are looking for the answer for the practical question:

what is the effect of the income increase on the food expenditure in case of households having low income or high income, respectively? The technique which can give the answers is the *quantile regression*.

13.2   Concept of Quantile

13.2.1   *Quantile as a Generalization of Median*

We shall see that the extension of the definition of the *mean* and *median* value can lead to the definition of the *quantile*. Let us consider 10,000 random uniformly distributed integer numbers from the interval $[-10, 10]$ (Fig. 13.7)

$\Rightarrow$ y = RandomReal[{$-10,10$},10000]; Histogram[y, ImageSize $\rightarrow$ 350]
The *mean value* (μ_{av}) of these numbers is

$$\mu_{av} = \frac{1}{n}\sum_{i=1}^{n}y_i.$$

$\Rightarrow$ Mean[y]
$\Leftarrow$ 0.06423866
    This mean value also can be defined as the minimum of the following objective function

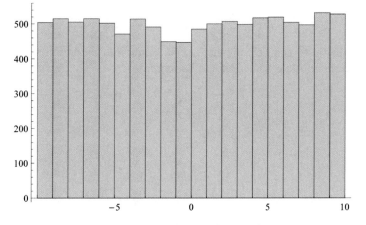

Fig. 13.7 Histogram of uniformly distributed integers from $[-10, 10]$

$$G(\mu) = \sum_{i=1}^{n}(\mu - y_i)^2.$$

Namely

$$\mu_{av} = \min_{\mu}(G(\mu)).$$

$\Rightarrow$ G = Total[Map[(# $-\mu_{av}$)^2&, y]];

$\Rightarrow$ NMinimize[G, μ_{av}][[2]]

$\Leftarrow \{\mu_{av} \rightarrow 0.0642386\}$

Similarly the *median* (μ_m), which is the "middle" value, separating the higher half of an *ordered data* sample, a population, or a probability distribution, from the lower half.

$\Rightarrow$ m = Median[y]

$\Leftarrow$ 0.198665

The number of the elements being greater or equal to the median is equal with the number of the elements that are smaller than the median. Indeed, the number of the greater elements is

$\Rightarrow$ Select[y, # $\geq$ m&]//Length

$\Leftarrow$ 5000

and that of the smaller elements is

$\Rightarrow$ Select[y, # $<$ m&]//Length

$\Leftarrow$ 5000

The median also can be computed as the solution of an optimization problem. The sum these absolute values should be minimized,

$$G(\mu) = \sum_{i=1}^{n}|(\mu - y_i)|.$$

Namely

$$\mu_m = \min_{\mu}(G(\mu)).$$

$\Rightarrow$ G = Total[Map[Abs[(# $-\mu_m$)]&, Y]];

$\Rightarrow$ qm = NMinimize[G, μ_m][[2]]

$\Leftarrow \{\mu_m \rightarrow 0.1989649\}$

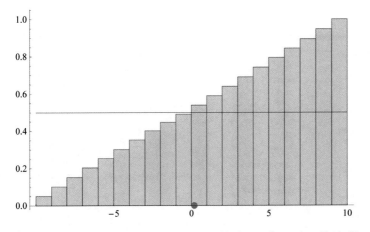

Fig. 13.8 The CDF of uniformly distributed integers with the median value (the ball)

This measure has a symmetrical characteristic; see the Cumulative Distribution Density Function (CDF) in Fig. 13.8.

We have seen that in the data set, there are as many values bigger than the median as there are smaller. Now, we *generalize* this idea! The quantile is a generalization of the median Koenker (2005).

The quantile p of a population or data set y is a member of the data set μ_q, $(\mu_q = Q(y, p))$ so that

$$\frac{\text{Length}\left(\{y_i : y_i > \mu_q\}\right)}{\text{Length}(y)} = 1 - p$$

and

$$\frac{\text{Length}\left(\{y_i : y_i \leq \mu_q\}\right)}{\text{Length}(y)} = p$$

If $p = 0.5$ we get $\mu_q = \mu_m$. For example

$\Rightarrow \text{p} = 0.5;$

$\Rightarrow \text{Quantile}[\text{y}, 0.5]$

$\Rightarrow \text{Quantile}[\text{y}, 0.5]$

So we see that the median of a set is its quantile at 0.5, namely $\mu_m = Q(0.5)$.

Now, one may ask what is the value separating the higher 80% of an *ordered data* sample, from the lower 20%? See Fig. 13.9,

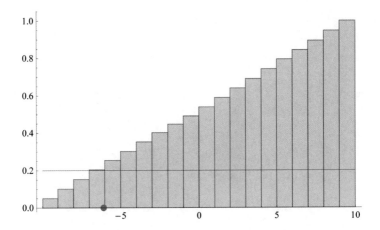

Fig. 13.9 The CDF of uniformly distributed integers with the quantile value μ_q (the ball)

$\Rightarrow$ p = 0.2;

$\Rightarrow \mu_q$ = Quantile[y, p] Indeed

$\Leftarrow$ -6.0761

$\Rightarrow$ nG = Select [y, #$\leq \mu_q$&]//Length

$\Leftarrow$ 8000

$\Rightarrow$ nS = Select [y, #$\leq \mu_q$&]//Length

$\Leftarrow$ 8000

The task to find the $q = Q(y, p)$ can be defined as an optimization problem, too. The objective function is

$$G(y, p, \mu) = \sum_{i \in \{i : y_i < \mu\}} (1 - p)|(\mu - y_i)| + \sum_{i \in \{i : y_i \geq \mu\}} p|(\mu - y_i)|$$

and

$$\min_{\mu} G(y, p, \mu) \rightarrow \mu_q.$$

$\Rightarrow$ G = Total[Map[If [# < μ_q, (1 − p)Abs [# − μ_q], pAbs [# − μ_q]] &, y]];

$\Rightarrow$ NMinimize [G, μ_q] [2]

$\Leftarrow \{\mu_q \rightarrow -6.07417\}$

13.2.2 *Quantile for Probability Distributions*

Instead of sets of discrete values, now we consider continuous *distributions*. First, consider the *uniform distribution*.

⇒ U = UniformDistribution[{−10, 10}];
The probability density function is Fig. 13.10.
Now let us compute the value of $Q(U, 0.2)$

⇒ Quantile[U, 0.2]
⇐ −6
The cumulative density function is Fig. 13.11.
Indeed

⇒ NSolve[CDF[U, x] == p, x]//Quiet//Flatten
⇐ {x → −6.}
Which means, that the quantile p of the distribution D, $\mu_q = Q(D, p)$ satisfies the following relation

$$\text{CDF}\,(\mu_q) = p.$$

Namely the value of the CDF of the distribution D at μ_q is equal to p.
Now, let us see a *normal distribution* $N(\mu, \sigma) = N(1, 5)$

⇒ D = NormalDistribution[1, 5];
The probability density function is Fig. 13.12

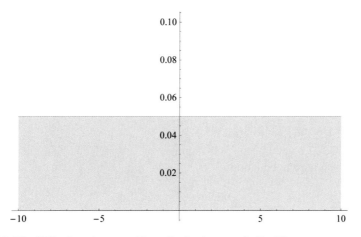

Fig. 13.10 The PDF of continuous uniform distribution over [−10, 10]

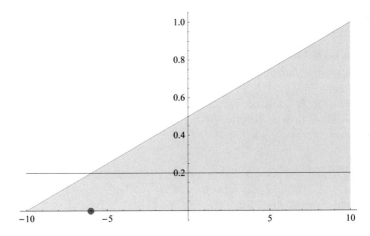

Fig. 13.11 The CDF of continuous uniform distribution over $[-10, 10]$ with the quantile value (gray ball)

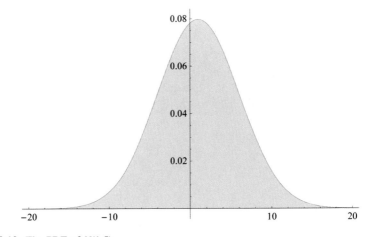

Fig. 13.12 The PDF of $N(1.5)$

Calculate quantile for $p = 0.2$, namely $Q(N (1, 5), 0.2) = ?$

$\Rightarrow \mu_q = \texttt{Quantile[D, 0.2]//N}$
$\Leftarrow -3.20811$
Indeed, solving the equation

$$\text{CDF} (\mu_q) - p = 0.$$

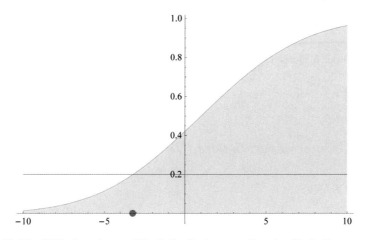

Fig. 13.13 The CDF of continuous $N(1, 5)$ distribution quantile value (the ball)

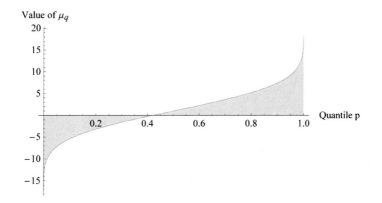

Fig. 13.14 The μ_q values of $N(1, 5)$ for $p \in [0, 1]$

$\Rightarrow \mu_q = //\text{Quiet}$

$\Rightarrow \text{NSolve}[\text{CDF}[\text{D},\mu_q] == p, \mu_q]//\text{Quiet}//\text{Flatten}$

$\Leftarrow \{\mu_q \rightarrow --3.20811\}$

Let us visualize the result in Fig. 13.13,

Now let us compute the quantiles for all of the values of $p \in [0, 1]$.

This curve is not symmetric, since the mean value is not zero, but 1. This function on Fig. 13.14 is the inverse of the function of the CDF on Fig. 13.13, see Davino et al (2014).

13.3   Linear Quantile Regression

Let us consider a linear regression model, namely

$$y = \beta_0 + \beta_1 x.$$

13.3.1   Ordinary Least Square (OLS)

The objective function is

$$G(\beta_0, \beta_1,) = \sum_{i=1}^{n} (\beta_0 + \beta_1 x_i - y_i)^2$$

where the data pairs of the observation are $\{x_i, y_i\}$, $i = 1, 2, \ldots, n$.

13.3.2   Median Regression (MR)

The objective function is

$$G(\beta_0, \beta_1,) = \sum_{i=1}^{n} |(\beta_0 + \beta_1 x_i - y_i)|$$

where the data pairs of the observation are $\{x_i, y_i\}$, $i = 1, 2, \ldots, n$.

Now let us compute the model parameter for MR,

$$\Rightarrow \texttt{G = Total[Map[Abs[} \beta_0 + \beta_1 \texttt{]\#[[1]]} - \texttt{\#[[2]]]\&, eng]};$$
$$\Rightarrow \texttt{NMinimize[} G, \{\beta_0 + \beta_1\} \texttt{][[2]]}$$
$$\Leftarrow \{\beta_0 \rightarrow 81.4825, \beta_1 \rightarrow 0.56018\}$$

Figure 13.15 shows some improvement of the quality of the estimation in case of the MR approach. We can also see this improvement in case of MR, when we consider the predicted food expenditure vs. observed food expenditure in both cases; see Fig. 13.16.

The performance of the MR is better than that of the OLS in the low food expenditure region.

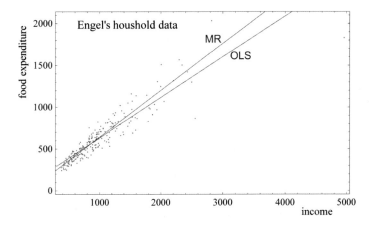

Fig. 13.15 The regression lines of the OLS and the MR approach

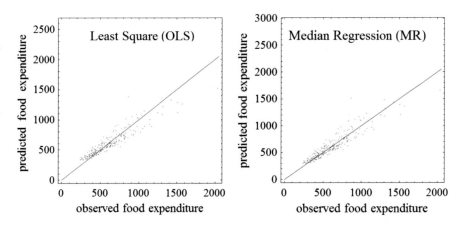

Fig. 13.16 The predicted versus observed food expenditure in case of the two different approach

13.3.3 Quantile Regression (QR)

We can generalize the median regression similarly to that of the generalization of the median as quantile! The objective function is

$$G(x, y, p, \beta_0, \beta_1) = \sum_{i \in \{i: y_i < \beta_0 + \beta_1 y_i\}} (1 - p)|(\beta_0 + \beta_1 x_i - y_i)| +$$
$$\sum_{i \in \{i: y_i \geq \beta_0 + \beta_1 x\}} p|(\beta_0 + \beta_1 x_i - y_i)|$$

and

$$\min_{\beta_0,\beta_1}(G(x,y,p,\beta_0,\beta_1) \rightarrow \beta_0,\beta_1.$$

Let us compute the regression line for the region of data representing high food expenditure, $p = 0.95$

$\Rightarrow p = 0.95;$
The result is (Fig. 13.17)

$\Leftarrow \{\beta_0 \rightarrow 64.1042, \beta_1 \rightarrow 0.709068\}$

What does this regression line ($p = 0.95$) represent? Let us consider the distribution of the output (food expenditure) with the value of the food expenditure belonging to $p = 0.95$,

$\Rightarrow \{X, Y\} = \texttt{Transpose[eng]};$
$\Rightarrow \texttt{Quantile[Y,p]}$
$\Leftarrow 1143.42$

In Fig. 13.18 we see that the regression line for $p = 0.5$ represents the relation between income and food expenditure when the output data (food expenditure) greater than or equal to 1143.42 are weighted with 0.95, and the output data less than 1143.2 are weighted with 0.05. This means that the food expenditure data greater than 1143.42 are strongly over-weighted. Therefore, this regression line essentially represents the relation between input and output at high output values!

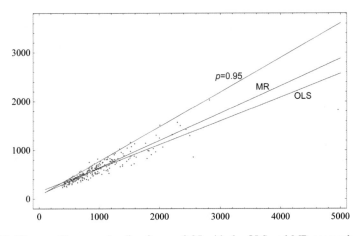

Fig. 13.17 The quantile regression line for $p = 0.95$ with the OLS and MR approach

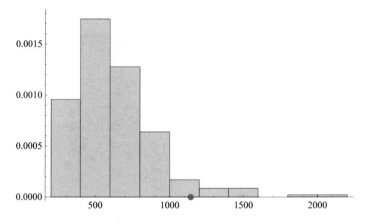

Fig. 13.18 The histogram of the output value (the food expenditure) with the quantile value 1143.42 belonging to $p = 0.95$

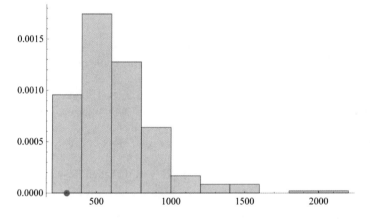

Fig. 13.19 The histogram of the output value (the food expenditure) with the quantile value 301 belonging to $p = 0.05$

Now let us consider a low quantile value $p = 0.05$ (Fig. 13.19).

⇒ p = 0.05;

⇒ Quantile[Y, p]

⇐ 301

Now the quantile regression line represents the relation between income and food expenditure when the output data greater than or equal to 301 are weighted with 0.05, while output data less than 301 are weighted with 0.95. Thus the food expenditure data greater than 301 are strongly under-weighted. Therefore this regression line essentially represents the relation between input and output at low output values.

We can use a built-in function to compute the quantile regression for a set of p values

$\Rightarrow$ qs $= \{0.05, 0.1, 0.25, 0.5, 0.75, 0.9, 0.95\}$;
The result is

$\Leftarrow 124.880 + 0.343361x$
$\quad 110.142 + 0.401766x$
$\quad 95.4835 + 0.474103x$
$\quad 81.4822 + 0.560181x$
$\quad 62.3966 + 0.644014x$
$\quad 67.3509 + 0.686299x$
$\quad 64.1040 + 0.709069x$

We also apply **Fit** to the data and the model functions in order to compare the regression quantiles with the least-squares regression fit:

$\Leftarrow 147.475 + 0.485178 \, x$

Here is a plot that combines the computed regression quantiles and the least squares fit (Fig. 13.20):

For example, the line for quantile $p = 0.5$ just above the green line (OLS) represents the regression where all data are equally weighted. So this line represents the median regression (MR).

Similarly to Fig. 13.14 we can compute the slope of the regression lines belonging to the different quantiles.

This figure shows the change of the slope with the quantile value (Fig. 13.21).

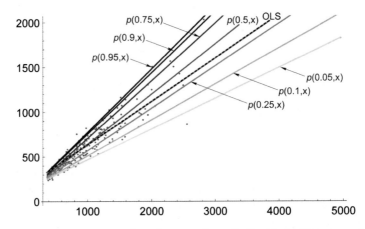

Fig. 13.20 The quantile regression lines for a set of $p \in (0, 1)$ with the OLS approach

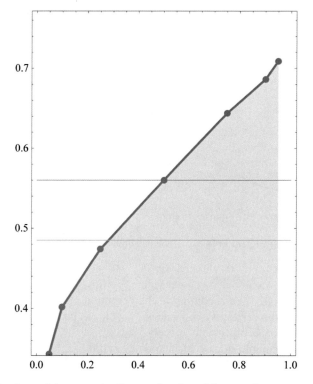

Fig. 13.21 The slope of the regression lines as function of the quantile p

Figure 13.20 also can be understood as a contour plot representing a 3D graph, with food expenditure and income on the y and x axis, respectively. The third dimension arises from the *conditional probability density* of the output values, i.e., the values of the food expenditure, variable y.

Let us consider following Fig. 13.22,

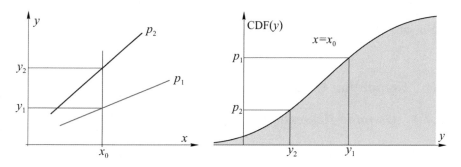

Fig. 13.22 The computation of the conditional probability density function $(y|x)$ at $x = x_0$

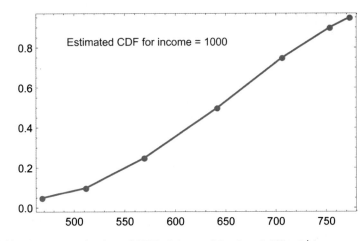

Fig. 13.23 The computed points of CDF of the conditional probability $(y|x)$ at $x = x_0 = 1000$

We compute the values of the output values, y_i with the different regression lines at $x = x_0$, $y_i = y_{pi}(x_0)$. Then according to the definition of the quantile (Figs. 13.9, 13.11 and 13.13), the corresponding values $\{y_i, p_i\}$ will provide the points of CDF of the conditional probability of y, namely $y|x_0$.

Let us see an example in case of income $x_0 = 1000$.

$$\Rightarrow \texttt{x0} = \texttt{1000};$$

The function values of the regression lines at different quantiles (see Fig. 13.20).

$$\Rightarrow \texttt{yi} = \texttt{qrFuncs/.x} \rightarrow \texttt{x0};$$

Now let us display the $\{y_i, p_i\}$ pairs (Fig. 13.23).

Since we want to also derive $PDF_{x_0}(y)$ from the estimated $CDF_{x_0}(y)$, via differentiation, it is better to use an interpolation object of $CDF_{x_0}(y)$:

Here is a plot of the derived $PDF_{x_0}(y)$ (Fig. 13.24).

These last two figures provide more insight into the input—output relation, and they can be used for *Monte-Carlo simulation*—namely generating further artificial data points.

13.4   Computing Quantile Regression

13.4.1   *Quantile Regression via Linear Programming*

Employing direct global minimization is a very time consuming technique. It is desirable, therefore, to transform the optimization problem of the quantile

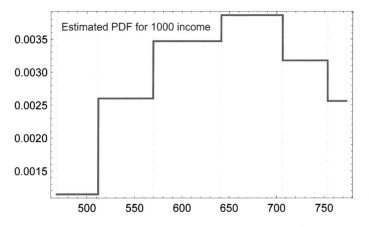

Fig. 13.24 The computed points of PDF of the conditional probability $(y|x)$ at $x = x_0 = 1000$

regression into a linear programming task of a linear regression model. As we have seen the linear quantile regression can be formulated as the following optimization problem, see Koenker (2005):

$$G(x, y, p, \beta_0, \beta_1) = \sum_{i \in \{i: y_i < \beta_0 + \beta_1 x_i\}} (1 - p)|(\beta_0 + \beta_1 x_i - y_i)| +$$

$$\sum_{i \in \{i: y_i \geq \beta_0 + \beta_1 x\}} p|(\beta_0 + \beta_1 x_i - y_i)|$$

and

$$\min_{\beta_0, \beta_1}(G(x, y, p, \beta_0, \beta_1)) \rightarrow \beta_0, \beta_1$$

In order to convert this problem into a linear programming task, let us introduce new non-negative variables u_i and v_i. Then the following linear equations as constraints can be given,

$$y_i - (\beta_0 + \beta_1 x_i) + u_i = 0 \quad \text{for } i \in \{i : y_i \geq \beta_0 + \beta_1 x_i\}$$

$$u_i = 0 \quad \text{for } i \notin \{i : y_i \geq \beta_0 + \beta_1 x_i\}$$

$$(\beta_0 + \beta_1 x_i) - y_i + v_i = 0 \quad \text{for } i \in \{i : y_i < \beta_0 + \beta_1 x_i\}$$

$$v_i = 0 \quad \text{for } i \notin \{i : y_i < \beta_0 + \beta_1 x_i\}$$

The linear objective function is

$$G(x,y,p,\beta_0,\beta_1,u,v) = \sum_{i\in\{i:y_i<\beta_0+\beta_1 x_i\}} (1-p)u_i + \sum_{i\in\{i:y_i\geq\beta_0+\beta_1 x_i\}} pv_i$$

which should be minimized

$$\min_{\beta_0,\beta_1,u,v} (G(x,y,p,\beta_0,\beta_1)) \rightarrow \beta_0,\beta_1,u,v$$

So we have now a linear problem, but we pay for it with having a system with a large number of variables. To compute the further illustration examples, we employed the *Mathematica* functions developed by Antonov (2013).

13.4.2   Boscovich's Problem

Between 1735 and 1754 the French Academy carried out four measurements of the length of an arc of a meridian at widely different latitudes with the purpose of determining the figure of the Earth, expressed as its ellipticity. Pope Benedict XIV contributing this project commissioned *Boscovich* and *Christopher Maire* to measure an arc of the meridian near Rome and constract a new map, see Hald (2007). The five measurements appearing in Table 13.1 had been made. It was clear from these measurements that *Arc Length* was increasing as one moved towards the pole from the equator, thus qualitatively confirming Newton's conjecture that the earth's rotation could be expected to make it bulge at the equator with a corresponding flattening at the poles, yielding an oblate spheroid, "more like a grapefruit than a lemon."

But how the five measurements should be combined to produce one estimate of earth's ellipticity was unclear. The form of the approximation is suggested by Koenker (2005)

$$y = \beta_0 + \beta_1 \sin^2 \lambda,$$

where y is the *Arc Length* and λ is the latitude.

Table 13.1 Measured data used by Boscovich

Location (i)	λ_i = Latitude	X_i = $\sin^2$(Latitude)	y_i = Arc Length
Quito	0° 00′	0	56,751
Cape of Good Hope	33° 18′	0.2987	57,037
Rome	42° 59′	0.4648	56,979
Paris	49° 23′	0.5762	57,074
Lapland	66° 19′	0.8386	57,422

Then the ellipticity could be computed as

$$\frac{1}{\text{ellipticity}} = \frac{3\beta_0}{\beta_1}$$

Now let us formulate the quantile regression problem for this data as a linear regression task. First we illustrate the solution for quantile value $p = 0.5$.

Input data for $X_i = \sin^2(\lambda_i)$ are

$$\Rightarrow X = \{0, 0.2987, 0.4648, 0.5762, 0.8386\};$$
and for $y_i = $ Arc Length

$$\Rightarrow y = \{56751, 57037, 56979, 57074, 57422\};$$

Introducing new variables

$$\Rightarrow U = \text{Table}[u_i, \{i, 1, 5\}]$$
$$\Leftarrow \{u_1, u_2, u_3, u_4, u_5\}$$

$$\Rightarrow V = \text{Table}[v_i, \{i, 1, 5\}]$$
$$\Leftarrow \{v_1, v_2, v_3, v_4, v_5\}$$
Constraints with equations are

$$\Leftarrow 56751 + u_1 - v_1 - \beta_0 == 0$$
$$57037 + u_2 - v_2 - \beta_0 - 0.2987\beta_1 == 0$$
$$56979 + u_3 - v_3 - \beta_0 - 0.4648\beta_1 == 0$$
$$57074 + u_4 - v_4 - \beta_0 - 0.5762\beta_1 == 0$$
$$57422 + u_5 - v_5 - \beta_{10} - 0.8386\beta_1 == 0$$
Constraints with inequalities are

$$\Leftarrow u_1 \geq 0$$
$$u_2 \geq 0$$
$$u_3 \geq 0$$
$$u_4 \geq 0$$
$$u_5 \geq 0$$
$$v_1 \geq 0$$
$$v_2 \geq 0$$
$$v_3 \geq 0$$
$$v_4 \geq 0$$
$$v_5 \geq 0$$

All constrains can be joined together,

$$\Leftarrow 56751 + u_1 - v_1 - \beta_0 == 0$$
$$57037 + u_2 - v_2 - \beta_D - 0.2987\beta_1 == 0$$
$$56979 + u_3 - v_3 - \beta_0 - 0.4648\beta_1 == 0$$
$$57074 + u_4 - v_4 - \beta_0 - 0.5762\beta_1 == 0$$
$$57422 + u_5 - v_5 - \beta_{\bar{0}} - 0.8386\beta_1 == 0$$

$$\Leftarrow u_1 \geq 0$$
$$u_2 \geq 0$$
$$u_3 \geq 0$$
$$u_4 \geq 0$$
$$u_5 \geq 0$$
$$v_1 \geq 0$$
$$v_2 \geq 0$$
$$v_3 \geq 0$$
$$v_4 \geq 0$$
$$v_5 \geq 0$$

The linear objective function is

$$\Leftarrow (1-p)u_1 + (1-p)u_2 + (1-p)u_3 + (1-p)u_4 + (1-p)u_5 + pv_1 + pv_2 + pv_3 + pv_4 + pv_5$$

The variables of the problem,

$$= \{\beta_0, \beta_1, u_1, u_2, u_3, u_4, u_5, v_1, v_2, v_3, v_4, v_5\}$$

When we minimize these we get

$$\Leftarrow \{164.473, \{\beta_0 \rightarrow 56751., \beta_1 \rightarrow 800.143, u_1 \rightarrow 0.,$$
$$u_2 \rightarrow 0., u_3 \rightarrow 143.907, u_4 \rightarrow 138.042, u_5 \rightarrow 0.,$$
$$v_1 \rightarrow 0., v_2 \rightarrow 46.9973, v_3 \rightarrow 0., v_4 \rightarrow 0., v_5 \rightarrow 0.\}\}$$

Now let us compute the parameters β's for different quantile values $p \in (0, 1)$.
Figure 13.25 shows the distinct regression quantile solutions for the slope
parameter, β_1.

This figure shows an interesting step-wise feature! It turns out that these steps
correspond to the pairwise solution of the regression problem. For example, the
interval $(0, 0.21)$ yields as a unique solution the line passing through the towns
Quito and *Rome*, indeed,

$$\Leftarrow \{\{\beta_0 \rightarrow 56751\cdot, \beta_1 \rightarrow 490.534\}\}$$

At $p = 0.21$ the solution jumps, and throughout the interval $(0.21, 0.48)$, we have
a solution characterized by the line passing through *Quito* and *Paris*. The process

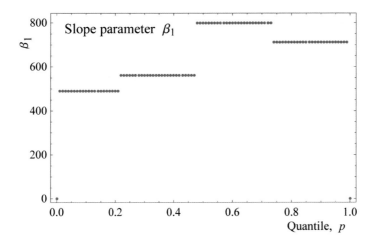

Fig. 13.25 The computed slope parameter as function of the quantile value p

continues until we get to $p = 0.78$, where the solution through *Lapland* and *Cape of Good Hope* prevails up to $p = 1$.

The pairs of points now play the role of order statistics and serve to define the estimated linear conditional quantile functions. Again, in the terminology of linear programming, such solutions are "*basic*" and constitute extreme points of the polyhedral constraint set. If we imagine the plane represented by the objective function (F) rotating as p increases, we may visualize the solutions of F as passing from one vertex of the constraint set to another, see Fig. 13.26. Each vertex represents an exact fit of a line to *a pair of sample observations*. At a few isolated points, the plane will make contact with an entire edge of the constraint set and we will get a set-valued solution. One occasionally encounters the view that quantile regression estimators must "ignore sample information" since they are inherently determined by a small subset of the observations.

Fig. 13.26 Explanation of the stepwise solutions of the problem

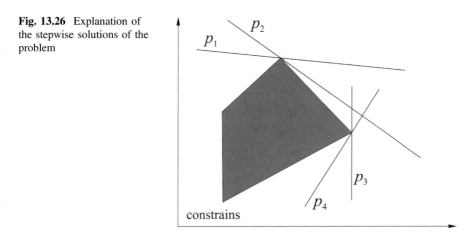

13.4.3   Extension to Linear Combination of Nonlinear Functions

The quantile regression formulation as a linear programming problem can be done for any model consisting of a linear combination of nonlinear functions, namely

$$y(x) = \sum_{i=1}^{n} \beta_i g_i(x)$$

where $g_i(x)$ nonlinear functions and β_i coefficients to be computed. As an illustration, let the data to be considered in Fig. 13.27.

The histogram of the output values y shows anomalies of the data. Let us consider the following quantiles:

$$\Rightarrow qs = \{0.05, 0.25, 0.5, 0.75, 0.95\};$$

We want to find curves that separate the data according the quantiles. Those curves are called "regression quantiles".

Pretending that we do not know how the data is generated, just by looking at the plot we assume that the model for the data is

$$y = \beta_0 + \beta_1 x + \beta_2 \sqrt{x} + \beta_3 \log(x)$$

Here we find the regression quantiles:

$$\Leftarrow -0.675199 + 1.28437\sqrt{x} + 0.15953x + 1.17151\log[x]$$
$$-0.675199 + 2.03544\sqrt{x} + 0.114855x + 0.67658\log[x]$$
$$0.523285 + 1.37319\sqrt{x} + 0.170873x + 1.26068\log[x]$$
$$3.01222 + 4.83723 * 10^{-7}\sqrt{x} + 0.252894x + 2.21582\log[x]$$
$$1.53156 + 3.12814\sqrt{x} + 0.015158x + 1.59654 * 10^{-7}\log[x]$$

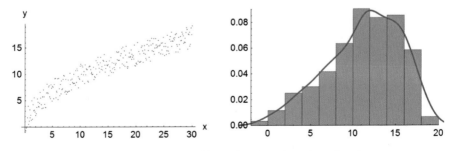

Fig. 13.27 Data for quantile regression with combination of nonlinear functions

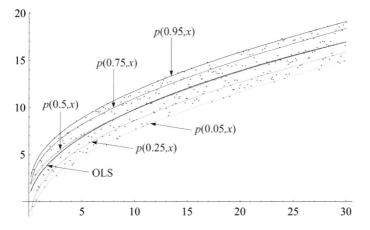

Fig. 13.28 Application of quantile regression with linear combination of nonlinear functions

We also apply **Fit** to the data and the model functions to compare the regression quantiles with the least-squares regression fit:

Therefore the OLS results in

$$\Leftarrow -0.522654 + 3.39992\sqrt{x} - 0.0329017x - 0.0430765\text{Log}[x]$$

Here is a plot that combines the found regression quantiles and least squares fit (Fig. 13.28).

Let us check how good the regression quantiles are for separating the data according to the quantiles they were computed for. In case of the ith quantile (p_i), we compute the fraction of the data,

$$f_i = \sum_{k=j}^{n} y_k \bigg/ \sum_{k=1}^{n} y_k, j \in \left\{ j : y_j \geq \beta_0^{(i)} + \beta_1^{(i)} x_j + \beta_2^{(i)} \sqrt{x_j} + \beta_3^{(i)} \log(x_j) \right\}.$$

and $p_i \approx f_i$ should be true (Table 13.2).

$$p_i + f_i \approx 1$$

Table 13.2 Quantile values (P_i) versus fractions above (P_i)

Quantile (P_i)	Fraction above (P_i)
0.05	0.949833
0.25	0.745819
0.50	0.505017
0.75	0.254181
0.95	0.046823

Fig. 13.29 Generated data for B-spline quantile regression

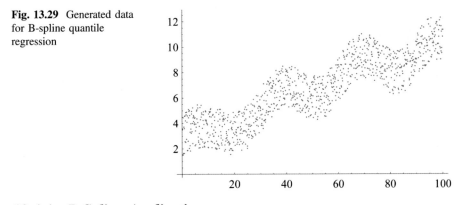

13.4.4 B-Spline Application

Splines, as basic functions in the linear combination, provide more flexibility of the regression line. Here we use B-splines.

We are going to consider two implementations of the Quantile Regression (QR) calculation with B-spline bases. The first implementation is based on the Linear Programming (LP) formulation of the quantile minimization problem. The second implementation is a direct translation of the non-LP minimization formulation, see Antonov (2013).

Let us see an example. We generate data with a sinusoidal function and add some random noise (Fig. 13.29).

If the number of knots of our spline is 15, then the quantile B-spline functions can be computed as third order polynomials,

$$
\Leftarrow 0. + 8.95237 \left(\left\{ \begin{array}{ll} -2752.84 + 88.4778\#1 - 0.9479088\#1^2 + 0.00338515\#1^3 & 93.34 \le \#1 \le 100.0 \\ 0 & \text{True} \end{array} \right. \right)
$$

$$
+ 9.19095 \left(\left\{ \begin{array}{ll} 4954.52 - 157.88\#1 + 1.67575\#1^2 - 0.005924\#1^3 & 93.34 \le \#1 \le 100.0 \\ -551.155 + 19.0755\#1 - 0.220068\#1^2 + 0.000846286\#1^3 & 86.68 \le \#1 < 93.34 \\ 0 & \text{True} \end{array} \right. \right)
$$

$$
+ 2.79210 \left(\left\{ \begin{array}{ll} 2.04538 - 0.457239\#1 + 0.0340715\#1^2 - 0.000846286\#1^3 & 6.76 \le \#1 \le 13.42 \\ -0.0460655 + 0.470919\#1 - 0.10323\#1^2 + 0.005924\#1^3 & 0.1 \le \#1 < 6.76 \ \text{True} \\ 0 & \textit{True} \end{array} \right. \right)
$$

$$
+ 2.00006 \left(\left\{ \begin{array}{ll} 0.00034128 - 0.0068566\#1 + 0.034749\#1^2 - 0.00310305\#1^3 & 0.1 \le \#1 < 6.76 \\ 4.56791 - 0.682456\#1 + 0.0339869\#1^2 - 0.000564191\#1^3 & 13.42 \le \#1 < 20.08 \\ -1.56825 + 0.689262\#1 - 0.0682276\#1^2 + 0.00197467\#1^3 & 6.76 \le \#1 < 13.42 \\ 0 & \text{True} \end{array} \right. \right)
$$

$$
+ 8.82548 \left(\left\{ \begin{array}{ll} 1364.38 - 46.3887\#1 + 0.524765\#1^2 - 0.00197467\#1^3 & 86.68 \le \#1 < 93.34 \\ -289.082 + 10.8379\#1 - 0.13544\#1^2 + 0.000564191\#1^3 & 80.02 \le \#1 < 86.68 \\ -2764.87 + 86.328\#1 - 0.897097\#1^2 + 0.00310305\#1^3 & 93.34 \le \#1 \le 100.0 \\ 0 & \text{True} \end{array} \right. \right)
$$

$$
+ 5.71623 \left(\left\{ \begin{array}{ll} 285.724 - 15.597\#1 + 0.282321\#1^2 - 0.00169257\#1^3 & 53.38 \le \#1 < 60.04 \\ 222.743 - 9.1089\#1 + 0.124167\#1^2 - 0.000564191\#1^3 & 66.7 \le \#1 \le 73.36 \\ -57.5353 + 3.69448\#1 - 0.079077\#1^2 + 0.000564191\#1^3 & 46.72 \le \#1 < 53.38 \\ -446.931 + 21.0114\#1 - 0.327411\#1^2 + 0.00169257\#1^3 & 60.04 \le \#1 < 66.70 \\ 0 & \text{True} \end{array} \right. \right)
$$

$$+ \ 4.11158 \left(\!\!\left(\begin{cases} 193.87 - 12.0617\#1 + 0.248504\#1^2 - 0.00169257\#1^3 & 46.72 \le \#1 < 53.38 \\ 167.419 - 7.53007\#1 + 0.112895\#1^2 - 0.000564191\#1^3 & 60.04 \le \#1 \le 66.70 \\ -36.2709 + 2.71625\#1 - 0.0678045\#1^2 + 0.000564191\#1^3 & 40.06 \le \#1 < 46.72 \\ -321.018 + 16.8755\#1 - 0.293594\#1^2 + 0.00169257\#1^3 & 53.38 \le \#1 < 60.04 \\ 0 & \text{True} \end{cases}\!\!\right)\right.$$

$$+ \ 7.71789 \left(\!\!\left(\begin{cases} 402.622 - 19.5827\#1 + 0.316139\#1^2 - 0.00169257\#1^3 & 60.04 \le \#1 < 66.70 \\ 289.082 - 10.8379\#1 + 0.13544\#1^2 - 0.000564191\#1^3 & 73.36 \le \#1 \le 80.02 \\ -85.8147 + 4.82286\#1 - 0.0903495\#1^2 + 0.000564191\#1^3 & 53.38 \le \#1 < 60.04 \\ -601.889 + 25.5977\#1 - 0.361229\#1^2 + 0.00169257\#1^3 & 66.7 \le \#1 < 73.36 \\ 0 & \text{True} \end{cases}\!\!\right)\right.$$

$$+ \ 2.60176 \left(\!\!\left(\begin{cases} 16.908 - 2.425\#1 + 0.113233\#1^2 - 0.00169257\#1^3 & 20.08 \le \#1 < 26.74 \\ 36.2709 - 2.71625\#1 + 0.0678045\#1^2 - 0.000564191\#1^3 & 33.4 \le \#1 \le 40.06 \\ -1.36359 + 0.304826\#1 - 0.0227143\#1^2 + 0.000564191\#1^3 & 13.42 \le \#1 < 20.08 \\ -47.8154 + 4.83642\#1 - 0.158323\#1^2 + 0.00169257\#1^3 & 26.74 \le \#1 < 33.40 \\ 0 & \text{True} \end{cases}\!\!\right)\right.$$

$$+ \ 4.52253 \left(\!\!\left(\begin{cases} 124.062 - 8.97682\#1 + 0.214686\#1^2 - 0.00169257\#1^3 & 40.06 \le \#1 < 46.72 \\ 122.109 - 6.10139\#1 + 0.101622\#1^2 - 0.000564191\#1^3 & 53.38 \le \#1 \le 60.04 \\ -21.0216 + 1.88817\#1 - 0.0565319\#1^2 + 0.000564191\#1^3 & 33.4 \le \#1 < 40.06 \\ -221.15 + 13.19\#1 - 0.259776\#1^2 + 0.00169257\#1^3 & 46.72 \le \#1 < 53.38 \\ 0 & \text{True} \end{cases}\!\!\right)\right.$$

$$+ \ 6.45381 \left(\!\!\left(\begin{cases} 933.587 - 34.2426\#1 + 0.417592\#1^2 - 0.00169257\#1^3 & 80.02 \le \#1 < 86.68 \\ 564.191 - 16.9257\#1 + 0.169257\#1^2 - 0.000564191\#1^3 & 93.34 \le \#1 \le 100.0 \\ -222.743 + 9.1089\#1 - 0.124167\#1^2 + 0.000564191\#1^3 & 73.36 \le \#1 < 80.02 \\ -1271.03 + 42.0595\#1 - 0.462682\#1^2 + 0.00169257\#1^3 & 86.68 \le \#1 < 93.34 \\ 0 & \text{True} \end{cases}\!\!\right)\right.$$

$$+ \ 4.61042 \left(\!\!\left(\begin{cases} 38.581 - 4.15849\#1 + 0.147051\#1^2 - 0.00169257\#1^3 & 26.74 \le \#1 < 33.40 \\ 57.5353 - 3.69448\#1 + 0.079077\#1^2 - 0.000564191\#1^3 & 40.06 \le \#1 \le 46.72 \\ -4.56791 + 0.682456\#1 - 0.0339869\#1^2 + 0.000564191\#1^3 & 20.08 \le \#1 < 26.74 \\ -87.5485 + 7.17051\#1 - 0.192141\#1^2 + 0.00169257\#1^3 & 33.40 \le \#1 < 40.06 \\ 0 & \text{True} \end{cases}\!\!\right)\right.$$

$$+ \ 2.06603 \left(\!\!\left(\begin{cases} 0.697149 - 0.309369\#1 + 0.0455979\#1^2 - 0.00169257\#1^3 & 6.76 \le \#1 < 13.42 \\ 10.7872 - 1.21024\#1 + 0.0452594\#1^2 - 0.000564191\#1^3 & 20.08 \le \#1 \le 26.74 \\ -5.642 \times 10^{-7} + 0.000017\#1 - 0.0001693\#1^2 + 0.000564\#1^3 & 0.1 \le \#1 < 6.76 \\ -7.48439 + 1.51959\#1 - 0.090688\#1^2 + 0.00169257\#1^3 & 13.42 \le \#1 < 20.08 \\ 0 & \text{True} \end{cases}\!\!\right)\right.$$

$$+ \ 1.66144 \left(\!\!\left(\begin{cases} 5.28007 - 1.14196\#1 + 0.0794155\#1^2 - 0.00169257\#1^3 & 13.42 \le \#1 < 20.08 \\ 21.0216 - 1.88817\#1 + 0.0565319\#1^2 - 0.000564191\#1^3 & 26.74 \le \#1 \le 33.40 \\ -0.174287 + 0.077346\#1 - 0.011442\#1^2 + 0.000564191\#1^3 & 6.76 \le \#1 < 13.42 \\ -22.1274 + 2.95278\#1 - 0.124506\#1^2 + 0.00169257\#1^3 & 20.08 \le \#1 < 26.74 \\ 0 & \text{True} \end{cases}\!\!\right)\right.$$

$$+ \ 5.93117 \left(\!\!\left(\begin{cases} 73.2991 - 6.34243\#1 + 0.180868\#1^2 - 0.00169257\#1^3 & 33.4 \le \#1 < 40.06 \\ 85.8147 - 4.82286\#1 + 0.0903495\#1^2 - 0.000564191\#1^3 & 46.72 \le \#1 \le 53.38 \\ -10.7872 + 1.21024\#1 - 0.0452594\#1^2 + 0.000564191\#1^3 & 26.74 \le \#1 < 33.40 \\ -144.327 + 9.95505\#1 - 0.225958\#1^2 + 0.00169257\#1^3 & 40.06 \le \#1 < 46.72 \\ 0 & \text{True} \end{cases}\!\!\right)\right.$$

$$+ \ 6.35067 \left(\!\!\left(\begin{cases} 723.553 - 28.9055\#1 + 0.383774\#1^2 - 0.00169257\#1^3 & 73.36 \le \#1 < 80.02 \\ 458.806 - 14.7463\#1 + 0.157985\#1^2 - 0.000564191\#1^3 & 86.68 \le \#1 \le 93.34 \\ -167.419 + 7.53007\#1 - 0.112895\#1^2 + 0.000564191\#1^3 & 66.7 \le \#1 < 73.36 \\ -1010.94 + 36.1218\#1 - 0.428864\#1^2 + 0.00169257\#1^3 & 80.02 \le \#1 < 86.68 \\ 0 & \text{True} \end{cases}\!\!\right)\right.$$

$$+ \ 8.01989 \left(\!\!\left(\begin{cases} 547.565 - 24.0189\#1 + 0.349956\#1^2 - 0.00169257\#1^3 & 66.7 \le \#1 < 73.36 \\ 367.437 - 12.717\#1 + 0.146712\#1^2 - 0.000564191\#1^3 & 80.02 \le \#1 \le 86.68 \\ -122.109 + 6.10139\#1 - 0.101622\#1^2 + 0.000564191\#1^3 & 60.04 \le \#1 < 66.7 \\ -788.893 + 30.6345\#1 - 0.395046\#1^2 + 0.00169257\#1^3 & 73.36 \le \#1 < 80.02 \\ 0 & \text{True} \end{cases}\!\!\right)\right. \&$$

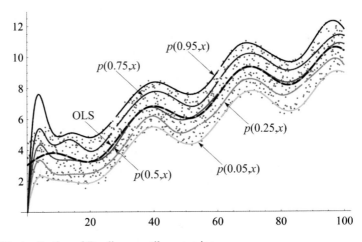

Fig. 13.30 Application of B-spline quantile regression

To compare them with the OLS method, here we employ trigonometric polynomials with a linear tail (Fig. 13.30).

$$y(x) = \sum_{i=5}^{10} \beta_i \sin\left(\frac{x}{i}\right) + \beta_1 x + \beta_0.$$

$\Leftarrow$ y = 3.03703 + 0.0711411x − 3.90944Sin[x/10] + 7.64373Sin[x/9]−
    6.6894Sin[x/8] + 3.14463Sin[x/7] − 1.33602Sin[x/6] + 1.14319Sin[x/5]

Let us visualize the result,

At small x values, $x < 20$, the values of the OLS method are somewhat worse than that of the B-spline approximation.

Let us demonstrate the robustness of the regression quantiles with the data of the example from Sect. 12.4.3, but suppose that for some reason 50% of the data y-values greater than 10.5 are altered by multiplying them with some factor greater than 1, say, $\alpha = 2.5$. Then the altered data looks like Fig. 13.31. This figure even shows us the histogram of the data on the right hand side.

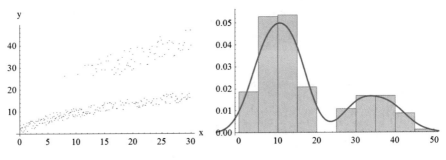

Fig. 13.31 Data generated with outliers

⇒ data = Map[{#, (Sqrt[10 x]) + RandomReal[{−2.6, 1.9}]/.x → #}&,
   Range[0.2, 30, 0.1]];

⇒ α = 2.5;

⇒ dataAlt = Map[If[Randominteger[{0, 1}] == 1 && #[[2]] > 10.5,
   {#[[1]], α#[[2]]}, #]&, data];

The histogram of the output values are indicating outliers. Let us employ the following quantiles,

⇒ qs = {0.05, 0.25, 0.5, 0.75, 0.95};

The basic functions are,

⇒ funcs = {1, x, $\sqrt{x}$, Log[x]};

Let us compute the five (qs) regression quantiles for the altered data:

⇐ $1.29453 \times 10^{-9} + 0.396145\sqrt{x} + 0.225675x + 1.80078\text{Log}[x]$

   $2.16463\sqrt{x} + 0.14238x + 0.204086\text{Log}[x]$

   $0.698279 + 2.49018\sqrt{x} + 0.168249x + 3.3233 * 10^{-9}\text{Log}[x]$

   $1.73343 + 1.45241x$

   $7.82056\sqrt{x} + 0.0819461x + 0.547324\text{Log}[x]$

We also compute the least squares fit of the model consisting of the basic functions (OLS),

⇐ $-16.1728 + 19.0888\sqrt{x} - 0.958974 x - 9.26759\text{ Log}[x]$

Here is a plot of the altered data and all fitted functions, see Fig. 13.32.

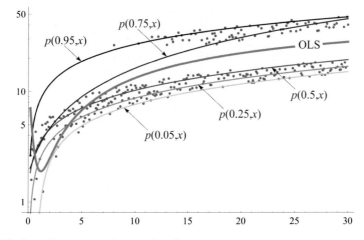

Fig. 13.32 Quantile regression in case of outliers

Figure 13.32 shows that regression lines with quantile $p < 0.5$ can reject outliers, while OLS provides a misleading result.

13.5   Applications

13.5.1   *Separate Outliers in Cloud Points*

LiDAR (Light Detection and Ranging) point clouds contain abundant spatial (3D) information. A dense distribution of scanned points on an object's surface strongly implies surface features. In particular, plane features commonly appear in a typical LiDAR dataset of artificial structures. Plane fitting is the key process for extracting plane features from LiDAR data.

Let's consider a *synthetic dataset* generated from a multivariate Gauss distribution. The 3D points have means of $(2, 8, 6)$ and variances $(5, 5, 0.01)$. The ideal plane is $z = 6$; see Fig. 13.33.

In this case let us consider $N = 20{,}000$ points.

Figure 13.33 shows the data points with the ideal plane. Now, let us generate 3000 outliers randomly, see Fig. 13.34.

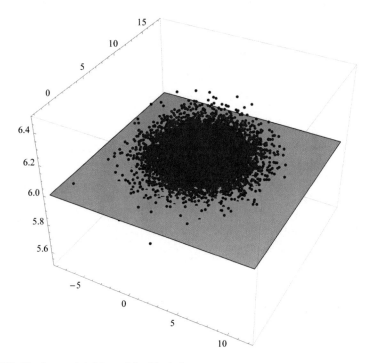

Fig. 13.33 Random point data and the ideal plane

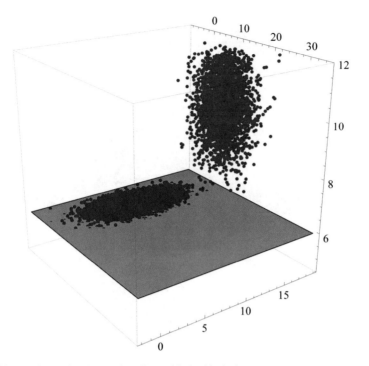

Fig. 13.34 Random point data and outliers with the ideal plane

Let us see the histogram of all the data points, see Fig. 13.35,

⇒ XYZC = Join[XYZ, XYZOutLiers];

Figure 13.35 indicates the anomaly caused by outliers. In order to eliminate the outlier points from the plane fitting process, let us use $p = 0.9$ quantile value (Fig. 13.36).

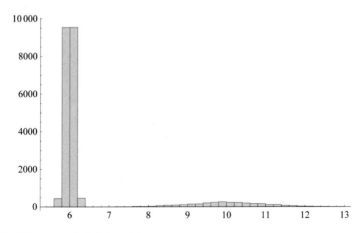

Fig. 13.35 Histogram of all data points

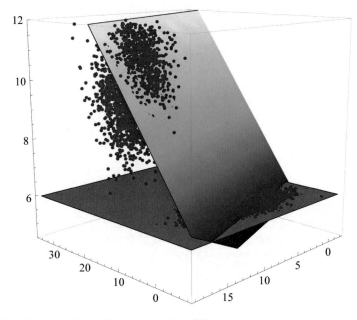

Fig. 13.36 The result of quantile regression ($p = 0.9$)

We get:

$$\Leftarrow \{2887.81, \{\alpha \rightarrow 0.218001, \beta \rightarrow 0.15376, \gamma \rightarrow 5.20335\}\}$$

Here we used the perpendicular distance of points from the plane as the error definition.

Now let us see the result of the case of $p = 0.1$.

$$\Rightarrow \mathtt{p} = 0.1;$$

The objective function to be minimized in order to set the plane parameters is, see Sect. 13.3.3,

$$\Rightarrow \mathtt{G} = \mathtt{Total}\left[\mathtt{MapThread}\left[\mathtt{If}\left[\frac{\#[[3]] - (\alpha\#[[1]] + \beta\#[[1]] + \gamma)}{\sqrt{1 + \alpha^2 + \beta^2}} < 0,\right.\right.\right.$$

$$(1 - \mathtt{p})\,\mathtt{Abs}\left[\frac{\#[[3]] - (\alpha\#[[1]] + \beta\#[[1]] + \gamma)}{\sqrt{1 + \alpha^2 + \beta^2}}\right],$$

$$\left.\left.\mathtt{p}\,\mathtt{Abs}\left[\frac{\#[[3]] - (\alpha\#[[1]] + \beta\#[[2]] + \gamma)}{\sqrt{1 + \alpha^2 + \beta^2}}\right]\right]\right.\&, \{\mathtt{X}, \mathtt{Y}, \mathtt{Z}\}]];$$

$$\Rightarrow \mathtt{sol} = \mathtt{NMinimize}[\mathtt{G}, \{\alpha, \beta, \gamma\}]$$

$$\Leftarrow \{1535.13, \{\alpha \rightarrow 0.0241206, \beta \rightarrow 0.0128114, \gamma \rightarrow 5.71022\}\}$$

Figure 13.37 shows the resulting plane.

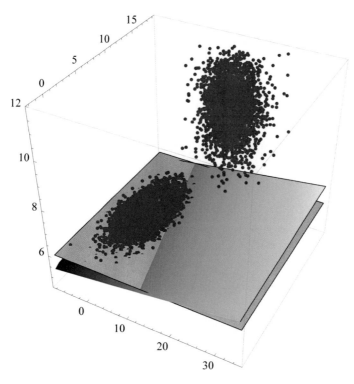

Fig. 13.37 The result of quantile regression ($p = 0.1$)

How can one find the proper quantile value? Let us compute the histogram of this plane fitting using $p = 0.1$; see Fig. 13.38.

We see that the left hand side of the histogram represents the errors of the inliers. By inspection we can determine the maximum error (Abs(Error) < 0.5), or more precisely and automatically one can do it with the *Expectation Maximization* (EM) algorithm. Let us select data points which give smaller error for the model fitted with $p = 0.1$.

$\Rightarrow$ XYZS =

$$\text{Select}\left[\text{XYZC},\text{Abs}\left[\left(\frac{\#[[3]] - \alpha\#[[1]] - \beta\#[[2]] - \gamma}{\sqrt{1 + \alpha^2 + \beta^2}}\right)/\text{sol}[[2]]\right] < 0.5\&\right];$$

In this way, most of the outliers could be eliminated. The number of points remaining is 19979. After minimizing

$$\Rightarrow G = \text{Total}\left[\text{Map}\left[\left(\frac{\#[[3]] - \alpha\#[[1]] - \beta\#[[2]] - \gamma}{\sqrt{1 + \alpha^2 + \beta^2}}\right)^2 \&, \text{XYZS}\right]\right];$$

we get

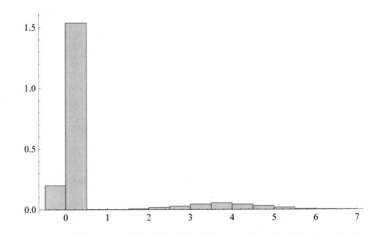

Fig. 13.38 The error histogram of the plane fitted by quantile regression ($p = 0.1$)

$\Leftarrow \{198.172, \{\alpha \to 0.000232049, \beta \to 0.0000462067, \gamma \to 5.99977\}\}$
See Fig. 13.39.

Now, let us introduce another example with *real LiDAR measurement*. We have data points of the downhill covered by bushes as vegetation. We are going to find the slope of the hill by fitting a plane to the data points. In this context the bushes represent outliers, see Fig. 13.40.

The number of the points is 41,392. But we remove some corrupted data and have 33,292 left.

Let us see the histogram of the data in Fig. 13.41.

Employing quantile regression, we can successfully find the proper plane; see Fig. 13.42. Let

$\Rightarrow \mathtt{p} = 0.05;$

$$\Rightarrow \mathtt{G} = \mathtt{Total}\left[\mathtt{MapThread}\left[\mathtt{If}\left[\frac{\#[[3]] - (\alpha\#[[1]] + \beta\#[[2]] + \gamma)}{\sqrt{1 + \alpha^2 + \beta^2}} < 0,\right.\right.\right.$$

$$(1 - \mathtt{p})\,\mathtt{Abs}\left[\frac{\#[[3]] - (\alpha\#[[1]] + \beta\#[[2]] + \gamma)}{\sqrt{1 + \alpha^2 + \beta^2}}\right],$$

$$\left.\left.\left[\frac{\#[[3]] - (\alpha\#[[1]] + \beta\#[[2]] + \gamma)}{\sqrt{1 + \alpha^2 + \beta^2}}\right]\right]\&, \{\mathtt{X}, \mathtt{Y}, \mathtt{Z}\}\right]\right];$$

The parameters of the fitted plane are

$\Leftarrow \{1652.71, \{\alpha \to 0.107359, \beta \to 0.508726, \gamma \to 202.536\}\}$
The fitting is quite successful, see Fig. 13.42.

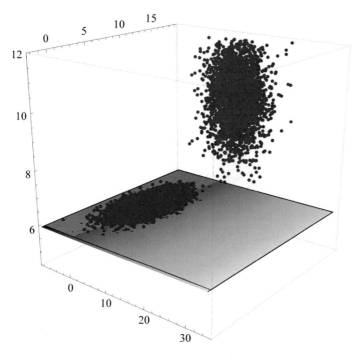

Fig. 13.39 The corrected plane via separating in- and outliers

Fig. 13.40 Original LiDAR data

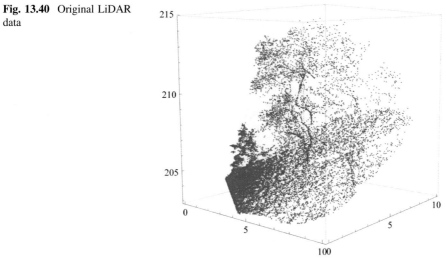

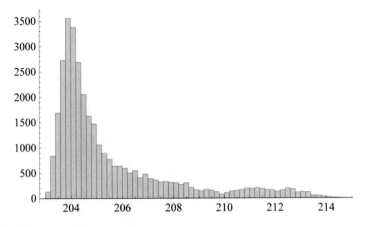

Fig. 13.41 Histogram of the output data

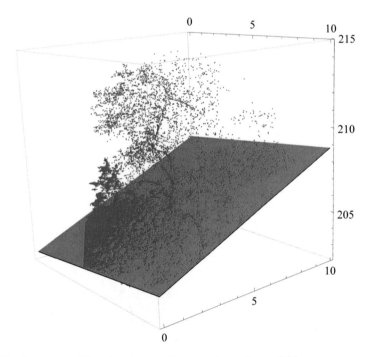

Fig. 13.42 Estimation of the plane by quantile regression with $p = 0.05$

13.5.2   Modelling Time-Series

Let us consider an application of the quantile regression to time series. Here we employ the daily temperature data of Melbourne from 1 Jan 2010 up to 21 Dec 2014, see Fig. 13.43.

First we are looking for the conditional (at a certain day) cumulative density function of the daily temperature. For example we should like to estimate the probability that temperature is less than a given value, or between two given values. Let us consider the following quantile values:

$\Rightarrow$ qs $= \{0.05, 0.25, 0.5, 0.75, 0.95\}$;

Let us carry out the quantile regression with B-splines having 20 knots.
Plot regression quantiles and time series data; see Fig. 13.44.
Let us select a day, $i = 123$

$\Rightarrow$ x0 $= 123$;

$\Rightarrow$ yi $=$ tempq$[[x0]]$;

Then using the five different quantile values, we get the p1-temperature diagram, which is the discrete points of the CDF of the temperature as condition ($i = 123$) random variable; see Fig. 13.45.

Let us employ linear interpolation for this discrete CDF,

$\Rightarrow$ tempq$[[123]]$

$\Leftarrow \{6.22348, 11.6218, 15.6562, 16.1488, 20.0468\}$

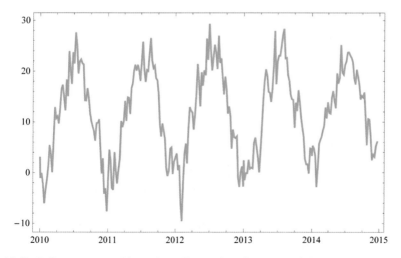

Fig. 13.43 Daily temperature history in Melbourne in a five year period

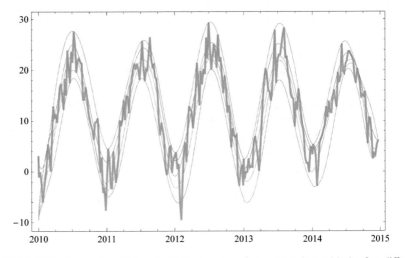

Fig. 13.44 Daily temperature history in Melbourne in a five year period, with the five different quantile B-splines

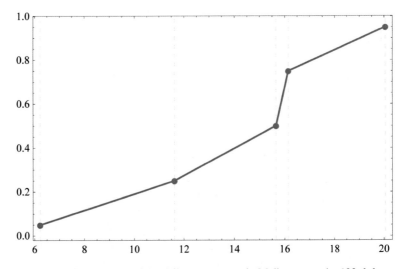

Fig. 13.45 The estimated CDF of the daily temperature in Melbourne at the 123rd day

Then for example the average temperature can be computed as

$\Rightarrow$ NSolve[qCDFInt[x] == 0.5, x]//Quiet

$\Leftarrow$ {{x → 15.6562}}

The probability that the temperature will be less than temperature is 6.22 is

$\Rightarrow$ tempq$[[123,1]]$

$\Leftarrow$ 6.22348

or

$\Rightarrow$ qCDFInt$[\%]$

$\Leftarrow$ 0.05

The probability that the temperature will be between

$\Rightarrow$ tempq$[[123,1]]$

$\Leftarrow$ 6.22348

and

$\Rightarrow$ tempq$[[123,2]]$

$\Leftarrow$ 11.6218

is

$\Rightarrow$ qCDFInt$[\%]$ $-$ qCDFInt$[\%\%]$

$\Leftarrow$ 0.2

and so on. We can visualize the PDF too, via differentiating CDF; see Fig. 13.46.

Secondly, we consider a more difficult forecasting problem. Assume that we have temperature data for a given location. We want to predict today's temperature at that location using yesterday's temperature. More generally, the problem discussed in this section can be stated as "How to estimate the conditional density of the predicted variable given a value of the conditioning covariate?".

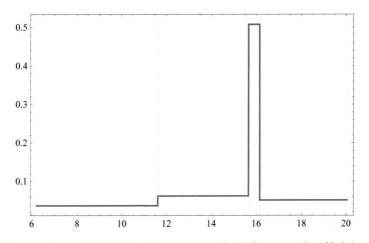

Fig. 13.46 The estimated PDF of the daily temperature in Melbourne at the 123rd day

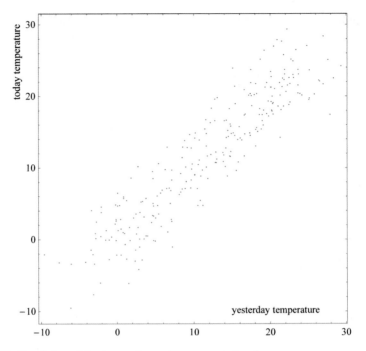

Fig. 13.47 Correlation of the temperatures of two consecutive days

One way to answer this question is to provide a family of regression quantiles and from them to estimate the Cumulative Distribution Function (CDF) for a given value of the covariate. In other words, given data pairs $\{x_i, y_i\}_{i=1}^{n}$ and a value x_0 we find $CDF_{x_0}(y)$. From the estimated CDF we can estimate the Probability Density Function (PDF). We can go further and use a Monte Carlo type of simulation with the obtained PDF's (this will be discussed elsewhere).

Using the time series data let us make pairs of yesterday and today data. The resulting pairs are in Fig. 13.47.

Let us use quantile regression with the following quantile values, qsi

```
⇒ qs = Join[{0.02}, FindDivisions[{0, 1}, 10][[2; ; − 2]], {0.98}]//N
⇐ {0.02, 0.1, 0.2, 0.3, 0.4, 0.5, 0.6, 0.7, 0.8, 0.9, 0.98}
```

We apply $n = 11$ B-spline with 5 knots (Fig. 13.48).

Given yesterday's temperature value, let us say $t_0 = 5$ °C, we can estimate $CDF_{t_0}(t)$ using the values of the quantile at t_0 of the corresponding regression quantile functions.

```
⇒ t0 = 5;
⇒ xs = Through[qFuncs[t0]];
```

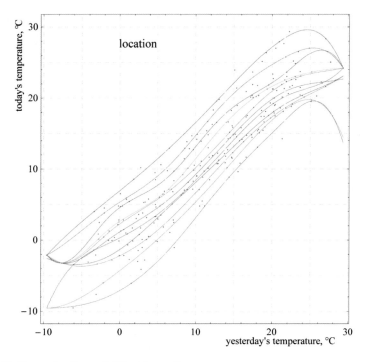

Fig. 13.48 Regression lines of different quantiles

We can get a first order (linear) approximation by simply connecting the consecutive points of $\left\{x_i^{t_0}, q_i\right\}_{i=1}^{|qs|}$ as it is shown on the following plot, Fig. 13.49.

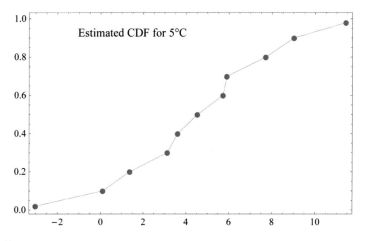

Fig. 13.49 Constructed conditional CDF of the today temperature based on the yesterday temperature of 5 °C

Since we want to also derive $PDF_{t_0}(t)$ from the estimated $CDF_{t_0}(t)$, make plots, and do other manipulations, it is better to make and use an interpolation object for $CDF_{t_0}(t)$.

$\Rightarrow$ qCDFInt = Interpolation[cdfPairs, InterpolationOrder → 1];

Here is a plot with both $CDF_{t_0}(t)$ and the $PDF_{t_0}(t)$ derived from it; see Fig. 13.50.

On the plot the dashed vertical grid lines are for the quantiles {0.05, 0.25, 0.5, 0.75, 0.95}; the solid vertical gray grid line is for t_0 (Fig. 13.51).

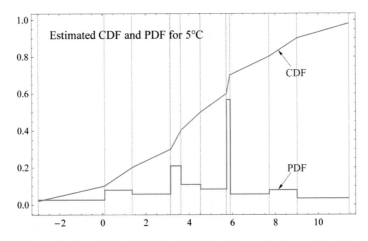

Fig. 13.50 Constructed conditional PDF and CDF of the today temperature based on the yesterday temperature of 5 °C

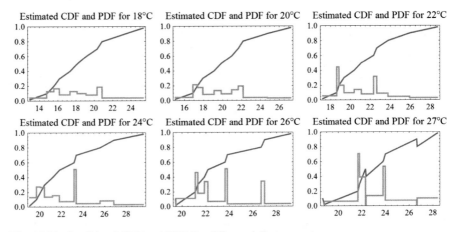

Fig. 13.51 Conditional CDF and PDF for different daily temperatures

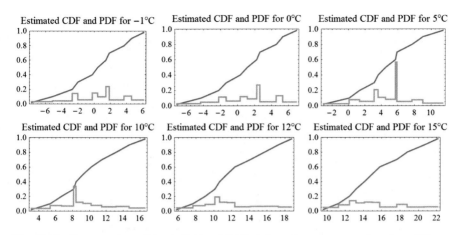

Fig. 13.52 Constructed conditional CDF and PDF of the today temperature based on different yesterday temperatures

Using this technique let us plot the estimated CDF's and PDF's for a collection of temperatures. Note that the quantiles are given with vertical dashed grid lines; the median is colored with green. The values of the conditioning covariate are given on the plot labels (Fig. 13.52).

13.6 Exercise

13.6.1 Regression of Implicit-Functions

Let us consider the following point cloud; see Fig. 13.53.

```
⇒ npoints = 12000
  data = {RandomReal[{0, 2π}, npoints]
  RandomVariate[SkewNormalDistribution[0.6, 0.3, 4], npoints]};
  data = MapThread[#2 * {Cos[#1], Sin[#1]}&, data];
  rmat = RotationMatrix[−π/2 · 5].DiagonalMatrix[{2, 1}];
  data = Transpose[rmat.Transpose[data]];
  data = TranslationTransform[{−Norm[Abs[#]/3], 0}][#]&/ddata;
  sndData = Standardize[data];
⇒ hup1 = ListPlot[sndData, PlotRange → All, AspectRatio → 1,
  PlotTheme → Detailed, GridLines → Map[Quantile[#,
  Range[0, 1, 1/19]]&, Transpose[sndData]], ImageSize → 300]
```

This cloud of points cannot be represented by a single explicit function. Let us try to carry out quantile regression with B-splines having 15 knots. The quantile values are:

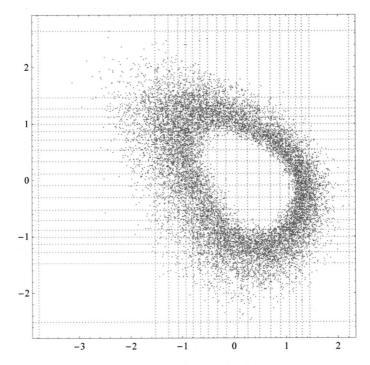

Fig. 13.53 Point cloud may be represented by implicit function

$\Rightarrow$ qs = Range[0.05, 0.95, 0.1]

$\Leftarrow$ {0.05, 0.15, 0.25, 0.35, 0.45, 0.55, 0.65, 0.75, 0.85, 0.95}

$\Rightarrow$ AbsoluteTiming[

  qFuncs = QuantileRegression[sndData, 15, qs,

  Method $\rightarrow$ {LinearProgramming, Method $\rightarrow$ CLP}];]

Let us visualize the results (Fig. 13.54).

$\Rightarrow$ SS = Through[qFuncs[x]];

$\Rightarrow$ hup2 = Plot[Evaluate[SS], {x, $-2.1, 2$}, PlotPoints $\rightarrow$ 130,

  ImageSize $\rightarrow$ 300, AspectRatio $\rightarrow$ 1];

$\Rightarrow$ Show[{hup2, hup1}]

Directional quantile envelope technique

However we can employ a so called *directional quantile envelope* technique, see Antonov (2013). The idea of directional quantile envelopes is conceptually simple and straightforward to derive. The calculation is also relatively simple: over a set of uniformly distributed directions we find the lines that separate the data according to

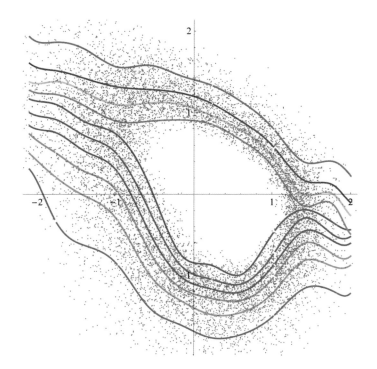

Fig. 13.54 Point cloud represented by explicit functions (B-splines)

a quantile parameter q, $0 < q < 1$, and with those lines we approximate the enveloping curve for data that corresponds to q. The quantile regression is carried out from different angles using rotation of the data; see Fig. 13.55.

Now let us employ our function. The quantiles are

$\Rightarrow$ qs $= \{0.7, 0.75, 0.8, 0.85, 0.90, 0.95, 0.98, .99\}$;

One can use the quantile regression function with the option, see Antonov (2013)

$\Rightarrow$ Options[QuantileEnvelope}

$\Leftarrow$ {Tangents $\rightarrow$ True}

$\Rightarrow$?QuantileEnvelope

```
QuantileRegression[data_?MatrixQ,qs:(_?NumberQ|{_?NumberQ..}),n_Integer]
    experimental implementation of quantile envelopes points finding.
```

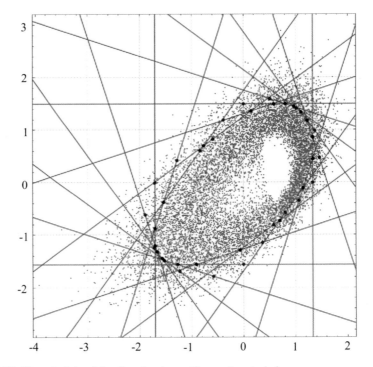

Fig. 13.55 The principle of the directional quantile envelope technique

Let us visualize the results in Fig. 13.56.

⇒ Dimensions[qsPoints]
⇐ {8, 60, 2}

⇒ Block[{data = sndData},
  Show[{ListPlot[data, AspectRatio → 1, PlotTheme{Detailed},
  GridLines → Map[Quantile[#, Range[0, 1, 1/(20 − 1)]]&,
  Transpose[data]], PlotLegends → SwatchLegend[
  Blend[{Red, Orange}, Rescale[#1, {Min[qs], Max[qs]},
  {0, 1}]]&/@qs, qs]], Graphics[{PointSize[0.005],
  Thickness[0.0035], MapThread[{Blend[{Red, Orange}, Rescale[#1,
  {Min[qs], Max[qs]}, {0, 1}]], Tooltip[Line[Append[#2, #2[[1]]]], #1],
  Point[#2]}&, {qs, qsPoints}]}]}, ImageSize → 300]]

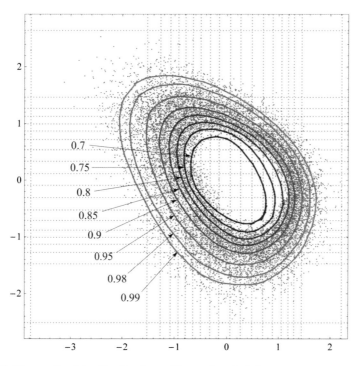

Fig. 13.56 The results of the directional quantile envelope technique

References

Antonov A (2013) Quantile regression Mathematica package, source code at GitHub, https://github.com/antononcube/MathematicaForPrediction, package QuantileRegression.m,

Davino C, Furno M, Vistocco D (2014) Quantile regression—theory and applications. Wiley Series in Probability and Statistics

Hald A (2007) A history of parametric statistical inference from Bernoulli to Fisher 1713–1935. Springer Verlag, Berlin, Heisenberg, New York

Koenker R (2005) Quantile regression. Econometric Society Monograph, Cambridge University Press, Cambridge. http://www2.sas.com/proceedings/sugi30/213-30.pdf

Logan J, Petscher Y (2013) An introduction to quantile regression, presentation for modern modeling methods conference: 5-23-2013. http://www.modeling.uconn.edu/m3c/assets/File/Logan_Quantile%20regression.pdf

Chapter 14
Robust Regression

14.1 Basic Methods in Robust Regression

14.1.1 Concept of Robust Regression

In many fields such as robotics (Poppinga et al. 2006), computer vision (Mitra and Nguyen 2003), digital photogrammetry (Yang and Förtsner 2010), surface reconstruction (Nurunnabi, Belton and West 2012), computational geometry (Lukács et al. 1998) as well as in the increasing applications of laser scanning (Stathas et al 2003), it is a fundamental task for extracting features from 3D point cloud. Since the physical limitations of the sensors, the occlusions, multiple reflectance and noise can produce off-surface points, robust fitting techniques are required. *Robust regression (fitting)* means an estimation technique which is able to estimate accurate model parameters not only despite small-scale noise in the data set but occasionally large scale measurement errors (outliers). *Outliers definition* is not easy. Perhaps considering the problem from the practical point of the view, we can say that data points, which appearance in the data set causes dramatically change in the result of the parameter estimation can be labeled as outliers.

Basically there are two different methods to handle outliers:

(a) weighting out outliers
(b) discarding outliers.

Weighting outliers means that we do not kick out certain data points labeled as outlier but during the parameter estimation process we take them with a low weight into consideration in the objective function. Such a technique is the good old *Danish method*.

Supplementary Information The online version contains supplementary material available at https://doi.org/10.1007/978-3-030-92495-9_14.

J. L. Awange et al., *Mathematical Geosciences*,
https://doi.org/10.1007/978-3-030-92495-9_14

The other technique will try to identify data points, which make "troubles" during the parameter estimation process. Troubles mean that their existence in the data set change the result of the parameter estimation considerably. One of the representative of this technique is the RANdom SAmple Consensus (RANSAC) method.

Both these techniques eliminate outliers in a way. However there are softer methods used in case of presence of outliers, too.

The simplest methods of estimating parameters in a regression model that are less sensitive to outliers than the least squares estimates, is to use least absolute deviations. Even then, gross outliers can still have a considerable impact on the model, motivating research into even more robust approaches.

In 1973, Huber introduced M-estimation for regression. The M in M-estimation stands for "maximum likelihood type". The method is robust to outliers in the response variables, but turned out not to be resistant to outliers in the explanatory variables (leverage points). In fact, when there are outliers in the explanatory variables, the method has no advantage over least squares. This handicap maybe eliminated employing total least square technique.

First let us see the application of the maximum likelihood estimation technique to fitting cloud data to a plane.

Remark This chapter contains many lines of codes, therefore please do consult with the electronic supplement.

14.1.2   Maximum Likelihood Method

14.1.2.1   Symbolic Solution

Let a point be P_i with coordinates $\{x_i, y_i, z_i\}$ and d_i is its distance from a general point $\{x, y, z\}$ of the plane $z = \alpha x + \beta y + \gamma$, then

$$\Rightarrow d_i = (z_i - \alpha x - \beta y - \gamma)^2 + (x_i - x)^2 + (y_i - y)^2$$

The distance of P_i from its perpendicular projection to the plane is the shortest distance, which represents its regression error ε_{MI}. From the necessary conditions we get

$$\Rightarrow \text{deq1} = D[d_i, x]//\text{Expand}$$
$$\Leftarrow 2x + 2x\alpha^2 + 2y\alpha\beta + 2\alpha\gamma - 2x_i - 2\alpha z_i$$

$$\Rightarrow \text{deq2} = D[d_i, y]//\text{Expand}$$
$$\Leftarrow 2y + 2x\alpha\beta + 2y\beta^2 + 2\beta\gamma - 2y_i - 2\beta z_i$$

$\Rightarrow \text{xy} = \text{Solve}[\{\text{deq1} == 0, \text{deq2} == 0\}, \{x, y\}]//\text{Flatten}$

$$\Leftarrow \left\{x \rightarrow -\frac{\alpha\gamma - x_i - \beta^2 x_i + \alpha\beta y_i - \alpha z_i}{1 + \alpha^2 + \beta^2}, y \rightarrow -\frac{\beta\gamma + \alpha\beta x_i - y_i - \alpha^2 y_i - \beta z_i}{1 + \alpha^2 + \beta^2}\right\}$$

Then the square of the error is

$\Rightarrow \Delta_i = d_i /.\text{xy}//\text{Simplify}$

$$\Leftarrow \frac{(\gamma + \alpha x_i + \beta y_i - z_i)^2}{1 + \alpha^2 + \beta^2}$$

Consequently our error model is

$$\Leftarrow \varepsilon = \frac{z_i - x_i \alpha - y_i \beta - \gamma}{\sqrt{1 + \alpha^2 + \beta^2}} ;$$

Now let us develop the likelihood with this error model and assuming that the model error has a Gaussian distribution with zero mean.

Then the Maximum Likelihood (ML) estimator for Gaussian-type noise is

$$\Rightarrow \text{L} = \prod_{i=1}^{N} \text{PDF}[\text{NormalDistribution}[0,\sigma], e_{M\,i}]$$

$$\Rightarrow \prod_{i=1}^{N} \frac{e^{-\frac{(e_M)_i^2}{2\sigma^2}}}{\sqrt{2\pi}\sigma}$$

Then the likelihood function—considering its logarithm—should be maximized

$$\Rightarrow \text{LogL} = \log[\text{L}/.e_{M\,i} \rightarrow w_i \varepsilon]//\text{Simplify}$$

$$\Leftarrow -\frac{1}{2}N\log[2] - \frac{1}{2}N\log[\pi] - N\log[\sigma] + \sum_{i=1}^{N} -\frac{\gamma^2 w_i^2}{2(1 + \alpha^2 + \beta^2)\sigma^2}$$

$$+ \sum_{i=1}^{N} -\frac{\alpha\gamma w_i^2 x_i}{(1 + \alpha^2 + \beta^2)\sigma^2} + \sum_{i=1}^{N} -\frac{\alpha^2 w_i^2 x_i^2}{2(1 + \alpha^2 + \beta^2)\sigma^2} + \sum_{i=1}^{N} -\frac{\beta\gamma w_i^2 y_i}{(1 + \alpha^2 + \beta^2)\sigma^2}$$

$$+ \sum_{i=1}^{N} -\frac{\alpha\beta w_i^2 x_i y_i}{(1 + \alpha^2 + \beta^2)\sigma^2} + \sum_{i=1}^{N} -\frac{\beta^2 w_i^2 y_i^2}{2(1 + \alpha^2 + \beta^2)\sigma^2} + \sum_{i=1}^{N} \frac{\gamma w_i^2 z_i}{(1 + \alpha^2 + \beta^2)\sigma^2}$$

$$+ \sum_{i=1}^{N} \frac{\alpha w_i^2 x_i z_i}{(1 + \alpha^2 + \beta^2)\sigma^2} + \sum_{i=1}^{N} \frac{\beta w_i^2 y_i z_i}{(1 + \alpha^2 + \beta^2)\sigma^2} + \sum_{i=1}^{N} -\frac{w_i^2 z_i^2}{2(1 + \alpha^2 + \beta^2)\sigma^2}$$

Here the weighted error with weights w_i is used. In order to avoid global maximization we transform the problem of the global solution of a multivariate polynomial system. From the necessary conditions one can obtain

$\Rightarrow$ eq1 = D[LogL,α]

$$\Leftarrow \sum_{i=1}^{N} \frac{\alpha \gamma^2 w_i^2}{\left(1+\alpha^2+\beta^2\right)^2 \sigma^2} + \sum_{i=1}^{N}\left(\frac{2\alpha^2 \gamma w_i^2 x_i}{\left(1+\alpha^2+\beta^2\right)^2 \sigma^2} - \frac{\gamma w_i^2 x_i}{\left(1+\alpha^2+\beta^2\right)\sigma^2}\right)$$

$$+ \sum_{i=1}^{N}\left(\frac{\alpha^3 w_i^2 x_i^2}{\left(1+\alpha^2+\beta^2\right)^2 \sigma^2} - \frac{\alpha w_i^2 x_i^2}{\left(1+\alpha^2+\beta^2\right)\sigma^2}\right) + \sum_{i=1}^{N} \frac{2\alpha\beta\gamma w_i^2 y_i}{\left(1+\alpha^2+\beta^2\right)^2 \sigma^2}$$

$$+ \sum_{i=1}^{N} \frac{\alpha\beta^2 w_i^2 y_i^2}{\left(1+\alpha^2+\beta^2\right)^2 \sigma^2} + \sum_{i=1}^{N}\left(\frac{2\alpha^2\beta w_i^2 x_i y_i}{\left(1+\alpha^2+\beta^2\right)^2 \sigma^2} - \frac{\beta w_i^2 x_i y_i}{\left(1+\alpha^2+\beta^2\right)\sigma^2}\right)$$

$$+ \sum_{i=1}^{N} -\frac{2\alpha\gamma w_i^2 z_i}{\left(1+\alpha^2+\beta^2\right)^2 \sigma^2} + \sum_{i=1}^{N} -\frac{2\alpha\beta w_i^2 y_i z_i}{\left(1+\alpha^2+\beta^2\right)^2 \sigma^2}$$

$$+ \sum_{i=1}^{N} \frac{\alpha w_i^2 z_i^2}{\left(1+\alpha^2+\beta^2\right)^2 \sigma^2} + \sum_{i=1}^{N}\left(-\frac{2\alpha^2 w_i^2 x_i z_i}{\left(1+\alpha^2+\beta^2\right)^2 \sigma^2} + \frac{w_i^2 x_i z_i}{\left(1+\alpha^2+\beta^2\right)\sigma^2}\right)$$

$\Rightarrow$ eq2 = D[LogL,β]

$$\Rightarrow \sum_{i=1}^{N} \frac{\beta \gamma^2 w_i^2}{\left(1+\alpha^2+\beta^2\right)^2 \sigma^2} + \sum_{i=1}^{N} \frac{2\alpha\beta\gamma w_i^2 x_i}{\left(1+\alpha^2+\beta^2\right)^2 \sigma^2}$$

$$+ \sum_{i=1}^{N} \frac{\alpha^2 \beta w_i^2 x_i^2}{\left(1+\alpha^2+\beta^2\right)^2 \sigma^2} + \sum_{i=1}^{N}\left(\frac{2\beta^2 \gamma w_i^2 y_i}{\left(1+\alpha^2+\beta^2\right)^2 \sigma^2} - \frac{\gamma w_i^2 y_i}{\left(1+\alpha^2+\beta^2\right)\sigma^2}\right)$$

$$+ \sum_{i=1}^{N}\left(\frac{2\alpha\beta^2 w_i^2 x_i y_i}{\left(1+\alpha^2+\beta^2\right)^2 \sigma^2} - \frac{\alpha w_i^2 x_i y_i}{\left(1+\alpha^2+\beta^2\right)\sigma^2}\right)$$

$$+ \sum_{i=1}^{N}\left(\frac{\beta^3 w_i^2 y_i^2}{\left(1+\alpha^2+\beta^2\right)^2 \sigma^2} - \frac{\beta w_i^2 y_i^2}{\left(1+\alpha^2+\beta^2\right)\sigma^2}\right)$$

$$+ \sum_{i=1}^{N} -\frac{2\beta\gamma w_i^2 z_i}{\left(1+\alpha^2+\beta^2\right)^2 \sigma^2} + \sum_{i=1}^{N} -\frac{2\alpha\beta w_i^2 x_i z_i}{\left(1+\alpha^2+\beta^2\right)^2 \sigma^2}$$

$$+ \sum_{i=1}^{N} \frac{\beta w_i^2 z_i^2}{\left(1+\alpha^2+\beta^2\right)^2 \sigma^2} + \sum_{i=1}^{N}\left(-\frac{2\beta^2 w_i^2 y_i z_i}{\left(1+\alpha^2+\beta^2\right)^2 \sigma^2} + \frac{w_i^2 y_i z_i}{\left(1+\alpha^2+\beta^2\right)\sigma^2}\right)$$

$\Rightarrow$ eq3 = D[LogL,γ]

$$\Leftarrow \sum_{i=1}^{N} -\frac{\gamma w_i^2}{\left(1+\alpha^2+\beta^2\right)\sigma^2} + \sum_{i=1}^{N} -\frac{\alpha w_i^2 x_i}{\left(1+\alpha^2+\beta^2\right)\sigma^2}$$

$$+ \sum_{i=1}^{N} -\frac{\beta w_i^2 y_i}{\left(1+\alpha^2+\beta^2\right)\sigma^2} + \sum_{i=1}^{N} \frac{w_i^2 z_i}{\left(1+\alpha^2+\beta^2\right)\sigma^2}$$

We get a multivariate algebraic system for the unknown parameters α, β and γ. In compact form

$$\Rightarrow \text{eq1} = j\alpha\gamma^2 + 2\alpha^2\gamma a - \gamma a(1+\alpha^2+\beta^2) + \alpha^3 b - \alpha b(1+\alpha^2+\beta^2)$$
$$+ 2\alpha\beta\gamma c + \alpha\beta^2 d + 2\alpha^2\beta e - \beta e(1+\alpha^2+\beta^2) - 2\alpha\gamma f - 2\alpha\beta g + \alpha h$$
$$- 2\alpha^2 i + i(1+\alpha^2+\beta^2)//\text{Expand}$$
$$\Leftarrow i - b\alpha + h\alpha - i\alpha^2 - e\beta - 2g\alpha\beta + e\alpha^2\beta + i\beta^2 - b\alpha\beta^2$$
$$+ d\alpha\beta^2 - e\beta^3 - a\gamma - 2f\alpha\gamma + a\alpha^2\gamma + 2c\alpha\beta\gamma - a\beta^2\gamma + j\alpha\gamma^2$$

$$\Rightarrow \text{eq2} = j\beta\gamma^2 + 2\alpha\beta\gamma a + \alpha^2\beta b + 2\beta^2\gamma c - \gamma c(1+\alpha^2+\beta^2) + 2\alpha\beta^2 e$$
$$- \alpha e(1+\alpha^2+\beta^2) + \beta^3 d - \beta d(1+\alpha^2+\beta^2) - 2\beta\gamma f - 2\alpha\beta i + \beta h$$
$$- 2\beta^2 g + g(1+\alpha^2+\beta^2)//\text{Expand}$$
$$\Leftarrow g - e\alpha + g\alpha^2 - e\alpha^3 - d\beta + h\beta - 2i\alpha\beta + b\alpha^2\beta - d\alpha^2\beta$$
$$- g\beta^2 + e\alpha\beta^2 - c\gamma - c\alpha^2\gamma - 2f\beta\gamma + 2a\alpha\beta\gamma + c\beta^2\gamma + j\beta\gamma^2$$

$$\Rightarrow \text{eq3} = f - j\gamma - \alpha a - \beta c$$
$$\Leftarrow f - a\alpha - c\beta - j\gamma$$

where the constants are

$$\Leftarrow a = \sum_{i=1}^{N} w_i^2 x_i, \quad b = \sum_{i=1}^{N} w_i^2 x_i^2, \quad c = \sum_{i=1}^{N} w_i^2 y_i, \quad d = \sum_{i=1}^{N} w_i^2 y_i^2, \quad e = \sum_{i=1}^{N} w_i^2 x_i y_i,$$

$$f = \sum_{i=1}^{N} w_i^2 z_i, \quad g = \sum_{i=1}^{N} w_i^2 y_i z_i, \quad h = \sum_{i=1}^{N} w_i^2 z_i^2, \quad i = \sum_{i=1}^{N} w_i^2 x_i z_i, \quad j = \sum_{i=1}^{N} w_i^2$$

Similarly the compact form of objective function, which will be required later

$$\Leftarrow L = -\frac{j\gamma^2}{2(1+\alpha^2+\beta^2)\sigma^2} - \frac{1}{2}N\log[2] - \frac{1}{2}N\log[\pi] - N\log[\sigma]$$
$$- a\frac{\alpha\gamma}{2(1+\alpha^2+\beta^2)\sigma^2} - \frac{\alpha^2 b}{2(1+\alpha^2+\beta^2)\sigma^2} - \frac{\beta\gamma c}{(1+\alpha^2+\beta^2)\sigma^2}$$
$$- \frac{\alpha\beta e}{(1+\alpha^2+\beta^2)\sigma^2} - \frac{\beta^2 d}{2(1+\alpha^2+\beta^2)\sigma^2} + \frac{\gamma f}{(1+\alpha^2+\beta^2)\sigma^2}$$
$$+ \frac{\alpha i}{(1+\alpha^2+\beta^2)\sigma^2} + \frac{\beta g}{(1+\alpha^2+\beta^2)\sigma^2} - \frac{h}{2(1+\alpha^2+\beta^2)\sigma^2}$$

$$\Leftarrow -\frac{h}{2\left(1+\alpha^2+\beta^2\right)\sigma^2} + \frac{i\alpha}{\left(1+\alpha^2+\beta^2\right)\sigma^2} - \frac{b\alpha^2}{2\left(1+\alpha^2+\beta^2\right)\sigma^2}$$

$$+ \frac{g\beta}{\left(1+\alpha^2+\beta^2\right)\sigma^2} - \frac{e\alpha\beta}{\left(1+\alpha^2+\beta^2\right)\sigma^2} - \frac{d\beta^2}{2\left(1+\alpha^2+\beta^2\right)\sigma^2}$$

$$+ \frac{f\gamma}{\left(1+\alpha^2+\beta^2\right)\sigma^2} - \frac{a\alpha\gamma}{\left(1+\alpha^2+\beta^2\right)\sigma^2} - \frac{c\beta\gamma}{\left(1+\alpha^2+\beta^2\right)\sigma^2}$$

$$- \frac{j\gamma^2}{2\left(1+\alpha^2+\beta^2\right)\sigma^2} - \frac{1}{2}N\log[2] - \frac{1}{2}N\log[\pi] - N\log[\sigma]$$

This technique is similar to finding regression parameters of a straight line, see Paláncz (2016).

This polynomial system can be solved in symbolic way via Gröbner basis and Sylvester resultant. The polynomial system can be reduced to monomials of higher order. First let us eliminate γ via Gröbner basis, we get

$\Leftarrow$ gb$\alpha\beta$= GroebnerBasis[{eq1, eq2, eq3}, {α, β, γ}, {γ},

   MonomialOrder $\to$ EliminationOrder]

$\Leftarrow$ {$-af\alpha + ij\alpha + a^2\alpha^2 - f^2\alpha^2 - bj\alpha^2 + hj\alpha^2 + af\alpha^3 - ij\alpha^3 - cf\beta + gj\beta + 2ac\alpha\beta - 2ej\alpha\beta$
$+ cf\alpha^2\beta - gj\alpha^2\beta + c^2\beta^2 - f^2\beta^2 - dj\beta^2 + hj\beta^2 + af\alpha\beta^2 - ij\alpha\beta^2 + cf\beta^3 - gj\beta^3$,

$-af + ij + a^2\alpha - f^2\alpha - bj\alpha + hj\alpha + af\alpha^2 - ij\alpha^2 + ac\beta - ej\beta + 2cf\alpha\beta - 2gj\alpha\beta$
$-ac\alpha^2\beta + ej\alpha^2\beta - af\beta^2 + ij\beta^2 + a^2\alpha\beta^2 - c^2\alpha\beta^2 - bj\alpha\beta^2 + dj\alpha\beta^2 + ac\beta^3 - ej\beta^3$,

$cf - gj - ac\alpha + ej\alpha + cf\alpha^2 - gj\alpha^2 - ac\alpha^3 + ej\alpha^3 - c^2\beta + f^2\beta + dj\beta - hj\beta - 2af\alpha\beta$
$+ 2ij\alpha\beta + a^2\alpha^2\beta - c^2\alpha^2\beta - bj\alpha^2\beta + dj\alpha^2\beta - cf\beta^2 + gj\beta^2 + ac\alpha\beta^2 - ej\alpha\beta^2$,

$-ag + ci - bc\alpha + ae\alpha - fg\alpha + ch\alpha + ef\alpha^2 - ag\alpha^2 - bc\alpha^3 + ae\alpha^3 - fg\alpha^3 + ch\alpha^3 + ef\alpha^4 - ci\alpha^4$
$+ ad\beta - ce\beta - ah\beta + fi\beta - bf\alpha\beta + df\alpha\beta - cg\alpha\beta + ai\alpha\beta + ad\alpha^2\beta - ce\alpha^2\beta - ah\alpha^2\beta + fi\alpha^2\beta$

$-bf\alpha^3\beta + df\alpha^3\beta - cg\alpha^3\beta + ai\alpha^3\beta - ef\beta^2 + ci\beta^2 - bc\alpha\beta^2 + ae\alpha\beta^2 - fg\alpha\beta^2 + ch\alpha\beta^2 + ag\alpha^2\beta^2$
$-ci\alpha^2\beta^2 + ad\beta^3 - ce\beta^3 - ah\beta^3 + fi\beta^3 - bf\alpha\beta^3 + df\alpha\beta^3 - cg\alpha\beta^3 + ai\alpha\beta^3 - ef\beta^4 + ag\beta^4$,

$-afg + 2cfi - gij - bcf\alpha + 2aef\alpha - f^2g\alpha + cfh\alpha - aci\alpha - a^2e\alpha + 2ef^2\alpha^2 - afg\alpha^2 + cfi\alpha^2$
$+ bej\alpha^2 - ehj\alpha^2 - gij\alpha^2 - bcf\alpha^3 - f^2g\alpha^3 + cfh\alpha^3 - aci\alpha^3 + 2eij\alpha^3 + ef^2\alpha^4 - cfi\alpha^4$

$+ adf\beta - afh\beta - c^2i\beta + 2f^2i\beta - egj\beta + dij\beta - hij\beta - 2ace\alpha\beta - bf^2\alpha\beta + df^2\alpha\beta - cfg\alpha\beta$
$+ 2e^2j\alpha\beta + i^2j\alpha\beta + adf^2\beta - 2cef^2\beta - afh\alpha^2\beta - c^2i\alpha^2\beta + 2f^2i\alpha^2\beta + egj\alpha^2\beta + dij\alpha^2\beta - hij\alpha^2\beta$
$-bf^2\alpha^3\beta + df^2\alpha^3\beta - cfg\alpha^3\beta + i^2j\alpha^3\beta - c^2e\beta^2 + cfi\beta^2 + dej\beta^2 - ehj\beta^2 - bcf\alpha\beta^2 - f^2g\alpha\beta^2 + cfh\alpha\beta^2$
$-aci\alpha\beta^2 + 2eij\alpha\beta^2 + afg\alpha^2\beta^2 - 2cfi\alpha^2\beta^2 + gij\alpha^2\beta^2 + adf\beta^3 - 2cef\beta^3 - afh\beta^3 - c^2i\beta^3 + 2f^2i\beta^3$

$+ egj\beta^3 + dij\beta^3 - hij\beta^3 - bf^2\alpha\beta^3 + df^2\alpha\beta^3 - cfg\alpha\beta^3 + i^2j\alpha\beta^3 - ef^2\beta^4 + afg\beta^4 - cfi\beta^4 + gij\beta^4$,
$-bcf + aef - \alpha^2g + cfh + aci + bgj - ghj - eij + ef^2\alpha - 2afg\alpha - cfi\alpha + 2gij\alpha - bcf\alpha^2$
$+ aef\alpha^2 - f^2g\alpha^2 + cfh\alpha^2 + aci\alpha^2 - eij\alpha^2 + ef^2\alpha^3 - cfi\alpha^3 + bc^2\beta + a^2\alpha\beta - 2ace\beta - bf^2\beta - 2cfg\beta$
$-a^2h\beta - c^2h\beta + f^2h\beta + 2afi\beta - bdj\beta + e^2j\beta + 2g^2j\beta + bhj\beta + dhj\beta - h^2j\beta - i^2j\beta + 2adf\alpha\beta$

$$-cef\alpha\beta + 2acg\alpha\beta - 2afh\alpha\beta + c^2 i\alpha\beta - 2egj\alpha\beta - 2dij\alpha\beta + 2hij\alpha\beta + bc^2\alpha^2\beta - 2ace\alpha^2\beta$$
$$-bf^2\alpha^2\beta + df^2\alpha^2\beta - c^2ha^2\beta + 2afi\alpha^2\beta + e^2j\alpha^2\beta - i^2j\alpha^2\beta - cef\alpha^3\beta + c^2i\alpha^3\beta + cdf\beta^2 + 2c^2g\beta^2$$
$$-2f^2g\beta^2 - cfh\beta^2 - 3dgj\beta^2 + 3ghj\beta^2 - 2acd\alpha\beta^2 + ef^2\alpha\beta^2 + 2afg\alpha\beta^2 + 2ach\alpha\beta^2 - cfi\alpha\beta^2$$
$$+2dej\alpha\beta^2 - 2ehj\alpha\beta^2 - 2gij\alpha\beta^2 + bcf\alpha^2\beta^2 - cdf\alpha^2\beta^2 - aef\alpha^2\beta^2 + c^2g\alpha^2\beta^2 - aci\alpha^2\beta^2$$

$$+eij\alpha^2\beta^2 + bc^2\beta^3 + a^2d\beta^3 - c^2d\beta^3 - 2ace\beta^3 - bf^2\beta^3 + df^2\beta^3 + 2cfg\beta^3 - a^2h\beta^3 + 2afi\beta^3$$
$$-bdj\beta^3 + d^2j\beta^3 + e^2j\beta^3 - 2g^2j\beta^3 + bhj\beta^3 - dhj\beta^3 - i^2j\beta^3 - cef\alpha\beta^3 - 2acg\alpha\beta^3 + c^2i\alpha\beta^3$$
$$+2egj\alpha\beta^3 + bcf\beta^4 - cdf\beta^4 - aef\beta^4 + a^2g\beta^4 - aci\beta^4 - bgj\beta^4 + dgj\beta^4 + eij\beta^4, -cdf + cfh$$
$$+dgj - ghj + acd\alpha + 2afg\alpha - ach\alpha - 2cfi\alpha - dej\alpha + ehj\alpha + bcf\alpha^2 - cdf\alpha^2 - 2aef\alpha^2 - a^2g\alpha^2$$

$$+2f^2g\alpha^2 + aci\alpha^2 + bgj\alpha^2 + dgj\alpha^2 - 2ghj\alpha^2 + acd\alpha^3 + a^2e\alpha^3 - 2ef^2\alpha^3 - ach\alpha^3 - cfi\alpha^3 - bej\alpha^3$$
$$-dej\alpha^3 + 2ehj\alpha^3 + 2gij\alpha^3 + bcf\alpha^4 + f^2g\alpha^4 - cfh\alpha^4 + aci\alpha^4 - 2eij\alpha^4 - ef^2\alpha^5 + cfi\alpha^5 + c^2d\beta$$
$$-df^2\beta - c^2h\beta + f^2h\beta - d^2j\beta + 2dhj\beta - h^2j\beta - acg\alpha\beta + c^2i\alpha\beta - 2f^2i\alpha\beta + 2egj\alpha\beta - 2dij\alpha\beta$$
$$+2hij\alpha\beta + c^2d\alpha^2\beta + 2ace\alpha^2\beta + bf^2\alpha^2\beta - 2df^2\alpha^2\beta - cfg\alpha^2\beta - c^2ha^2\beta + f^2ha^2\beta - d^2j\alpha^2\beta$$

$$-2e^2j\alpha^2\beta + 2g^2j\alpha^2\beta + 2dhj\alpha^2\beta - h^2j\alpha^2\beta - i^2j\alpha^2\beta + 2cef\alpha^3\beta + acg\alpha^3\beta + c^2i\alpha^3\beta - 2f^2i\alpha^3\beta$$
$$-2egj\alpha^3\beta - 2dij\alpha^3\beta + 2hij\alpha^3\beta + bf^2\alpha^4\beta - df^2\alpha^4\beta + cfg\alpha^4\beta - i^2j\alpha^4\beta + acd\alpha\beta^2 + c^2e\alpha\beta^2$$
$$-ach\alpha\beta^2 - cfi\alpha\beta^2 - 2dej\alpha\beta^2 + 2ehj\alpha\beta^2 + bcf\alpha^2\beta^2 + cdf\alpha^2\beta^2 + c^2g\alpha^2\beta^2 - 2cfh\alpha^2\beta^2$$
$$+aci\alpha^2\beta^2 - 2dgj\alpha^2\beta^2 + 2ghj\alpha^2\beta^2 - 2eij\alpha^2\beta^2 + 2cfi\alpha^3\beta^2 - 2gij\alpha^3\beta^2 + c^2d\beta^3 - df^2\beta^3 - cfg\beta^3$$

$$-c^2h\beta^3 + f^2h\beta^3 - d^2j\beta^3 + g^2j\beta^3 + 2dhj\beta^3 - h^2j\beta^3 + 2cef\alpha\beta^3 + acg\alpha\beta^3 + c^2i\alpha\beta^3 - 2f^2i\alpha\beta^3$$
$$-2egj\alpha\beta^3 - 2dij\alpha\beta^3 + 2hij\alpha\beta^3 + bf^2\alpha^2\beta^3 - df^2\alpha^2\beta^3 + 2cfg\alpha^2\beta^3 - g^2j\alpha^2\beta^3 - i^2j\alpha^2\beta^3 + cdf\beta^4$$
$$+c^2g\beta^4 - f^2g\beta^4 - cfh\beta^4 - 2dgj\beta^4 + 2ghj\beta^4 + ef^2\alpha\beta^4 + cfi\alpha\beta^4 - 2gij\alpha\beta^4 + cfg\beta^5 - g^2j\beta^5\}$$

The basis consists of 7 polynomials of variables α and β. The lengths of these polynomials are

$$\Rightarrow \texttt{Map}[\texttt{Length}[\#]\&, \texttt{gb}\alpha\beta]$$
$$\Leftarrow \{22, 22, 22, 48, 75, 102, 120\}$$

Let us consider the first and the third one and eliminate β via Sylvester resultant,

$$\Rightarrow \texttt{alfa} = \texttt{Resultant}[\texttt{gb}\alpha\beta[[1]], \texttt{gb}\alpha\beta[[3]], \beta]$$

```
a² c⁵ d f j α² - a c⁶ e f j α² + 2 a² c³ d f³ j α² -
3 a c⁴ e f³ j α² + ▨ ...4832... ▨ + 4 b e² g i j⁵ α⁹ - 4 d e² g i j⁵ α⁹ -
b² e i² j⁵ α⁹ + 2 b d e i² j⁵ α⁹ - d² e i² j⁵ α⁹ - 4 e³ i² j⁵ α⁹

large output     show less     show more     show all     set size limit...
```

so we get a univariate polynomial for α. Similarly eliminate α from the first and second basis,

$$\Rightarrow \texttt{beta} = \texttt{Resultant[gba}\beta\texttt{[[1]],gba}\beta\texttt{[[2]],}\alpha\texttt{]}$$

$-a^5 b c^2 f j \beta^2 + a^6 c e f j \beta^2 - 2 a^3 b c^2 f^3 j \beta^2 +$
$3 a^4 c e f^3 j \beta^2 - a b c^2 f^5 j \beta^2 + 3 a^2 c e f^5 j \beta^2 +$
⬚⬚⬚ 4829 ⬚⬚⬚ $+ 4 b e^2 g i j^5 \beta^9 - 4 d e^2 g i j^5 \beta^9 -$
$b^2 e i^2 j^5 \beta^9 + 2 b d e i^2 j^5 \beta^9 - d^2 e i^2 j^5 \beta^9 - 4 e^3 i^2 j^5 \beta^9$

| large output | **show less** | **show more** | **show all** | **set size limit...** |

which leads to a univariate polynomial of β.
The coefficients of the univariate polynomial for α are

$$\Rightarrow \texttt{ca0} = \texttt{Coefficient[alfa,}\alpha,0\texttt{]}$$
$$\Leftarrow 0$$

$\Rightarrow \texttt{ca1} = \texttt{Coefficient[alfa,}\alpha,1\texttt{]}$
$\Leftarrow 0$
$\Rightarrow \texttt{ca2} = \texttt{Coefficient[alfa,}\alpha,2\texttt{]}$
$\Leftarrow a^2c^5dfj - ac^6efj + 2a^2c^3df^3j - 3ac^4ef^3j + a^2cdf^5j - 3ac^2ef^5j - aef^7j - a^2c^6gj - a^2c^4f^2gj$
$+ a^2c^2f^4gj + a^2f^6gj - a^2c^5fhj - 2a^2c^3f^3hj - a^2cf^5hj + ac^7ij + 3ac^5f^2ij + 3ac^3f^4ij + acf^6ij$
$- 2a^2c^3d^2fj^2 + ac^4defj^2 + c^5e^2fj^2 + 2a^2cd^2f^3j^2 - 2ac^2def^3j^2 + 2c^3e^2f^3j^2 - 3adef^5j^2 + ce^2f^5j^2$
$+ 2a^2c^4dgj^2 + 2ac^5egj^2 - 12a^2c^2df^2gj^2 + 12ac^3ef^2gj^2 + 2a^2df^4gj^2 + 10acef^4gj^2 + 8a^2c^3fg^2j^2$

$- 8a^2cf^3g^2j^2 + 4a^2c^3dfhj^2 - ac^4efhj^2 - 4a^2cdf^3hj^2 + 2ac^2ef^3hj^2 + 3aef^5hj^2 - 2a^2c^4ghj^2$
$+ 12a^2c^2f^2ghj^2 - 2a^2f^4ghj^2 - 2a^2c^3fh^2j^2 + 2a^2cf^3h^2j^2 - 3ac^5dij^2 - c^6eij^2 - 2ac^3df^2ij^2$
$- c^4ef^2ij^2 + acdf^4ij^2 + c^2ef^4ij^2 + ef^6ij^2 - 10ac^4fgij^2 - 12ac^2f^3gij^2 - 2af^5gij^2 + 3ac^5hij^2$
$+ 2ac^3f^2hij^2 - acf^4hij^2 - c^5fi^2j^2 - 2c^3f^3i^2j^2 - cf^5i^2j^2 + a^2cd^3fj^3 + ac^2d^2efj^3 - 2c^3de^2fj^3$

$- 3ad^2ef^3j^3 + 2cde^2f^3j^3 - a^2c^2d^2gj^3 - 4ac^3degj^3 - c^4e^2gj^3 + a^2d^2f^2gj^3 + 12acdef^2gj^3$
$- 10c^2e^2f^2gj^3 - e^2f^4gj^3 + 4a^2cdfg^2j^3 - 20ac^2efg^2j^3 - 4aef^3g^2j^3 - 4a^2c^2g^3j^3 + 4a^2f^2g^3j^3$
$- 3a^2cd^2fhj^3 - 2ac^2defhj^3 + 2c^3e^2fhj^3 + 6adef^3hj^3 - 2ce^2f^3hj^3 + 2a^2c^2dghj^3 + 4ac^3eghj^3$
$- 2a^2df^2ghj^3 - 12acef^2ghj^3 - 4a^2cfg^2hj^3 + 3a^2cdfh^2j^3 + ac^2efh^2j^3 - 3aef^3h^2j^3 - a^2c^2gh^2j^3$

$+ a^2f^2gh^2j^3 - a^2cfh^3j^3 + 3ac^3d^2ij^3 + 3c^4deij^3 - acd^2f^2ij^3 - 2c^2def^2ij^3 + 3def^4ij^3$
$+ 12ac^2dfgij^3 + 8c^3efgij^3 - 4adf^3gij^3 - 8cef^3gij^3 + 4ac^3g^2ij^3 + 20acf^2g^2ij^3 - 6ac^3dhij^3$
$- 3c^4ehij^3 + 2acdf^2hij^3 + 2c^2ef^2hij^3 - 3ef^4hij^3 - 12ac^2fghij^3 + 4af^3ghij^3 + 3ac^3h^2ij^3$
$- acf^2h^2ij^3 + 2c^3dfi^2j^3 - 2cdf^3i^2j^3 + c^4gi^2j^3 + 10c^2f^2gi^2j^3 + f^4gi^2j^3 - 2c^3fhi^2j^3 + 2cf^3hi^2j^3$

$- ad^3efj^4 + cd^2e^2fj^4 + 2acd^2egj^4 + 2c^2de^2gj^4 - 2de^2f^2gj^4 - 4adefg^2j^4 + 12ce^2fg^2j^4$
$+ 8aceg^3j^4 + 3ad^2efhj^4 - 2cde^2fhj^4 - 4acdeghj^4 - 2c^2e^2ghj^4 + 2e^2f^2ghj^4 + 4aefg^2hj^4$
$- 3adefh^2j^4 + ce^2fh^2j^4 + 2acegh^2j^4 + aefh^3j^4 - acd^3ij^4 - 3c^2d^2eij^4 + 3d^2ef^2ij^4 - 2ad^2fgij^4$
$- 8cdefgij^4 - 4acdg^2ij^4 - 4c^2eg^2ij^4 + 4ef^2g^2ij^4 - 8afg^3ij^4 + acd^2hij^4 + 6c^2dehij^4$

$- 6def^2hij^4 + 4adfghij^4 + 8cefghij^4 + 4acg^2hij^4 - 3acdh^2ij^4 - 3c^2eh^2ij^4 + 3ef^2h^2ij^4$
$- 2afgh^2ij^4 + ach^3ij^4 - cd^2fi^2j^4 - 2c^2dgi^2j^4 + 2df^2gi^2j^4 - 12cfg^2i^2j^4 + 2cdfhi^2j^4$
$+ 2c^2ghi^2j^4 - 2f^2ghi^2j^4 - cfh^2i^2j^4 - d^2e^2gj^5 - 4e^2g^3j^5 + 2de^2ghj^5 - e^2gh^2j^5 + d^3eij^5$
$+ 4deg^2ij^5 - 3d^2ehij^5 - 4eg^2hij^5 + 3deh^2ij^5 - eh^3ij^5 + d^2gi^2j^5 + 4g^3i^2j^5 - 2dghi^2j^5 + gh^2i^2j^5$

$\Rightarrow \texttt{ca3} = \texttt{Coefficient[alfa,}\alpha,3\texttt{]; Short[ca3,10]}$
$\Leftarrow -abc^7j - a^3c^5dj + 2a^2c^6ej - 3abc^5f^2j - 6a^3c^3df^2j + 2ac^5df^2j + 10a^2c^4ef^2j - c^6ef^2j$
$- 3abc^3f^4j - 5a^3cdf^4j + 4ac^3df^4j + 14a^2c^2ef^4j - 3c^4ef^4j - abcf^6j + 2acdf^6j$
$+ 6a^2ef^6j - 3c^2ef^6j - ef^8j \ll 550 \gg + 3deh^3j^5 - eh^4j^5 - 2bd^2gij^5 + d^3gij^5 - 12de^2gij^5$
$- 8bg^3ij^5 + 4dg^3ij^5 + 4bdghij^5 - d^2ghij^5 + 12e^2ghij^5 + 4g^3hij^5 - 2bgh^2ij^5$
$- dgh^2ij^5 + gh^3ij^5 + 5d^2ei^2j^5 - 4eg^2i^2j^5 - 10dehi^2j^5 + 5eh^2i^2j^5 + 4dgi^3j^5 - 4ghi^3j^5$

$\Rightarrow$ ca4 = Coefficient[alfa,α, 4]; Short[ca4, 10]

$\Leftarrow 5a^2bc^5fj - bc^7fj + 6a^4c^3dfj - 11a^3c^4efj + ac^6efj + 10a^2bc^3f^3j - 3bc^5f^3j + 10a^4cdf^3j$
$\quad -10a^2c^3df^3j + c^5df^3j - 26a^3c^2ef^3j + 8ac^4ef^3j + 5a^2bcf^5j - 3bc^3f^5j - 10a^2cdf^5j + 2c^3df^5j$
$\quad -15a^3ef^5j + 13ac^2ef^5j - bcf^7j + cdf^7j + \ll 665 \gg + d^3eij^5 + 8de^3ij^5 + 12beg^2ij^5$
$\quad -16deg^2ij^5 + 14bdehij^5 + 4d^2ehij^5 - 8e^3hij^5 + 4eg^2hij^5 - 7beh^2ij^5 - 11deh^2ij^5 + 6eh^3ij^5$
$\quad -10bdgi^2j^5 + 5d^2gi^2j^5 - 8e^2gi^2j^5 + 10bghi^2j^5 - 5gh^2i^2j^5 + 8dei^3j^5 - 8ehi^3j^5 + 4gi^4j^5$

$\Rightarrow$ ca5 = Coefficient[alfa,α, 5]; Short[ca5, 10]

$\Leftarrow -2a^3bc^5j - 2abc^7j - 2a^5c^3dj - 2a^3c^5dj + 4a^4c^4ej + 4a^2c^6ej - 12a^3bc^2f^2j$
$\quad + abc^5f^2j - 10a^5cdf^2j + 6a^3c^3df^2j + 3ac^5df^2j + 24a^4c^2ef^2j - 3a^2c^4ef^2j - c^6ef^2j$
$\quad -10a^3bcf^4j + 8abc^3f^4j + 20a^3cdf^4j - 2ac^3df^4j + \ll 769 \gg + 2bdghij^5 + 6d^2ghij^5$
$\quad -12e^2ghij^5 + 8g^3hij^5 + 7bgh^2ij^5 - 7dgh^2ij^5 - 14bdei^2j^5 + d^2ei^2j^5 + 8e^3i^2j^5$
$\quad -8eg^2i^2j^5 + 14behi^2j^5 + 12dehi^2j^5 - 13eh^2i^2j^5 - 12bgi^3j^5 + 4dgi^3j^5 + 8ghi^3j^5 + 4ei^4j^5$

$\Rightarrow$ ca6 = Coefficient[alfa,α, 6]; Short[ca6, 10]

$\Leftarrow 6a^4bc^3fj + 4a^2bc^5fj - 2bc^7fj + 5a^6cdfj + 2a^4c^3dfj - 3a^2c^5dfj$
$\quad -11a^5c^2efj - 6a^3c^4efj + 5ac^6efj + 10a^4bcf^3j - 6a^2bc^3f^3j - 4bc^5f^3j - 20a^4cdf^3j$
$\quad -6a^2c^3df^3j + 2c^5df^3j - 15a^5ef^3j + 18a^3c^2ef^3j \ll 770 \gg + 9d^2ehij^5 + 4e^3hij^5$
$\quad -4eg^2hij^5 + 8beh^2ij^5 - 8deh^2ij^5 + 13b^2gi^2j^5 - 12bdgi^2j^5 - d^2gi^2j^5 + 8e^2gi^2j^5$
$\quad -8g^3i^2j^5 - 14bghi^2j^5 + 14dghi^2j^5 - 8bei^3j^5 - 4dei^3j^5 + 12ehi^3j^5 - 4gi^4j^5$

$\Rightarrow$ ca7 = Coefficient[alfa,α, 7]; Short[ca7, 10]

$\Leftarrow -a^5bc^3j - 2a^3bc^5j - abc^7j - a^7cdj - 2a^5c^3dj - a^3c^5dj + 2a^6c^2ej + 4a^4c^4ej$
$\quad + 2a^2c^6ej - 5a^5bcf^2j + 5abc^5f^2j + 10a^5cdf^2j + 10a^3c^3df^2j + 6a^6ef^2j - 7a^4c^2ef^2j$
$\quad -12a^2c^4ef^2j + c^6ef^2j + 10a^3bcf^4j + \ll 669 \gg + 11b^2dgij^5 - 4bd^2gij^5 - d^3gij^5 - 4be^2gij^5$
$\quad + 16de^2gij^5 + 8bg^3ij^5 - 8dg^3ij^5 + 7b^2ghij^5 - 14bdghij^5 + 7d^2ghij^5 - 12e^2ghij^5$
$\quad + 5b^2ei^2j^5 - 5d^2ei^2j^5 + 8eg^2i^2j^5 - 10behi^2j^5 + 10dehi^2j^5 + 8bgi^3j^5 - 8dgi^3j^5 - 4ei^4j^5$

$\Rightarrow$ ca8 = Coefficient[alfa,α, 8]; Short[ca8, 10]

$\Leftarrow a^6bcfj + a^4bc^3fj - a^2bc^5fj - bc^7fj - 2a^6cdfj - 4a^4c^3dfj - 2a^2c^5dfj - a^7efj + a^5c^2efj$
$\quad + 5a^3c^4efj + 3ac^6efj - 5a^4bcf^3j - 6a^2bc^3f^3j - bc^5f^3j + 5a^4cdf^3j + 6a^2c^3df^3j$
$\quad + c^5df^3j + 6a^5ef^3j + 4a^3c^2ef^3j - 2ac^4ef^3j + \ll 554 \gg + 4de^2ghj^5 - b^3eij^5 + b^2deij^5$
$\quad + bd^2eij^5 - d^3eij^5 - 4be^3ij^5 - 4de^3ij^5 - 12beg^2ij^5 + 12deg^2ij^5 + 2b^2ehij^5 - 4bdehij^5$
$\quad + 2d^2ehij^5 + 8e^3hij^5 - 5b^2gi^2j^5 + 10bdgi^2j^5 - 5d^2gi^2j^5 + 4e^2gi^2j^5 + 4bei^3j^5 - 4dei^3j^5$

$\Rightarrow$ ca9 = Coefficient[alfa,α, 9]; Short[ca9, 10]

$\Leftarrow a^5bcf^2j + 2a^3bc^3f^2j + abc^5f^2j - a^5cdf^2j - 2a^3c^3df^2j - ac^5df^2j - a^6ef^2j - a^4c^2ef^2j + a^2c^4ef^2j$
$\quad + c^6ef^2j + a^7fgj + 3a^5c^2fgj + 3a^3c^4fgj + ac^6fgj - a^5cfij - 3a^4c^3fij - 3a^2c^5fij - c^7fij$
$\quad -2a^3b^2cf^2j^2 + 2ab^2c^3f^2j^2 + \ll 211 \gg + acd^2i^2j^4 + 2a^2bei^2j^4 - 2bc^2ei^2j^4 - 2a^2dei^2j^4$
$\quad + 2c^2dei^2j^4 + 12ace^2i^2j^4 + b^2eg^2j^5 - 2bdeg^2j^5 + d^2eg^2j^5 + 4e^3g^2j^5 + b^3gij^5 - 3b^2dgij^5$
$\quad + 3bd^2gij^5 - d^3gij^5 + 4be^2gij^5 - 4de^2gij^5 - b^2ei^2j^5 + 2bdei^2j^5 - d^2ei^2j^5 - 4e^3i^2j^5$

$\Rightarrow$ ca10 = Coefficient[alfa,α, 10]

$\Leftarrow 0$

Therefore our polynomial for α can be written as

$\Rightarrow A = (ca2 + ca3\alpha + ca4\alpha^2 + ca5\alpha^3 + ca6\alpha^4 + ca7\alpha^5 + ca8\alpha^6 + ca9\alpha^7)\alpha^2$;

Let us check these coefficients

```
⇒ (alfa − A)//Simplify
⇐ 0
```

Now, the coefficients of the polynomial of β

```
⇒ cb0 = Coefficient[beta,β, 0]
⇐ 0
```

```
⇒ cb1 = Coefficient[beta,β, 1]
⇐ 0
```

```
⇒ cb2 = Coefficient[beta,β, 2]; Short[cb2, 10]
```
$\Leftarrow -a^5bc^2fj + a^6cefj - 2a^3bc^2f^3j + 3a^4cef^3j - abc^2f^5j + 3a^2cef^5j + cef^7j$
$-a^7cgj - 3a^5cf^2gj - 3a^3cf^4gj - acf^6gj + a^5c^2fhj + 2a^3c^2f^3hj + ac^2f^5hj + a^6c^2ij$
$+a^4c^2f^2ij - a^2c^2f^4ij - c^2f^6ij + \ll 214 \gg +12afg^2i^2j^4 - 4cefhi^2j^4 - 4acghi^2j^4$
$-8acei^3j^4 + 8cfgij^3j^4 - b^3egj^5 + 3b^2eghj^5 - 3begh^2j^5 + egh^3j^5 + b^2e^2ij^5 - b^2g^2ij^5$
$-2be^2hij^5 + 2bg^2hij^5 + e^2h^2ij^5 - g^2h^2ij^5 - 4begi^2j^5 + 4eghi^2j^5 + 4e^2i^3j^5 - 4g^2i^3j^5$

```
⇒ cb3 = Coefficient[beta,β, 3]; Short[cb3, 10]
```
$\Leftarrow a^5bc^3j + a^7cdj - 2a^5bc^2ej - 2a^5bcf^2j + 6a^3bc^3f^2j + 3a^5cdf^2j + a^6ef^2j - 10a^4c^2ef^2j$
$-4a^3bcf^4j + 5abc^3f^4j + 3a^3cdf^4j + 3a^4ef^4j - 14a^2c^2ef^4j - 2abcf^6j + acdf^6j + 3a^2ef^6j$
$-6c^2ef^6j + ef^8j - a^7fgj + 4a^5c^2fgj - 3a^5f^3gj + \ll 553 \gg +12be^2gij^5 - 4bg^2ij^5 + b^2ghij^5$
$-4bdghij^5 - 12e^2ghij^5 + 4g^2hij^5 + bgh^2ij^5 + 2dgh^2ij^5 - gh^3ij^5 + 2b^2ei^2j^5 + 4bdei^2j^5$
$-12e^3i^2j^5 + 4eg^2i^2j^5 - 8behi^2j^5 - 4dehi^2j^5 + 6eh^2i^2j^5 - 4bgi^3j^5 + 8dgi^3j^5 - 4ghi^3j^5 + 8ei^4j^5$

```
⇒ cb4 = Coefficient[beta,β, 4]; Short[cb4, 10]
```
$\Leftarrow -6a^3bc^4fj + a^7dfj - 5a^5c^2dfj - a^6cefj + 11a^4c^3efj - a^5bf^3j + 10a^3bc^2f^3j - 10abc^4f^3j$
$+3a^5df^3j - 10a^3c^2df^3j - 8a^4cef^3j + 26a^2c^3ef^3j - 2a^3bf^5j + 10abc^2f^5j + 3a^3df^5j$
$-5ac^2df^5j - 13a^2cef^5j + 15c^3ef^5j - abf^7j + \ll 665 \gg +10de^2hij^5 - 10dg^2hij^5 + 2b^2h^2ij^5$
$-bdh^2ij^5 - d^2h^2ij^5 - 10e^2h^2ij^5 + 5g^2h^2ij^5 - bh^3ij^5 + dh^3ij^5 + 16begi^2j^5 - 12degi^2j^5$
$-4eghi^2j^5 + b^2i^3j^5 + 4bdi^3j^5 - 4d^2i^3j^5 - 20e^2i^3j^5 - 6bhi^3j^5 + 4dhi^3j^5 + h^2i^3j^5 + 4i^5j^5$

```
⇒ cb5 = Coefficient[beta,β, 5]; Short[cb5, 10]
```
$\Leftarrow 2a^5bc^3j + 2a^3bc^5j + 2a^7cdj + 2a^5c^3dj - 4a^6c^2ej - 4a^4c^4ej - 3a^5bcf^2j - 6a^3bc^3f^2j$
$+10abc^5f^2j - a^5cdf^2j + 12a^3c^3df^2j + a^6ef^2j + 3a^4c^2ef^2j - 24a^2c^4ef^2j + 2a^3bcf^4j$
$-20abc^3f^4j - 8a^3cdf^4j + 10ac^3df^4j + 2a^4ef^4j + \ll 771 \gg +8d^2ghij^5 + 12e^2ghij^5$
$-8g^3hij^5 + 7bgh^2ij^5 - 7dgh^2ij^5 + 3b^2ei^2j^5 - 8bdei^2j^5 + 8d^2ei^2j^5 + 12e^3i^2j^5$
$+8eg^2i^2j^5 + 2behi^2j^5 - 8dehi^2j^5 + 3eh^2i^2j^5 + 4bgi^3j^5 + 4dgi^3j^5 - 8ghi^3j^5 - 4ei^4j^5$

```
⇒ cb6 = Coefficient[beta,β, 6]; Short[cb6, 10]
```
$\Leftarrow 3a^5bc^2fj - 2a^3bc^4fj - 5abc^6fj + 2a^7dfj - 4a^5c^2dfj - 6a^3c^4dfj - 5a^6cefj + 6a^4c^3efj$
$+11a^2c^5efj - 2a^5bf^3j + 6a^3bc^2f^3j + 20abc^4f^3j + 4a^5df^3j + 6a^3c^2df^3j - 10ac^4df^3j$
$-9a^4cef^3j - 18a^2c^3ef^3j + \ll 768 \gg +8be^2hij^5 + 8de^2hij^5 - 14bg^2hij^5 + 14dg^2hij^5$
$+2b^2h^2ij^5 - 4bdh^2ij^5 + 2d^2h^2ij^5 - 8e^2h^2ij^5 + 8begi^2j^5 - 12degi^2j^5 + 4eghi^2j^5$
$+2b^2i^3j^5 + 2bdi^3j^5 - 4d^2i^3j^5 - 12e^2i^3j^5 + 8g^2i^3j^5 - 6bhi^3j^5 + 6dhi^3j^5 + 4i^5j^5$

⇒ cb7 = Coefficient[beta,β,7]; Short[cb7,10]

⇐ $a^5bc^3j + 2a^3bc^5j + abc^7j + a^7cdj + 2a^5c^3dj + a^3c^5dj - 2a^6c^2ej - 4a^4c^4ej - 2a^2c^6ej$
$-10a^3bc^3f^2j - 10abc^5f^2j - 5a^5cdf^2j + 5ac^5df^2j - a^6ef^2j + 12a^4c^2ef^2j + 7a^2c^4ef^2j$
$-6c^6ef^2j + 6a^3bcf^4j + \ll 668 \gg + 6d^3gij^5 - 16be^2gij^5 + 4de^2gij^5 + 8bg^3ij^5$
$-8dg^3ij^5 - 7b^2ghij^5 + 14bdghij^5 - 7d^2ghij^5 + 12e^2ghij^5 - 10bdei^2j^5 + 10d^2ei^2j^5$
$+20e^3i^2j^5 - 8eg^2i^2j^5 + 10behi^2j^5 - 10dehi^2j^5 + 8bgi^3j^5 - 8dgi^3j^5 - 12ei^4j^5$

⇒ cb8 = Coefficient[beta,β,8]; Short[cb8,10]

⇐ $2a^5bc^2fj + 4a^3bc^4fj + 2abc^6fj + a^7dfj + a^5c^2dfj - a^3c^4dfj - ac^6dfj - 3a^6cefj - 5a^4c^3efj$
$-a^2c^5efj + c^7efj - a^5bf^3j - 6a^3bc^2f^3j - 5abc^4f^3j + a^5df^3j + 6a^3c^2df^3j + 5ac^4df^3j$
$+2a^4cef^3j - 4a^2c^3ef^3j - 6c^5ef^3j + a^7cgj + \ll 550 \gg + 8bde^2ij^5 - 6d^2e^2ij^5 - 8e^4ij^5$
$+5b^2g^2ij^5 - 10bdg^2ij^5 + 5d^2g^2ij^5 - 4e^2g^2ij^5 - b^3hij^5 + 3b^2dhij^5 - 3bd^2hij^5 + d^3hij^5$
$-4be^2hij^5 + 4de^2hij^5 - 12begi^2j^5 + 12degi^2j^5 + b^2i^3j^5 - 2bdi^3j^5 + d^2i^3j^5 + 12e^2i^3j^5$

⇒ cb9 = Coefficient[beta,β,9]; Short[cb9,10]

⇐ $a^5bcf^2j + 2a^3bc^3f^2j + abc^5f^2j - a^5cdf^2j - 2a^3c^3df^2j - ac^5df^2j - a^6ef^2j - a^4c^2ef^2j$
$+a^2c^4ef^2j + c^6ef^2j + a^7fgj + 3a^5c^2fgj + 3a^3c^4fgj + ac^6fgj - a^6cfij - 3a^4c^3fij - 3a^2c^5fij$
$-c^7fij - 2a^3b^2cf^2j^2 + 2ab^2c^3f^2j^2 + \ll 211 \gg + acd^2i^2j^4 + 2a^2bei^2j^4 - 2bc^2ei^2j^4 - 2a^2dei^2j^4$
$+2c^2dei^2j^4 + 12ace^2i^2j^4 + b^2eg^2j^5 - 2bdeg^2j^5 + d^2eg^2j^5 + 4e^3g^2j^5 + b^3gij^5 - 3b^2dgij^5$
$+3bd^2gij^5 - d^3gij^5 + 4be^2gij^5 - 4de^2gij^5 - b^2ei^2j^5 + 2bdei^2j^5 - d^2ei^2j^5 - 4e^3i^2j^5$

⇒ cb10 = Coefficient[beta,β,10]

⇐ 0

then

⇒ B = (cb2 + cb3β + cb4β² + cb5β³ + cb6β⁴ + cb7β⁵ + cb8β⁶ + cb9β⁷)β² ;
⇒ (beta − B)//Simplify
⇐ 0

As illustration let us consider a synthetic dataset generated from a multivariate Gauss distribution. Regular 3D points have means of (2, 8, 6) and variances $(5, 5, 0.01)$, see Nurunnabi et al. (2012). The ideal plane is $z = 6$. In this case let us consider here $N = 20,000$ points. There are additional 3000 points, which represent outliers.

⇒ Clear[XYZ]
⇒ N = 20000;
⇒ XYZ = Table[{Null, Null, Null}, {i, 1, N}];
⇒ Do[XYZ[[i]] = RandomVariate[MultinormalDistribution[{2., 8, 6},
  {{5., 0, 0}, {0, 5., 0}, {0, 0, 0.01}}]], {i, 1, N}];
⇒ p1 = Plot3D[6, {x, −8, 12}, {y, −2, 18}, BoxRatios → {1, 1, 0.5},
  ClippingStyle → None, Mesh → None, PlotRange → {5.5, 6.5}];
⇒ p2 = ListPointPlot3D[XYZ, PlotStyle → Blue];
⇒ NOutLiers = 3000;

Now, let us generate 3000 outliers randomly (Fig. 14.1).

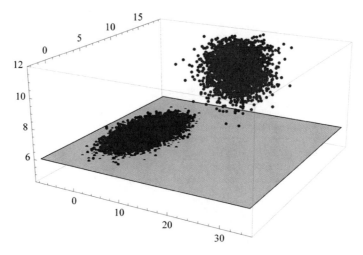

Fig. 14.1 The plane and the artificial measured points with outliers (red)

⇒ XYZOutLiers = Table[{Null, Null, Null}, {i, 1, NOutLiers}];
⇒ Do[XYZOutLiers[[i]] =
RandomVariate[MultinormalDistribution[{15., 15, 10},
{{10., 0, 0}, {0, 2., 0}, {0, 0, 1}}]], {i, 1, NOutLiers}];
⇒ p1 = Plot3D[6, {x, −8, 35}, {y, −2, 19}, BoxRatios → {1, 1, 0.5},
ClippingStyle → None, Mesh → None, PlotRange → {4.5, 12}];
⇒ p3 = ListPointPlot3D[XYZOutLiers, PlotStyle → Red];
⇒ Show[{p1, p2, p3}]

⇒ XYZC = Join[XYZ, XYZOutLiers];
⇒ X = XYZC[[All, 1]]; Y = XYZC[[All, 2]]; Z = XYZC[[All, 3]];

The number of data

⇒ N = Length[XYZC]
⇐ 23000

We use the same weight for all data, $W_i = 1$.

⇒ W = Table[1, {N}];
⇒ W2 = Map[#^{2}&, W];

The coefficients are

⇒ {a, b, c, d, e, f, g, h, i, j} = Total/@Map[MapThread[#1#2&, {#, W2}]&,
{X, X^2, Y, Y^2, XY, Z, YZ, Z^2, XZ, W2}];

The monomial for α

$\Rightarrow$ AbsoluteTiming [pα=
ca2 + ca3α + ca4α^2 + ca5α^3 + ca6α^4 + ca7α^5 + ca8α^6 + ca9α^7 ;]
$\Leftarrow \{0.00911141, \text{Null}\}$

$\Rightarrow$ pα
$\Leftarrow 1.23806 \times 10^48 - 7.11611 \times 10^48\alpha + 6.94251 \times 10^48\alpha^2 - 2.76003 \times 10^49\alpha^3$
$\quad + 3.84629 \times 10^49\alpha^4 - 2.98505 \times 10^49\alpha^5 - 8.5202 \times 10^48\alpha^6 + 5.1252 \times 10^47\alpha^7$

The real roots of this monomial are

$\Rightarrow$ AbsoluteTiming[solα= Reduce[pα== 0,α, Reals];]
$\Leftarrow \{0.00498473, \text{Null}\}$

$\Rightarrow$ solα
$\Leftarrow \alpha== -3.94057 || \alpha== 0.188435 || \alpha== 19.4301$

We select the smaller positive one

$\Rightarrow$ ToRules[solα[[2]]]
$\Leftarrow \{\alpha \rightarrow 0.188435\}$

Similarly the monomial for β and γ, and their solutions are

$\Rightarrow$ AbsoluteTiming[pβ=
cb2 + cb3β + cb4β^2 + cb5β^3 + cb6β^4 + cb7β^5 + cb8β^6 + cb9β^7 ;]
$\Leftarrow \{0.00841855, \text{Null}\}$

$\Rightarrow$ pβ
$\Leftarrow -5.02172 \times 10^{48} + 3.87559 \times 10^{49}\beta - 3.22113 \times 10^{47}\beta^2 + 6.86247 \times 10^{49}\beta^3$
$\quad + 1.75699 \times 10^{49}\beta^4 + 2.78598 \times 10^{49}\beta^5 + 1.96989 \times 10^{49}\beta^6 + 5.1252 \times 10^{47}\beta^7$

$\Rightarrow$ solβ= Reduce[pβ== 0,β, Reals]
$\Leftarrow \beta== -36.9881 || \beta== -2.04298 || \beta== 0.126022$

$\Rightarrow$ ToRules[solβ[[3]]]
$\Leftarrow \{\beta \rightarrow 0.126022\}$

$\Rightarrow$ pγ= Solve[eq3 == 0,γ]//Flatten
$\Leftarrow \{\gamma \rightarrow -0.0000434783 \quad (-149980. + 84667.8\alpha + 204930.\beta)\}$

$\Rightarrow$ %/.ToRules[solα[[2]]]/.ToRules[solβ[[3]]]
$\Leftarrow \{\gamma \rightarrow 4.70437\}$

14.1.2.2   Solution Via Numerical Gröbner Basis

Instead of solving the equation system developed from the necessary conditions in symbolic way, we can solve this algebraic system in numeric form, too.

$\Rightarrow$ eq1

$\Leftarrow 688361.2 + 138096.3\alpha - 688361.2\alpha^2 - 993513.5\beta - 2819366.6\alpha\beta$
$+ 993513.5\alpha^2\beta + 688361.2\beta^2 + 1175652.4\alpha\beta^2 - 993513.5\beta^3 - 84667.86\gamma$
$- 299960.98\alpha\gamma + 84667.8\alpha^2\gamma + 409859.5\alpha\beta\gamma - 84667.8\beta^2\gamma + 23000\alpha\gamma^2$

$\Rightarrow$ eq2

$\Leftarrow 1409683.3 - 993513.5\alpha + 1409683.3\alpha^2 - 993513.5\alpha^3 - 1037556.1\beta$
$- 1376722.4\alpha\beta - 1175652.4\alpha^2\beta - 1409683.3\beta^2 + 993513.5\alpha\beta^2 - 204929.78\gamma$
$- 204929.7\alpha^2\gamma - 299960.9\beta\gamma + 169335.6\alpha\beta\gamma + 204929.7\beta^2\gamma + 23000\beta\gamma^2$

$\Rightarrow$ eq3

$\Leftarrow 149980.5 - 84667.8\alpha - 204929.7\beta - 23000\gamma$

$\Rightarrow$ AbsoluteTiming[sol$\alpha\beta\gamma$= NSolve[{eq1, eq2, eq3}, {α, β, γ}, Reals];]
$\Leftarrow \{0.0869599, \text{Null}\}$

The running time is acceptably short.

$\Rightarrow$ sol$\alpha\beta\gamma$
$\Leftarrow \{\{\alpha \to -1.10422 \times 10^{15}, \beta \to -5.71182 \times 10^{14}, \gamma \to 9.15411 \times 10^{15}\},$
$\quad \{\alpha \to 2.16013 \times 10^{13}, \beta \to -4.17602 \times 10^{13}, \gamma \to 2.92564 \times 10^{14}\},$
$\quad \{\alpha \to 19.4301, \beta \to -36.9881, \gamma \to 264.558\},$
$\quad \{\alpha \to -3.94057, \beta \to -2.04298, \gamma \to 39.2299\},$
$\quad \{\alpha \to 0.1884345, \beta \to 0.126022, \gamma \to 4.70437\}\}$

How can we select the proper real solution? Let us consider the values of the formal objective function based on the different real solutions

$\Rightarrow$ L$\alpha\beta\gamma$= Map[L/.#&, sol$\alpha\beta\gamma$]/.$\sigma \to 1$//N
$\Leftarrow \{-369601., -76614.4, -76645., -387030., -26180.4\}$

The real solution will be selected according to which maximizes the objective function. We seek for the real global maximum of the maximum likelihood function. The value of σ will not influence the relative comparison of the real solutions, therefore we consider $\sigma = 1$. The different real solutions of the polynomial system represent the different real local maximums. Therefore the optimal solution, which gives the highest value for the maximum likelihood function is

```
⇒ solopt = solαβγ[[Position[Lαβγ,max[Lαβγ]]//Flatten//First]]
⇐ {α → 0.188435, β → 0.126022, γ → 4.70437}
```

It is reasonable to encapsulate these statements in a function, namely

```
⇒ PlaneTLSW[XYZ_,W_] :=
Module[{X,Y,Z,a,b,c,d,e,f,g,h,i,j,N,W2,eq1,eq2,eq3,solαβγ,
Lαβγ,σ,L,solopt},
X = XYZ[[All,1]]; Y = XYZ[[All,2]]; Z = XYZ[[All,3]];

N = Length[XYZ];

W2 = Map[#²&,W];
{a,b,c,d,e,f,g,h,i,j} =
Total/@Map[MapThread[#1#2&,{#,W2}]&,

X,X²,Y,Y²,XY,Z,YZ,Z²,XZ,W2];
```

$$eq1 = j\alpha\gamma^2 + 2\alpha^2\gamma a - \gamma a(1+\alpha^2+\beta^2) + \alpha^3 b - \alpha b(1+\alpha^2+\beta^2) + 2\alpha\beta\gamma c$$
$$+ \alpha\beta^2 d + 2\alpha^2\beta e - \beta e(1+\alpha^2+\beta^2) - 2\alpha\gamma f - 2\alpha\beta g + \alpha h - 2\alpha^2 i$$
$$+ i(1+\alpha^2+\beta^2);$$

$$eq2 = j\beta\gamma^2 + 2\alpha\beta\gamma a + \alpha^2\beta b + 2\beta^2\gamma c - \gamma c(1+\alpha^2+\beta^2) + 2\alpha\beta^2 e$$
$$- \alpha e(1+\alpha^2+\beta^2) + \beta^3 d - \beta d(1+\alpha^2+\beta^2) - 2\beta\gamma f - 2\alpha\beta i + \beta h - 2\beta^2 g$$
$$+ g(1+\alpha^2+\beta^2);$$

$$eq3 = f - j\gamma - \alpha a - \beta c;$$

```
solαβγ= NSolve[{eq1,eq2,eq3},{α,β,γ},Reals];
```

$$L = -\frac{j\gamma^2}{2(1+\alpha^2+\beta^2)\sigma^2} - \frac{1}{2}N\log[2] - \frac{1}{2}N\log[\pi] - N\log[\sigma]$$

$$-a\frac{\alpha\gamma}{(1+\alpha^2+\beta^2)\sigma^2} - \frac{\alpha^2 b}{2(1+\alpha^2+\beta^2)\sigma^2} - \frac{\beta\gamma c}{(1+\alpha^2+\beta^2)\sigma^2}$$

$$-\frac{\alpha\beta e}{(1+\alpha^2+\beta^2)\sigma^2} - \frac{\beta^2 d}{2(1+\alpha^2+\beta^2)\sigma^2} + \frac{\gamma f}{(1+\alpha^2+\beta^2)\sigma^2}$$

$$+\frac{\alpha i}{(1+\alpha^2+\beta^2)\sigma^2} + \frac{\beta g}{(1+\alpha^2+\beta^2)\sigma^2} - \frac{h}{2(1+\alpha^2+\beta^2)\sigma^2};$$

```
Lαβγ= Map[L/.#&,solαβγ]/.σ → 1//N;
solopt = solαβγ[[Position[Lαβγ,max[Lαβγ]]//Flatten//First]]]
```

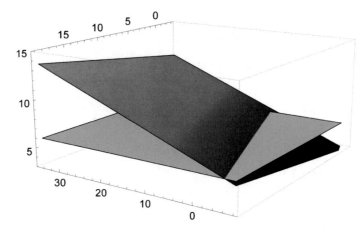

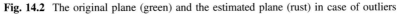

Fig. 14.2 The original plane (green) and the estimated plane (rust) in case of outliers

Now, let us employ it,

⇒ PlaneTLSW[XYZC, W]

⇐ {α → 0.188435, β → 0.126022, γ → 4.70437}

Then the estimated plane can be visualized (Fig. 14.2),

⇒ p4 = Plot3D[αx + βy + γ /.%, {x, −8, 35}, {y, −3, 19},
  BoxRatios → {1, 1, 0.5}, ClippingStyle → None, Mesh → None,
  PlotRange → {3, 15}, ColorFunction → "RustTones"];

⇒ Show[{p4, p1}]

The result of this soft method is not good. Although it would have been possible to give ad hoc values for the weights, it is difficult to find their proper ones.

In the next section we shall demonstrate how the Danish method can provide the proper weights.

14.1.3 Danish Algorithm

14.1.3.1 Basic Idea

The Danish method was proposed by Krarup et al. (1980) and is purely heuristic with no rigorous statistical theory. The method works with adaptive weights, which are altered iteratively. The weight of an observation is computed according to its error (residual), which is different from 1 if the error of this observation is greater than the standard deviation of the error distribution, σ_ϵ, namely

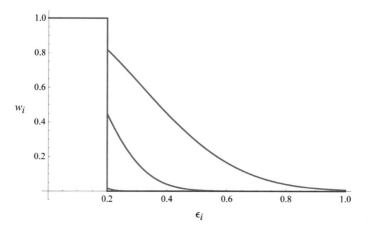

Fig. 14.3 The actual weights of the observations in case of $\sigma_\epsilon = 0.1$ and with different c_1 values

$$w_i^{(k+1)} = \exp\left(-c_1\left(\varepsilon_i^{(k)}\right)^2\right) \quad \text{if} \;\; \varepsilon_i^{(k)} \geq \frac{1}{c_2}\sigma_\epsilon, \quad \text{otherwise} \;\; w_i^{(k+1)} = 1.$$

where $\varepsilon_i^{(k)}$ is the error of the i-th observation computed with weight $w_i^{(k)}$ in the k-th iteration step. The error definition is the same as before,

$$\varepsilon_i = \frac{|z_i - \alpha x_i - \beta y_i - \gamma_s|}{\sqrt{1 + \alpha^2 + \beta^2}}.$$

Here c_2 is a suitable constant. For example in this study, let $\sigma_\varepsilon = 0.1$ and $c_2 = 1/2$. Therefore $(1/c_2)\sigma_\varepsilon = 0.2$, so $\varepsilon_i^{(k)} \geq 0.2$. In this study c_1 will be 5, 20, 100, 1000; see Fig. 14.3.

```
⇒ Plot[Map[If[ε³ᵢ0.2, exp[−#ε²ᵢ], 1]&, {5, 20, 100, 1000}],
    {εᵢ, 0, 1}, PlotRange → All, AxesLabel → {"ε″ᵢ″", "w″ᵢ″"}]
```

So when $c_1 = 1000$, we have $w_i \approx 0$ weight for outliers.

14.1.3.2 Danish Algorithm with Gröbner Basis

We want to *combine numerical Gröbner basis solution* into with the solution of the maximum likelihood estimation with *Danish algorithm* therefore we need to introduce a certain initialization phase. First we define the maximum number of the iterations,

```
⇒ N = 10; (*Number of iterations*)
```

The number of observations,

```
⇒ NC = Length[XYZC]
⇐ 23000
```

The initial weights of the observations

```
⇒ W = Table[1, {NC}]; (*initialization of weights *)
```

In order to follow the change of weights during the iteration, these weights will be collected in every iteration step,

```
⇒ CollectorW = {W}; (*initialization of the weights collertor*)
```

Similarly, we should like to follow the convergence of the solution of the estimated parameters. These values will be collected in every iteration step, too.

```
⇒ CollectorSols = {};
```

The parameter c_1 is

```
⇒ c1 = 1000;
```

Then the cycle for the iteration can be written as

```
⇒ AbsoluteTiming[For[J = 1, J ≤ N, J + +,
 sols = PlaneTLSW[XYZC, W];
 Error = Map[((#[[3]]−α#[[1]]−β#[[2]]−γ)/(√(1+α² + β²)))/.
 sols&, XYZC];
 StD = StandardDeviation[Error];
 W = Table[If[(Abs[Error[[i]]] < StD), 1,
 Exp[−c1 Error[[i]]²]], {i, 1, NC}];
 CollectorW = Append[CollectorW, W];
 CollectorSols = Append[CollectorSols, {α, β, γ}/.sols];
 ]; ]
⇐ {9.55748, Null}
```

The next Figs. 14.4, 14.5 and 14.6 show the convergence of the different estimated parameters as a function of the number of the iterations.

```
⇒ ListPlot[Transpose[CollectorSols][[1]],
 Joined → True, Joined → True, PlotRange → All,
 AxesLabel → {"Number of iterations", "α"}]
```

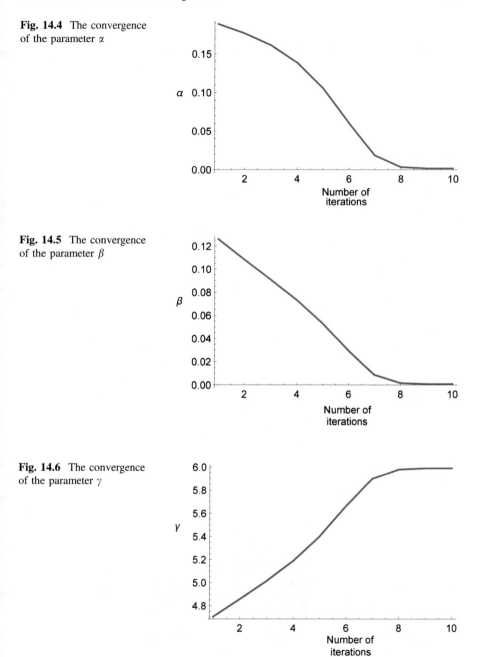

Fig. 14.4 The convergence of the parameter α

Fig. 14.5 The convergence of the parameter β

Fig. 14.6 The convergence of the parameter γ

⇒ ListPlot[Transpose[CollectorSols][[2]],
   Joined → True, Joined → True, PlotRange → All,
   AxesLabel → {"Number of iterations", "β"}]

⇒ ListPlot[Transpose[CollectorSols][[3]],
   Joined → True, Joined → True, PlotRange → All,
   AxesLabel → {"Number of iterations", "γ"}]

The result of the iteration

⇒ W = Last[CollectorW];
⇒ sols = PlaneTLSW[XYZC, W]
⇐ {α → 0.00169951, β → 0.000787971, γ → 5.98921}

The observations, the inlier data points which remained active (Fig. 14.7),

⇒ inliers = Select[MapThread[If[#1 < 0.1, 0, #2]&, {W, XYZC}],
   Length[#] == 3&];
⇒ Length[inliers]
⇐ 20011

⇒ p5 = ListPointPlot3D[inliers, PlotStyle → Brown];
⇒ Show[{p1, p5}]

Only eleven points remained from the outliers.

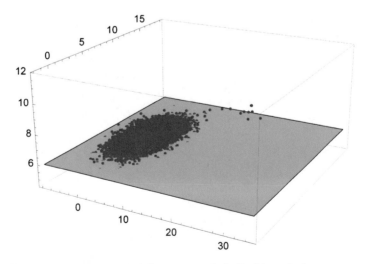

Fig. 14.7 The original plane and the inliers computed via Danish method

14.1.4 Danish Algorithm with PCA

The Danish algorithm can cure other soft algorithms too, and make them proper for robust estimation. Let us consider the Principal Component Analysis (PCA), which is a popular statistical technique that describes the covariance structure of data by means of a small number of components. These components are linear combinations of the original variables that rank the variability in the data through the variances, and produce directions using the eigenvectors of the covariance matrix. The eigenvector corresponding to the smallest eigenvalue is exactly the normal of the best-fitted plane.

In order to employ PCA in the Danish algorithm, one should extend the traditional PCA algorithm to weighted PCA (PCAW), see Weingarten et al. (2004).

First, we compute the weighted center of gravity,

$$c_0 = \frac{\sum_{i=1}^{N} \tilde{w}_i \begin{pmatrix} x_i \\ y_i \\ z_i \end{pmatrix}}{\sum_{i=1}^{N} \tilde{w}_i}.$$

Let us shift the data of observations to this point,

$$C_{c_0 i} = \begin{pmatrix} x_i \\ y_i \\ z_i \end{pmatrix} - c_0, \quad i = 1, 2, \ldots N.$$

The weighted covariance matrix is symmetric,

$$A = \begin{pmatrix} \sum_i \tilde{w}_i (C_{c_0 i})_1^2 & \sum_i \tilde{w}_i (C_{c_0 i})_1 (C_{c_0 i})_2 & \sum_i \tilde{w}_i (C_{c_0 i})_1 (C_{c_0 i})_3 \\ \cdot & \sum_i \tilde{w}_i (C_{c_0 i})_2^2 & \sum_i \tilde{w}_i (C_{c_0 i})_2 (C_{c_0 i})_3 \\ \cdot & \cdot & \sum_i \tilde{w}_i (C_{c_0 i})_3^2 \end{pmatrix}.$$

Considering the Hesse form, the weighted residual to be minimized can be written,

$$R(n_x, n_y, n_z, d) = \sum_{i=1}^{N} \tilde{w}_i (n_x x_i + n_y y_i + n_z z_i - d)^2.$$

Remark Remember that in the estimator based on the maximization of the likelihood function we applied $w_i = \sqrt{\tilde{w}_i}$.

Therefore considering that

$$\frac{\partial R}{\partial d} = 0$$

we get

$$n_x \sum_{i=1}^{N} \tilde{w}_i x_i + n_y \sum_{i=1}^{N} \tilde{w}_i y_i + n_z \sum_{i=1}^{N} \tilde{w}_i z_i - d \sum_{i=1}^{N} \tilde{w}_i = 0.$$

Since the plane normal can be computed as the eigenvectors of the weighted covariance matrix A, the surface parameter d is determined. Consequently the surface is

$$z = -\frac{n_x}{n_z} x - \frac{n_y}{n_z} y + \frac{d}{n_z}.$$

Now, let us check our algorithm with the weights provided by the computation done above in Sect. 13.1.3.2.

First let us compute $\tilde{w}_i = w_i^2$,

$$\Rightarrow \text{Ws} = \text{Map}\left[\#^2 \&, \text{W}\right];$$

The weighted data of the observations,

$$\Rightarrow \text{XYZW} = \text{MapThread}[\#1\ \#2\&, \{\text{XYZC}, \text{Ws}\}];$$

therefore the weighted center of gravity,

$$\Rightarrow \text{c0} = \text{Map}[\text{Total}[\#]\&, \text{Transpose}[\text{XYZW}]]/\text{Total}[\text{Ws}]$$
$$\Leftarrow \{1.98748, 8.01922, 6.00089\}$$

Then the shifted data

$$\Rightarrow \text{Cc0} = \text{Map}[\# - \text{c0}\&, \text{XYZC}];$$

Preparation of the shifted data in order to compute their weighted covariance matrix,

$$\Rightarrow \text{Cc0W} = \text{MapThread}[\#1\ \sqrt{\#2}\ \&, \{\text{Cc0}, \text{Ws}\}];$$

Then the covariance matrix

$$\Rightarrow \text{AbsoluteTiming}[\text{A} = \text{Covariance}[\text{Cc0W}]; \text{MatrixForm}[\text{A}]]$$

$$\Leftarrow \left\{ 0.0903777, \begin{pmatrix} 4.40013 & 0.0429781 & 0.00899076 \\ 0.0429781 & 4.39533 & 0.002346 \\ 0.00899076 & 0.002346 & 0.0093104 \end{pmatrix} \right\}$$

The eigenvector having $nz = 1$ is choosen, see the expression of $z = z(x, y)$ above,

$\Rightarrow \{nx, ny, nz\} = Eigenvectors[A][[3]]$

$\Leftarrow \{-0.00204257, -0.000514862, 0.999998\}$

Therefore the parameter d can be computed

$\Rightarrow sold = Solve[\{nx, ny, nz\}.Total[XYZW] - d\ Total[Ws] == 0, d]//$
Flatten

$\Leftarrow \{d \rightarrow 5.99269\}$

Consequently the coefficients of surface model are

$\Rightarrow \left\{-\dfrac{nx}{nz}, -\dfrac{ny}{nz}, \dfrac{d}{nz}\right\}/.sold$

$\Leftarrow \{0.00204258, 0.000514863, 5.9927\}$

We want to implement the weighted PCA into the Danish method.
First let us summarize the steps of computations of the PCAW algorithm in a
Mathematica function,

```
⇒ PlanePCAW[XYZ_, W_] :=
    Module[{Ws, XYZW, co, Cco, CcoW, nx, ny, nz, d, sold},
    Ws = Map[#^2&, W];
    XYZW = MapThread[#1#2&, {XYZ, Ws}];
    co = Map[Total[#]&, Transpose[XYZW]]/Total[Ws];
    Cco = Map[# - co&, XYZ];
    CcoW = MapThread[#1 #2&, {Cc0, W}];
    {nx, ny, nz} = Eigenvectors[Covariance[CcoW]][[3]];
    sold =
    Solve [{nx, ny, nz}.Total[XYZW] - d Total[Ws] == 0, d]//Flatten;
```

$\left\{\alpha \rightarrow -\dfrac{nx}{nz}, \beta \rightarrow -\dfrac{ny}{nz}, \gamma \rightarrow \dfrac{d}{nz}\right\}/.sold]$

Let us check it

$\Rightarrow PlanePCAW[XYZC, W]$

$\Leftarrow \{\alpha \rightarrow 0.00204258, \beta \rightarrow 0.000514863, \gamma \rightarrow 5.9927\}$

Now, the integration of PCAW into the Danish algorithm is similar to that of the
numerical Gröber basis solution, see above

$\Rightarrow N = 10;\ (*Number\ of\ iterations*)$

The number of observations,

⇒ NC = Length[XYZC]

⇐ 23000

The initial weights of the observations

⇒ W = Table[1, {NC}]; (∗initialization of weights∗)

In order to follow the change of weights during the iteration, these weights will be collected in every iteration step,

⇒ CollectorW = {W}; (∗initialization of the weights collertor∗)

Similarly, we should like to follow the convergence of the solution of the estimated parameters. These values will be collected in every iteration step, too.

⇒ CollectorSols = {};

The parameter c_1 is

⇐ c1 = 1000;

Then the cycle for the iteration can be written as

⇒ AbsoluteTiming[For[j = 1, j ≤ N, j++,
    sols = PlanePCAW[XYZC, W];

Error = Map[((#[[3]]−α#[[1]]−β#[[2]]−γ)/($\sqrt{1+α^2+β^2}$))/.
sols&, XYZC];

StD = StandardDeviation[Error];

W = Table[If[(Abs[Error[[i]]] < StD), 1, Exp[−c1]Error[[i]]^2]],
{i, 1, NC}];

CollectorW = Append[CollectorW, W];

CollectorSols = Append[CollectorSols, {α, β, γ}/.sols];
];]

⇐ {7.23155, Null}

The computation time is somewhat longer than the Gröbner solution' time was. The next figures show the convergence of the different parameters as a function of the number of the iterations (Figs. 14.8, 14.9 and 14.10).

⇒ ListPlot[Transpose[CollectorSols][[1]], Joined → True,
Joined → True, PlotRange → All,
AxesLabel → {"Number of iterations", "α"}]

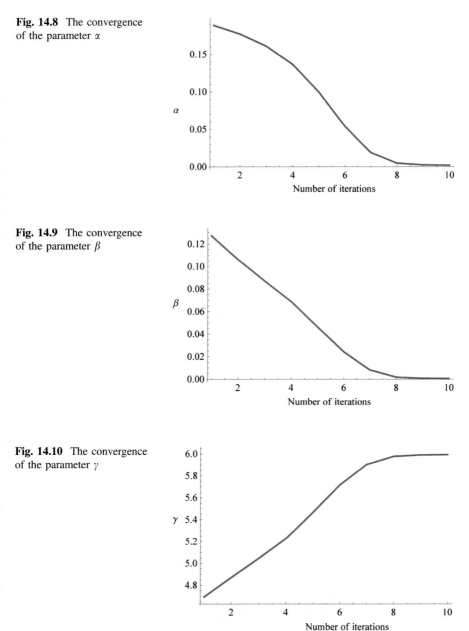

Fig. 14.8 The convergence of the parameter α

Fig. 14.9 The convergence of the parameter β

Fig. 14.10 The convergence of the parameter γ

⇒ ListPlot[Transpose[CollectorSols] [[2]], Joined → True,
  Joined → True, PlotRange → All,
  AxesLabel → {"Number of iterations", "β"}]

⇒ ListPlot[Transpose[CollectorSols] [[3]], Joined → True,
  Joined → True, PlotRange → All,
  AxesLabel → {"Number of iterations", "γ"}]

The result of the iteration

⇒ W = Last[CollectorW];
⇒ sols = PlanePCAW[XYZC, W]
⇐ {α → 0.00204258, β → 0.000514863, γ → 5.9927}

The observations which remained active (inliers) (Fig. 14.11),

⇒ inliers = Select[MapThread[If[#1 < 0.1, 0, #2]&, {W, XYZC}],
  Length[#] == 3&];
⇒ Length[inliers]
⇐ 20010

⇒ p6 = ListPointPlot3D[inliers, PlotStyle → Green];
⇒ Show[{p1, p6}]

The result is the same as that of the Danish method combined with M-estimation.

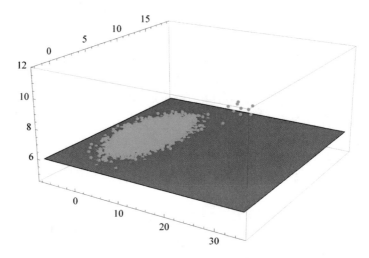

Fig. 14.11 The original plane and the inliers computed via Danish method with weighted PCA

14.1.5 RANSAC Algorithm

14.1.5.1 Basic Idea of RANSAC

Let us try to apply the RANSAC method, given in Zuliani (2012), which has proven to be successful for detecting outliers. The basic RANSAC algorithm is as follows:

(1) Pick up a model type (M)
(2) Input data as

> dataQ—data corrupted with outliers (cardinality (dataQ) = n).
> s—number of data elements required per subset.
> N—number of subsets to draw from the data.
> τ—threshold which defines if data element, $d_i \in$ dataQ, agrees with the model M

Remarks In special case s can be the minimal number of the data which results in a determined system for the unknown parameters of the model.

The number of subsets to draw from the data, N is chosen high enough to ensure that at least one of the subsets of the random examples does not include an outlier (with the probability p, which is usually set to 0.99). Let u represent the probability that any selected data point is an inlier and $v = 1 - u$ the probability of observing an outlier. Then the iterations N can be computed as

$$N = \frac{\log(1 - p)}{\log(1 - (1 - v)^s)}$$

(3) maximalConsensusSet $\leftarrow \varnothing$
(4) Iterate N times:

  (a) ConsensusSet $\leftarrow \varnothing$
  (b) Randomly draw a subset containing s elements and estimate the parameters of the model M
  (c) For each data element, $d_i \in$ dataQ:
    if agree (d_i, M, τ), ConsensusSet $\leftarrow d_i$
  (d) if cardinality (maximalConsensusSet) < cardinality(ConsensusSet),
    maximalConsensusSet $\leftarrow$ ConsensusSet

(5) Estimate model parameters using maximalConsensusSet.

14.1.5.2   The Necessary Number of Iterations

In our case we have 3 parameters of the plane to determine, so we need minimum 3 equations, therefore

$\Rightarrow s = 3;$

In this case the parameters can be computed directly from a linear system. However, in general if $s > 3$ we need linear least squares solution. Let

$\Rightarrow p = 0.99;$

In addition let us assume that the probability of observing outlier is

$\Rightarrow v = 1/2;$

Then the number of the necessary iterations is

$\Rightarrow N = \text{Round}[\text{Log}[1 - p]/\text{Log}[1 - (1 - n)^s]]$
$\Leftarrow 34$

14.1.5.3   Threshold

Let the threshold τ

$\Rightarrow \tau = 10;$

An element $d_i \in \text{dataQ}$ will be accepted if the local error is less than the prespecified threshold

$$\frac{|z_i - \alpha_s x_i - \beta_s y_i - \gamma_s|}{\sqrt{1 + \alpha_s^2 + \beta_s^2}} < \tau$$

Remarks d_i is represented by the triplet (x_i, y_i, z_i) and the parameters $(\alpha_s, \beta_s, \gamma_s)$ computed from a subset of s elements.

Here ordinary least square (OLS_z) estimation could be also applied as maximum support, see Fischler and Bolles (1981).

14.1.5.4   Computation of the Method Step by Step

$\Rightarrow \texttt{inliers} = \texttt{Table}[\{\}, \{\texttt{N}\}](* \text{ the consensus sets}, \texttt{j} = 1, 2, ...\texttt{N}*)$
$\Leftarrow \{\{\}, \{\}, \{\}, \{\}, \{\}, \{\}, \{\}, \{\}, \{\}, \{\}, \{\}, \{\}, \{\}, \{\}, \{\}, \{\}, \{\},$
$\quad \{\}, \{\}, \{\}, \{\}, \{\}, \{\}, \{\}, \{\}, \{\}, \{\}, \{\}, \{\}, \{\}, \{\}, \{\}, \{\}\}$

Let us consider a subset $(j = 1, 2, \ldots N)$.

```
⇒ j = 2 (*second iteration means second subset*)
⇐ 2
```

Assigning noisy data to dataQ.

```
⇒ dataQ = XYZC;
```

Let us select a triplet randomly

```
⇒ dataS = RandomSample[dataQ, s]
⇐ {{2.93055, 7.24291, 5.96181}, {1.26074, 5.85424, 5.93},
   {−0.126164, 7.10004, 6.10541}}
```

For further computation we need a function form of our algorithm.

14.1.5.5   RANSAC with Gröbner Basis

We want to integrate numerical Gröbner basis solution into the RANSAC algorithm, therefore it is reasonable to write a function for this suggested algorithm,

```
⇒ PlaneTLS[XYZ_] :=
  Module[{X, Y, Z, a, b, c, d, e, f, g, h, i, N, eq1, eq2, eq3, solαβγ,
  selsol, Lαβγ, L, solopt},
  X = XYZ[[All, 1]]; Y = XYZ[[All, 2]]; Z = XYZ[[All, 3]];
```

$$\{a, b, c, d, e, f, g, h, i\} = \text{Total}/@\{X, X^2, Y, Y^2, XY, Z, YZ, Z^2, XZ\};$$
$$N = \text{Length}[XYZ];$$
$$eq1 = N\alpha\gamma^2 + 2\alpha^2\gamma a - \gamma a(1 + \alpha^2 + \beta^2) + \alpha^3 b - \alpha b(1 + \alpha^2 + \beta^2) + 2\alpha\beta\gamma c$$
$$+ \alpha\beta^2 d + 2\alpha^2\beta e - \beta e(1 + \alpha^2 + \beta^2) - 2\alpha\gamma f - 2\alpha\beta g + \alpha h - 2\alpha^2 i + i(1 + \alpha^2 + \beta^2);$$

$$eq2 = N\beta\gamma^2 + 2\alpha\beta\gamma a + \alpha^2\beta b + 2\beta^2\gamma c - \gamma c(1 + \alpha^2 + \beta^2) + 2\alpha\beta^2 e - \alpha e(1 + \alpha^2$$
$$+ \beta^2) + \beta^3 d - \beta d(1 + \alpha^2 + \beta^2) - 2\beta\gamma f - 2\alpha\beta i + \beta h - 2\beta^2 g + g(1 + \alpha^2 + \beta^2);$$
$$eq3 = f - N\gamma - \alpha a - \beta c;$$
$$sol\alpha\beta\gamma = \text{NSolve}[\{eq1, eq2, eq3\}, \{\alpha, \beta, \gamma\}];$$

$$selsol = \text{Select}[sol\alpha\beta\gamma, \text{Im}[\#[[1]][[2]]] == 0\&];$$

$$L = -\frac{h}{2(1 + \alpha^2 + \beta^2)\sigma^2} + \frac{i\alpha}{(1 + \alpha^2 + \beta^2)\sigma^2} - \frac{b\alpha^2}{2(1 + \alpha^2 + \beta^2)\sigma^2}$$
$$+ \frac{g\beta}{(1 + \alpha^2 + \beta^2)\sigma^2} - \frac{e\alpha\beta}{(1 + \alpha^2 + \beta^2)\sigma^2} - \frac{d\beta^2}{2(1 + \alpha^2 + \beta^2)\sigma^2} + \frac{f\gamma}{(1 + \alpha^2 + \beta^2)\sigma^2}$$
$$- \frac{a\alpha\gamma}{(1 + \alpha^2 + \beta^2)\sigma^2} - \frac{c\beta\gamma}{(1 + \alpha^2 + \beta^2)\sigma^2} - \frac{N\gamma^2}{2(1 + \alpha^2 + \beta^2)\sigma^2}$$

$$-\frac{1}{2}\mathrm{Nlog}[2]-\frac{1}{2}\mathrm{Nlog}[\pi]-\mathrm{Nlog}[\sigma];$$

$$\mathtt{L\alpha\beta\gamma= Map[L/.\#\&, selsol]/.\sigma \rightarrow 1//N};$$

$$\mathtt{solopt = selsol[[Position[L\alpha\beta\gamma, max[L\alpha\beta\gamma]]//Flatten//First]]}$$

Now let us compute the parameters from the randomly selected subset

$\Rightarrow \mathtt{sols = PlaneTLS[dataS]}$

$\Leftarrow \{\alpha \rightarrow -0.0509108,\ \beta \rightarrow 0.0841216,\ \gamma \rightarrow 5.50172\}$

Then check for all elements of **dataQ** ($i = 1, 2, \ldots, n$) whether it can satisfy the threshold. For example let

$\Rightarrow \mathtt{i = 1131};$

$\Rightarrow \mathtt{dataQ[[i]]}$

$\Leftarrow \{0.320363, 4.46075, 5.90436\}$

$\Rightarrow \mathtt{If[(((1/\sqrt{1+\alpha^2+\beta^2})Abs[dataQ[[i]][[3]]-\alpha dataQ[[i]][[1]]}$
$\mathtt{-\beta dataQ[[i]][[2]]-\gamma])/.sols)<\tau,}$
$\mathtt{inliers[[j]] = Append[inliers[[j]], dataQ[[i]]]]}$

$\Leftarrow \{\{0.320363, 4.46075, 5.90436\}\}$

$\Rightarrow \mathtt{inliers}$

$\Leftarrow \{\{\}, \{\{0.320363, 4.46075, 5.90436\}\}, \{\}, \{\}, \{\}, \{\}, \{\}, \{\}, \{\}, \{\}, \{\}, \{\}, \{\},$
$\{\}, \{\}, \{\}, \{\}, \{\}, \{\}, \{\}, \{\}, \{\}, \{\}, \{\}, \{\}, \{\}, \{\}, \{\}, \{\}, \{\}, \{\}, \{\}\}$

Employing the following double **For** cycle

$\Rightarrow \mathtt{For\ j = 1,2,...N}$
$\quad \mathtt{For\ i = 1,2,...NC}$

we can get the maximal consensus whose size is between s and *NC*. Here *NC* is the number of observations, see Sect. 13.1.3.2. Then the computation is straightforward.

14.1.5.6 Application the RANSAC Method

Now let us use for threshold $\tau = 0.75$

$\Rightarrow \mathtt{\tau= 0.75};$

and let

$\Rightarrow \mathtt{s = 3};$

therefore

$$\Rightarrow N = \text{Round}[\text{Log}[1 - p]/\text{Log}[1 - (1 - n)^s]]$$
$$\Leftarrow 34$$

Preparation of the sets of inliers for every iteration step—at the beginning they are empty sets:

$$\Rightarrow \texttt{inliers} = \texttt{Table}[\{\}, \{N\}](* \text{ the consensus sets, } j = 1, 2, ...N*);$$

Then let us carry out the double **For** cycle to find the maximum consensus set

```
⇒ AbsoluteTiming[For[j = 1, j ≤ N, j + +,
  (*get four random points*)
  dataS = RandomSample[dataQ, s];
  (*determine model parameters*)
  sols = PlaneTLS[dataS];
  (* determine inliers*)
  For[i = 1, i ≤ NC, i + +,
    If[((((1/√(1 + α² + β²))Abs[dataQ[[i]][[3]]−αdataQ[[i]][[1]]
      −βdataQ[[i]][[2]]−γ])/.sols) < τ,
      inliers[[j]] = Append[inliers[[j]], dataQ[[i]]];
    ]; ]; ]
⇐ {39.9638, Null}
```

Select the consensus set which has the highest number of inliers (Fig. 14.12).

$$\Rightarrow \texttt{trimmeddata} = \texttt{inliers}[[\texttt{Ordering}[\texttt{inliers}, -1]]][[1]];$$
$$\Rightarrow \texttt{Length}[\texttt{trimmeddata}]$$
$$\Leftarrow 20048$$

$$\Rightarrow \texttt{p5} = \texttt{ListPointPlot3D}[\texttt{trimmeddata}, \texttt{PlotStyle} \rightarrow \texttt{Green}];$$
$$\Rightarrow \texttt{Show}[\{\texttt{p1}, \texttt{p5}\}]$$

Then we can compute the parameters

$$\Rightarrow \texttt{PlaneTLS}[\texttt{trimmeddata}]$$
$$\Leftarrow \{\alpha \rightarrow 0.0137893, \beta \rightarrow 0.00655942, \gamma \rightarrow 5.92509\}$$

Having a multicore machine, one can evaluate **For** cycles in parallel. Let us compute the internal **For** cycle parallel, then `ParallelTable` should be used used instead of **For**. Now we launch twice as many kernels as the number of cores.

Let us refresh the sets of inliers for every iteration step—at the beginning they are empty sets:

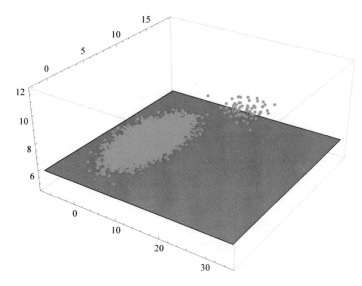

Fig. 14.12 The original plane and the inliers computed via RANSAC (green)

⇒ inliers= Table[{}, {N}](*the consensus sets, j = 1, 2,...N*);

⇒ AbsoluteTiming[For[j = 1, j ≤ N, j + +,

(*get three random points*)

dataS = RandomSample[dataQ, s];

(*determine model parameters*)

 sols = PlaneTLS[dataS];

(*determine inliers*)

S = ParallelTable[

  If[(((1/√(1 + a² + b²))Abs[dataQ[[i]][[3]] − adataQ[[i]][[1]]

  −bdataQ[[i]][[2]] − g])/.sols) < t, dataQ[[i]]];

{i, 1, NC}];

inliers[[j]] = Append[inliers[[j]],

Select[S, (Head[#] == List)&]][[1]];

];]

⇐ {39.9638, Null}

Select the consensus set which contains the highest number of inliers
(Fig. 14.13).

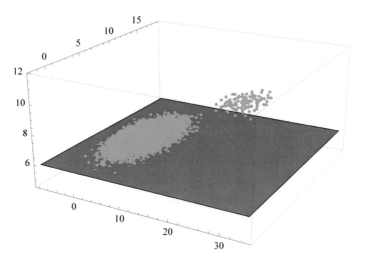

Fig. 14.13 The original plane and the inliers computed via RANSAC

⇒ trimmeddata = inliers[[Ordering[inliers, −1]]][[1]];
⇒ Length[trimmeddata]
⇐ 20077

⇒ p5 = ListPointPlot3D[trimmeddata, PlotStyle → Green];
⇒ Show[{p1, p5}]

Then we can compute the parameters

⇐ PlaneTLS[trimmeddata]
⇐ {α → 0.0277224, β → 0.0114123, γ → 5.86191}

We can improve the method—increase the precision and reduce the running time —by preparing the data set via a soft computing technique, like neural network algorithm, see in the next section.

14.1.5.7   Application of Self-Organizing Map (SOM) to the RANSAC Algorithm

The Self-Organizing Map (SOM) is a neural network algorithm, which uses a competitive learning technique to train itself in an unsupervised manner. SOMs are different from other artificial neural networks in the sense that they use a neighborhood function to preserve the topological properties of the input space and they are used to create an ordered representation of multi-dimensional data that simplifies complexity and reveals meaningful relationships.

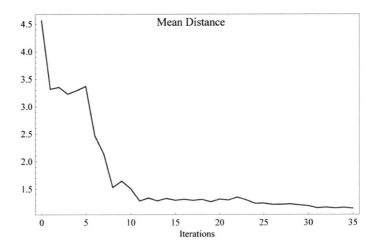

Fig. 14.14 The error of the Kohonen-map with a lattice of SOM of size 6 × 6 code-book vectors during the training of the network

In the last decade a lot of research has been carried out to develop surface reconstruction methods. A more recent approach to the problem of surface reconstruction is that of learning based methods see e.g. DalleMole et al. (2010). Learning algorithms are able to process very large or noisy data, such as point clouds obtained from 3D scanners and are used to construct surfaces. Following this approach some studies have been employed SOM and their variants for surface reconstruction. SOM is suitable for this problem because it can form topological maps representing the distribution of input data. In our case this mapping occurs from 3D space to 2D, see Kohonen (1998).

Therefore we try to represent this cloud of observed data points by considerably fewer points using the Kohonen-map with a lattice of SOM of size 6 × 6 code-book vectors with symmetrical neighborhood (Fig. 14.14).

```
⇒ < <NeuralNetworks′
⇒ AbsoluteTiming[{som, fitrecord} =
  UnsupervisedNetFit[XYZC, 36, 35, SOMR{6, 6}]; //Quiet]
```

Figure 14.15 shows the original plane with the 6 × 6 = 36 codebook vector points.

```
⇒ dataSOM = som[[1]];
⇒ p6 = ListPointPlot3D[dataSOM,
  PlotStyle → Directive[Magenta, PointSize[0.015]]];
⇒ Show[{p1, p6}]
```

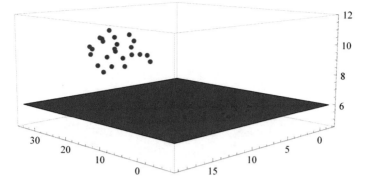

Fig. 14.15 The original plane and the code-book vectors of the SOM of 6 × 6

Now let us apply RANSAC to these 36 code-book vectors with a low threshold

```
⇒ τ = 0.05;
⇒ NSOM = Length[dataSOM];
⇒ inliers = Table[{}, {N}] (* the consensus sets, j = 1, 2, ...N*);

⇒ AbsoluteTiming[For[j = 1, j£N, j + +,
  (*get three random points*)
  dataS = RandomSample[dataSOM, s];
  (*determine model parameters*)
  sols = PlaneTLS[dataS];
  (* determine inliers*)
  For[i = 1, i → NSOM, i + +,
   If[(((1/√(1 + α² + β²)Abs[dataQ[[i]][[3]]−αdataQ[[i]][[1]]
    −βdataQ[[i]][[2]]−γ])/.sols) < τ,
    inliers[[j]] = Append[inliers[[j]], dataQ[[i]]];
  ]; ]; ]
⇐ {1.46515, Null}
```

Select the consensus set which has the highest number of inliers

```
⇒ trimmeddata = inliers[[Ordering[inliers, −1]]][[1]];
⇒ Length[trimmeddata]
⇐ 14
```

Figure 14.16 shows the result. These 14 codevectors are practically on the plane.

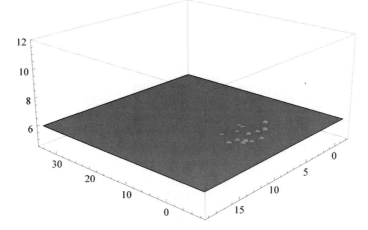

Fig. 14.16 The original plane and the inliers of RANSAC (magenta)

⇒ p7 = ListPointPlot3D[trimmeddata, PlotStyle → Directive[Green,
  PointSize[0.015]]];
⇒ Show[{p1, p7}]

Then we can compute the parameters

⇒ PlaneTLS[trimmeddata]
⇐ {α → −0.0119509, β → −0.00791939, γ → 6.05779}

The error of the method is quite acceptable. Now the total computation time of SOM + RANSAC parallel combination is just the same as that of the direct RANSAC with the original number of points. However having a more than four core machine, this is not the case, since parallel processing can reduce the computation time further.

14.2   Application Examples

14.2.1   Fitting a Sphere to Point Cloud Data

In order to understand the geometry of an object as a function of its depth values relative to the sensor, it is vital to have both a smooth and accurate estimate of the depth map. Due to the limitations of the depth sensing technology used by inexpensive devices, a significant loss occurs when capturing depth information. Unlike depth measuring technologies, like laser based LIDAR or range finding cameras, theses sensors contain much missing data and are incapable of measuring depth for shiny objects or objects of certain orientation. On the flip side, traditional depth capturing devices cost in the order of several thousand dollars, while low resolution sensors costs around $100–200. These sensors mainly are intended for close distance human pose recognition, like Microsoft's Kinect, and are operated in an environment with a conspicuous foreground (human) and background (room), so a clean and smooth depth map is not necessary.

However, if the ability of such sensors is to be extended to a more general setting, such as the inclusion in consumer laptops for human–computer interaction, robotics for robot motion planning, or graphics for environment reconstruction, both smooth and complete depth information are essential and could significantly enhance the accuracy of object detection.

In our example the measurement has been carried out with *Microsoft Kinect XBOX* see Fig. 14.17. Microsoft's Kinect contains a diverse set of sensors, most notably a depth camera based on *PrimeSense's* infrared structured light technology. With a proper calibration of its color and depth cameras, the Kinect can capture detailed color point clouds at up to 30 frames per second. This capability uniquely positions the Kinect for use in fields such as robotics, natural user interfaces, and three-dimensional mapping, see i.e. Draelos (2012).

This device provides 2D RGB color imaging and RGB data representing the depth—the distance of the object—in 11 bits (0.2048).

This low resolution causes a discontinuity effect, which can be even 10 cm above a four meter object distance, and is about 2 cm when the object distance is less than 2.5 m. In our case a spherical object, having radius $R = 0.152$ m was placed in the real world position $x = 0$, $y = 0$ and the object distance $z = 3$ m. Since the intensity threshold resolution is low, one may have quantized levels caused by round-off processes. Figure 14.18 represents the quantized depth data points simulated as synthetic data for illustration.

Real measurements suffer not only from quantization but from random errors. Let us load a real measured data set adopted from Molnár et al. (2012).

$\Rightarrow$ dataQ = Import$\left["\text{H} :\backslash\backslash \text{f\_03\_05.dat}"\right]/1000;$

The number of the data points is,

$\Rightarrow$ n = Length[dataQ]

$\Leftarrow$ 2710

Fig. 14.17 Microsoft Kinect XBOX

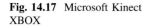

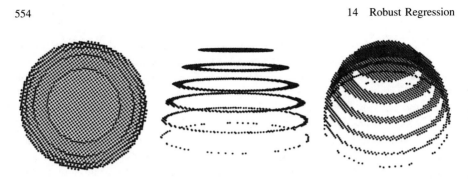

Fig. 14.18 Simulated synthetic quantized depth data points in case of a sphere from 3 different view points

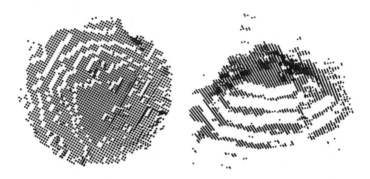

Fig. 14.19 Real measured data points from different view points

The cloud of measured points is shown in Fig. 14.19.

There are three basic approaches for estimating the position (a, b, c) and the radius (R) of a sphere from data point clouds, namely algebraic, geometric and directional least squares estimate.

14.2.1.1   Algebraic Least Square When R is Unknown

In this case, we want to fit given n points to a sphere such that the sum of the squared algebraic distance is minimized then the local error is defined as

$$\delta_i = (x_i - a)^2 + (y_i - b)^2 + (z_i - c)^2 - R^2$$

where (x_i, y_i, z_i) are the coordinates of the measured point. First let us suppose that the position as well as the radius of the sphere is unknown.

The parameters to be estimated are R, a, b and c. Let us introduce a new parameter,

$$d = a^2 + b^2 + c^2 - R^2.$$

Then δ_i can be expressed as

$$\delta_i = (-2x_i, -2y_i, -2z_i, 1) \begin{pmatrix} a \\ b \\ c \\ d \end{pmatrix} + (x_i^2 + y_i^2 + z_i^2).$$

Consequently the parameter estimation problem is linear, namely one should solve an overdetermined system. In matrix form

$$M p = h$$

where

$$M = \begin{pmatrix} -2x_1 & -2y_1 & -2z_1 & 1 \\ & & \vdots & \\ -2x_i & -2y_i & -2z_i & 1 \\ & & \vdots & \\ -2x_n & -2y_n & -2z_n & 1 \end{pmatrix}, \quad p = \begin{pmatrix} a \\ b \\ c \\ d \end{pmatrix} \quad and \quad h = \begin{pmatrix} -(x_1^2 + y_1^2 + z_1^2) \\ -(x_i^2 + y_i^2 + z_i^2) \\ \vdots \\ -(x_n^2 + y_n^2 + z_n^2) \end{pmatrix}.$$

Then the solution in least square sense

$$\begin{pmatrix} a \\ b \\ c \\ d \end{pmatrix} = M^{-1} h$$

where M^{-1} is the pseudoinverse of M.

The attractive feature of the algebraic parameter estimation is that its formulation results in a linear least squares problem which has non-iterative closed-form solution. Therefore the algebraic distance based fitting can be used as a quick way to calculate approximate parameter values without requiring an initial guess.

First we create the M matrix.

```
⇒ M = Map[{-2 #[[1]], -2 #[[2]], -2 #[[3]], 1}&, dataQ];
⇒ Dimensions[M]
⇐ {2710, 4}
```

The right hand side vector is

```
⇒ h = Map[-(#[[1]]² + #[[2]]² + #[[3]]²)&, dataQ];
```

Then we can compute the parameters a, b, c and d,

$\Rightarrow \{a, b, c, d\} = \text{PseudoInverse}[M].h$

$\Leftarrow \{-0.0413512, -0.00851136, 3.00166, 8.99215\}$

and the estimated radius of the sphere is

$\Rightarrow R = \sqrt{a^2 + b^2 + c^2 - d}$

$\Leftarrow 0.139948$

Since we want to use these parameter values as initial values for other methods, we assign them to the following variables,

$\Rightarrow a0 = a; \ b0 = b; \ c0 = c; \ d0 = d; \ R0 = R;$

Let us display this estimated sphere with the data points, see Fig. 14.20.

The disadvantages of this method are its sensitivity for noise, outliers and partial occlusion, namely when the data points cover only a part of the surface of the sphere.

14.2.1.2   Algebraic Least Square When R is Known

Sometimes the radius of sphere is known i.e. in case of calibration, and we want to find only the position of the sphere. Therefore there are 3 parameters a, b and c to be estimated. Now we have an overdetermined polynomial system to be solved,

$$(x_i - a)^2 + (y_i - b)^2 + (z_i - c)^2 - R^2 = 0 \quad i = 1, 2, \dots n.$$

Fig. 14.20 The measured data with the estimated object

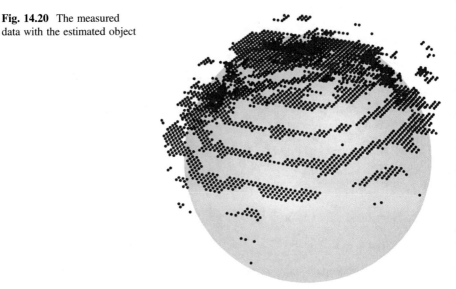

The equations can be easily generated:

$\Rightarrow$ Clear[a, b, c]; R = 0.152;

the prototype of the equation,

$\Rightarrow$ G = $(x - a)^2 + (y - b)^2 + (z - c)^2 - R^2$;

Then the system can be created,

$\Rightarrow$ eqs = Map[G/.{x $\rightarrow$ #[[1]], y $\rightarrow$ #[[2]], z $\rightarrow$ #[[3]]}&, dataQ];

For example the first equation is,

$\Rightarrow$ eqs[[1]]

$\Leftarrow$ $-0.023104 + (-0.1958 - a)^2 + (-0.0309 - b)^2 + (2.967 - c)^2$

The number of the equations is,

$\Rightarrow$ Length[eqs]

$\Leftarrow$ 2710

The solution of this problem is not unique since the reduced Gröbner basis for the determined polynomial subsystems ($n = 3$) are second order polynomials, see Appendix A1 at the end of this section. To solve the overdetermined system one can use local or global techniques. The local methods need an initial guess for the parameter values, which can be the solution of the linear problem. One of the most effective local methods is the *Extended Newton method*. Extended Newton method can handle nonlinear overdetermined systems by using the pseudoinverse of the Jacobian.

Let us employ the result of the algebraic solution as initial guess

$\Rightarrow$ NewtonExtended[eqs, {a, b, c}, {a0, b0, c0}]

$\Leftarrow$ {$-0.0403983, -0.00727025, 3.0184$}

Another local method is based on the *direct minimization of the squares of the residuals of the equations*,

$$\rho = \sum_{i=1}^{n} \delta_i^2 = \sum_{i=1}^{n} \left((x_i - a)^2 + (y_i - b)^2 + (z_i - c)^2 - R \right)^2.$$

Now we employ *local minimization* with the initial guess again computed for the linear problem. The objective function is

$\Rightarrow$ W = Apply[Plus, Map[#²&, eqs]];

Then

$$\Rightarrow \texttt{FindMinimum}[\texttt{W}, \{\{\texttt{a}, \texttt{a0}\}, \{\texttt{b}, \texttt{b0}\}, \{\texttt{c}, \texttt{c0}\}\}]$$
$$\Leftarrow \{0.0355584, \{\texttt{a} \to -0.0403983, \texttt{b} \to -0.00727025, \texttt{c} \to 3.0184\}\}$$

The solution of the two methods is the same, but the computation time of the local minimization is somewhat shorter.

Although as we mentioned, the reduced Gröbner basis of the determined system ($n = 3$) can be computed in symbolic form represented by second order polynomials, the high number of the subsystems, namely

$$\Rightarrow \texttt{Binomial}[\texttt{n}, 3]$$
$$\Leftarrow 3313414020$$

hinders the realization of the Gauss-Jacobi combinatorial computation as a global method. However, later we shall demonstrate, that if one can represent the cloud of points with considerably fewer points, for example via SOM (Self Organizing Map), then the symbolic solution is a reasonable alternative to the local methods.

Appendix A1 Algebraic approach: Computation of the Gröbber basis, when R is known.

If we have three exact noiseless measurement points, then the prototype system for 3 points:

$$\Rightarrow \texttt{Clear}[\texttt{a}, \texttt{b}, \texttt{c}, \texttt{R}];$$

The prototype of the equations,

$$\Rightarrow \texttt{G} = (\texttt{x} - \texttt{a})^2 + (\texttt{y} - \texttt{b})^2 + (\texttt{z} - \texttt{c})^2 - \texttt{R}^2;$$
$$\Rightarrow \texttt{proto} = \texttt{Table}[(\texttt{G}/.\{\texttt{x} \to \eta_i, \texttt{y} \to \xi_i, \texttt{z} \to \chi_i\}), \{\texttt{i}, 1, 3\}]$$
$$\Leftarrow \{-\texttt{R}^2 + (-\texttt{a} + \eta_1)^2 + (-\texttt{b} + \xi_1)^2 + (-\texttt{c} + \chi_1)^2,$$
$$-\texttt{R}^2 + (-\texttt{a} + \eta_2)^2 + (-\texttt{b} + \xi_2)^2 + (-\texttt{c} + \chi_2)^2, -\texttt{R}^2 + (-\texttt{a} + \eta_3)^2 + (-\texttt{b} + \xi_3)^2 + (-\texttt{c} + \chi_3)^2\}$$

The polynomial for *a* is

$$\Rightarrow \texttt{grba} = \texttt{GroebnerBasis}[\texttt{proto}, \{\texttt{a}, \texttt{b}, \texttt{c}\}, \{\texttt{b}, \texttt{c}\}]//\texttt{Simplify};$$
$$\Rightarrow \texttt{Coefficient}[\texttt{grba}, \texttt{a}, 2]//\texttt{Simplify}$$

$$\Leftarrow \{4(\eta_1^2 \xi_2^2 - 2\eta_1^2 \xi_2 \xi_3 + \eta_1^2 \xi_3^2 + \xi_2^2 \chi_1^2 - 2\xi_2 \xi_3 \chi_1^2 + \xi_3^2 \chi_1^2 + \eta_3^2(\xi_1^2 - 2\xi_1 \xi_2 + \xi_2^2 + (\chi_1 - \chi_2)^2) - 2\xi_1 \xi_2 \chi_1 \chi_2$$
$$+ 2\xi_1 \xi_3 \chi_1 \chi_2 + 2\xi_2 \xi_3 \chi_1 \chi_2 - 2\xi_3^2 \chi_1 \chi_2 + \eta_1^2 \chi_2^2 + \xi_1^2 \chi_2^2 - 2\xi_1 \xi_3 \chi_2^2 + \xi_3^2 \chi_2^2 + \eta_2^2(\xi_1^2 - 2\xi_1 \xi_3 + \xi_3^2 + (\chi_1 - \chi_3)^2)$$

$$+ 2\eta_1 \eta_3(-\xi_2^2 + \xi_1(\xi_2 - \xi_3) + \xi_2 \xi_3 + (\chi_1 - \chi_2)(\chi_2 - \chi_3)) + 2\xi_1 \xi_2 \chi_1 \chi_3 - 2\xi_2^2 \chi_1 \chi_3 - 2\xi_1 \xi_3 \chi_1 \chi_3 + 2\xi_2 \xi_3 \chi_1 \chi_3$$
$$- 2\eta_1^2 \chi_2 \chi_3 - 2\xi_1^2 \chi_2 \chi_3 + 2\xi_1 \xi_2 \chi_2 \chi_3 + 2\xi_1 \xi_3 \chi_2 \chi_3 - 2\xi_2 \xi_3 \chi_2 \chi_3 + \eta_1^2 \chi_3^2 + \xi_1^2 \chi_3^2 - 2\xi_1 \xi_2 \chi_3^2 + \xi_2^2 \chi_3^2 - 2\eta_2(\eta_3(\xi_1^2$$
$$+ \xi_2 \xi_3 - \xi_1(\xi_2 + \xi_3) + (\chi_1 - \chi_2)(\chi_1 - \chi_3)) + \eta_1(\xi_1(\xi_2 - \xi_3) - \xi_2 \xi_3 + \xi_3^2 + \chi_1 \chi_2 - \chi_1 \chi_3 - \chi_2 \chi_3 + \chi_3^2))))\}$$

$$\Rightarrow \texttt{Coefficient}[\texttt{grba}, \texttt{a}, 3]//\texttt{Simplify}$$
$$\Leftarrow \{0\}$$

This means we have a second order polynomial indicating non-unique solution.

14.2.1.3 Geometric Fitting When R is Unknown

In this case, we want to minimize the sum of squared geometric distances to the point set. Then the local error is defined as

$$\delta_i = \sqrt{(x_i - a)^2 + (y_i - b)^2 + (z_i - c)^2} - R.$$

Now we have an overdetermined *nonlinear equation system* to be solved for the parameter estimation problem even in case of unknown R. However in this case, the solution of the system is unique, since the Gröbner basis of the determined subsystems are first order polynomials. In general the use of geometric distance results in better solution compared to the algebraic distance. This is especially true in the presence of noise and small outliers. Similarly to the algebraic approach, local methods like Extended Newton method as well as direct local minimization can be employed.

Let us create the system of the equations. First we reset a, b, c and R to be symbolic variables again,

$\Rightarrow$ Clear[a, b, c, R]

The prototype of the equation,

$\Rightarrow$ G = $\sqrt{(x - a)^2 + (y - b)^2 + (z - c)^2}$ −R

Then the system can be created as,

$\Rightarrow$ eqs = Map[G/.{x $\rightarrow$ #[[1]], y $\rightarrow$ #[[2]], z $\rightarrow$ #[[3]]}&, dataQ];

For example the first equation is

$\Rightarrow$ eqs[[1]]

$\Leftarrow$ $\sqrt{(-0.1958-a)^2 + (-0.0309-b)^2 + (2.967-c)^2}$ − R

Let us employ the Extended Newton method with the result of the algebraic solution as initial guess. We get:

$\Leftarrow$ {−0.0411326, −0.00844824, 3.01198, 0.145917}

Indeed, the geometric approach provides better estimation than algebraic one, however it needs a good initial guess.

The other possible solution is minimization of the squares of the residuals of the equations,

$$\rho = \sum_{i=1}^{n} \delta_i^2 = \sum_{i=1}^{n} \left(\sqrt{(x_i - a)^2 + (y_i - b)^2 + (z_i - c)^2} - R \right)^2.$$

Now again we can employ *local minimization* with the initial guess computed from the linear problem,

```
⇒ W = Apply[Plus, Map[#² &, eqs]];
⇒ FindMinimum[W, {{a, a0}, {b, b0}, {c, c0}, {R, R0}}]
⇐ {0.351, {a → −0.0411326, b → −0.0084482, c → 3.01198, R → 0.145917}}
```

Again this computation required less time than the Newton method.

14.2.1.4   Geometric Fitting When R is Known

Again one can use the Extended Newton method,

```
⇒ NewtonExtended[eqs/.R → 0.152, {a, b, c}, {a0, b0, c0}]
⇐ {−0.040658, −0.00793428, 3.02032}
```

The solution with the direct local minimization

```
⇒ FindMinimum[W/.R → 0.152, {{a, a0}, {b, b0}, {c, c0}}]
⇐ {0.360787, {a → −0.040658, b → −0.00793428, c → 3.02032}}
```

In this case the nonlinear system has more solutions since the Gröbner basis of the determined subsystems are second order polynomials, see Appendix A2. Generally speaking, the algebraic as well as the geometric approach with known radius always lead to nonlinear problems, where in case of geometric model the solution is not unique.

Appendix A2 Geometric approach: Computation of the Gröbber basis, when R is known.

If we have three exact noiseless measurement points, then the prototype system for 3 points:

```
⇒ Clear[a, b, c, R];
```

The prototype of the equations,

```
⇒ G = √((x − a)² + (y − b)² + (z − c)²) − R;
⇒ proto = Table[(G/.{x → ηᵢ, y → ξᵢ, z → cᵢ}), {i, 1, 3}]
```

$$\Leftarrow \{-R^2 + \sqrt{(-a+\eta_1)^2 + (-b+\xi_1)^2 + (-c+\chi_1)^2},$$

$$-R^2 + \sqrt{(-a+\eta_2)^2 + (-b+\xi_2)^2 + (-c+\chi_2)^2}, -R^2 + \sqrt{(-a+\eta_3)^2 + (-b+\xi_3)^2 + (-c+\chi_3)^2}\}$$

The polynomial for a is

```
⇒ grba = GroebnerBasis[proto, {a, b, c}, {b, c}]//Simplify;
⇒ Length[grba]
⇐ 1
⇒ ca2 = Coefficient[grba[[1]], a, 2]//Simplify
```

$\Leftarrow \{4(n_1^2\xi_2^2 - 2n_1^2\xi_2\xi_3 + n_1^2\xi_3^2 + \xi_2^2x_1^2 - 2\xi_2\xi_3x_1^2 + \xi_3^2x_1^2 + n_3^2(\xi_1^2 - 2\xi_1\xi_2 + \xi_2^2 + (x_1-x_2)^2) - 2\xi_1\xi_2x_1x_2$

$+ 2\xi_1\xi_3x_1x_2 + 2\xi_2\xi_3x_1x_2 - 2\xi_3^2x_1x_2 + n_1^2x_2^2 + \xi_1^2x_2^2 - 2\xi_1\xi_3x_2^2 + \xi_3^2x_2^2 + n_2^2(\xi_1^2 - 2\xi_1\xi_3 + \xi_3^2 + (x_1-x_3)^2)$

$+ 2n_1n_3(-\xi_2^2 + \xi_1(\xi_2-\xi_3) + \xi_2\xi_3 + (x_1-x_2)(x_2-x_3)) + 2\xi_1\xi_2x_1x_3 - 2\xi_2^2x_1x_3 - 2\xi_1\xi_3x_1x_3 + 2\xi_2\xi_3x_1x_3$

$- 2n_1^2x_2x_3 - 2\xi_1^2x_2x_3 + 2\xi_1\xi_2x_2x_3 + 2\xi_1\xi_3x_2x_3 - 2\xi_2\xi_3x_2x_3 + n_1^2x_3^2 + \xi_1^2x_3^2 - 2\xi_1\xi_2x_3^2 + \xi_2^2x_3^2 - 2n_2(n_3(\xi_1^2$

$+ \xi_2\xi_3 - \xi_1(\xi_2+\xi_3) + (x_1-x_2)(x_1-x_3)) + n_1(\xi_1(\xi_2-\xi_3) - \xi_2\xi_3 + \xi_3^2 + x_1x_2 - x_1x_3 - x_2x_3 + x_3^2)))\}$

```
⇒ ca3 = Coefficient[grba[[1]], a, 3]//Simplify
⇐ 0
```

Again, we have a second order polynomial indicating non-unique solution.

14.2.1.5 Directional Fitting

For directional fitting, the error function is based on the distance between the measured point $P_i = P\{x_i, y_i, z_i\}$ and its projection onto the surface of the sphere closer to the instrument along the direction of P_i, see Fig. 14.21.

In Cartesian coordinates, the expressions for r_i, p_i and q_i (see Fig. 14.21) can be written as, see i.e. Franaszek et al. (2009)

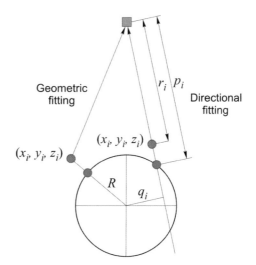

Fig. 14.21 Comparison of the geometric and directional fitting

$$r_i = \sqrt{(x_c - x_i)^2 + (y_c - y_i)^2 + (z_c - z_i)^2},$$

where x_c, y_c and z_c are the coordinates of the center point. Let,

$$p_i = au_i + bv_i + cw_i$$

and

$$q_i = \sqrt{(v_i c - w_i b)^2 + (w_i a - u_i c)^2 + (u_i b - v_i a)^2},$$

where

$$u_i = \frac{x_i}{r_i}, \quad v_i = \frac{y_i}{r_i} \quad and \quad w_i = \frac{z_i}{r_i}.$$

The systems of the nonlinear equations are

$$\left(p_i - \sqrt{R^2 - q_i^2} - r_i\right)^2 = 0 \quad if \quad q_i < R$$

$$(p_i - r_i)^2 + (q_i - R)^2 = 0 \quad if \quad q_i \geq R.$$

In our case the camera is in the origin,

$$x_c = 0, \quad y_c = 0 \quad and \quad z_c = 0.$$

Now, let us create the nonlinear system of the problem,

$\Rightarrow$ Eqs = Res[a, b, c, R, dataQ];

The number of equations are

$\Rightarrow$ Length[Eqs]

$\Leftarrow$ 2710

For example the first equation is,

$\Rightarrow$ Eqs[[1]]

$\Leftarrow$ If[Sqrt[((0.0103914a − 0.0658458b)² + (−0.997776b − 0.0103914c)²
$\qquad$ + (0.997776a + 0.0658458c)²)] < R, ((−2.97361 − 0.0658458a − 0.0103914b
$\qquad$ + 0.997776c − Sqrt[(−(0.0103914a − 0.0658458b)² − (−0.99778b − 0.0103914c)²
$\qquad$ −(0.997776a + 0.0658458c)² + R²)])²), (−2.97361 − 0.0658458a − 0.0103914b
$\qquad$ + 0.997776c)² + [Sqrt[((0.0103914a − 0.0658458b)²
$\qquad$ + (−0.997776b − 0.0103914c)² + (0.997776a + 0.0658458c)²)] − R)²]

The Extended Newton method cannot handle these equations properly because the derivatives cannot be compute symbolically, therefore local minimization is used.

The objective function is the sum of the residual of the equations,

$\Rightarrow$ W = Apply[Plus, Eqs];

Then the local minimization is employed with a derivative free method,

$\Rightarrow$ FindMinimum[W, {{a, a0}, {b, b0}, {c, c0}, {R, R0}},
    Method $\Rightarrow$ " PrincipalAxis"]

$\Leftarrow$ {1.56588, {a $\rightarrow$ -0.036743, b $\rightarrow$ -0.00738915, c $\rightarrow$ 3.00028,
    R $\rightarrow$ 0.140142}}

Now let us assume that R is known. Then we have the same equation system where $R = 0.152$.

Then the local minimization gives

$\Rightarrow$ FindMinimum[W/.R $\rightarrow$ 0.152, {{a, a0}, {b, b0}, {c, c0}},
    Method $\rightarrow$ " PrincipalAxis"]

$\Leftarrow$ {1.38047, {a $\rightarrow$ -0.0369308, b $\rightarrow$ -0.00800386, c $\rightarrow$ 3.0196}}

In the following sections we shall demonstrate the application of Self-Organizing Map (SOM) which can reduce the number of data points and open the way for the application of the Gauss-Jacobi combinatoric solution.

14.2.1.6   Application Self-Organizing Map to Reducing Data

Using a Kohonen-map with a lattice of SOM (Self Organizing Map) of size 5×5 code-book vectors with symmetrical neighborhood. (Fig. 14.22)

The 25 representing code-book vectors are

$\Rightarrow$ dataSOM = som[[1]];

Let us display the code-book vectors with the sphere to be estimated, see Fig. 14.23.

At this time the application of the Gauss-Jacobi algorithm as a global method to solve polynomial systems is reasonable, since now the number of the subsets is "only"

$\Rightarrow$ Binomial[25, 4]

$\Leftarrow$ 12650

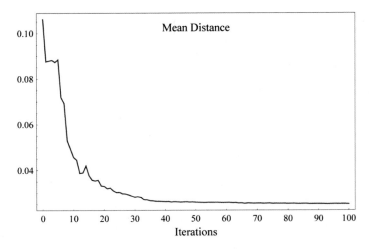

Fig. 14.22 The error of the Kohonen-map with a lattice of SOM of size 5 × 5 code-book vectors

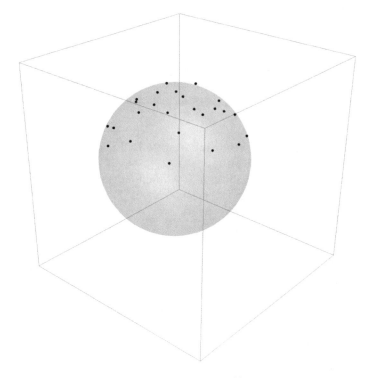

Fig. 14.23 The code-book vectors with the exact target object

when R is unknown, and

$\Rightarrow$ Binomial$[25, 3]$

$\Leftarrow 2300$

when R is known.

But before we do that, let us compute the parameters from these code-book vectors as input data in order to see if precise results can be expected at all with a reduced data set. We use the earlier methods. First we consider geometric fitting with unknown and then with known radius.

The parameters when R is unknown:

$\Rightarrow$ Clear$[a, b, c, R]$

Then the system can be created

$\Rightarrow$ eqs = Map$[G/.\{x \rightarrow \#[[1]], y \rightarrow \#[[2]], z \rightarrow \#[[3]]\}\&, \text{dataSOM}]$;

Let us employ the Extended Newton method with the result of the algebraic solution as initial guess

$\Rightarrow$ NewtonExtended$[\text{eqs}, \{a, b, c, R\}, \{a0, b0, c0, R0\}]$

$\Leftarrow \{-0.040915, -0.00897106, 3.00846, 0.141147\}$

Then for known radius, we get

$\Rightarrow$ NewtonExtended$[\text{eqs}/.R \rightarrow 0.152, \{a, b, c\}, \{a0, b0, c0\}]$

$\Leftarrow \{-0.0397396, -0.00766308, 3.02364\}$

It means that the code-book vectors provide a quite good representation of the point cloud.

Let us compute the parameters from the directional approach, too. Creating the nonlinear system of the problem for R is unknown,

$\Rightarrow$ Eqs = Res$[a, b, c, R, \text{dataSOM}]$;

The number of equations is

$\Rightarrow$ Length$[\text{Eqs}]$

$\Leftarrow 25$

For example the first equation is,

$\Rightarrow$ Eqs[[1]]

$\Leftarrow$ If[Sqrt[$((0.0333223a - 0.0476156b)^2 + (-0.99831b - 0.0333223c)^2$

$+ (0.99831a + 0.0476156c)^2)$] $< R, (-2.97169 - 0.0476156a - 0.0333223b$

$+ 0.99831c - $ Sqrt[$(-(0.0333223a - 0.0476156b)^2$

$-(-0.99831b - 0.0333223c)^2 - (0.99831a + 0.0476156c)^2 + R^2)])^2,$

$((-2.97169 - 0.0476156a - 0.0333223b + 0.99831c)^2)$

$+ $ Sqrt[$((0.0333223a - 0.0476156b)^2 + (-0.99831b - 0.0333223c)^2$

$+ (0.99831a + 0.0476156c)^2)$] $- R)^2)$]

Again let us employ local minimization

$\Rightarrow$ W = Apply[Plus, Eqs];

$\Rightarrow$ FindMinimum[W, {{a, a0}, {b, b0}, {c, c0}, {R, R0}},

Method $\rightarrow$ "PrincipalAxis"]

$\Leftarrow$ {0.00139774, {a $\rightarrow$ -0.0398059, b $\rightarrow$ -0.0102369, c $\rightarrow$ 3.0147,

R $\rightarrow$ 0.144784}}

However, now SOM provides a smooth and well arranged data point set, therefore Newton method can be also successful,

$\Rightarrow$ NewtonExtended[Eqs, {a, b, c, R}, {a0, b0, c0, R0}]

$\Leftarrow$ {$-0.0398873, -0.00998192, 3.01548, 0.145522$}

Now let us compute the Gröbner basis for the equation system resulted from the geometric approach, when R is unknown.

14.2.1.7   Symbolic Computation of Gröbner Basis for Geometric Fitting

Considering 4 exact noiseless measurement points, then the prototype system for these 4 points is,

$\Rightarrow$ proto = Table[(G/.{x $\rightarrow$ η_i, y $\rightarrow$ ξ_i, z $\rightarrow$ χ_i}), {i, 1, 4}]

$\Leftarrow$ {$-R + \sqrt{(-a + \eta_1)^2 + (-b + \xi_1)^2 + (-c + \chi_1)^2}$,

$-R + \sqrt{(-a + \eta_2)^2 + (-b + \xi_2)^2 + (-c + \chi_2)^2}$,

$-R + \sqrt{(-a + \eta_3)^2 + (-b + \xi_3)^2 + (-c + \chi_3)^2}$,

$-R + \sqrt{(-a + \eta_4)^2 + (-b + \xi_4)^2 + (-c + \chi_4)^2}$}

One can realize, that the elimination of R reduces the system

$\Rightarrow$ protoR = Take[proto, {2, 4}]/.R $\rightarrow$ $\sqrt{(-a+\eta_1)^2+(-b+\xi_1)^2+(-c+\chi_1)^2}$

$\Leftarrow \{-\sqrt{(-a+\eta_1)^2+(-b+\xi_1)^2+(-c+\chi_1)^2}+\sqrt{(-a+\eta_2)^2+(-b+\xi_2)^2+(-c+\chi_2)^2},$

$-\sqrt{(-a+\eta_1)^2+(-b+\xi_1)^2+(-c+\chi_1)^2}+\sqrt{(-a+\eta_3)^2+(-b+\xi_3)^2+(-c+\chi_3)^2},$

$-\sqrt{(-a+\eta_1)^2+(-b+\xi_1)^2+(-c+\chi_1)^2}+\sqrt{(-a+\eta_4)^2+(-b+\xi_4)^2+(-c+\chi_4)^2}\}$

To get univariate polynomial for parameter a let us eliminate b and c from the basis

$\Rightarrow$ grba = GroebnerBasis[protoR, {a, b, c}, {b, c}]//Simplify;

This basis contains

$\Rightarrow$ Length[grba]

$\Leftarrow 1$

polynomials. We take the first one

$\Rightarrow$ grba[[1]]

$\Leftarrow 2a\eta_2\xi_3\chi_1-\eta_2^2\xi_3\chi_1-\xi_2^2\xi_3\chi_1+\xi_2\xi_3^2\chi_1-2a\eta_2\xi_4\chi_1+\eta_2^2\xi_4\chi_1+\xi_2^2\xi_4\chi_1-\xi_3^2\xi_4\chi_1-\xi_2\xi_4^2\chi_1+\xi_3\xi_4^2\chi_1-2a\eta_1\xi_3\chi_2$
$+\eta_1^2\xi_3\chi_2+\xi_1^2\xi_3\chi_2-\xi_1\xi_3^2\chi_2+2a\eta_1\xi_4\chi_2-\eta_1^2\xi_4\chi_2-\xi_1^2\xi_4\chi_2+\xi_3^2\xi_4\chi_2+\xi_1\xi_4^2\chi_2-\xi_3\xi_4^2\chi_2+\xi_3\chi_1^2\chi_2-\xi_4\chi_1^2\chi_2$
$-\xi_3\chi_1\chi_2^2+\xi_4\chi_1\chi_2^2-2a\eta_2\xi_1\chi_3+\eta_2^2\xi_1\chi_3+2a\eta_1\xi_2\chi_3-\eta_1^2\xi_2\chi_3-\xi_1^2\xi_2\chi_3+\xi_1\xi_2^2\chi_3-2a\eta_1\xi_4\chi_3+\eta_1^2\xi_4\chi_3$

$+2a\eta_2\xi_4\chi_3-\eta_2^2\xi_4\chi_3+\xi_1^2\xi_4\chi_3-\xi_2^2\xi_4\chi_3-\xi_1\xi_4^2\chi_3+\xi_2\xi_4^2\chi_3-\xi_2\chi_1^2\chi_3+\xi_4\chi_1^2\chi_3+\xi_1\chi_2^2\chi_3-\xi_4\chi_2^2\chi_3+\xi_2\chi_1\chi_3^2$
$-\xi_4\chi_1\chi_3^2-\xi_1\chi_2\chi_3^2+\xi_4\chi_2\chi_3^2+\eta_4^2(\xi_3(\chi_1-\chi_2)+\xi_1(\chi_2-\chi_3)+\xi_2(-\chi_1+\chi_3))+2a\eta_4(\xi_3(-\chi_1+\chi_2)$
$+\xi_2(\chi_1-\chi_3)+\xi_1(-\chi_2+\chi_3))+2a\eta_1\xi_4-\eta_1^2\xi_4-2a\eta_1\xi_2\chi_4+\eta_1^2\xi_2\chi_4+\xi_1^2\xi_2\chi_4-\xi_1\xi_2^2\chi_4+2a\eta_1\xi_3\chi_4$

$-\eta_1^2\xi_3\chi_4-2a\eta_2\xi_3\chi_4+\eta_2^2\xi_3\chi_4-\xi_1^2\xi_3\chi_4+\xi_2^2\xi_3\chi_4+\xi_1\xi_3^2\chi_4-\xi_2\xi_3^2\chi_4+\xi_2\chi_1^2\chi_4-\xi_3\chi_1^2\chi_4-\xi_1\chi_2^2\chi_4+\xi_3\chi_2^2\chi_4$
$+\xi_1\chi_3^2\chi_4-\xi_2\chi_3^2\chi_4-\xi_2\chi_1\chi_4^2+\xi_3\chi_1\chi_4^2+\xi_1\chi_2\chi_4^2-\xi_3\chi_2\chi_4^2-\xi_1\chi_3\chi_4^2+\xi_2\chi_3\chi_4^2-2a\eta_3(\xi_4(-\chi_1+\chi_2)$
$+\xi_2(\chi_1-\chi_4)+\xi_1(-\chi_2+\chi_4))+\eta_3^2(\xi_4(-\chi_1+\chi_2)+\xi_2(\chi_1-\chi_4)+\xi_1(-\chi_2+\chi_4))$

It turns out that this is a first order polynomial, since

$\Rightarrow$ ca2 = Coefficient[grba[[1]], a, 2]

$\Leftarrow 0$

Therefore the solution for parameter a is simple,

$\Rightarrow$ aG = $-\dfrac{\text{Coefficient[grba[[1]], a, 0]}}{\text{Coefficient[grba[[1]], a, 1]}}$//Simplify

$\Leftarrow (-\eta_2^2\xi_3\chi_1-\xi_2^2\xi_3\chi_1+\xi_2\xi_3^2\chi_1+\eta_2^2\xi_4\chi_1+\xi_2^2\xi_4\chi_1-\xi_3^2\xi_4\chi_1-\xi_2\xi_4^2\chi_1+\xi_3\xi_4^2\chi_1+\eta_1^2\xi_3\chi_2+\xi_1^2\xi_3\chi_2-\xi_1\xi_3^2\chi_2-\eta_1^2\xi_4\chi_2$
$-\xi_1^2\xi_4\chi_2+\xi_3^2\xi_4\chi_2+\xi_1\xi_4^2\chi_2-\xi_3\xi_4^2\chi_2+\xi_3\chi_1^2\chi_2-\xi_4\chi_1^2\chi_2-\xi_3\chi_1\chi_2^2+\xi_4\chi_1\chi_2^2+\eta_2^2\xi_1\chi_3-\eta_1^2\xi_2\chi_3-\xi_1^2\xi_2\chi_3+\xi_1\xi_2^2\chi_3$

$+\eta_1^2\xi_4\chi_3-\eta_2^2\xi_4\chi_3+\xi_1^2\xi_4\chi_3-\xi_2^2\xi_4\chi_3-\xi_1\xi_4^2\chi_3+\xi_2\xi_4^2\chi_3-\xi_2\chi_1^2\chi_3+\xi_4\chi_1^2\chi_3+\xi_1\chi_2^2\chi_3-\xi_4\chi_2^2\chi_3+\xi_2\chi_1\chi_3^2-\xi_4\chi_1\chi_3^2$
$-\xi_1\chi_2\chi_3^2+\xi_4\chi_2\chi_3^2+\eta_4^2(\xi_3(\chi_1-\chi_2)+\xi_1(\chi_2-\chi_3)+\xi_2(-\chi_1+\chi_3))-\eta_2^2\xi_1\chi_4+\eta_1^2\xi_2\chi_4+\xi_1^2\xi_2\chi_4-\xi_1\xi_2^2\chi_4$

$$-n_1^2 \xi_3 x_4 + n_2^2 \xi_3 x_4 - \xi_1^2 \xi_3 x_4 + \xi_2^2 \xi_3 x_4 + \xi_1 \xi_3^2 x_4 - \xi_2 \xi_3^2 x_4 + \xi_2 x_1^2 x_4 - \xi_3 x_1^2 x_4 - \xi_1 x_2^2 x_4 + \xi_3 x_2^2 x_4 + \xi_1 x_3^2 x_4 - \xi_2 x_3^2 x_4$$
$$-\xi_2 x_1 x_4^2 + \xi_3 x_1 x_4^2 + \xi_1 x_2 x_4^2 - \xi_3 x_2 x_4^2 - \xi_1 x_3 x_4^2 + \xi_2 x_3 x_4^2 + n_3^2(\xi_4(-x_1 + x_2) + \xi_2(x_1 - x_4) + \xi_1(-x_2 + x_4)))$$
$$/(2(-n_2 \xi_3 x_1 + n_2 \xi_4 x_1 + n_1 \xi_3 x_2 - n_1 \xi_4 x_2 + n_2 \xi_1 x_3 - n_1 \xi_2 x_3 + n_1 \xi_4 x_3 - n_2 \xi_4 x_3 + n_4(\xi_3(x_1 - x_2) + \xi_1(x_2 - x_3)$$
$$+ \xi_2(-x_1 + x_3)) - n_2 \xi_1 x_4 + n_1 \xi_2 x_4 - n_1 \xi_3 x_4 + n_2 \xi_3 x_4 + n_3(\xi_4(-x_1 + x_2) + \xi_2(x_1 - x_4) + \xi_1(-x_2 + x_4))))$$

This means that from the coordinates of the corresponding 4 points the parameter can be directly computed. Similar expressions can be developed for the other two parameters b and c, too.

For parameter b:

$$\Rightarrow \texttt{grbb} = \texttt{GroebnerBasis[protoR}, \{a, b, c, R\}, \{a, c\}];$$

$$\Rightarrow \texttt{bG} = -\frac{\texttt{Coefficient[grbb[[1]]}, b, 0]}{\texttt{Coefficient[grbb[[1]]}, b, 1]} //\texttt{Simplify}$$

$$\Leftarrow (n_4 \xi_2^2 x_1 - n_4 \xi_3^2 x_1 - n_1^2 n_4 x_2 + n_1 n_4^2 x_2 - n_4 \xi_1^2 x_2 - n_1 \xi_3^2 x_2 + n_4 \xi_3^2 x_2 + n_1 \xi_4^2 x_2 - n_4 x_1^2 x_2 + n_4 x_1 x_2^2 + n_1^2 n_4 x_3 - n_1 n_4^2 x_3$$
$$+ n_4 \xi_1^2 x_3 + n_1 \xi_2^2 x_3 - n_4 \xi_2^2 x_3 - n_1 \xi_4^2 x_3 + n_4 x_1^2 x_3 + n_1 x_2^2 x_3 - n_4 x_2 x_3^2 - n_4 x_1 x_3^2 - n_1 x_2 x_3^2 + n_4 x_2 x_3^2 - n_1 \xi_2^2 x_4 + n_1 \xi_3^2 x_4$$

$$- n_1 x_2^2 x_4 + n_1 x_3^2 x_4 + n_1 x_2 x_4^2 - n_1 x_3 x_4^2 + n_2(\xi_3^2 x_1 - \xi_4^2 x_1 - n_1^2 x_3 - \xi_1^2 x_3 + \xi_4^2 x_3 - x_1^2 x_3 + x_1 x_3^2 + n_4^2(-x_1 + x_3)$$
$$+ n_3^2(x_1 - x_4) + n_1^2 x_4 + \xi_1^2 x_4 - \xi_3^2 x_4 + x_1^2 x_4 - x_3^2 x_4 - x_1 x_4^2 + x_3 x_4^2) + n_4^2(n_4(x_1 - x_3) + n_1(x_3 - x_4) - n_3(x_1 - x_4))$$

$$+ n_3(\xi_4^2 x_1 + n_4^2(x_1 - x_2) + n_1^2 x_2 + \xi_1^2 x_2 - \xi_4^2 x_2 + x_1^2 x_2 - x_1 x_2^2 - n_1^2 x_4 - \xi_1^2 x_4 - x_1^2 x_4 + x_2^2 x_4 + x_1 x_4^2 - x_2 x_4^2$$
$$+ \xi_2^2(-x_1 + x_4)) + n_3^2(n_4(-x_1 + x_2) + n_1(-x_2 + x_4)))/(2(n_2 \xi_3 x_1 - n_2 \xi_4 x_1 - n_1 \xi_3 x_2 + n_1 \xi_4 x_2 - n_2 \xi_1 x_3$$
$$+ n_1 \xi_2 x_3 - n_1 \xi_4 x_3 + n_2 \xi_4 x_3 + n_4(\xi_3(-x_1 + x_2) + \xi_2(x_1 - x_3) + \xi_1(-x_2 + x_3)) + n_2 \xi_1 x_4 - n_1 \xi_2 x_4 + n_1 \xi_3 x_4$$
$$- n_2 \xi_3 x_4 + n_3(\xi_4(x_1 - x_2) + \xi_1(x_2 - x_4) + \xi_2(-x_1 + x_4))))$$

For parameter c:

$$\Rightarrow \texttt{grbc} = \texttt{GroebnerBasis[protoR}, \{a, b, c, R\}, \{a, b\}];$$

$$\Rightarrow \texttt{cG} = -\frac{\texttt{Coefficient[grbc[[1]]}, c, 0]}{\texttt{Coefficient[grbc[[1]]}, c, 1]} //\texttt{Simplify}$$

$$\Leftarrow (n_1^2 n_4 \xi_2 - n_1 n_4^2 \xi_2 + n_4 \xi_1^2 \xi_2 - n_4 \xi_1 \xi_2^2 - n_1^2 n_4 \xi_3 + n_1 n_4^2 \xi_3 - n_4 \xi_1^2 \xi_3 - n_1 \xi_2^2 \xi_3 + n_4 \xi_2^2 \xi_3 + n_4 \xi_1 \xi_3^2 + n_1 \xi_2 \xi_3^2 - n_4 \xi_2 \xi_3^2$$
$$+ n_3^2(n_4(\xi_1 - \xi_2) + n_1(\xi_2 - \xi_4)) + n_1 \xi_2^2 \xi_4 - n_1 \xi_3^2 \xi_4 - n_1 \xi_2 \xi_4^2 + n_1 \xi_3 \xi_4^2 + n_2^2(n_4(-\xi_1 + \xi_3) + n_3(\xi_1 - \xi_4))$$

$$+ n_1(-\xi_3 + \xi_4)) + n_4 \xi_2 x_1^2 - n_4 \xi_3 x_1^2 - n_4 \xi_1 x_2^2 - n_1 \xi_3 x_2^2 + n_4 \xi_3 x_2^2 + n_1 \xi_4 x_2^2 + n_4 \xi_1 x_3^2 + n_1 \xi_2 x_3^2 - n_4 \xi_2 x_3^2 - n_1 \xi_4 x_3^2$$
$$- n_1 \xi_2 x_4^2 + n_1 \xi_3 x_4^2 + n_3(-\xi_1^2 \xi_2 + \xi_1 \xi_2^2 + n_4^2(-\xi_1 + \xi_2) + \xi_1^2 \xi_4 - \xi_2^2 \xi_4 - \xi_1 \xi_4^2 + \xi_2 \xi_4^2 + n_1^2(-\xi_2 + \xi_4) - \xi_2 x_1^2 + \xi_4 x_1^2$$

$$+ \xi_1 x_2^2 - \xi_4 x_2^2 - \xi_1 x_4^2 + \xi_2 x_4^2) + n_2(n_4^2(\xi_1 - \xi_3) + n_1^2 \xi_3 + \xi_1^2 \xi_3 - \xi_1 \xi_3^2 - n_1^2 \xi_4 - \xi_1^2 \xi_4 + \xi_3^2 \xi_4 + \xi_1 \xi_4^2 - \xi_3 \xi_4^2 - n_3^2(\xi_1 - \xi_4)$$
$$+ \xi_3 x_1^2 - \xi_4 x_1^2 - \xi_1 x_3^2 + \xi_4 x_3^2 + \xi_1 x_4^2 - \xi_3 x_4^2))/(2(n_2 \xi_3 x_1 - n_2 \xi_4 x_1 - n_1 \xi_3 x_2 + n_1 \xi_4 x_2 - n_2 \xi_1 x_3 - n_1 \xi_4 x_3$$
$$+ n_2 \xi_4 x_3 + n_4(\xi_3(-x_1 + x_2) + \xi_2(x_1 - x_3) + \xi_1(-x_2 + x_3)) + n_2 \xi_1 x_4 - n_1 \xi_2 x_4 + n_1 \xi_3 x_4 - n_2 \xi_3 x_4$$
$$+ n_3(\xi_4(x_1 - x_2) + \xi_1(x_2 - x_4) + \xi_2(-x_1 + x_4))))$$

14.2.1.8   Geometric Fitting Via Gauss-Jacobi Method with SOM Data

We can get a good approximation to compute the arithmetic average of the solution of the $\binom{25}{4} = 12\,650$ subsets. Let us generate the indices of the equations belonging to the same subset,

$$\Rightarrow \texttt{index} = \texttt{Partition[Map[\#\&, Flatten[Subsets[Range[25], \{4\}]]], 4];}$$

For example the 10th subset consists of the following equations

$\Rightarrow$ index[[10]]

$\Leftarrow \{1, 2, 3, 13\}$

Now we apply parallel computation to reduce computation time. The coordinates of the points of the subsets are

$\Rightarrow$ dataSub = ParallelMap[Map[dataSOM[[#]]&, #]&, index];

For example

$\Rightarrow$ dataSub = [[10]]

$\Leftarrow \{\{-0.141499, -0.0990233, 2.96667\}, \{-0.0517075, -0.136479, 2.9448\},$
$\{-0.00368423, -0.136832, 2.94386\}, \{-0.089566, -0.0862473, 2.91522\}\}$

In order to avoid ill-posed subsets, we compute the product of the distances of every vertex of a quadrant representing a subset,

$\Rightarrow$ avgSide[x_] := Apply[Times, Map[Norm[x[[#[[1]]]] − x[[#[[2]]]]]&,
$\{\{1, 2\}, \{1, 3\}, \{1, 4\}, \{2, 3\}, \{2, 4\}, \{3, 4\}\}]]$

Let us select the only subsets for which this product is higher than 2.5×10^{-5}

$\Rightarrow$ dataSubS = Parallelize[Select[dataSub, avgSide[#] > 0.000025&]];

Generating a prototype for the equations of the subsets

$\Rightarrow$ psi = Table[($\{\eta_i, \xi_i, \chi_i\}), \{i, 1, 4\}];$

Substituting numerical values

$\Rightarrow$ dataS =
ParallelMap[Flatten[MapThread[MapThread[#1 $\rightarrow$ #2&, {#1, #2}]&,
$\{psi, \#\}]]\&, dataSubS];$

Then the number of the equations to be considered is

$\Rightarrow$ Length[dataS]

$\Leftarrow$ 1231

instead of the 12,650.

Employing the result of the Gröbner basis the parameter a is

$\Rightarrow$ aS = ParallelMap[aG/.#&, dataS];

Then the average is

$\Rightarrow$ aaS = Mean[aS]

$\Leftarrow$ −0.0382357

The parameters b and c can be computed in the same way. b is -0.0108443 and c is 3.01347.

Now employing these averages of computed parameters to compute the radius of the sphere, we get 0.143481.

14.2.1.9   Application of RANdom SAmple Consensus (RANSAC)

In order to see how this data reduction technique can work in our case, let us select the simplest model, namely the algebraic least squares estimation when R is known.

The number of data elements required per subset s is

$$\Rightarrow \mathsf{s} = 4;$$

The solution of a subsystem of size s (which is a determined system) is

$$M = \begin{pmatrix} -2x_1 & -2y_1 & -2z_1 & 1 \\ & & \cdot & \\ -2x_i & -2y_i & -2z_i & 1 \\ & & \cdot & \\ -2x_s & -2y_s & -2z_s & 1 \end{pmatrix}, \quad p = \begin{pmatrix} a \\ b \\ c \\ d \end{pmatrix} \quad and \quad h = \begin{pmatrix} -\left(x_1^2 + y_1^2 + z_1^2\right) \\ -\left(x_i^2 + y_i^2 + z_i^2\right) \\ -\left(x_s^2 + y_s^2 + z_s^2\right) \end{pmatrix}$$

and

$$Mp = h$$

Remark Although it is a determined system, subsets representing ill-posed configuration can arise. Therefore the best way to avoid this situation is to restrict the condition number of M. Let's say if this condition number is higher than 1000 we leave out the computation of the subset and draw another one. Here we use another less efficient method, namely we compute pseudoinverse.

The total number of the data is,

$$\Rightarrow \mathsf{n}$$
$$\Leftarrow 2710$$

Let

$$\Rightarrow \mathsf{p} = 0.99;$$

In addition let us assume that the probability of observing an outlier is

$$\Rightarrow \mathsf{v} = 1/2;$$

Then the number of the necessary iterations is

$$\Rightarrow N = Round[Log[1 - p]/Log[1 - (1 - v)^s]]$$
$$\Leftarrow 71$$

Remark The higher this probability, the large the number of necessary iterations.

Thinking of Gauss-Jacobi technique, which can also be employed to kick out outliers, the possible subset of which can be drawn from **dataQ**:

$$\Rightarrow Binomial[n, s]$$
$$\Leftarrow 2242352938035$$

Let the threshold τ be 1.

$$\Rightarrow \tau = 1;$$

An element $d_i \in$ **dataQ** will be accepted if

$$\left| (x_i - a_s)^2 + (y_i - b_s)^2 + (z_i - c_s)^2 - R_s^2 \right| < \tau^2$$

Remark d_i is represented by the triplet (x_i, y_i, z_i) and the parameters (a_s, b_s, c_s, R_s) computed from a subset of s elements. We get four random points:

$$\Leftarrow \{\{0.0201, -0.0553, 2.893\}, \{-0.0766, -0.1431, 2.942\},$$
$$\{0.0153, 0.092, 2.942\}, \{-0.0703, 0.0804, 2.893\}\}$$

First we compute the parameters from this subset. Let us create the M matrix.

$$\Leftarrow \begin{pmatrix} -0.0402 & 0.1106 & -5.786 & 1 \\ 0.1532 & 0.2862 & -5.884 & 1 \\ -0.0306 & -0.184 & -5.884 & 1 \\ 0.1406 & -0.1608 & -5.786 & 1 \end{pmatrix}$$

Then we can compute the parameters a, b, c and d, by multiplying the pseudoinverse of M by h:

$$\Leftarrow \{-0.0631953, -0.0128281, 3.0033, 9.00307\}$$

and the estimated radius of the sphere is

$$\Rightarrow R = \sqrt{a^2 + b^2 + c^2 - d}$$
$$\Leftarrow 0.144597$$

Then check each element of **dataQ** ($i = 1, 2, \ldots, n$) to see whether it can satisfy the threshold. We get one inlier:

$$\Leftarrow \{-0.1958, -0.0412, 2.967\}$$

Now let us use for threshold $\tau = 1.5$ cm.

$\Rightarrow \tau = 0.015;$

$\Leftarrow 173$

As a last step we apply algebraic least squares with the maximal consensus set to checkout previous results. The estimated radius of the sphere is

$\Leftarrow 0.1414112$

The selected points—the elements of the maximal consensus set—are close to the surface of the sphere, see Fig. 14.24.

This seems to be a better result than the direct algebraic approach without RANSAC, and quite close to the result of the directional approach.

To find the sphere parameters, different methods—algebraic, geometrical and directional—have been employed. The algebraic approach results in a linear least square problem which does not require initial guess. The geometric as well as the directional approach lead to a nonlinear problem that requires iteration with an initial guess. This initial guess can be the result of the algebraic method. One also may employ direct global minimization of the sum of the residual of the geometrical model equations; however, it may need restrictions for the search space.

As an alternative method, Gauss-Jacobi combinatorial method based on Gröbner basis solution utilizing SOM representation of the measured data points has been demonstrated. This method, similar to the algebraic solution, does not require an initial guess. Parallel computation on a multi-core machine can be utilized decreasing the computation time considerably. The results of the different methods are in the following Tables 14.1 and 14.2.

14.2.2 Fitting a Cylinder

Fitting real circular cylinders is an important problem in the representation of geometry of man-made structures such as industrial plants Vosselman et al. (2004), deformation analysis of tunnels, Stal et al. (2012), detecting and monitoring the deformation in deposition holes, Carrea et al. (2014), positioning of femur pieces for surgical fracture reduction, Winkelbach et al. (2003), estimating tree stems, Khameneh (2013) and so on. Since planes and cylinders compose up to 85% of all objects in industrial scenes, research in 3D reconstruction and modeling—see CAD-CAM applications—have largely focused on these two important geometric primitives, e.g. Petitjean (2002).

In general, to define a cylinder we need 5 parameters: 4 for the axis and 1 for the radius, so one requires at last 5 points to determine parameters. It goes without saying, that in special cases, like the cylinder is parallel to an axis or to a plane, fewer parameters are enough; see Beder and Förstner (2006).

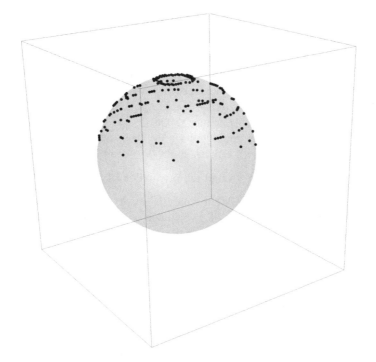

Fig. 14.24 Elements of the maximal consensus set

Table 14.1 Estimated parameters employing different methods (R is unknown)

Type of estimation	a [mm]	b [mm]	c [mm]	R [mm]
Algebraic	−41.4	−8.5	300.2	140.0
Geometric	−41.1	−8.5	301.2	145.9
Directional	−36.7	−7.4	300.0	140.1
Geometric based on SOM	−41.0	−8.9	300.6	141.0
Directional based on SOM	−40.3	−10.1	301.5	144.8
Gauss-Jacobi with Gröbner based on SOM	−38.2	−11.0	301.3	144.0
RANSAC with Algebraic	–34.0	–5.7	300.8	141.0

Table 14.2 Estimated parameters employing different methods ($R = 152$ mm)

Type of estimation	a [mm]	b [mm]	c [mm]
Algebraic	−40.4	−7.3	301.8
Geometric	−40.7	−7.9	302.0
Directional	−36.7	−7.9	300.0
Geometric based on SOM	−39.7	−7.7	302.4

In case of more than 5 points, one has to find the 5 parameters of the cylinder so that the sum of the distances of data points from the cylinder surface is minimum in the least squares sense. Basically, two approaches can be followed:

- find the direction vector of the cylinder center-axis and transform the data points into vertical position to the $x - y$ plane, then the remaining 3 parameters of the cylinder oriented parallel to z axis—2 shifting parameters and the radius—can be computed; see Beder and Förstner (2006).
- all of the 5 parameters are computed via the least squares method using local optimization techniques, e.g. Lukacs et al (1998) and Lichtblau (2006, 2012).

There are different methods to find the orientation of the cylinder center axis:

- one may use Principal Component Analysis (PCA),
- considering the general form of a cylinder as a second order surface, the direction vector can be partially extracted, and computed via linear least squares; see Khameneh (2013),
- the PCA method can be modified such as instead of single points, a local neighborhood of randomly selected points are employed; see Ruiz et al (2013).
- employing Hough transformation; see Su and Bethel (2010), Rabbani and van den Heuvel (2005).

All of these methods consider a least squares method assuming that model error has normal Gaussian distribution with zero mean value. In this study we consider realistic model error distribution applying maximization likelihood technique for parameter estimation, where this distribution is represented by a Gaussian mixture of the in- and outliers error, and identified by an expectation maximization algorithm. For geometric modeling of a general cylinder, a vector algebraic approach is followed; see Lichtblau (2012) and Paláncz et al. (2016).

14.2.2.1 Vector Algebraic Definition

Given a cylinder with axis line L in R^3. We may parametrize it using five model parameters (a, b, c, d, r), where r stands for the radius of the cylinder. If $P(x, y, z)$ is a point of the cylinder and the vector of the axis line of L is $vec = \{1, a, c\}$, let us consider L' to pass through the origin of the coordinate system and parallel to L. The translation vector is $offset = \{0, b, d\}$. The result of this translation is $P' \to P$ and $L' \to L$. We project this P' point onto L' and denote the length of the orthogonal projection $perp$. It is computed as follows.

The vector of P' is projected onto L' and the projection will be subtracted from the vector of P'. Then for the magnitude of $perp$, its norm, one can write

$$\|perp\|^2 - r^2 = 0.$$

Let us apply this definition to set the equation of the cylinder. Consider a point

Fig. 14.25 Explanation of
perp function

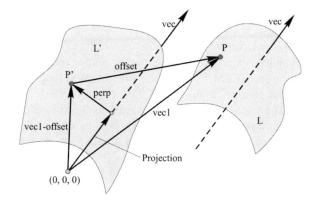

$\Rightarrow \mathrm{P} = \{x, y, z\};$

The vector of the locus of this point on the cylinder having model parameters
(a, b, c, d, r) can be computed as follows, see Fig. 14.25.
The vector of the axis line

$\Rightarrow \mathrm{vec} = \{1, a, c\};$

The vector translating the axis line to pass the origin is

$\Rightarrow \mathrm{offset} = \{0, b, d\};$

The function computing the orthogonal projection *perp*, carries out projection
onto the translated axis line *L'* and subtracts it from the vector of locus *P'*:

$\Rightarrow \mathrm{perp}[\mathrm{vec1\_}, \mathrm{vec\_}, \mathrm{offset\_}] :=$
$\quad \mathrm{vec1} - \mathrm{offset} - \mathrm{Projection}[\mathrm{vec1} - \mathrm{offset}, \mathrm{vec}, \mathrm{Dot}]$

Applying it to point *P:*

$\Rightarrow \mathrm{perp}[\mathrm{P}, \mathrm{vec}, \mathrm{offset}]//\mathrm{Simplify}$
$\Leftarrow \{x - (x + a(-b + y) + c(-d + z))/(1 + a^2 + c^2),$
$\quad -b + y - (a(x + a(-b + y) + c(-d + z)))/(1 + a^2 + c^2),$
$\quad -d + z - (c(x + a(-b + y) + c(-d + z)))/(1 + a^2 + c^2)\}$

Clearing denominators and formulating of the equation for *perp*

$\Rightarrow \mathrm{vector} = \mathrm{Numerator}[\mathrm{Together}[\%.\% - r^2]]$
$\Leftarrow b^2 + b^2 c^2 - 2abcd + d^2 + a^2 d^2 - r^2 - a^2 r^2 - c^2 r^2 + 2abx + 2cdx + a^2 x^2 + c^2 x^2 - 2by$
$\quad -2bc^2 y + 2acdy - 2axy + y^2 + c^2 y^2 + 2abcz - 2dz - 2a^2 dz - 2cxz - 2acyz + z^2 + a^2 z^2$

This is the implicit equation of the cylinder with model parameters (a, b, c, d, r).
It is important to realize that creating this equation, the algebraic error definition,
$\|perp\|^2 - r^2$ has been used.

14.2.2.2 Parametrized Form of the Cylinder Equation

In order to employ these model parameters to define the parametric equation of the cylinder, let us develop the usual parametric equation in form of $x = x\,(u,\,v)$, $y = y\,(u,\,v)$ and $z = z\,(u,\,v)$ with actual scaling parameters $(u,\,v)$ and using the model parameters $(a,\,b,\,c,\,d,\,r)$.

The locus of point P is obtained as the sum of a vector on L' plus a vector of length r perpendicular to L. Let v be a parameter along the length of axis L'. Then the projection of $P\,(u,\,v)$ on L is a vector

$$\mu = v\,vec + offset.$$

All vectors perpendicular to L are spanned by any independent pair. We can obtain an orthonormal pair $\{w_1,\,w_2\}$ in the standard way by finding the null space to the matrix whose one row is the vector along the axial direction, that is vec, and then using Gram-Schmidt to orthogonalize that pair. Using parameter u, this vector having length r and perpendicular to L can be written as

$$\rho = r\cos(u)w_1 + r\sin(u)w_2.$$

Then the locus vector of a general point of the cylinder is, see Fig. 14.26,

$$\lambda = \mu + \rho.$$

Let us carry out this computation step by step in symbolic way employing *Mathematica*,

```
⇒ pair = NullSpace[{vec}];
⇒ {w1,w2} = Orthogonalize[pair, Dot];
```

Then the parametric equation of a general circular cylinder using $\{a,\,b,\,c,\,d,\,r\}$ parameters is

Fig. 14.26 Explanation of general locus vector λ

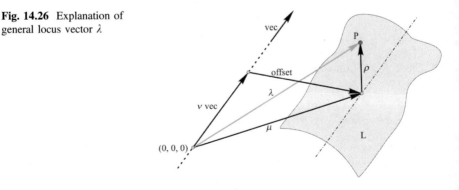

$$\Leftarrow \{v - \frac{cr\cos[u]}{\sqrt{1+c^2}} - \frac{ar\sin[u]}{(1+c^2)\sqrt{1+a^2/(1+c^2)}}, b + av + \frac{r\sin[u]}{\sqrt{1+a^2/(1+c^2)}},$$

$$d + cv + \frac{r\cos[u]}{\sqrt{1+c^2}} - \frac{acr\sin[u]}{(1+c^2)\sqrt{1+a^2/(1+c^2)}}\}$$

where u and v the actual scaling parameters.

14.2.2.3 Implicit Equation from the Parametric One

One can develop the implicit equation from the parametric one, too. This implicitazion requires that $\sin(u)$, $\cos(u)$ and v all be eliminated from the algebraic system.

$$x = v - \frac{cr\cos(u)}{\sqrt{1+c^2}} - \frac{ar\sin(u)}{(1+c^2)\sqrt{1+\frac{a^2}{1+c^2}}},$$

$$y = b + av + \frac{r\sin(u)}{\sqrt{1+\frac{a^2}{1+c^2}}},$$

$$z = d + cv + \frac{r\cos(u)}{\sqrt{1+c^2}} - \frac{acr\sin(u)}{(1+c^2)\sqrt{1+\frac{a^2}{1+c^2}}},$$

and

$$\sin^2(u) + \cos^2(u) = 1.$$

In order to carry out this computation in *Mathematica,* for practical reasons let $\sin(u) = U$ and $\cos(u) = V$. Then our system is

```
⇒ polys =
  Append[vvec + offset + rVw1 + rUw2 − x, y, z, U² + V²−1]//Simplify
```

$$\Leftarrow \{-\frac{arU}{(1+c^2)\sqrt{(1+a^2+c^2)/(1+c^2)}} + v - \frac{crV}{\sqrt{1+c^2}} - x, b + \frac{rU}{\sqrt{(1+a^2+c^2)/(1+c^2)}} + av - y,$$

$$d - \frac{acrU}{(1+c^2)\sqrt{(1+a^2+c^2)/(1+c^2)}} + cv + \frac{rV}{\sqrt{1+c^2}} - z, -1 + U^2 + V^2\}$$

Let us clear denominators:

$$\Leftarrow \{-a\sqrt{1+c^2}\,rU-(1+c^2)\sqrt{\frac{1+a^2+c^2}{1+c^2}}\left(-\sqrt{1+c^2}\,v+crV+\sqrt{1+c^2}\,x\right),$$

$$b\sqrt{\frac{1+a^2+c^2}{1+c^2}}+rU+\sqrt{\frac{1+a^2+c^2}{1+c^2}}(av-y),\,(1+c^2)^{3/2}\sqrt{\frac{1+a^2+c^2}{1+c^2}}\,d-ac\sqrt{1+c^2}\,rU$$

$$+(1+c^2)\sqrt{\frac{1+a^2+c^2}{1+c^2}}\left(c\sqrt{1+c^2}\,v+rV-\sqrt{1+c^2}\,z\right),-1+U^2+V^2\}$$

and get

$$\Leftarrow \{-a\sqrt{1+c^2}\,rU+\sqrt{1+a^2+c^2}(v+c^2v-c\sqrt{1+c^2}\,rV-x-c^2x),$$

$$b\sqrt{1+a^2+c^2}+\sqrt{1+c^2}\,rU+\sqrt{1+a^2+c^2}(av-y),$$

$$(1+c^2)\sqrt{1+a^2+c^2}\,d-ac\sqrt{1+c^2}\,rU+\sqrt{1+a^2+c^2}(cv+c^3v+\sqrt{1+c^2}\,rV-z-c^2z),-1+U^2+V^2\}$$

Applying Gröbner basis to eliminate U and V and v, the implicit form is:

$$\Leftarrow b^2+b^2c^2-2abcd+d^2+a^2d^2-r^2-a^2r^2-c^2r^2+2abx+2cdx+a^2x^2+c^2x^2-2by$$

$$-2bc^2y+2acdy-2axy+y^2+c^2y^2+2abcz-2dz-2a^2dz-2cxz-2acyz+z^2+a^2z^2$$

This is the same equation, which was computed in Sect. 13.2.2.1.

14.2.2.4 Computing Model Parameters

Let us consider five triples of points

$$\Rightarrow \texttt{points5 = Table[}\{x_i,y_i,z_i\},\{i,1,5\}]$$

$$\Leftarrow \{\{x_1,y_1,z_1\},\{x_2,y_2,z_2\},\{x_3,y_3,z_3\},\{x_4,y_4,z_4\},\{x_5,y_5,z_5\}\}$$

Substituting these points into the implicit form and setting $r^2 = rsqr$, we get

$$\Rightarrow \texttt{eqs = Map[implicit/.x} \rightarrow \#[[1]], \texttt{y} \rightarrow \#[[2]], \texttt{z} \rightarrow \#[[3]]\&,$$

$$\texttt{points5]/.r}^2 \rightarrow \texttt{rsqr}$$

$$\Leftarrow \{b^2+b^2c^2-2abcd+d^2+a^2d^2-rsqr-a^2rsqr-c^2rsqr+2abx_1+2cdx_1+a^2x_1^2+c^2x_1^2-2by_1$$

$$-2bc^2y_1+2acdy_1-2ax_1y_1+y_1^2+c^2y_1^2+2abcz_1-2dz_1-2a^2dz_1-2cx_1z_1-2acy_1z_1+z_1^2+a^2z_1^2,$$

$$b^2+b^2c^2-2abcd+d^2+a^2d^2-rsqr-a^2rsqr-c^2rsqr+2abx_2+2cdx_2+a^2x_2^2+c^2x_2^2-2by_2$$

$$-2bc^2y_2+2acdy_2-2ax_2y_2+y_2^2+c^2y_2^2+2abcz_2-2dz_2-2a^2dz_2-2cx_2z_2-2acy_2z_2+z_2^2+a^2z_2^2,$$

$$b^2+b^2c^2-2abcd+d^2+a^2d^2-rsqr-a^2rsqr-c^2rsqr+2abx_3+2cdx_3+a^2x_3^2+c^2x_3^2-2by_3$$

$$-2bc^2y_3+2acdy_3-2ax_3y_3+y_3^2+c^2y_3^2+2abcz_3-2dz_3-2a^2dz_3-2cx_3z_3-2acy_3z_3+z_3^2+a^2z_3^2,$$

$$b^2+b^2c^2-2abcd+d^2+a^2d^2-rsqr-a^2rsqr-c^2rsqr+2abx_4+2cdx_4+a^2x_4^2+c^2x_4^2-2by_4$$

$$-2bc^2y_4+2acdy_4-2ax_4y_4+y_4^2+c^2y_4^2+2abcz_4-2dz_4-2a^2dz_4-2cx_4z_4-2acy_4z_4+z_4^2+a^2z_4^2,$$

$$b^2+b^2c^2-2abcd+d^2+a^2d^2-rsqr-a^2rsqr-c^2rsqr+2abx_5+2cdx_5+a^2x_5^2+c^2x_5^2-2by_5$$

$$-2bc^2y_5+2acdy_5-2ax_5y_5+y_5^2+c^2y_5^2+2abcz_5-2dz_5-2a^2dz_5-2cx_5z_5-2acy_5z_5+z_5^2+a^2z_5^2\}$$

This is a polynomial system for the parameters based on the algebraic error definition. In order to create the determined system for the geometric error model, let us consider *geometric error model*, namely $\|perp\| - r$,

$$\Delta_i = \|perp\| - r \rightarrow \Delta_i^2 = \left(\sqrt{\#.\#} - r\right)^2$$

which can be applied to the **points5** data, then

$\Rightarrow$ eqs $=$ Map[Numerator[Together[Sqrt[#.#] $-$ r]]&,

Map[perp[#, vec, offset]&, points5]]

$\Leftarrow \{-r + \mathrm{Sqrt}[\left(x_1 - \dfrac{x_1 + a(-b+y_1) + c(-d+z_1)}{1+a^2+c^2}\right)^2$

$+ \left(-b + y_1 - \dfrac{a(x_1 + a(-b+y_1) + c(-d+z_1))}{1+a^2+c^2}\right)^2 + \left(-d + z_1 - \dfrac{c(x_1 + a(-b+y_1) + c(-d+z_1))}{1+a^2+c^2}\right)^2]$,

$-r + \mathrm{Sqrt}[\left(x_2 - \dfrac{x_2 + a(-b+y_2) + c(-d+z_2)}{1+a^2+c^2}\right)^2$

$+ \left(-b + y_2 - \dfrac{a(x_2 + a(-b+y_2) + c(-d+z_2))}{1+a^2+c^2}\right)^2 + \left(-d + z_2 - \dfrac{c(x_2 + a(-b+y_2) + c(-d+z_2))}{1+a^2+c^2}\right)^2]$,

$-r + \mathrm{Sqrt}[\left(x_3 - \dfrac{x_3 + a(-b+y_3) + c(-d+z_3)}{1+a^2+c^2}\right)^2$

$+ \left(-b + y_3 - \dfrac{a(x_3 + a(-b+y_3) + c(-d+z_3))}{1+a^2+c^2}\right)^2 + \left(-d + z_3 - \dfrac{c(x_3 + a(-b+y_3) + c(-d+z_3))}{1+a^2+c^2}\right)^2]$,

$-r + \mathrm{Sqrt}[\left(x_4 - \dfrac{x_4 + a(-b+y_4) + c(-d+z_4)}{1+a^2+c^2}\right)^2$

$+ \left(-b + y_4 - \dfrac{a(x_4 + a(-b+y_4) + c(-d+z_4))}{1+a^2+c^2}\right)^2 + \left(-d + z_4 - \dfrac{c(x_4 + a(-b+y_4) + c(-d+z_4))}{1+a^2+c^2}\right)^2]$,

$-r + \mathrm{Sqrt}[\left(x_5 - \dfrac{x_5 + a(-b+y_5) + c(-d+z_5)}{1+a^2+c^2}\right)^2$

$+ \left(-b + y_5 - \dfrac{a(x_5 + a(-b+y_5) + c(-d+z_5))}{1+a^2+c^2}\right)^2 + \left(-d + z_5 - \dfrac{c(x_5 + a(-b+y_5) + c(-d+z_5))}{1+a^2+c^2}\right)^2]\}$

14.2.2.5 Computing Model Parameters in Overdetermined Case

Let us consider an *algebraic error model* first. This means to minimize the residual of the implicit form

$$G(a, b, c, d, r) = \sum_{i=1}^{n} \Delta_i^2$$

where the algebraic error is, see **implicit** or **vector**,

$\Leftarrow \Delta_i = b^2 + b^2c^2 - 2abcd + d^2 + a^2d^2 - rsqr - a^2rsqr - c^2rsqr + 2abx_i + 2cdx_i + a^2x_i^2 + c^2x_i^2 - 2by_i$
$- 2bc^2y_i + 2acdy_i - 2ax_iy_i + y_i^2 + c^2y_i^2 + 2abcz_i - 2dz_i - 2a^2dz_i - 2cx_iz_i - 2acy_iz_i + z_i^2 + a^2z_i^2,$

while the geometric error is,

$$\Rightarrow \Delta_i = \texttt{Map[Numerator[Together[}\sqrt{\texttt{\#.\#}}-\texttt{r]]\&,Map[perp[\#,vec,offset]\&,}$$
$$\texttt{\{\{x_i,y_i,z_i\}\}]]}$$
$$\Leftarrow \{-r+\texttt{Sqrt}[\left(x_i-\frac{x_i+a(-b+y_i)+c(-d+z_i)}{1+a^2+c^2}\right)^2$$
$$+\left(-b+y_i-\frac{a(x_i+a(-b+y_i)+c(-d+z_i))}{1+a^2+c^2}\right)^2+\left(-d+z_i-\frac{c(x_i+a(-b+y_i)+c(-d+z_i))}{1+a^2+c^2}\right)^2]\}$$

In order to carry out this minimization problem via local minimization, since that is much faster than the global one, we may solve the determined system ($n = 5$) for randomly selected data points to get initial guess values.

14.2.2.6   Application to Estimation of Tree Stem Diameter

Outdoor laser scanning measurements have been carried out in the backyard of Budapest University of Technology and Economics, see Fig. 14.27 left. In order to get a simple fitting problem rather than a segmentation one, the test object, the lower part of the trunk of a tree was preselected by segmentation, see Fig. 14.27 right.

The experiment has been carried out with a Faro Focus 3D terrestrial laser scanner, see Fig. 14.28.

Fig. 14.27 Test environment (to left) and test object (to right)

Fig. 14.28 Faro Focus 3D scanner

The scanning parameters were set to 1/2 resolution that equals to 3 mm/10 m point spacing. The test data set was cropped from the point cloud; moreover, further resampling was applied in order to reduce the data size. The final data set is composed of 16,434 points in ASCII format, and only the x, y, z coordinates were kept (no intensity values). Let us load the measured data of 16,434 points (Fig. 14.29).

We shall compute the parameters from the geometric error model employing local minimization of the sum of the squares of the residual. To do that we compute the initial guess values from five randomly chosen points of data using the algebraic error model, since the resulting polynomial system can be solved easily via numerical Gröbner basis. In the deterministic case, the number of real solutions must be an even number (0, 2, 4, or 6). According to our numerical experiences, to ensure reliable initial values, one needs such a five points that provide at least four real solutions.

$\Rightarrow$ dataPR = RandomSample[dataP, 5]

$\Leftarrow$ {{1.5402, −12.2018, 140.461}, {1.5536, −12.1603, 141.285},
   {1.4512, −12.319, 141.43}, {1.6294, −12.3794, 140.952},
   {1.6821, −12.2508, 140.86}}

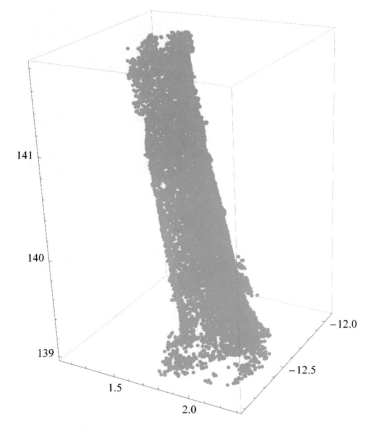

Fig. 14.29 The points of cloud of data

The equations are

$$\Rightarrow \mathtt{perps = Map[perp[\#, vec, offset]\&, dataPR]}$$
$$\Leftarrow \{\{1.5402 - (1.5402 + a(-12.2018 - b) + c(140.461 - d))/(1 + a^2 + c^2),$$
$$-12.2018 - b - (a(1.5402 + a(-12.2018 - b) + c(140.461 - d)))/(1 + a^2 + c^2),$$
$$140.461 - (c(1.5402 + a(-12.2018 - b) + c(140.461 - d)))/(1 + a^2 + c^2) - d\},$$
$$\{1.5536 - (1.5536 + a(-12.1603 - b) + c(141.285 - d))/(1 + a^2 + c^2),$$
$$-12.1603 - b - (a(1.5536 + a(-12.1603 - b) + c(141.285 - d)))/(1 + a^2 + c^2),$$
$$141.285 - (c(1.5536 + a(-12.1603 - b) + c(141.285 - d)))/(1 + a^2 + c^2) - d\},$$
$$\{1.4512 - (1.4512 + a(-12.319 - b) + c(141.43 - d))/(1 + a^2 + c^2),$$
$$-12.319 - b - (a(1.4512 + a(-12.319 - b) + c(141.43 - d)))/(1 + a^2 + c^2),$$
$$141.43 - (c(1.4512 + a(-12.319 - b) + c(141.43 - d)))/(1 + a^2 + c^2) - d\},$$
$$\{1.6294 - (1.6294 + a(-12.3794 - b) + c(140.952 - d))/(1 + a^2 + c^2),$$
$$-12.3794 - b - (a(1.6294 + a(-12.3794 - b) + c(140.952 - d)))/(1 + a^2 + c^2),$$
$$140.952 - (c(1.6294 + a(-12.3794 - b) + c(140.952 - d)))/(1 + a^2 + c^2) - d\},$$
$$\{1.6821 - (1.6821 + a(-12.2508 - b) + c(140.86 - d))/(1 + a^2 + c^2),$$
$$-12.2508 - b - (a(1.6821 + a(-12.2508 - b) + c(140.86 - d)))/(1 + a^2 + c^2),$$
$$140.86 - (c(1.6821 + a(-12.2508 - b) + c(140.86 - d)))/(1 + a^2 + c^2) - d\}\}$$

In order to avoid round off error, one may use higher precision or alternatively employ integer coefficients.

$\Rightarrow$ exprs = Map[Numerator[Together[Rationalize[#.#, 0] − rsqr]]&, perps]

$\Leftarrow$ {$248477205553 + 469830309a + 246645809213a^2 + 305045000b + 38505000ab + 12500000b^2$
$-5408450805c + 42846925745ac + 3511525000abc + 1890701741c^2 + 305045000bc^2$
$+12500000b^2c^2 - 3511525000d - 3511525000a^2d + 38505000cd - 305045000acd - 25000000abcd$
$+12500000d^2 + 12500000a^2d^2 - 12500000rsqr - 12500000a^2rsqr - 12500000c^2rsqr,$
$2010932412109 + 3778448416a + 1996386489796a^2 + 2432060000b + 310720000ab$
$+100000000b^2 - 43900075200c + 343613597100ac + 28257000000abc + 15028656905c^2$
$+2432060000bc^2 + 100000000b^2c^2 - 28257000000d - 28257000000a^2d + 310720000cd$
$-2432060000acd - 200000000abcd + 100000000d^2 + 100000000a^2d^2 - 100000000rsqr$
$-100000000a^2rsqr - 100000000c^2rsqr,$
$503855066525 + 893866640a + 500113772036a^2 + 615950000b + 72560000ab + 25000000b^2$
$-10262160800c + 87113808500ac + 7071500000abc + 3846593561c^2 + 615950000bc^2$
$+25000000b^2c^2 - 7071500000d - 7071500000a^2d + 72560000cd - 615950000acd - 50000000abcd$
$+25000000d^2 + 25000000a^2d^2 - 25000000rsqr - 25000000a^2rsqr - 25000000c^2rsqr,$
$500517896209 + 1008549718a + 496753031209a^2 + 618970000b + 81470000ab + 25000000b^2$
$-11483359440c + 87245059440ac + 7047600000abc + 3897612218c^2 + 618970000bc^2$
$+25000000b^2c^2 - 7047600000d - 7047600000a^2d + 81470000cd - 618970000acd - 50000000abcd$
$+25000000d^2 + 25000000a^2d^2 - 25000000rsqr - 25000000a^2rsqr - 25000000c^2rsqr,$
$1999162170064 + 4121414136a + 1984436906041a^2 + 2450160000b + 336420000ab$
$+100000000b^2 - 47388121200c + 345129537600ac + 28172000000abc + 15291156105c^2$
$+2450160000bc^2 + 100000000b^2c^2 - 28172000000d - 28172000000a^2d + 336420000cd$
$-2450160000acd - 200000000abcd + 100000000d^2 + 100000000a^2d^2 - 100000000rsqr$
$-100000000a^2rsqr - 100000000c^2rsqr$}

Solving the system via numerical Gröbner basis yields six solutions:

$\Rightarrow$ sol5 = NSolve[exprs, {a, b, c, d, rsqr}]
$\Leftarrow$ {{$a \rightarrow -0.745866, b \rightarrow -11.1109, c \rightarrow -22.1205, d \rightarrow 175.724, rsqr \rightarrow 0.0124972$},
{$a \rightarrow -0.674735, b \rightarrow -11.2742, c \rightarrow -3.34268, d \rightarrow 145.884, rsqr \rightarrow 0.0316293$},
{$a \rightarrow -0.0533048 + 0.95706i, b \rightarrow -12.1959 - 1.50844i, c \rightarrow 0.372737 + 0.0412819i,$
$d \rightarrow 140.563 - 0.0688756i, rsqr \rightarrow 0.097419 - 0.00128866i$},
{$a \rightarrow -0.0533048 - 0.95706i, b \rightarrow -12.1959 + 1.50844i, c \rightarrow 0.372737 - 0.0412819i,$
$d \rightarrow 140.563 + 0.0688756i, rsqr \rightarrow 0.097419 + 0.00128866i$},
{$a \rightarrow 2.43015, b \rightarrow -14.9782, c \rightarrow -1.54214, d \rightarrow 142.734, rsqr \rightarrow 0.268222$},
{$a \rightarrow -2.03283, b \rightarrow -9.11152, c \rightarrow 5.59169, d \rightarrow 132.397, rsqr \rightarrow 0.0299727$}}

The real solutions are:

$\Rightarrow$ solnsR = Select[sol5, Im[#[[1, 2]]] == 0&]
$\Leftarrow$ {{$a \rightarrow -0.745866, b \rightarrow -11.1109, c \rightarrow -22.1205, d \rightarrow 175.724, rsqr \rightarrow 0.0124972$},
{$a \rightarrow -0.674735, b \rightarrow -11.2742, c \rightarrow -3.34268, d \rightarrow 145.884, rsqr \rightarrow 0.0316293$},
{$a \rightarrow 2.43015, b \rightarrow -14.9782, c \rightarrow -1.54214, d \rightarrow 142.734, rsqr \rightarrow 0.268222$},
{$a \rightarrow -2.03283, b \rightarrow -9.11152, c \rightarrow 5.59169, d \rightarrow 132.397, rsqr \rightarrow 0.0299727$}}

We compute the corresponding r:

$\Leftarrow$ {{$a \rightarrow -0.745866, b \rightarrow -11.1109, c \rightarrow -22.1205, d \rightarrow 175.724, r \rightarrow 0.111791$},
{$a \rightarrow -0.674735, b \rightarrow -11.2742, c \rightarrow -3.34268, d \rightarrow 145.884, r \rightarrow 0.177846$},
{$a \rightarrow 2.43015, \rightarrow -14.9782, c \rightarrow -1.54214, d \rightarrow 142.734, r \rightarrow 0.517901$},
{$a \rightarrow > -2.03283, b \rightarrow -9.11152, c \rightarrow 5.59169, d \rightarrow 132.397, r \rightarrow 0.173126$}}

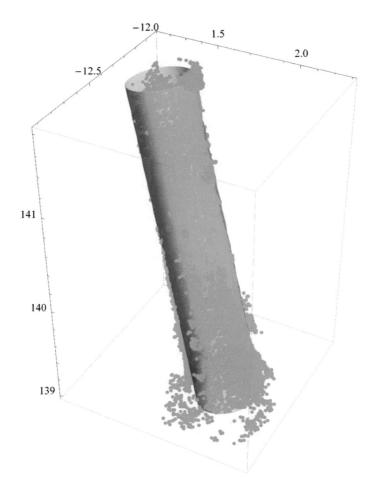

Fig. 14.30 The fitted cylinder via geometrical error model

From these solutions we can select the one that gives the smallest residual. It is:

$\Leftarrow \{a \rightarrow -0.674735, b \rightarrow -11.2742, c \rightarrow -3.34268, d \rightarrow 145.884, r \rightarrow 0.177846\}$

This solution will be used as initial guess values for the local minimization. The result, with plot, is (Fig. 14.30):

$\Leftarrow \{25.041, \{a \rightarrow -0.604342, b \rightarrow -11.3773, c \rightarrow -3.68704, d \rightarrow 146.479, r \rightarrow 0.178783\}\}$

The distribution of the model error

$\Rightarrow \text{error} = \text{exprs}/.\text{sol}[[2]];$

$\Rightarrow \text{p5} = \text{Histogram}[\text{error}, \text{Automatic}, "\text{PDF}"]$

⇒ Mean[error]

⇐ 2.08444 × 10⁻¹⁶

⇒ Min[error]

⇐ −0.0813939

⇒ Max[error]

⇐ 0.224278

⇒ StandardDeviation[error]

⇐ 0.0390362

It is clear from Fig. 14.31 that the assumption for the model error is not true. Therefore one should employ a likelihood function for the parameter estimation. The real model error distribution as well as the model parameters can be computed in an iterative way. We can consider the distribution shown in Fig. 14.31 as a first guess for this iteration. In order to handle an empirical distribution in the likelihood function there are at least three ways:

- Gaussian kernel functions can be applied to the values of different bins,
- Employing a known type of distribution to approximate the empirical histogram,
- Considering the empirical distribution as a Gaussian mixture of errors corresponding to inliers and outliers.

Here, we considered the third approach, and employed an expectation maximization algorithm to compute the parameters of the component distributions.

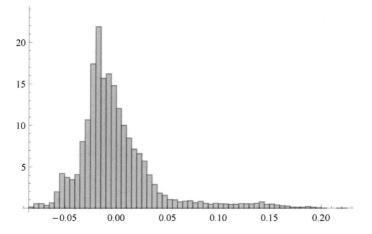

Fig. 14.31 Histogram of the model error of the standard approach assuming $N(0, \sigma)$

14.2.2.7   Expectation Maximization

Let us consider a two-component Gaussian mixture represented by the mixture model in the following form,

$$\mathcal{N}_{12}(x) = \eta_1 \mathcal{N}(\mu_1, \sigma_1, x) + \eta_2 \mathcal{N}(\mu_2, \sigma_2, x)$$

where

$$\mathcal{N}(\mu_i, \sigma_i, x) = \frac{e^{-\frac{(x-\mu_i)^2}{2\sigma_i^2}}}{\sqrt{2\pi}\sigma_i}, \quad i = 1, 2$$

and η_i's are the membership weights constrained with

$$\eta_1 + \eta_2 = 1.$$

We are looking for the parameter values (μ_1, σ_1) and (μ_2, σ_2). The log-likelihood function in case of N samples is

$$Log\mathcal{L}(x_i, \theta) = \sum_{i=1}^{N} \log(\mathcal{N}_{12}(x_i, \theta))$$

$$= \sum_{i=1}^{N} \log(\eta_1 \mathcal{N}(\mu_1, \sigma_1, x_i) + \eta_2 \mathcal{N}(\mu_2, \sigma_2, x_i)),$$

where $\theta = (\mu_1, \sigma_1, \mu_2, \sigma_2)$ the parameters of the normal densities. The problem is the direct maximization of this function, because of the sum of terms inside the logarithm. In order to solve this problem let us introduce the following alternative log-likelihood function

$$Log\mathcal{L}(x_i, \theta, \Delta) = \sum_{i=1}^{N}(1 - \Delta_i) \log(\mathcal{N}(\mu_1, \sigma_1, x_i)) + \Delta_i \log(\mathcal{N}(\mu_2, \sigma_2, x_i))$$

$$+ \sum_{i=1}^{N}(1 - \Delta_i) \log(\alpha_1) + \Delta_i \log(\alpha_2)$$

Here Δ_i's are considered as unobserved latent variables taking values 0 or 1. If x_i belongs to the first component then $\Delta_i = 0$, so

$$Log\mathcal{L}(x_i, \theta, \Delta) = \sum_{i \in N_1(\Delta)} \log(\mathcal{N}(\mu_1, \sigma_1, x_i)) + N_1 \log(\alpha_1),$$

otherwise x_i belongs to the second component then $\Delta_i = 1$, therefore

$$LogL(x_i, \theta, \Delta) = \sum_{i \in N_2(\Delta)} \log(\mathcal{N}(\mu_2, \sigma_2, x_i)) + N_2 \log(\alpha_2),$$

where N_1 and N_2 are the number of the elements of the mixture, which belong to the first and to the second component, respectively.

Since the values of the Δ_i's are actually unknown, we proceed in an iterative fashion, substituting for each Δ_i its expected value,

$$\xi_i(\theta) = E(\Delta_i | \theta, x) = \Pr(\Delta_i = 1 | \theta, x)$$

$$\approx \frac{\eta_2 \mathcal{N}(\mu_2, \sigma_2, x_i)}{(1 - \eta_2)\mathcal{N}(\mu_1, \sigma_1, x_i) + \eta_2 \mathcal{N}(\mu_2, \sigma_2, x_i)},$$

This expression is also called the responsibility of component two for observation i. Then the procedure, called the EM algorithm for two-component Gaussian mixture, is the following:

- Take initial guess for the parameters: $\theta = (\tilde{\mu}_1, \tilde{\sigma}_1, \tilde{\mu}_2, \tilde{\sigma}_2)$ and for $\tilde{\eta}_2$.
- *Expectation Step*: compute the responsibilities:

$$\tilde{\xi}_2 = \frac{\tilde{\eta}_2 \mathcal{N}(\tilde{\mu}_2, \tilde{\sigma}_2, x_i)}{(1 - \tilde{\eta}_2)\mathcal{N}(\tilde{\mu}_1, \tilde{\sigma}_1, x_i) + \tilde{\eta}_2 \mathcal{N}(\tilde{\mu}_2, \tilde{\sigma}_2, x_i)}, \quad for \ i = 1, 2, \ldots, N,$$

- *Maximization Step*: compute the weighted means and variances for the two components:

$$\tilde{\mu}_1 = \sum_{i=1}^{N} \left(1 - \tilde{\xi}_i\right) x_i / \sum_{i=1}^{N} \left(1 - \tilde{\xi}_i\right),$$

$$\tilde{\sigma}_1 = \sum_{i=1}^{N} \left(1 - \tilde{\xi}_i\right)(x_i - \tilde{\mu}_1)^2 / \sum_{i=1}^{N} \left(1 - \tilde{\xi}_i\right),$$

$$\tilde{\mu}_2 = \sum_{i=1}^{N} \tilde{\xi}_i x_i / \sum_{i=1}^{N} \tilde{\xi}_i,$$

$$\tilde{\sigma}_2 = \sum_{i=1}^{N} \tilde{\xi}_i (x_i - \tilde{\mu}_1)^2 / \sum_{i=1}^{N} \tilde{\xi}_i,$$

and the mixing probability

$$\tilde{\eta}_2 = \sum_{i=1}^{N} \tilde{\xi}_i / N,$$

- Iterate these steps until convergence.

This algorithm is implemented for *Mathematica* see, Fox et al. (2013) as a *Mathematica* Demonstration project. The code has been modified and applied here. In the next section we illustrate how this function works.

The result of the parameter values is:

$\Leftarrow$ {{0.081303, 0.0535765}, {−0.00940588, 0.0226878}, {0.103693, 0.896307}}

These can be displayed in a table form (Table 14.3).

Figure 14.32 shows the density functions of the two component (Figs. 14.33 and 14.34).

The following Figs. 14.35, 14.36 and 14.37 show the convergence of the different parameters.

Next, we separate the mixture of the samples into two clusters: cluster of outliers and cluster of inliers. Membership values of the first cluster.

Table 14.3 Parameters of the Gaussian mixture after zero iteration

η	μ	σ
0.103693	0.0813029	0.0535765
0.896307	−0.00940588	0.0226878

Fig. 14.32 The PDF of the two components

Fig. 14.33 The joint PDF of the mixture

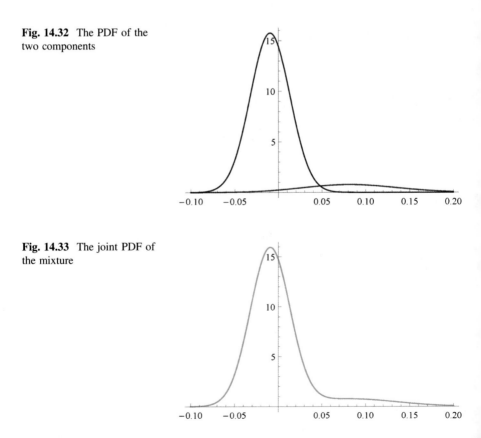

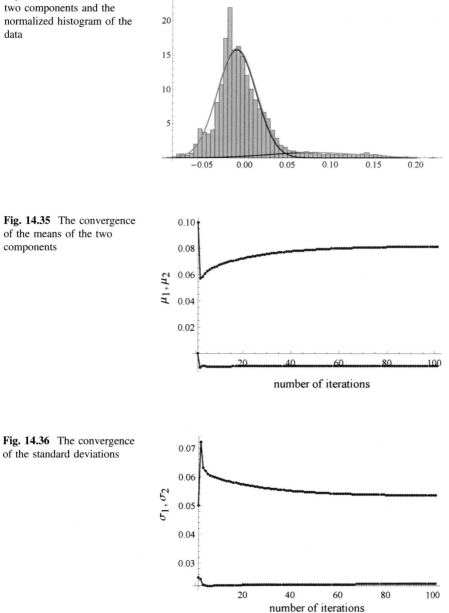

Fig. 14.34 The PDF of the two components and the normalized histogram of the data

Fig. 14.35 The convergence of the means of the two components

Fig. 14.36 The convergence of the standard deviations

Fig. 14.37 The convergence
of the membership weights

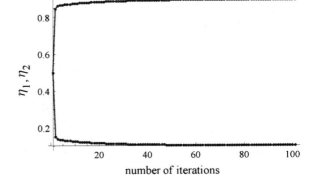

number of iterations

$\Rightarrow$ Short$[\Delta[[1]], 10]$
$\Leftarrow \{0.163639, 0.00848004, 0.0116534, 1., 0.00851251, 0.0156639, 0.00894375,$
$0.00945738, <<16418> >, 1., 1., 1., 1., 1., 1., 0.999997, 0.999995\}$

Membership values of the second cluster

$\Rightarrow$ Short$[\Delta[[2]], 10]$
$\Leftarrow 0.836361, 0.99152, 0.988347, 7.93708 \times 10^{-8}, 0.991487, 0.984336, 0.991056,$
$0.990543, <<16419> >, 5.63356 \times 10^{-11}, 4.26408 \times 10^{-11}, 1.24908 \times 10^{-10},$
$1.40798 \times 10^{-11}, 2.28746 \times 10^{-8}, 3.32431 \times 10^{-6}, 5.16318 \times 10^{-6}$

In order to get Boolean (crisp) clustering let us round the membership values

$\Rightarrow$ S1 = Round$[\Delta[[1]]]$; S2 = Round$[\Delta[[2]]]$;

The elements in the first cluster (outliers) are the corresponding elements of
those having value 1 in the set containing the rounded member values (S1)

$\Rightarrow$ XYZOut = Map$[dataP[[\#]]\&, $Position$[S1, 1]//$Flatten$]$;

The number of these elements (outliers) is 1318.

Similarly, the elements in the second cluster (inliers) are the corresponding
elements of those having value 1 in the set containing the rounded member values
(S2). There are 15,116.

Let us display the ouliers and the inliers (Fig. 14.38).

14.2.2.8   Maximum Likelihood for Gaussian Mixture

The likelihood function for a two-component Gaussian mixture can be written as,

Fig. 14.38 The inliers (blue) and outliers (red) as a first guess resulted standard minimization of the total residual

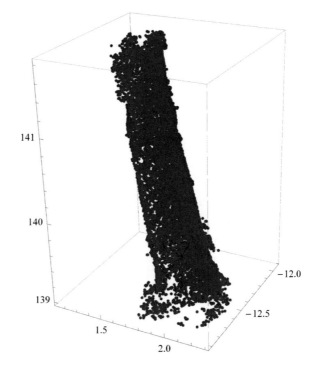

$$
Log\mathcal{L}(x_i, \theta) = \sum_{i \in N_1} \log(\mathcal{N}(\mu_1, \sigma_1, x_i)) + \sum_{i \in N_2} \log(\mathcal{N}(\mu_2, \sigma_2, x_i))
$$
$$
+ N_1 \log(\eta_1) + N_2 \log(\eta_2),
$$

where the likelihood function for one of the components can be developed as it follows.

We have seen that the geometric error is

$$
\Delta_i = -r + Sqrt\Bigg\{ \left(x_i - \frac{x_i + a(-b + y_i) + c(-d + z_i)}{1 + a^2 + c^2} \right)^2
$$
$$
+ \left(-b + y_i - \frac{a(x_i + a(-b + y_i) + c(-d + z_i))}{1 + a^2 + c^2} \right)^2
$$
$$
+ \left(-d + z_i - \frac{c(x_i + a(-b + y_i) + c(-d + z_i))}{1 + a^2 + c^2} \right)^2 \Bigg\}
$$

The probability density function for a single Gaussian

$$\frac{e^{-\frac{(x-\mu)^2}{2\sigma^2}}}{\sqrt{2\pi\sigma}}\,.$$

Let us substitute the expression of the geometric error, $(x \to \Delta_i)$

$$pdf = \frac{e^{-r-\mu+\sqrt{\left(x_i - \frac{x_i+a(-b+y_i)+c(-d+z_i)}{1+a^2+c^2}\right)^2+\left(-b+y_i-\frac{a(x_i+a(-b+y_i)+c(-d+z_i))}{1+a^2+c^2}\right)^2+\left(-d+z_i-\frac{c(x_i+a(-b+y_i)+c(-d+z_i))}{1+a^2+c^2}\right)^2}^{\;2}\Big/2\sigma^2}}{\sqrt{2\pi\sigma}}\,.$$

We apply a *Mathematica* function developed by Rose and Smith (2000) which is entitled as **SuperLog**. This function utilizes pattern-matching code that enhances *Mathematica*'s ability to simplify expressions involving the natural logarithm of a product of algebraic terms.

Then the likelihood function is

$$\mathcal{L} = \prod_{i=1}^{N} pdf$$

$$= \prod_{i=1}^{N} \frac{e^{-r-\mu+\sqrt{\left(x_i - \frac{x_i+a(-b+y_i)+c(-d+z_i)}{1+a^2+c^2}\right)^2+\left(-b+y_i-\frac{a(x_i+a(-b+y_i)+c(-d+z_i))}{1+a^2+c^2}\right)^2+\left(-d+z_i-\frac{c(x_i+a(-b+y_i)+c(-d+z_i))}{1+a^2+c^2}\right)^2}^{\;2}\Big/2\sigma^2}}{\sqrt{2\pi\sigma}}\,.$$

Considering its logarithm which should be maximized

$\Rightarrow \texttt{LogL} = Log[L]$

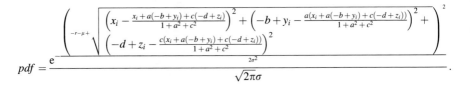

$$\Leftarrow -\frac{b^2N}{2\sigma^2} - \frac{a^2b^2N}{2(1+a^2+c^2)^2\sigma^2} - \frac{a^4b^2N}{2(1+a^2+c^2)^2\sigma^2} - \frac{a^2b^2c^2N}{2(1+a^2+c^2)^2\sigma^2} + \frac{a^2b^2N}{(1+a^2+c^2)\sigma^2}$$

$$-\frac{abcdN}{(1+a^2+c^2)^2\sigma^2} - \frac{a^3bcdN}{(1+a^2+c^2)^2\sigma^2} - \frac{abc^3dN}{(1+a^2+c^2)^2\sigma^2} + \frac{2abcdN}{(1+a^2+c^2)\sigma^2} - \frac{d^2N}{2\sigma^2}$$

$$-\frac{c^2d^2N}{2(1+a^2+c^2)^2\sigma^2} - \frac{a^2c^2d^2N}{2(1+a^2+c^2)^2\sigma^2} - \frac{c^4d^2N}{2(1+a^2+c^2)^2\sigma^2} + \frac{c^2d^2N}{(1+a^2+c^2)\sigma^2} - \frac{r^2N}{2\sigma^2} - \frac{rN\mu}{\sigma^2}$$

$$-\frac{N\mu^2}{2\sigma^2} - \frac{1}{2}N\log[2] - \frac{1}{2}N\log[\pi] - N\log[\sigma] + \sum_{i=1}^{N}\frac{abx_i}{(1+a^2+c^2)^2\sigma^2} + \sum_{i=1}^{N}\frac{a^3bx_i}{(1+a^2+c^2)^2\sigma^2}$$

$$+\sum_{i=1}^{N}\frac{abc^2x_i}{(1+a^2+c^2)^2\sigma^2} + \sum_{i=1}^{N}-\frac{2abx_i}{(1+a^2+c^2)\sigma^2} + \sum_{i=1}^{N}\frac{cdx_i}{(1+a^2+c^2)^2\sigma^2} + \sum_{i=1}^{N}\frac{a^2cdx_i}{(1+a^2+c^2)^2\sigma^2}$$

$$+\sum_{i=1}^{N}\frac{c^3dx_i}{(1+a^2+c^2)^2\sigma^2} + \sum_{i=1}^{N}-\frac{2cdx_i}{(1+a^2+c^2)\sigma^2} + \sum_{i=1}^{N}-\frac{x_i^2}{2\sigma^2} + \sum_{i=1}^{N}-\frac{x_i^2}{2(1+a^2+c^2)^2\sigma^2}$$

$$+ \sum_{i=1}^{N} -\frac{a^2 x_i^2}{2(1+a^2+c^2)^2 \sigma^2} + \sum_{i=1}^{N} -\frac{c^2 x_i^2}{2(1+a^2+c^2)^2 \sigma^2} + \sum_{i=1}^{N} \frac{x_i^2}{(1+a^2+c^2)\sigma^2} + \sum_{i=1}^{N} \frac{by_i}{\sigma^2}$$

$$+ \sum_{i=1}^{N} \frac{a^2 by_i}{(1+a^2+c^2)^2 \sigma^2} + \sum_{i=1}^{N} \frac{a^4 by_i}{(1+a^2+c^2)^2 \sigma^2} + \sum_{i=1}^{N} \frac{a^2 bc^2 y_i}{(1+a^2+c^2)^2 \sigma^2} + \sum_{i=1}^{N} -\frac{2a^2 by_i}{(1+a^2+c^2)\sigma^2}$$

$$+ \sum_{i=1}^{N} \frac{acdy_i}{(1+a^2+c^2)^2 \sigma^2} + \sum_{i=1}^{N} \frac{a^3 cdy_i}{(1+a^2+c^2)^2 \sigma^2} + \sum_{i=1}^{N} \frac{ac^3 dy_i}{(1+a^2+c^2)^2 \sigma^2} + \sum_{i=1}^{N} -\frac{2acdy_i}{(1+a^2+c^2)\sigma^2}$$

$$+ \sum_{i=1}^{N} -\frac{ax_i y_i}{(1+a^2+c^2)^2 \sigma^2} + \sum_{i=1}^{N} -\frac{a^3 x_i y_i}{(1+a^2+c^2)^2 \sigma^2} + \sum_{i=1}^{N} -\frac{ac^2 x_i y_i}{(1+a^2+c^2)^2 \sigma^2} + \sum_{i=1}^{N} \frac{2ax_i y_i}{(1+a^2+c^2)\sigma^2} + \sum_{i=1}^{N}$$

$$-\frac{y_i^2}{2\sigma^2} + \sum_{i=1}^{N} -\frac{a^2 y_i^2}{2(1+a^2+c^2)^2 \sigma^2} + \sum_{i=1}^{N} -\frac{a^4 y_i^2}{2(1+a^2+c^2)^2 \sigma^2} + \sum_{i=1}^{N} -\frac{a^2 c^2 y_i^2}{2(1+a^2+c^2)^2 \sigma^2} +$$

$$+ \sum_{i=1}^{N} \frac{a^2 y_i^2}{(1+a^2+c^2)\sigma^2} + \sum_{i=1}^{N} \frac{abcz_i}{(1+a^2+c^2)^2 \sigma^2} + \sum_{i=1}^{N} \frac{a^3 bcz_i}{(1+a^2+c^2)^2 \sigma^2} + \sum_{i=1}^{N} \frac{abc^3 z_i}{(1+a^2+c^2)^2 \sigma^2}$$

$$+ \sum_{i=1}^{N} -\frac{2abcz_i}{(1+a^2+c^2)\sigma^2} + \sum_{i=1}^{N} \frac{dz_i}{\sigma^2} + \sum_{i=1}^{N} \frac{c^2 dz_i}{(1+a^2+c^2)^2 \sigma^2} + \sum_{i=1}^{N} \frac{a^2 c^2 dz_i}{(1+a^2+c^2)^2 \sigma^2} + \sum_{i=1}^{N} \frac{c^4 dz_i}{(1+a^2+c^2)^2 \sigma^2}$$

$$+ \sum_{i=1}^{N} -\frac{2c^2 dz_i}{(1+a^2+c^2)\sigma^2} + \sum_{i=1}^{N} -\frac{cx_i z_i}{(1+a^2+c^2)^2 \sigma^2} + \sum_{i=1}^{N} -\frac{a^2 cx_i z_i}{(1+a^2+c^2)^2 \sigma^2} + \sum_{i=1}^{N} -\frac{c^3 x_i z_i}{(1+a^2+c^2)^2 \sigma^2}$$

$$+ \sum_{i=1}^{N} \frac{2cx_i z_i}{(1+a^2+c^2)\sigma^2} + \sum_{i=1}^{N} -\frac{acy_i z_i}{(1+a^2+c^2)^2 \sigma^2} + \sum_{i=1}^{N} -\frac{a^3 cy_i z_i}{(1+a^2+c^2)^2 \sigma^2} + \sum_{i=1}^{N} -\frac{ac^3 y_i z_i}{(1+a^2+c^2)^2 \sigma^2}$$

$$+ \sum_{i=1}^{N} \frac{2acy_i z_i}{(1+a^2+c^2)\sigma^2} + \sum_{i=1}^{N} -\frac{z_i^2}{2\sigma^2} + \sum_{i=1}^{N} -\frac{c^2 z_i^2}{2(1+a^2+c^2)^2 \sigma^2} + \sum_{i=1}^{N} -\frac{a^2 c^2 z_i^2}{2(1+a^2+c^2)^2 \sigma^2}$$

$$+ \sum_{i=1}^{N} -\frac{c^4 z_i^2}{2(1+a^2+c^2)^2 \sigma^2} + \sum_{i=1}^{N} \frac{c^2 z_i^2}{(1+a^2+c^2)\sigma^2}$$

$$+ \sum_{i=1}^{N} \frac{1}{\sigma^2} r \, \text{Sqrt}\left\{\left(x_i - \frac{x_i + a(-b+y_i) + c(-d+z_i)}{1+a^2+c^2}\right)^2\right.$$

$$+ \left(-b+y_i - \frac{a(x_i + a(-b+y_i) + c(-d+z_i))}{1+a^2+c^2}\right)^2 + \left(-d+z_i - \frac{c(x_i + a(-b+y_i) + c(-d+z_i))}{1+a^2+c^2}\right)^2\right\}$$

$$+ \sum_{i=1}^{N} \frac{1}{\sigma^2} \mu \, \text{Sqrt}\left\{\left(x_i - \frac{x_i + a(-b+y_i) + c(-d+z_i)}{1+a^2+c^2}\right)^2\right.$$

$$+ \left(-b+y_i - \frac{a(x_i + a(-b+y_i) + c(-d+z_i))}{1+a^2+c^2}\right)^2 + \left(-d+z_i - \frac{c(x_i + a(-b+y_i) + c(-d+z_i))}{1+a^2+c^2}\right)^2\right\}$$

It is a rather complicated expression, but fortunately *Mathematica* has a built in function to compute this expression numerically (**Likelihood[]**).

Now we can create the corresponding likelihood functions for the identified outliers and inliers. We do not display the *Mathematica* code here.

In order to carry out local maximization, the result of the standard parameter estimation (from above) can be considered

$$\Leftarrow \{a \to -0.604342, b \to -11.3773, c \to -3.68704, d \to 146.479,$$
$$r \to 0.178783\}$$

The result of the computation is:

$$\Rightarrow \{a \to -0.643829, b \to -11.3064, c \to -3.60796, d \to 146.386,$$
$$r \to 0.175172\}$$

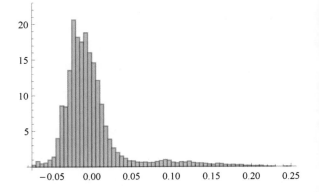

Fig. 14.39 Histogram of the model error employing maximum likelihood method

Now, let us compute the model error distribution and show the histogram:

This means that employing the model error distribution represented by Fig. 14.31, and using maximum likelihood method for estimating the parameters, with these parameters we get a model error distribution represented by Fig. 14.39. If the two distributions are the same, than the parameter estimation is correct. To answer this question let us employ EM for this later distribution.

The result of the parameter values is

$\Leftarrow \{\{0.0826151, 0.0596793\}, \{-0.0111856, 0.018304\}, \{0.118863, 0.881137\}\}$

These can be displayed in a Table 14.4 form.

Figure 14.40 shows the density functions of the two components (Figs. 14.41, 14.42, 14.43, 14.44 and 14.45).

We now separate the mixture of the samples into two clusters: cluster of outliers and cluster of inliers. Membership values of the first cluster:

Table 14.4 Parameters of the Gaussian mixture after zero iteration

η	μ	σ
0.118863	0.0826151	0.0596793
0.881137	−0.0111856	0.018304

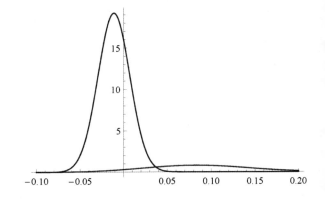

Fig. 14.40 The PDF of the two components

Fig. 14.41 The joint PDF of the mixture

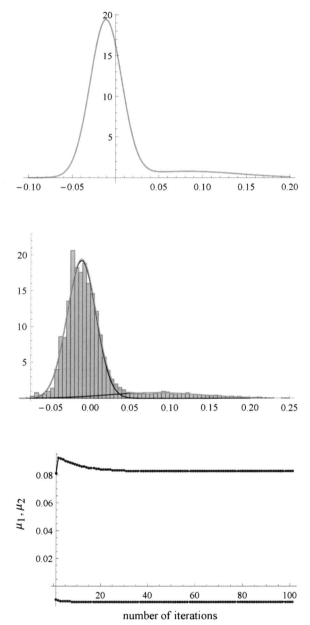

Fig. 14.42 The PDF of the two components and the normalized histogram of the data

Fig. 14.43 The convergence of the means of the two components

Fig. 14.44 The convergence
of the standard deviations

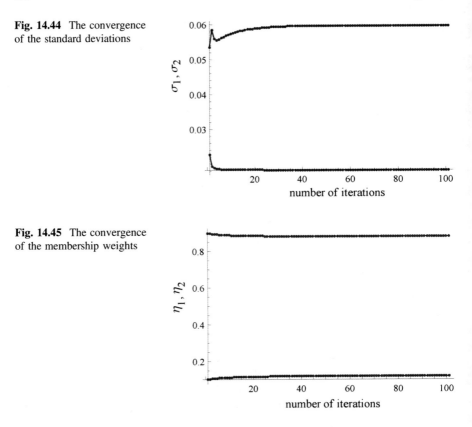

Fig. 14.45 The convergence
of the membership weights

⇒ Short[Δ[[1]], 10]
⇐ {0.0825756, 0.0117538, 0.0342882, 1., 0.0125805, 0.0147983, 0.0104937,
0.0196103, 0.0308094, 0.0161561, 0.0126288, < <16412 > > , 0.99999,
1., 1., 1., 1., 1., 1., 1., 1., 1., 1.}

Membership values of the second cluster:

⇒ Short[Δ[[1]], 10]
⇐ {0.917424, 0.988246, 0.965712, 1.78221 × 10⁻¹¹, 0.98742, 0.985202,
0.989506, < <16420 > > , 2.3251 × 10⁻²¹, 8.37015 × 10⁻²⁴, 4.22025 × 10⁻²¹,
1.97601 × 10⁻²⁴, 1.26222 × 10⁻¹⁷, 1.2341 × 10⁻¹³, 2.78243 × 10⁻¹³}

In order to get Boolean (crisp) clustering let us round the membership values.

⇒ S1 = Round[Δ[[1]]]; S2 = Round[Δ[[2]]];

The elements in the first cluster (outliers) are the corresponding elements of
those having value 1 in the set containing the rounded member values (S1).

⇒ XYZOut = Map[dataP[[#]]&, Position[S1, 1]//Flatten];

The number of these elements (outliers) is 1577.

Similarly, the elements in the second cluster (inliers) are the corresponding elements of those having value 1 in the set containing the rounded member values (S2).

⇒ XYZIn = Map[dataP[[#]]&, Position[S2, 1]//Flatten];

and there are 14,857 inliers.

Let us display the ouliers and the inliers (Fig. 14.46).

Since the two distributions are different, we need to carry out a further iteration steps. The results after three iteration steps can be seen on Fig. 14.47 and in Tables 14.5 and 14.6.

14.2.2.9   Application to Leafy Tree

In the example above there were only ∼ 10% outliers. Now let us see another example where the ratio of the outliers are much more higher, nearly 40%. This situation is closer to segmentation than to a simple fitting problem. Here the test object is the same tree, but with foliage, see Fig. 14.48.

Let us load the data, 91,089 points, and show their plot (Fig. 14.49).

The first guess will be the result, which we obtained without foliage (Fig. 14.50).

Fig. 14.46 The inliers (blue) and outliers (red) as a first guess resulted standard minimization of the total residual

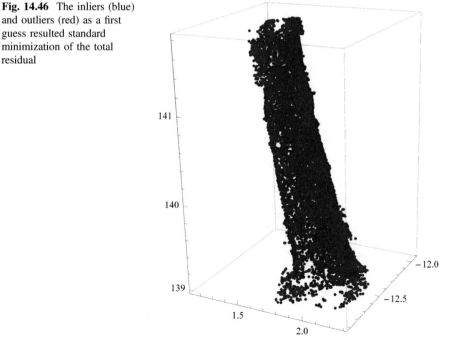

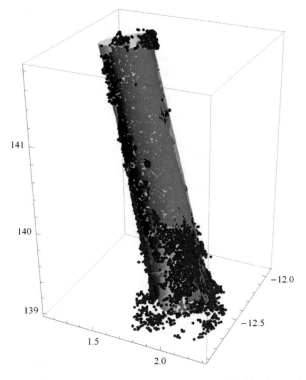

Fig. 14.47 The inliers (blue) and outliers (red) as a first guess resulted by the maximum likelihood technique

Table 14.5 The computed cylinder parameters

Iteration	a	b	c	d	r
0	−0.6043	−11.3773	−3.6870	146.479	0.1788
1	−0.6438	−11.3064	−3.6080	146.386	0.1752
2	−0.6556	−11.2872	−3.5783	146.343	0.1749
3	−0.6619	−11.2767	−3.5779	146.344	0.1750

Table 14.6 Parameters of the components of the Gaussian mixture

Iteration	μ_1	μ_2	σ_1	σ_2	η_1	η_2	N_1	N_2
0	0.0813	−0.010	0.0536	0.0227	0.1037	0.8962	1318	15116
1	0.0826	−0.011	0.0597	0.0183	0.1188	0.8811	1577	14857
2	0.0814	−0.012	0.0631	0.0176	0.1250	0.8749	1641	14793
3	0.0837	−0.012	0.0627	0.0176	0.1229	0.8771	1637	14797

Fig. 14.48 The inliers (blue) and outliers (red) as a first guess resulted standard minimization of the total residual

Fig. 14.49 The points of cloud of data

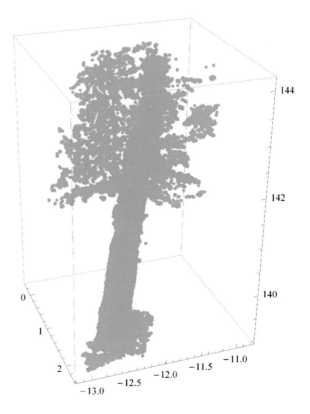

Fig. 14.50 The first
estimation of the cylinder
fitting to leafy tree

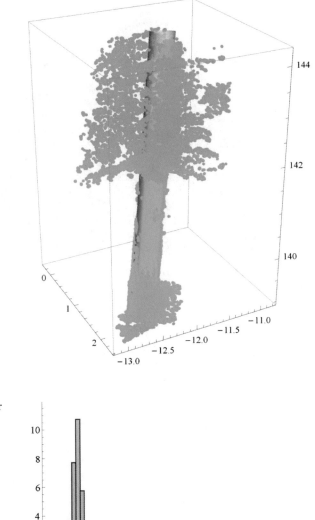

Fig. 14.51 The model error
distribution of the first
estimation

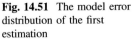

$\Leftarrow \{a \rightarrow -0.6619, b \rightarrow -11.2767, c \rightarrow -3.57789, d \rightarrow 146.344, r \rightarrow 0.175\}$

Applying the geometric error model, the model error distribution of this first
approach.

Using all of the points we have (Fig. 14.51).

Now let us identify the components of the Gaussian mixture.

$\Rightarrow \{\Delta, \texttt{param}\} = \texttt{ExpectationMaximization}[\texttt{error}, 100,$
$\quad \{\{0.6, 0.15\}, \{0, 0.1\}, \{0.5, 0.5\}\}];$

The result of the parameter values is

$\Leftarrow \{\{0.341165, 0.275897\}, \{-0.0109171, 0.0204849\}, \{0.463995, 0.536005\}\}$

These can be displayed in a Table 14.7 form.
The density functions of the two component (Figs. 14.52, 14.53, 14.54, 14.55, 14.56 and 14.57).

$\Rightarrow \texttt{Show}[\{\texttt{p5}, \texttt{p3}, \texttt{p4}\}]$

Now, we separate the mixture of the samples into two clusters: cluster of outliers and cluster of inliers. Membership values of the first cluster.

$\Rightarrow \texttt{Short}[\Delta[[1]], 10]$
$\Leftarrow \{1., 1., 1., 1., 1., 1., 1., 1., 1., 1., 1., 1., 1., 1., 1., 1., 1., 1., 1.,$
$\quad 1., 1., 1., 1., 1., 1., 1., 1., 1., 1., 1., 1., 1., 1., 1., 1., 1., 1., 1.,$
$\quad << 91009 > > , 1., 1., 1., 1., 1., 1., 1., 1., 1., 1., 1., 1., 1., 1., 1.,$
$\quad 1., 1., 1., 1., 1., 1., 1., 1., 1., 1., 1., 1., 1., 1., 1., 1., 1., 1., 1.,$
$\quad 1., 1., 1.\}$

Membership values of the second cluster:

$\Rightarrow \texttt{Short}[\Delta[[2]], 10]$
$\Leftarrow \{2.168186518114028 \times 10^{-569}, 5.715327153955006 \times 10^{-563},$
$\quad 4.439837476761972 \times 10^{-375}, 6.094575596255934 \times 10^{-374},$
$\quad 1.526341682623958 \times 10^{-585}, 2.869284522501804 \times 10^{-372},$
$\quad << 91078 > > , 9.51406 \times 10^{-40}, 6.5817 \times 10^{-32},$
$\quad 8.02675 \times 10^{-35}, 1.28721 \times 10^{-30}, 1.99815 \times 10^{-38}\}$

In order to get Boolean (crisp) clustering let us round the membership values.

$\Rightarrow \texttt{S1} = \texttt{Round}[\Delta[[1]]]; \texttt{S2} = \texttt{Round}[\Delta[[2]]];$

The elements in the first cluster (outliers) are the corresponding elements of those having value 1 in the set containing the rounded member values (S1). We get 39,982 outliers.

Similarly, the elements in the second cluster (inliers) are the corresponding elements of those having value 1 in the set containing the rounded member values (S2). There are 51,107.

Let us display the ouliers and the inliers (Fig. 14.58).

Now let us compute the first iteration employing maximum likelihood method. The parameters become (Fig. 14.59).

Table 14.7 Parameters of the Gaussian mixture after zero iteration

η	μ	σ
0.463995	0.341165	0.275897
0.536005	–0.0109171	0.0204849

Fig. 14.52 The PDF of the two components

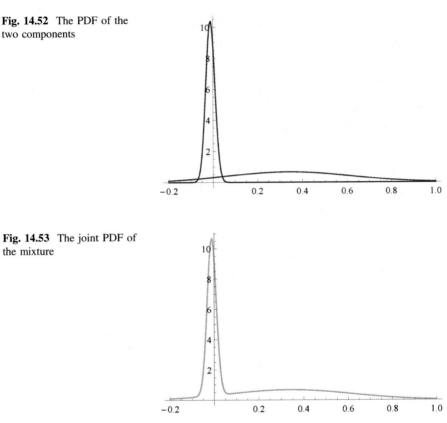

Fig. 14.53 The joint PDF of the mixture

$\Rightarrow \{a \rightarrow -0.660978, b \rightarrow -11.28, c \rightarrow -3.60862, d \rightarrow 146.398, r \rightarrow 0.17553\}$

Now the parameters of the Gaussian mixture:
The result of the parameter values:

$\Leftarrow \{\{0.343552, 0.275066\}, \{-0.0109826, 0.0207809\}, \{0.458853, 0.541147\}\}$

These can be displayed in a Table 14.8 form.

The density functions of the two components (Figs. 14.60, 14.61, 14.62, 14.63, 14.64 and 14.65).

Fig. 14.54 The PDF of the two components and the normalized histogram of the data

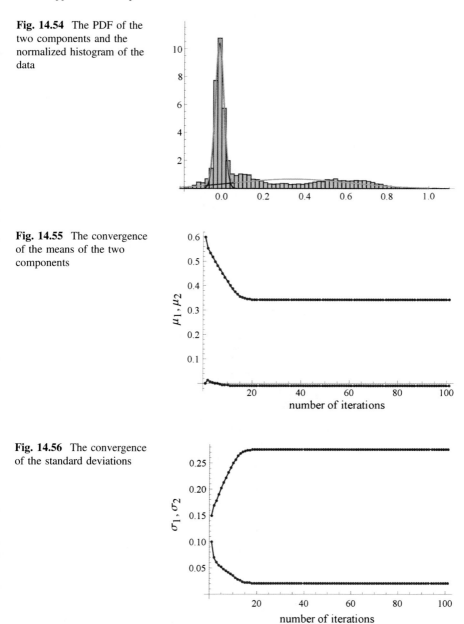

Fig. 14.55 The convergence of the means of the two components

Fig. 14.56 The convergence of the standard deviations

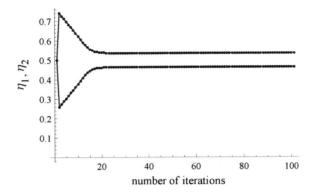

Fig. 14.57 The convergence of the membership weights

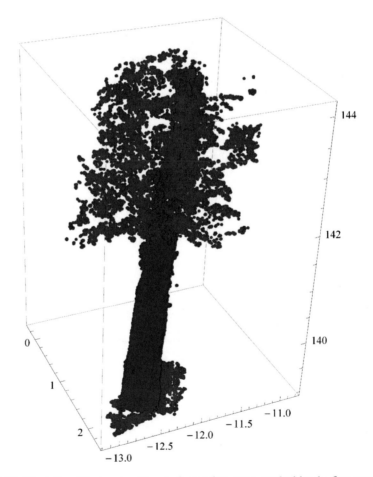

Fig. 14.58 The inliers (blue) and outliers (red) as a first guess resulted by the first approach

Fig. 14.59 The PDF of the two components and the normalized histogram of the data

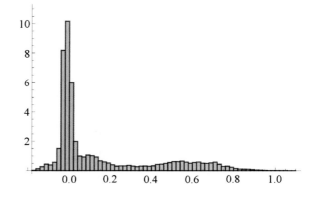

Table 14.8 Parameters of the Gaussian mixture after zero iteration

η	μ	σ
0.458853	0.343552	0.275066
0.541147	−0.0109826	0.0207809

Fig. 14.60 The PDF of the two components

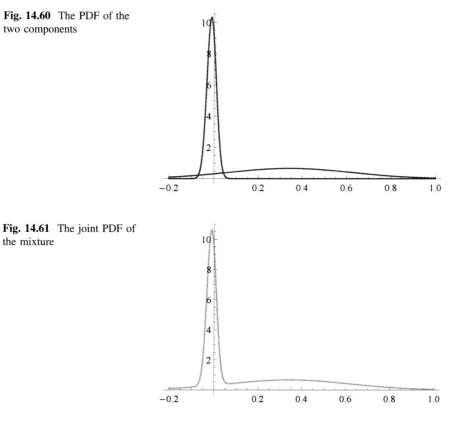

Fig. 14.61 The joint PDF of the mixture

Fig. 14.62 The PDF of the two components and the normalized histogram of the data

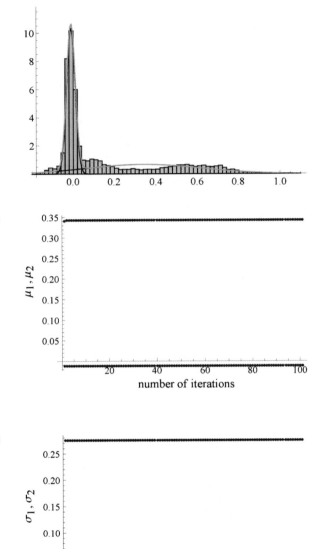

Fig. 14.63 The convergence of the means of the two components

Fig. 14.64 The convergence of the standard deviations

Fig. 14.65 The convergence of the membership weights

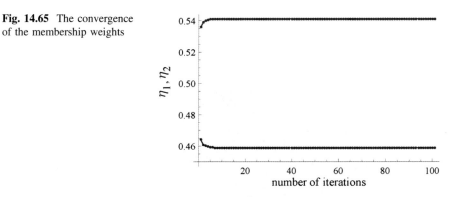

We separate the mixture of the samples into two clusters: cluster of outliers and cluster of inliers. Membership values of the first cluster:

⇒ Short[Δ[[1]], 10]

⇐ {1., 1., 1., 1., 1., 1., 1., 1., 1., 1., 1., 1., 1., 1., 1., 1., 1., 1., 1., 1., 1.,
   1., 1., 1., 1., 1., 1., 1., 1., 1., 1., 1., 1., 1., 1., 1., 1., 1., 1., 1.,
   1., 1., 1., 1., 1., 1., 1., 1., 1., 1., 1., 1., 1., 1., 1., 1., 1., <<91009>>,
   1., 1., 1., 1., 1., 1., 1., 1., 1., 1., 1., 1., 1., 1., 1., 1., 1., 1., 1.,
   1., 1., 1., 1., 1., 1., 1., 1., 1., 1., 1., 1., 1., 1., 1., 1., 1., 1., 1.,
   1., 1., 1., 1., 1., 1., 1., 1., 1., 1., 1., 1., 1., 1., 1., 1., 1.,}

Membership values of the second cluster:

⇒ Short[Δ[[2]], 10]

⇐ {5.197294751685767 × 10^{-563}, 9.62586576863179 × 10^{-557},
   2.44052537654221 × 10^{-371}, 3.138647064118679 × 10^{-370},
   7.696701760704474 × 10^{-579}, 1.319339597606332 × 10^{-368},
   <<91077>>, 5.26593 × 10^{-38}, 1.12943 × 10^{-38}, 4.7876 × 10^{-31},
   6.62412 × 10^{-34}, 7.8142 × 10^{-30}, 1.95385 × 10^{-37}}

In order to get Boolean (crisp) clustering let us round the membership values.

⇒ S1 = Round[Δ[[1]]]; S2 = Round[Δ[[2]]];

The elements in the first cluster (outliers) are the corresponding elements of those having value 1 in the set containing the rounded member values (S1). There are 39,404 outliers.

Similarly, the elements in the second cluster (inliers) are the corresponding elements of those having value 1 in the set containing the rounded member values (S2). There are 51,685 inliers.

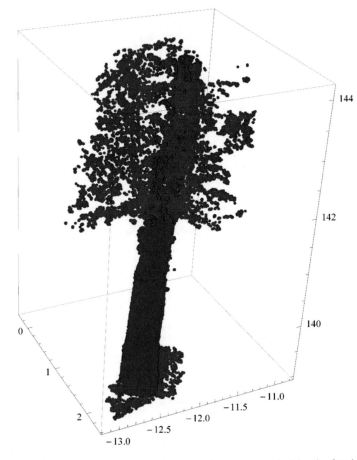

Fig. 14.66 The inliers (blue) and outliers (red) as a first guess resulted by the first iteration

Let us display the outliers and the inliers (Fig. 14.66).

The inlier points (Fig. 14.67),

We consider it the last iteration step. The parametric equation of the fitted cylinder is (Figs. 14.68 and 14.69; Table 14.9 and 14.10)

$\Leftarrow \{v + 0.168963\text{Cos}[u] + 0.00813893\text{Sin}[u], -11.28 - 0.660978\,v$
$+ 0.172661\text{Sin}[u], 146.398 - 3.60862\,v + 0.0468219\text{Cos}[u]$
$-0.0293703\text{Sin}[u]\}$

$\Rightarrow \text{Show}[\{p1, p2, pIn, pOut\}]$

Fig. 14.67 The inliers

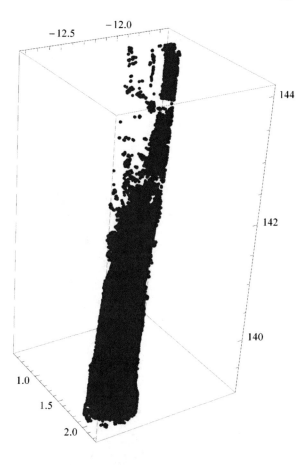

The results our computation illustrate that in case of noisy measurements, the frequently assumed $\mathcal{N}(0, \sigma)$ model error distribution is not true. Therefore to estimate the model parameters, iteration is necessary with the proper representation of the non-Gaussian distribution. In this study an expectation maximization algorithm was employed. In case of general cylinder geometry, the five model parameters can be computed via local maximization of the maximum likelihood function. The implicit equation based on an algebraic error definition, and solved easily by Gröbner basis, can provide a proper initial guess value for the maximization problem.

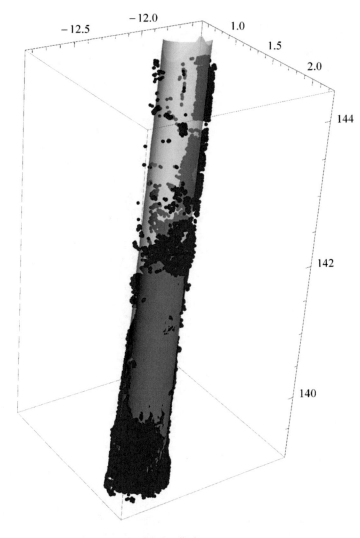

Fig. 14.68 The inliers (blue) and the fitted cylinder

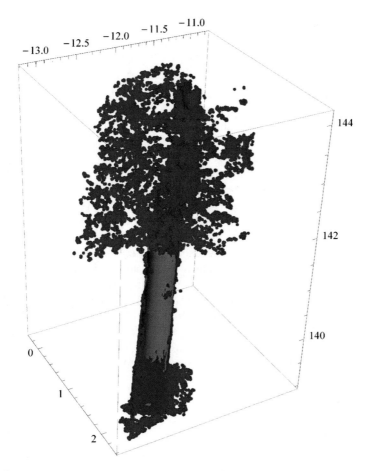

Fig. 14.69 The inliers (blue) and outliers (red) and the fitted cylinder

Table 14.9 The computed cylinder parameters

Iteration	a	b	c	d	r
0	−0.6619	−11.2767	−3.5779	146.344	0.1750
1	−0.6611	−11.2791	−3.6086	146.398	0.1753

Table 14.10 Parameters of the components of the Gaussian mixture

Iteration	μ_1	μ_2	σ_1	σ_2	η_1	η_2	N_1	N_2
0	0.3412	−0.011	0.2759	0.0205	0.4640	0.5360	39982	51107
1	0.3435	−0.010	0.2751	0.0208	0.4588	0.5418	39404	51685

14.3   Problem

14.3.1   Fitting a Plane to a Slope

14.3.1.1   Test Area, Equipment and Measured Data

Outdoor laser scanning measurements have been carried out in a hilly park of Budapest, see Fig. 14.70 The test area is on a steep slope, covered with dense but low vegetation.

The experiment has been carried out with a Faro Focus 3D terrestrial laser scanner, as before. The test also aimed to investigate the tie point detection capabilities of the scanner's processing software; different types of spheres were deployed all over the test area. In case of multiple scanning positions these spheres can be used for registering the point clouds.

The measurement range of the scanner is 120 m, the ranging error is ±2 mm, according to the manufacturer's technical specification (Fig. 14.71).

The scanning parameters were set to ″ resolution, which equals 3 mm/10 m point spacing. This measurement resulted in 178.8 million points that were acquired in five and half minutes. The test data set was cropped from the point cloud; moreover, further resampling was applied in order to reduce the data size. The final data set is composed of 38 318 points in ASCII format, and only the x, y, z coordinates were kept (no intensity values). Let us load the measured data (Figs. 14.72 and 14.73).

$\Rightarrow$ Clear[XYZ, N]

$\Rightarrow$ XYZ = Import$\left[''\text{H} : \backslash\backslash \text{Tamas\_adatok} // \text{output\_41\_392.dat}'' \right]$;

$\Rightarrow$ n = Length[XYZ]

$\Leftarrow$ 41392

Fig. 14.70 The test area in Budapest

Fig. 14.71 The scanner at the
top of the measured steep
slope with the different sizes
of white spheres as control
points in the background

⇒ XYZP = Select[XYZ, And[NumberQ[#[[1]]], NumberQ[#[[2]]],
NumberQ[#[[3]]]]&];

⇒ N = Length[XYZP]

⇐ 33292

⇒ p8 = ListPointPlot3D[XYZP, PlotStyle → Blue, BoxRatios$R${1, 1, 1},
ViewPoint → {1.8698922647815255′, −2.32400793103699′,
−0.1931120274143106′}, ViewVertical → {0.47714644303899606′,
−0.5674527451525903′, 0.6710653127036321′}]

Fig. 14.72 The measured colored LiDAR point cloud

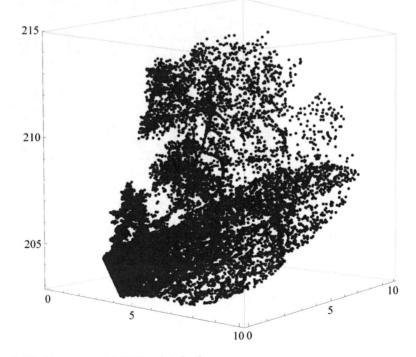

Fig. 14.73 The measured LiDAR point cloud

14.3.1.2 Application SVD

Skipping the details, after 9.1s SVD finds that the equation for the plane is:

$\Leftarrow \{z \rightarrow 198.698 + 0.495306x + 1.68771y\}$

14.3.1.3 Solution Via PCA

$\Rightarrow$ AbsoluteTiming[A = Covariance[Cc0]; MatrixForm[A]]

$$\Leftarrow \left\{ 0.0376042, \begin{pmatrix} 5.87736 & -0.0359188 & 1.96269 \\ -0.0359188 & 2.86103 & 1.78577 \\ 1.96269 & 1.78577 & 5.77838 \end{pmatrix} \right\}$$

$\Rightarrow \{nx, ny, nz\} = $ Eigenvectors[A][[3]]
$\Leftarrow \{-0.244802, -0.834143, 0.494245\}$

$\Rightarrow$ sold = Solve [$\{nx, ny, nz\}$.Total[XYZ] $-$ dN == 0, d]
$\Leftarrow \{d \rightarrow 98.2054\}$

Consequently the coefficients of surface model are

$\Leftarrow \{0.495306, 1.68771, 198.698\}$

14.3.1.4 Gröbner Basis Solution

$\Leftarrow \{\alpha \rightarrow 0.49530, \beta \rightarrow 1.68771, \gamma \rightarrow 198.698\}$

See Fig. 14.74.

14.3.1.5 Model Error Analysis

$\Rightarrow$ Error = Map $\left[\left(\dfrac{\#[[3]] - \alpha\#[[1]] - \beta\#[[2]] - \gamma}{\sqrt{1 + \alpha^2 + \beta^2}} \right) /.\text{solopt\&, XYZ} \right]$;

$\Rightarrow$ Min[Error]
$\Leftarrow -5.06393$
$\Rightarrow$ Max[Error]
$\Leftarrow 5.46962$
$\Rightarrow$ Mean[Error]
$\Leftarrow -1.27578 \times 10^{-14}$
$\Rightarrow$ StandardDeviation[Error]
$\Leftarrow 1.3388$
$\Rightarrow$ Histogram[Error, 40]

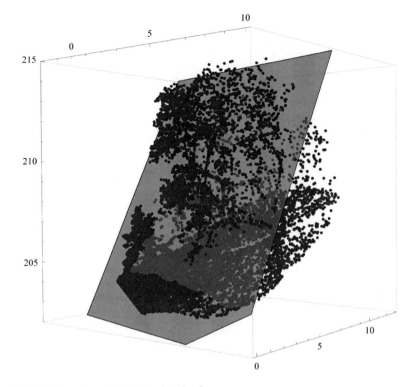

Fig. 14.74 The measured LiDAR point cloud

See Fig. 14.75.

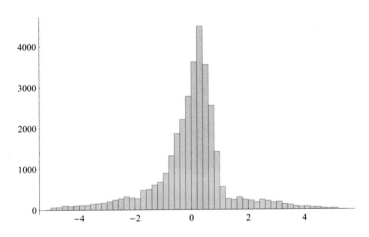

Fig. 14.75 The histogram of the errors

14.3.1.6   Application Danish Algorithm with Embedded Gröbner Basis Solution

We will use 20 iterations with 33,292 points.
  The initial weights of the observations.

  $\Rightarrow$ W $=$ Table[1, {N}]; (*initialization of weights *)

  The next figures show the convergence of the different parameters as a function of the number of the iterations (Figs. 14.76, 14.77 and 14.78).
  The result of the iteration

  $\Leftarrow \{\alpha \to 0.105602, \beta \to 0.504716, \gamma \to 202.664\}$

  The observations that remained active (inliers) are total 24,576. We plot them (Fig. 14.79).
  The inliers of the Danish solution (Fig. 14.80).

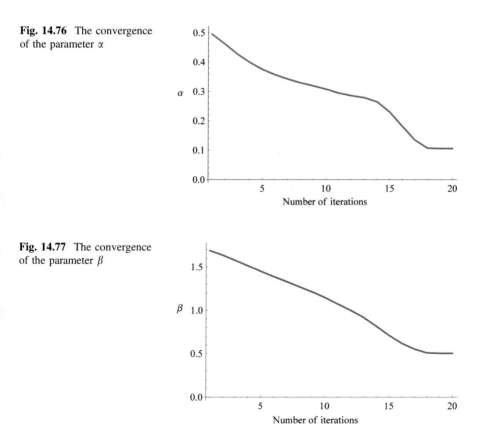

Fig. 14.76 The convergence of the parameter α

Fig. 14.77 The convergence of the parameter β

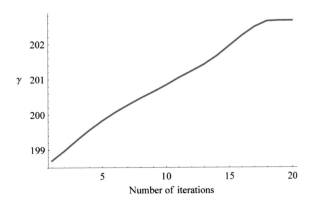

Fig. 14.78 The convergence of the parameter γ

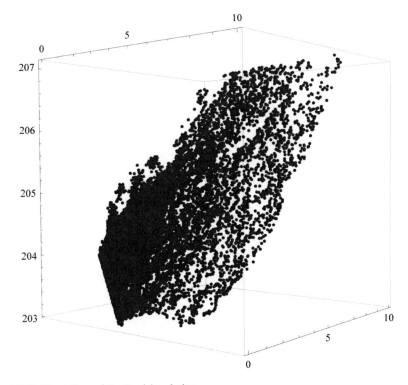

Fig. 14.79 The inliers of the Danish solution

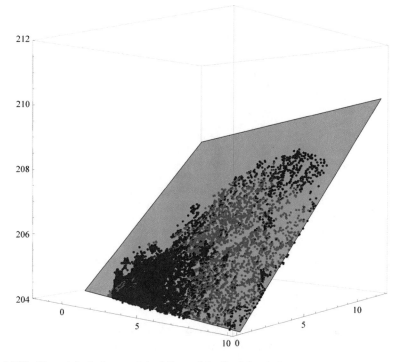

Fig. 14.80 The original plane and the inliers of the Danish solution

14.3.1.7   Model Error Analysis

$\Rightarrow$ Error = Map$\left[\left(\dfrac{\#[[3]]-\alpha\#[[1]]-\beta\#[[2]]-\gamma}{\sqrt{1+\alpha^2+\beta^2}}\right)/.\text{sols\&},\text{inliers}\right]$;

$\Rightarrow$ Min[Error]

$\Leftarrow -0.219962$

$\Rightarrow$ Max[Error]

$\Leftarrow 0.370434$

$\Rightarrow$ Mean[Error]

$\Leftarrow -9.44382 \times 10^{-15}$

$\Rightarrow$ StandardDeviation[Error]

$\Leftarrow 0.0703037$

$\Rightarrow$ pHT = Histogram[Error, 20]

  See Fig. 14.81.

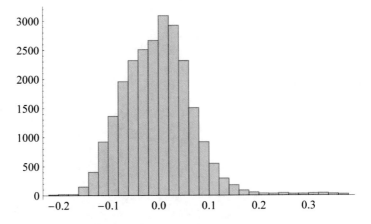

Fig. 14.81 The histogram of the errors

14.3.1.8   Application Danish Method with Embedded Weighted PCA (PCAW)

⇒ N = 50; (*Number of iterations*)

The next figures show the convergence of the different parameters as a function of the number of the iterations (Figs. 14.82, 14.83 and 14.84).

Fig. 14.82 The convergence of the parameter α

Fig. 14.83 The convergence of the parameter β

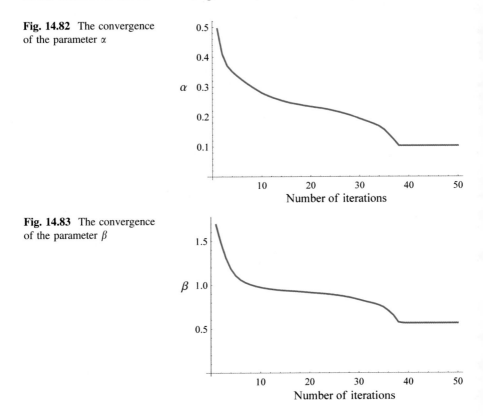

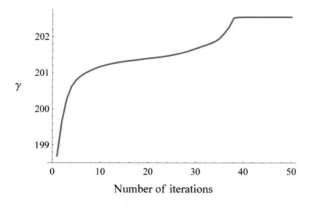

Fig. 14.84 The convergence of the parameter γ

The result of the iteration:

$$\Leftarrow \{\alpha \rightarrow 0.102968, \beta \rightarrow 0.566614, \gamma \rightarrow 202.536\}$$

The observations that remained active (inliers) are total 26,089 and they look like (Fig. 14.85).

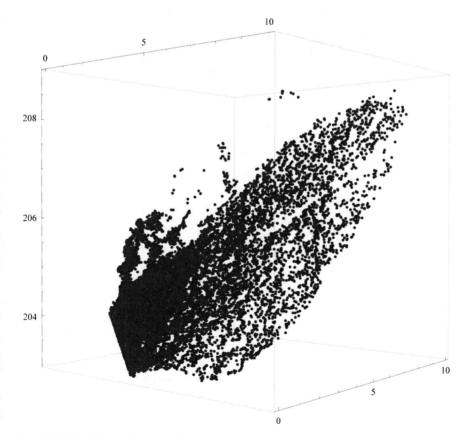

Fig. 14.85 The inliers of the Danish solution

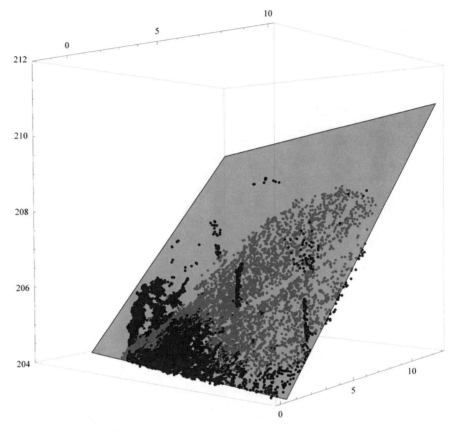

Fig. 14.86 The original plane and the inliers of the Danish solution

The plane that results from the Danish algorithm (Fig. 14.86).

14.3.1.9 Model Error Analysis

$\Rightarrow$ Error = Map $\left[\left(\dfrac{\#[[3]]-\alpha\#[[1]]-\beta\#[[2]]-\gamma}{\sqrt{1+\alpha^2+\beta^2}}\right)\middle/.\text{sols}\&, \text{inliers}\right];$

$\Rightarrow$ Min[Error]

$\Leftarrow -0.460056$

$\Rightarrow$ Max[Error]

$\Leftarrow 0.945997$

$\Rightarrow$ Mean[Error]

$\Leftarrow -1.68568 \times 10^{-14}$

$\Rightarrow$ StandardDeviation[Error]

$\Leftarrow 0.185912$

$\Rightarrow$ pHS = Histogram[Error, 20]

Fig. 14.87 The histogram of the errors in case of PCAW method

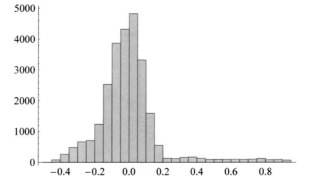

See Fig. 14.87.

In order to compare the solution without and with the Danish robust algorithm, let us consider the inliers selected by the Danish algorithm for the error computation of the result given *without robust estimation*,

$$\Rightarrow \text{Error} = \text{Map}\left[\left(\frac{\#[[3]] - \alpha\#[[1]] - \beta\#[[2]] - \gamma}{\sqrt{1 + \alpha^2 + \beta^2}}\right)/.\text{solopt\&}, \text{inliers}\right];$$

$\Rightarrow \text{Min}[\text{Error}]$

$\Leftarrow -5.06393$

$\Rightarrow \text{Max}[\text{Error}]$

$\Leftarrow 1.1246$

$\Rightarrow \text{Mean}[\text{Error}]$

$\Leftarrow -0.32325$

$\Rightarrow \text{StandardDeviation}[\text{Error}]$

$\Leftarrow 0.10649$

$\Rightarrow \text{pHW} = \text{Histogram}[\text{Error}, 20]$

See Fig. 14.88.

The 3 histograms are

$\Rightarrow \text{GraphicsGrid}[\{\{\text{pHT}, \text{pHS}, \text{pHW}\}\}]$

See Fig. 14.89.

In this section a new algorithm for plane fitting to large cloud of data set was presented.

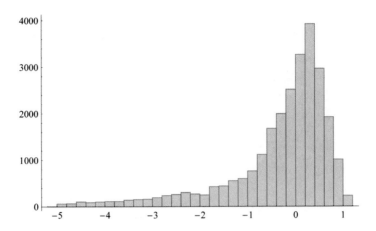

Fig. 14.88 The histogram of the errors in case of non-rubust method

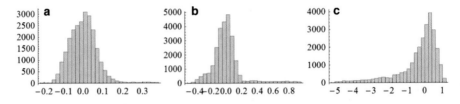

Fig. 14.89 Distributions of the model error in the 3 different cases: **a** Danish—Gröbner basis—maximum likelihood, **b** Danish—PCAW, **c** non-robust estimation

References

Carrea D, Jaboyedoff M, Derron MH (2014) Feasibility study of point cloud data from test deposition holes for deformation analysis. Working Report 2014-01, Universite de Lausanne (UNIL)

Chen CC, Stamos I (2007) Range image segmentation for modeling and object detection in urban scenes, 3DIM2007.1

DalleMole V, do Rego R, Araújo A (2010) The self-organizing approach for surface reconstruction from unstructured point clouds. In: Matsopoulos GK (ed) Self-organizing map, In Tech

Diebel JR, Thrun S, Brunig M (2006) A Bayesian method for probable surface reconstruction and decimation. ACM Trans Graph (TOG) 1

Fischler MA, Bolles RC (1981) Random sample consensus: a paradigm for model fitting with applications to image analysis and automated cartography. Commun ACM 24:381–395

Fox A, Smith J, Ford R, Doe J (2013) Expectation maximization for gaussian mixture distributions. Wolfram Demonstration Project

Huang CM, Tseng YH. Plane fitting methods of Lidar point cloud, Dept. of Geomatics, National Cheng Kung Uni. Taiwan, tseng@mail.ncku.edu.tw

Hubert M, Rousseeuw PJ, Van den Branden K (2005) ROBPCA: a new approach to robust principal component analysis. Technometrics 47(1):64–79

Khameneh M (2013) Tree detection and species identification using LiDAR Data, MSc. Thesis, KTH, Royal Institute of Technology, Stockholm

Kohonen T (1998) The self-organizing map. Neurocomputing 21:1–6

Krarup T, Kubik K, Juhl J (1980) Götterdammerung over least squares. In: Proceedings of international society for photogrammetry 14th congress, Hamburg, pp 370–378

Lakaemper R, Latecki LJ (2006) Extended EM for planar approximation of 3D data. In: IEEE international conference on robotics and automation (ICRa), Orlando, Florida

Lichtblau D (2007) Cylinders through five points: computational algebra and geometry. Autom Ded Geom Lect Notes Comp Sci 4869:80–97

Lukacs G, Martin R, Marshall D (1998) Faithful least-squares fitting of Spheres, cylinders, cones and Tori for reliable segmentation. Burkhardt H, Neumann B (eds) Computer vision-ECCV'98, Vol I. LNCS 1406, pp 671–686, Springer-Verlag

Mitra NJ, Nguyen (2003) SoCG'03, June 8–10, San Diego, California, USA, ACM 1-58113-663-3/03/0006, pp 322–328

Nievergelt Y (2000) A tutorial history of least square with applications to astronomy and geodesy. J Comp Appl Mathe 121:37–72

Nurunnabi A, Belton D, West G (2012) Diagnostic—robust statistical analysis for local surface fitting in 3D point cloud data. In: ISPRS annals of the photogrammetry, remote sensing and spatial information sciences, Vol 1–3, 2012 XXII ISPRS Congress, 25 Aug 2012, Melbourne, Australia, pp 269–275

Palancz B, Awange J, Somogyi A, Fukuda Y (2016) A robust cylindrical fitting to point cloud data. Australian J Earth Sci 63(5). https://doi.org/10.1080/08120099.2016.1230147

Project: Mathematical Geosciences—A hybrid (algebraic-numerical) solutionPaláncz B (2014) Fitting data with different error models. Mathe J. https://www.mathematica-journal.com/2014/04/30/fitting-data-with-different-error-models

Petitjean S (2002) A survey of methods for recovering quadrics in triangle meshes. ACM Comput Surv 34(2):211–262

Poppinga J, Vaskevicius N, Birk A, Pathak K (2006) Fast plane detection and polygonalization in noisy 3D range images. In: International conference on intelligent robots and systems (IROS), Nice, France, IEEE Press

Rose C, Smith D (2000) Symbolic maximum likelihood estimation with mathematica. Statistician 49:229–240

Ruiz O, Arroyave S, Acosta D (2013) Fitting of analytic surfaces to noisy point clouds. Am J Comput Mathe 3:18–26

Russeeuw PJ, Van Driessen K (1999) A fast algorithm for the minimum covariance determinant estimator. Technometrics 41(3):212–223

Stal C, Timothy Nuttens T, Constales D, Schotte K, De Backer H, De Wulf A (2012) Automatic filtering of terrestrial laser scanner data from cylindrical tunnels, TS09D—laser scanners III, 5812, FIG working week 2012, knowing to manage the territory, protect the environment, evaluate the cultural heritage. Rome, Italy, 6–10 May 2012

Stathas D, Arabatzi O, Dogouris S, Piniotis G, Tsini D, Tsinis D (2003) New monitoring techniques on the determination of structure deformation. In: Proceedings of the 11th FIG symposium on deformation measurements, Santorini, Greece

Su YT, Bethel J (2010) Detection and robust estimation of cylinder features in point clouds. In: ASPRS 2010 annual conference, San Diego, California, April 26–30

Vosselman G, Gorte B, Sithole G, Rabbani T (2004) Recognizing structure in laser scanner point clouds. Int Arch Photogr Rem Sens Spat Inform Sci 46:33–38

Weingarten JW, Gruener G, Siegwart R (2004) Probabilistic plane fitting in 3D and an application to robotic mapping. IEEE Int Conf 1:927–932

Winkelbach S, Westphal R, Goesling T (2003) Pose estimation of cylinder fragments for semi-automatic bone fracture reduction. In Pattern Recognition (DAGM 2003). Michaelis B and Krell G eds. Lecture Notes in Computer Science 2781, Springer Verlag, pp 566–573

Yang MY, Förtsner W (2010) Plane detection in point cloud data, TR-IGG-P-2010-01, Techn. Report. Nr. 1, Dept. of Photogrammetry Inst. of Geodesy and Geo-information, Uni., Bonn, Germany

Yaniv Z (2010) Random sample consensus (RANSAC) algorithm, a generic implementation. Georgetown University Medical Center, Washington, DC, USA, zivy@isis.georgetown.edu

Zuliani M (2012) RANSAC for Dummies, vision.ece.ucsb.edu/~zuliani

Chapter 15
Stochastic Modeling

15.1 Basic Stochastic Processes

15.1.1 Concept of Stochastic Processes

A *stochastic* or *random process* is a mathematical object usually defined as a collection of random variables. Historically, the random variables were associated with or indexed by a set of numbers, usually viewed as points in time, giving the interpretation of a stochastic process representing numerical values of some system randomly changing over time. From this point of view a random process can be discrete or continuous in time. Consequently stochastic processes are widely used as mathematical models of systems and phenomena that appear to vary in a random manner.

The term *random function* is also used to refer to a stochastic or random process, because a stochastic process can also be interpreted as a random element in a function space.

15.1.2 Examples for Stochastic Processes

Bernoulli process

One of the simplest stochastic processes is the *Bernoulli process*, which is a sequence of independent and identically distributed random variables, where each random variable takes either the value one with probability, say, p and value zero with probability $1 - p$. This process can be likened to somebody flipping a coin,

Supplementary Information The online version contains supplementary material available at https://doi.org/10.1007/978-3-030-92495-9_15.

where the probability of obtaining a head is p and its value is one, while the value of a tail is zero. In other words, a Bernoulli process is a sequence of Bernoulli random variables, where each coin flip is a Bernoulli trial. Using a *Mathematica* function for $p = 0.25$ could result in:

Using built in function

⇒ data = RandomFunction[BernoulliProcess[0.25], {0, 100}]

If we plot this from time 0 to time 100 we could get (Fig. 15.1).

Random walk

Random walks are stochastic processes that are usually defined as sums of random variables or random vectors in Euclidean space, so they are processes that change in discrete time. But some also use the term to refer to processes that change in continuous time.

A classic example of a random walk is known as the simple random walk, which is a stochastic process in discrete time with the integers as the state space, and is based on a Bernoulli process, where each Bernoulli variable takes either the value positive one or negative one. In other words, the simple random walk takes place on the integers, and its value increases by one with probability, say, p, or decreases by one with probability $1 - p$, so the index set of this random walk is the natural numbers, while its state space is the integers. If the $p = 0.5$, this random walk is called a symmetric random walk. Here is a graph with time 0–100 (Fig. 15.2).

A random process realization results a single trajectory. However, to get statistics of the process at single time stamp, one needs many realizations. Here are four (Fig. 15.3).

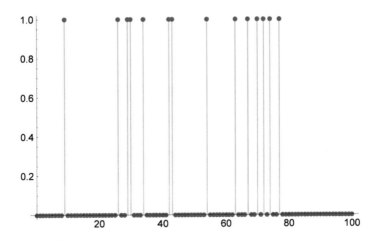

Fig. 15.1 Bernoulli process

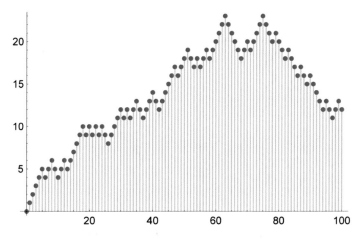

Fig. 15.2 Random walk process

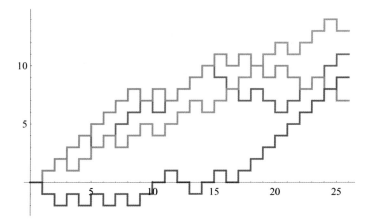

Fig. 15.3 Random walk process in two dimensions with four realizations

Wiener process

The Wiener process is a stochastic process with stationary and independent increments that are normally distributed based on the size of the increments. The Wiener process is named after Norbert Wiener, who proved its mathematical existence, but the process is also called the Brownian motion process or just Brownian motion due to its historical connection as a model for Brownian movement in liquids. It can be considered a continuous version of the simple random walk. It plays a central role in the stochastic calculus of the stochastic differential equations (Fig. 15.4).

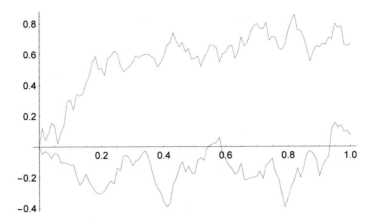

Fig. 15.4 Wiener process with drift and wfithout drift four realizations

$\Rightarrow$ data1 = Ran $-$ domFunction[WienerProcess[.8,.5], {0, 1, 0.01}];

$\Rightarrow$ data2 = RandomFunc $-$ tion[WienerProcess[0,.5], {0, 1, 0.01}];

15.1.3   Features of Stochastic Processes

Slice distribution

This is the distribution of the random process values at a fix time stamp.

Simulate 500 paths from a Wiener process and plot paths and histogram distribution of the slice distribution at $t = 1$ (Fig. 15.5).

Mean value function:

Mean value of a slice distribution of the random process.

Variance function:

Variance of a slice distribution of the random process.

Stationarity:

Stationary distribution is also known as steady-state distribution, its slice distribution is independent on the time.

Weak stationary:

A random process is weakly stationary if its mean function is independent of time, and its covariance function is independent of time translation.

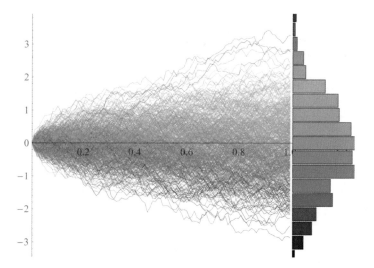

Fig. 15.5 Simulate 500 paths from a Wiener process and plot paths and histogram distribution of the slice distribution at $t = 1$

15.2 Time Series

15.2.1 Concept of Time Series

A time series is *a series of data points* indexed (or listed or graphed) in time order. Most commonly, a time series is a sequence taken at successive equally spaced points in time. Thus it is a sequence of discrete-time data. Examples of time series are heights of ocean tides, counts of sunspots.

Time series analysis comprises methods for *analyzing time series* data in order to extract meaningful statistics and other characteristics of the data. *Time series forecasting* is the use of a model to predict future values based on previously observed values.

For example here is the average temperature on the first day of a month in Chicago, IL. We have 133 data points from 2001 to 2012 (Fig. 15.6).

15.2.2 Models of Time Series

Models for time series data can have many forms and represent different stochastic processes. When modeling variations in the level of a process, three broad classes of practical importance are the autoregressive (AR) models, the integrated (I) models, and the moving average (MA) models. These three classes depend linearly on previous data points. Combinations of these ideas produce

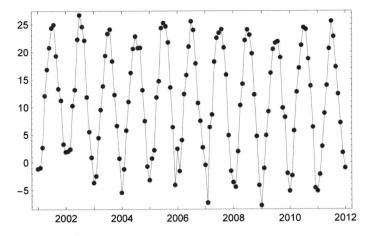

Fig. 15.6 The average temperature on the first day of a month in Chicago

autoregressive moving average (ARMA) and autoregressive integrated moving average (ARIMA) models. A further extension of the SRIMA model for seasonal time series is called as SARIMA model.

For example, an autoregressive (AR) model is a representation of a type of random process; as such, it is used to describe certain time-varying processes in nature, economics, etc. The autoregressive model specifies that the output variable depends linearly on its own previous values and on a stochastic term (an imperfectly predictable term); thus the model is in the form of a stochastic difference equation.

$$y_t = c + \sum_{i=1}^{p} \alpha_i y_{t-i} + \varepsilon_t$$

where p is the order of the autoregressive model, c is constant α_i are the parameters of the method and ε_t is white noise.

To demonstrate the application of this type of random process, let us fit an AR process to the average temperature of Chicago,

$\Rightarrow$ ARproc = TimeSeriesModelFit[temp,"AR"]

The fitted AR model of order six is,

$\Leftarrow$ ARProcess[11.2784, {0.607721, 0.0381793, −0.165581, −0.140076, −0.100544, −0.267987}, 8.05676]

The statistics of the model parameter can be seen in Table 15.1.
Now we may forecast the average temperature for one year ahead (Fig. 15.7),

$\Rightarrow$ forecast = TimeSeriesForecast[ARproc, {1, 12}];

Table 15.1 Statistics of the fitted AR(6) model

	Estimate	Standard error	t-statistic	P-value
α_1	0.607721	0.0835393	7.27467	1.34616×10^{-11}
α_2	0.0381793	0.0983854	0.388058	0.349297
α_3	-0.165581	0.0976889	-1.69498	0.0462095
α_4	-0.140076	0.0976889	-1.4339	0.076974
α_5	-0.100544	0.0983854	-1.02194	0.154332
α_6	-0.267987	0.0835393	-3.20792	0.00083767

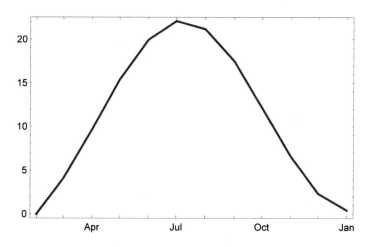

Fig. 15.7 The estimated of the average temperature for the first consecutive year

Figure 15.8 shows the known data with the forecasted temperature values.

Our model represents a random model consequently many representations of the process trajectory are possible. One hundred trajectories of the process with their mean can be seen on Fig. 15.9.

Let us calculate the standard error bands for the forecasted trajectory; see Fig. 15.10.

Similarly the 95% confidence band can be also computed, see Fig. 15.11.

Now we can try to employ different types of time series models and select the best one. In our case the SARIMA $\{(1, 0, 0), (2, 1, 1)\}$ model proved to be the model that fits with smaller error, see Table 15.2.

$\Leftarrow$ TimeSeriesModel[⊞ Family: SARIMA
Order: $\{\{1, 0, 0\}, \{2, 1, 1\}_{12}\}$]

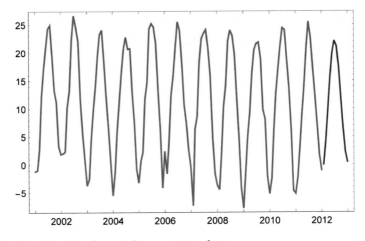

Fig. 15.8 The old and the forecasted temperature values

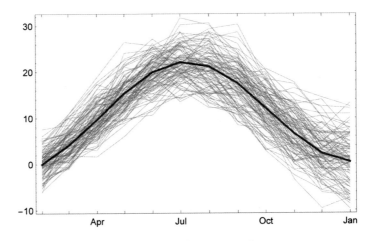

Fig. 15.9 Hundred generated trajectories with the average value

The best model is

$\Rightarrow$Normal[tsp]
$\Leftarrow$SARIMAProcess$[-0.21347, \{0.151342\}, 0, \{\},$
    $\{12, \{-0.292346, -0.347534\}, 1, \{-0.549841\}\}, 4.61513]$

To compare the different models we used the Akaike criterion. The Akaike information criterion (AIC) is a measure of the relative quality of statistical models for a given set of data. Given a collection of models for the data, AIC estimates the

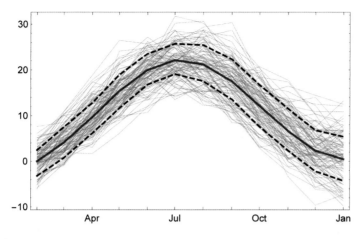

Fig. 15.10 The standard error band

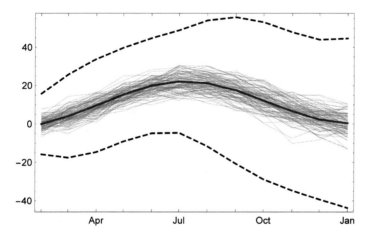

Fig. 15.11 The 95% confidence band

quality of each model, relative to each of the other models. Hence, AIC provides a means for model selection.

Suppose that we have a statistical model of some data. Let T be the maximum value of the likelihood function for the model (M) and let k be the number of estimated parameters in the model. Then the AIC value of the model is the following,

$$AIC = 2k - 2\ln(T)$$

Given a set of candidate models for the data, the preferred model is the one with the minimum AIC value. AIC rewards goodness of fit (as assessed by the likelihood function), but it also includes a penalty that is an increasing function of the number of estimated parameters. The penalty discourages over fitting, because increasing the number of parameters in the model almost always improves the goodness of the fit.

The AIC value of our previous selected AR model as well as that of the SARIMA model can be seen in Tables 15.2 and 15.3 respectively.

Then the probability that AR model is better than the SARIMA one is,

$$\Rightarrow \mathrm{Exp}[(215.402 - 293.506)/2]$$
$$\Leftarrow 1.09631 \times 10^{-17}.$$

The statistics of the main parameters of the selected SARIMA model can be seen in Table 15.4.

Figures 15.12, 15.13, 15.14, 15.15 and 15.16 shows the same features of the fitted time series in case of the SARIMA model, similarly to the ARIMA model.

Table 15.2 Comparison of the different SARIMA models

	Candidate	AIC
1	**SARIMAProcess[{1,0,0},{2,1,1}$_{12}$]**	**215.402**
2	SARIMAProcess[{1,0,0},{2,1,2}$_{12}$]	215.558
3	SARIMAProcess[{0,0,1},{2,1,2}$_{12}$]	215.656
4	SARIMAProcess[{0,0,0},{2,1,2}$_{12}$]	216.444
5	SARIMAProcess[{0,0,0},{2,1,1}$_{12}$]	216.526
6	SARIMAProcess[{1,0,0},{3,1,1}$_{12}$]	217.228
7	SARIMAProcess[{1,0,1},{2,1,2}$_{12}$]	217.423
8	SARIMAProcess[{1,0,1},{2,1,1}$_{12}$]	217.739
9	SARIMAProcess[{2,0,0},{2,1,2}$_{12}$]	217.889
10	SARIMAProcess[{0,0,0},{1,1,2}$_{12}$]	218.016

Table 15.3 The AIC measure of the AR(6) process

	Candidate	AIC
1	ARProcess[6]	293.506
2	ARProcess[5]	301.418

Table 15.4 Statistics of the selected SARIMA model

	Estimate	Standard error	t-staistic	P-value
α_1	0.151342	0.221397	0.683579	0.247715
α_1	−0.292346	0.121123	−2.41363	0.00857806
α_2	−0.347534	0.102341	−3.39583	0.000451239
β_1	−0.549841	0.1131	−4.86155	1.61578×10^{-6}

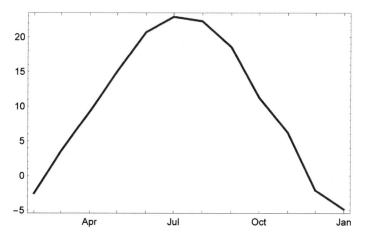

Fig. 15.12 The estimated average temperature for the first consecutive year in case of SARIMA model

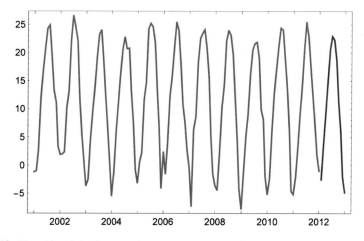

Fig. 15.13 The old and the forecasted temperature values of the SARIMA model

⇒ forecast = TimeSeriesForecast[tsp, {1, 12}];

Let's calculate the standard error bands for each time stamp:

⇒ sd1 = TimeSeriesThread[Mean[#] − StandardDeviation[#]&, simul];
⇒ sd2 = TimeSeriesThread[Mean[#] + StandardDeviation[#]&, simul];.

⇒ err = forecast[MeanSquaredErrors]

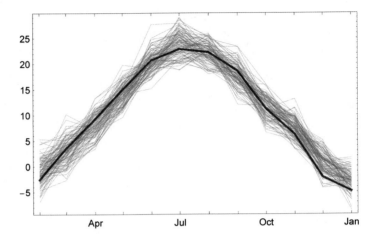

Fig. 15.14 One hundred generated trajectories with the average value

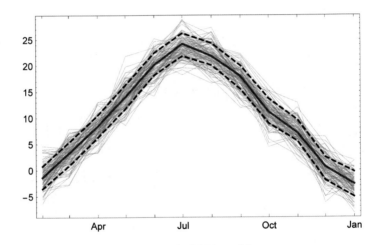

Fig. 15.15 The standard error band for the SARIMA model

⇐ TemporalData[]

These figures also demonstrate the superiority of the SARIMA model.

15.3   Stochastic Differential Equations (SDE)

The time series mainly represent *discrete* stochastic processes. Now we consider stochastic differential equations as a *continuous* model of the random processes.

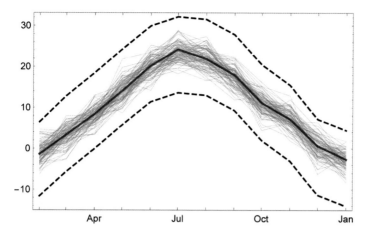

Fig. 15.16 The 95% confidence band in case of the SARIMA model

15.3.1 Ito Process

A typical form of the stochastic differential equations is

$$dy(t) = a(y(t), t)dt + b(y(t), t)dW(t)$$

where $dW(t)$ is a Wiener process, a continuous-time random walk.

15.3.2 Ito Numerical Integral

The solution of the SDE is

$$y(t) = y_0 + \int_0^t a(y(s), s)ds + \int_0^t b(y(s), s)dW(s)$$

In most practical situations there is no analytical solution for this integral, therefore numerical approximation is required.

15.3.3 Euler–Maruyama Method

To solve a model numerically, one needs—controversially—to discretize the model. The most simple discretized approximation is the *Euler–Maruyama* method,

$$y(t_{n+1}) = y(t_n) + a(y(t_n), t_n)\Delta t + b(y(t_n), t_n)\Delta W(t_n)$$

where $t_n = n\Delta t$ and $\Delta W(t_n) = W(t_n) - W(t_{n-1})$. The order of this schema is 1/2. To have higher order numerical approximation one may use the Stochastic Runge Kutta Method, too.

In the next section the numerical solution of SDE will be illustrated employing the stochastic form of the *FitzHugh–Nagumo* model.

15.4 Numerical Solution of (SDE)

The *FitzHugh–Nagumo* model for excitable media is a nonlinear model describing the reciprocal dependencies of the voltage $V(t)$ across an exon membrane and a recovery variable $R(t)$ summarizing outward currents. The model is general and is also to model excitable media, for example reaction–diffusion models. The deterministic ODE model (white-box model) is described by the following system of ordinary differential equations

$$\frac{dV(t)}{dt} = \gamma\left(V(t) - \frac{V(t)^3}{3} + R(t)\right),$$

$$\frac{dR(t)}{dt} = -\frac{1}{\gamma}(V(t) - \alpha + \beta R(t)),$$

with parameters α, β, and γ, initial condition $V(0) = -1$ and $R(0) = 1$.

White-box models are mainly constructed on the basis of knowledge of physics about the system. Solutions to ODE's are deterministic functions of time, and hence these models are built on the assumption that the future value of the state variables can be predicted exactly. An essential part of model validation is the analysis of the residual errors (the deviation between the true observations and the one-step predictions provided by the model). This validation method is based on the fact that a correct model leads to uncorrelated residuals. This is rarely obtainable for white-box models. Hence, in these situations, it is not possible to validate ODE models using standard statistical tools. However, by using a slightly more advanced type of equation, this problem can be solved by replacing ODE's with SDE's that incorporate the stochastic behavior of the system: modeling error, unknown disturbances, system noise, and so on.

The stochastic SDE gray-box model can be considered as an extension of the ODE model by introducing system noise:

$$dV(t) = \gamma\left(V(t) - \frac{V(t)^3}{3} + R(t)\right)dt + \sigma dW(t),$$

$$dR(t) = -\frac{1}{\gamma}(V(t) - \alpha + \beta R(t))dt + \sigma dW(t),$$

where $W(t)$ is a Wiener process (also known as Brownian motion), a continuous-time random walk. The next section carries out the numerical simulation of the SDE model using the parameter settings $\alpha = \beta = 0.2$, $\gamma = 3$ and $\sigma = 0.1$.

These equations represent an Ito-stochastic process that can be simulated with a stochastic Runge–Kutta method.

```
⇒procVR = ItoProcess[{dV[t] = γ(V[t] − V[t]³/3 + R[t])dt + σdwV[t],
   dR[t] = −(1/γ)(V[t] − α + βR[t])dt + σdwR[t]},
⇒{V[t], R[t]},{{V, R},{−1, 1}}, t,{wV > > WienerProcess[],
   wR > > WienerProcess[]}];
   param ={α → 0.2,β → 0.2,γ → 3,σ → 0.1};
```

15.4.1   Single Realization

First, a single realization is simulated in the time interval $0 \leq t \leq 20$ (Fig. 15.17),

```
⇒ dataVR = RandomFunction[procVR/.param, {0, 20, 0.1},
   Method → "StochasticRungeKutta"];
```

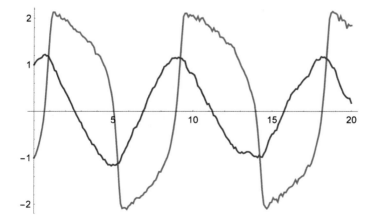

Fig. 15.17 The trajectories of the state variables $V(t)$ (blue) and $R(t)$ (brown), in case of a single realization of the Ito process

The values of V and R can be found at any time point, for example,

$\Rightarrow$ dataVR["SliceData", τ]/.$\tau \rightarrow$ 12.

$\Leftarrow \{\{1.52811, -0.477377\}\}$

15.4.2 Many Realizations

Now let us compute the trajectories for 100 realizations (Fig. 15.18).

Slice distributions of the state variables can be computed at any time point. First, let us simulate the trajectories in a slightly different and faster way, now using 1000 realizations.

$\Rightarrow$ td = RandomFunction[procVR/.param, $\{0., 20., 0.1\}, 1000$]

$\Rightarrow$ TemporalData[⊞ 〜 Time: 0. to 20. Data points: 201 000   Paths: 1000]

15.4.3 Slice Distribution

At the time point $\tau = 12$, we can compute the mean and standard deviation for both state variables $V(\tau)$ and $R(\tau)$ as well as their histograms (Fig. 15.19).

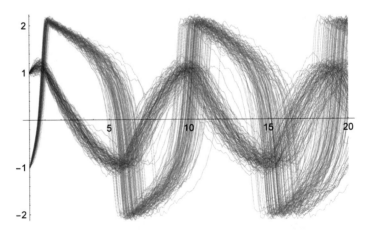

Fig. 15.18 The trajectories of the state variables $V(t)$ (blue) and $R(t)$ (brown), in case of 100 realizations of the Ito process

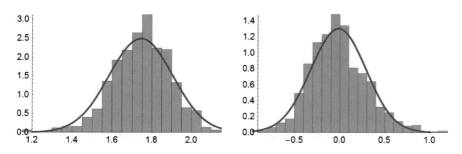

Fig. 15.19 The distribution of $V(\tau)$ (left) and $R(\tau)$ (right), in case of 100 realizations of the Ito process, at time $\tau = 12$

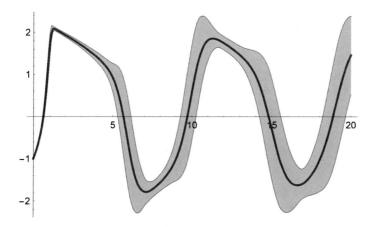

Fig. 15.20 The $\mu_V(t)$, the mean value of $V(t)$ (red) with its standard deviation, $\mu_V(t) \pm \sigma_V(t)$ (blue)

15.4.4 Standard Error Band

The mean value and the standard deviation of the trajectories along the simulation time, $0 \le t \le 20$ can be also computed and visualized. (This may take a few minutes.) (Fig. 15.20).

15.5 Parameter Estimation

Parameter estimation is critical since it decides how well the model compares to the measurement data. The measurement process itself may also have serially uncorrelated errors due to the imperfect accuracy and precision of the measurement equipment.

15.5.1   Measurement Values

Let us write the measurement equation as

$$y_k = V(t_k) + e_k,$$

where $e_k \sim N(0, \sigma_m)$. The voltage $V(t)$ is assumed to be sampled between $t = 0$ and $t = 20$ at discrete time points t_k, where $k = 0, 1, 2,..., N$, and $N = 40$ with an additive measurement noise σ_m. To get $V(t_k)$, we consider a single realization and sample it at time points t_k. Then we add white noise with $\sigma_m = 0.1$ (Fig. 15.21).

15.5.2   Likelihood Function

As we have seen, the solutions to SDE's are stochastic processes that are described by probability distributions. This property allows for maximum likelihood estimation. Let the deviation of the measurements from the model be

$$\varepsilon_k(\theta) = y_k - \mu_V(t_k, \theta),$$

where $\theta = \{\alpha, \beta, \gamma\}$. Assuming that the density function of ε_k can be approximated reasonably well by Gaussian density, the likelihood function to be maximized is

$$L(\theta) = \frac{1}{2\pi} \exp\left(-\sum_{k=1}^{N} \varepsilon_k(\theta)^2 \right),$$

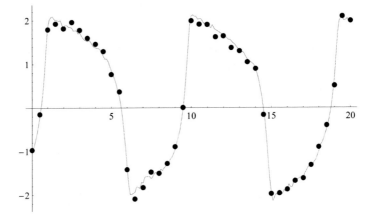

Fig. 15.21 A single realization of $V(t)$ (brown) and the simulated measured values y_k (black points)

For computation we use its logarithm. Now we accept the model parameter γ and estimate α and β from the SDE model employing maximum likelihood method.

$\Rightarrow \texttt{parm} = \{\gamma \rightarrow 3, \sigma \rightarrow 0.1\};$

Here are the values of the y_k.

$\Rightarrow \texttt{yk} = \texttt{Transpose[dataSV][[2]]};$

We use a logarithmic likelihood function.

15.5.3   Maximization of the Likelihood Function

The optimization of the likelihood function is not an easy task, since we have frequently a flat objective function, with non-differentiable terms and more local optimums. In addition there is a long evaluation time of the model. Instead of using direct global optimization, first we compute the values of the objective function on a 25×25 grid. In this way one can employ parallel computation in order to decrease the running time. We are looking for the parameters in a range $-0.1 \leq \alpha, \beta \leq 0.5$. Let us create the grid points

$\Rightarrow \texttt{XY} = \texttt{Table}[\{-0.1 + \texttt{i} \, 0.025, -0.1 + \texttt{j} \, 0.025\}, \{\texttt{i}, 0, 24\}, \{\texttt{j}, 0, 24\}];$

We compute the function values at the grid points in parallel. Then we apply interpolation for the grid point data and visualize the likelihood function (Fig. 15.22).

Employing different global optimization methods, we compute the parameters Table 15.5.

There are many local maximums. Let us choose the global one.

15.5.4   Simulation with the Estimated Parameters

The global maximum is provided by the Random Search method. Then the appropriate parameter list is

$\Rightarrow \texttt{paramN} = \{\alpha \rightarrow \texttt{u}, \beta \rightarrow \texttt{v}, \gamma \rightarrow 3, \sigma \rightarrow 0.1\}/.\texttt{uvN}[[2]]$

$\Rightarrow \{\alpha \rightarrow -0.125, \beta \rightarrow 0.0273485, \gamma \rightarrow 3, \sigma \rightarrow 0.1\}$

This computes 500 realizations and visualizes the result and the measurement data (Fig. 15.23).

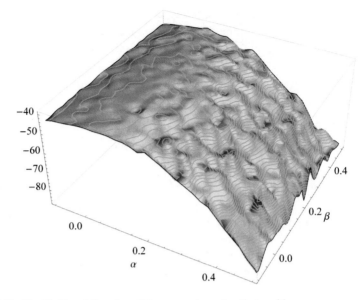

Fig. 15.22 The likelihood function of the parameter estimation problem

Table 15.5 Results of the different global methods

Differential evolution	Random search	Simulated annealing	Nelder Mead
−39.7162	−39.4906	−39.4907	−40.6119
u → 0.0500002	u → 0.125	u → 0.125001	u → −0.00364276
v → 0.25	v → 0.0273485	v → 0.027331	v → 0.225

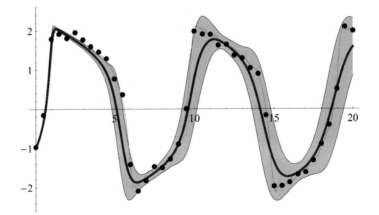

Fig. 15.23 The simulated $V(t)$ process with the estimated parameters

15.5.5 Deterministic Versus Stochastic Modeling

In the previous parameter estimation section, we illustrated the technique of the stochastic modeling. The measurement data was simulated with the correct model parameters ($\gamma = 3$), no assumed modeling error, existing system noise $\sigma = 0.1$, and measurement error $\sigma_m = 0.1$. The model to be fitted had two free parameters, α and β, since the system noise $\sigma = 0.1$ was adopted for the model.

Now let us consider a different situation. Suppose that we have modeling error, since the measured values are simulated with $\gamma = 2.5$; however, in the model to be fitted, we use $\gamma = 3$. In addition, let us increase the measure error to $\sigma_m = 0.2$. To compare the efficiencies of the deterministic and stochastic models, we should handle σ as a free parameter in the stochastic model to be fitted. So we have then three free parameters to be estimated: α, β and σ. Let us carry out the parameter estimation for different values of σ.

The results can be seen in Table 15.6; $\sigma = 0$ corresponds to the deterministic model. In order to demonstrate that the stochastic model can provide significant improvement compared with the deterministic model, the likelihood-ratio test can be applied. The likelihood-ratio test has been used to compare two nested models. In our case, one of the models is the ODE model having less parameters than the other one, the SDE model. The null hypothesis is that the two models are basically the same. The test statistic is,

$$R = 2\big((-\log L(\alpha, \beta))_D - ((-\log L(\alpha, \beta))_S\big),$$

where the subscripts D and S stand for the deterministic and stochastic model, respectively.

The distribution of R is $\chi^2(f)$, where f is the difference in the number of parameters between the two models; in our case, $f = 3 - 2 = 1$. Here is the critical value for $\chi^2(1)$ at confidence level of 95%.

$\Rightarrow$ `InverseCDF[ChiSquareDistribution[1], 0.95]`

$\Leftarrow 3.84146$

If the value of R is less than the critical value, then the null hypothesis can be accepted. In our case R = 2 × (61.060 − 52.761) = 16.598. Therefore we reject the null hypothesis, which means the SDE model can be considered as a different model providing a significant improvement compared with the ODE model.

Table 15.6 The results of the parameter estimation for different system noises

σ	$-\log L(\alpha,\ \beta)$	α	β
0.00	61.060	−0.037	0.230
0.10	54.938	−0.004	0.300
0.25	52.761	0.151	0.125
0.40	54.687	0.044	0.150

15.6   Application of Machine Learning

In the last decade for modeling stochastic processes the Machine Learning Methods
(ML) are frequently employed besides the traditional stochastic techniques, like
stochastic differential equations or time series. Even the combinations of stochastic
differential equations and ML methods, for example Neural Differential Equations
became popular. In this section we show the relation between the three different
methods and provide examples and demonstrate the application of these methods
for solving classification and regression problems.

In the previous sections the stochastic process was represented by stochastic
differential equations in Ito form and the model parameters were determined via
parameter estimation.

Now we shall employ machine learning method, namely Nearest Neighbors
method to describe the stocahstic process.

15.6.1   Applying Machine Learning

Let us employ a model $V_{n+1} = f(V_n, V_{n-1})$.

15.6.1.1   Input–Output Data Pairs

The measured data points are the same as they were in case of parameter estimation

```
⇒ n = Length[dataSV]
⇐ 41
```

Creating input–output pairs

```
⇒data = {};
⇒Do[
   AppendTo[data, {dataSV[[i − 1]][[2]], dataSV[[i]][[2]],
   dataSV[[i + 1]][[2]]}//Flatten], {i, 2, n − 1}];
⇒dataInOut = Map[{#[[1]], #[[2]]} → #[[3]]&, data]; ]
```

Here we employ the Nearest Neighbors method.

15.6.1.2 Training Process

⇒p = Predict[dataInOut, PerformanceGoal → "Quality",
    Method → {"NearestNeighbors","DistributionSmoothing" → 1,
    "NeighborsNumber" → 3,"NearestMethod" → "Octree"},*l*
  TimeGoal → 100]
⇐PredictorFunction

$Failed		Input type:NumericalVector (length:2)
		Method: NearestNeighbors

Information about the learning quality of the ML method

⇒ Information[p]

Predictor Information

⇐

Data type	Numerical vector (lenght: 2)
Method	NearestNeighbors
Standard deviation	0.899 ±0.0080
Loss	0.938 ± 0.075
Single evaluation time	3.06 ms/example
Batch evaluation speed	54.6 example/ms
Model memory	789. kB
Training examples used	39 examples
Training time	1 min 41 s

15.6.1.3 Using ML Method for Simulation

⇒ outml = Map[p[#[[1]]]&, dataInOut];
⇒ xoutml = Take[Map[#[[1]]&, dataSV], {3, n}];
⇒ u = Transpose[{xoutml, outml}]//N;
⇒ VML = Interpolation[u]
⇐ InterpolatingFunction

[$Failed		Domain: {{1.,20.}}	]
		Output: scalar	

⇒ pML1 = Plot[VML[s], {s, 1, 20}, PlotStyle → Red]

The standard deviations (Fig. 15.24)

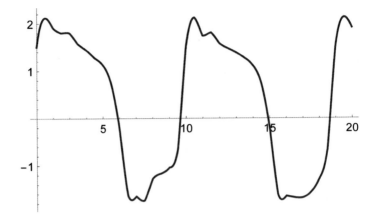

Fig. 15.24 The process mean values simulated by ML method

⇒ st = Map[StandardDeviation[p[#[[1]],"Distribution"]]&, dataInOut];
⇒ u = Trans − pose[{xoutml, st}]//N;
⇒ SDVML = Interpolation[u]
⇐ InterpolatingFunction

[$Failed  Domain: {{1.,20.}}
                       Output: scalar]

⇒pML2 = Plot[{VML[s] + SDVML[s], VML[s] − SDVML[s]},
   {s, 1, 20}, Filling → {1 → {2}},
     PlotStyle → {{Blue, Thin}, {Blue, Thin}}, PlotRange → All];
   Show[{pML2, pML1, pSV, pVm}]]

The distribution at τ = 12 (Figs. 15.25 and 15.26)

⇒ VML[12]
⇐ 1.57552
⇒ SDVML[12]
⇐ 0.279982
⇒ Plot[PDF[NormalDistribution[VML[12], SDVML[12]], s], {s, 0.75, 2.6}]

15.6.2 Machine Learning Differential Equation Model

As we can see one can modeling a process in two ways see below.

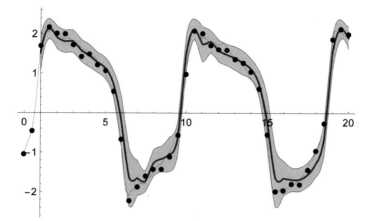

Fig. 15.25 The simulated $V(t)$ process with the nearest-neighbor ML method. The mean of the scientific model solution (brown)—the ML solution (red)

Fig. 15.26 The distribution of $V(\tau)$ at time $\tau = 12$ resulted by the ML method

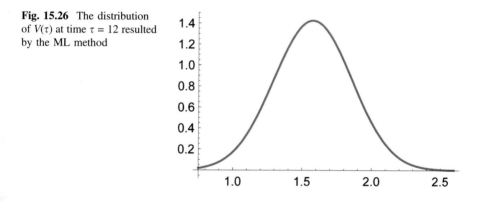

Differential Equation Model:

$$\frac{dy}{dt} = f(y(t)),$$

Machine Learning Model:

$$\frac{dy}{dt} = ML(y(t))$$

Nowadays a new approach gains high popularity namely the Neural Differential Equation, which is in certain extend a simplify label, since instead of Neural Network one may use any other ML method. The basic idea is to approximate the function derivate instead of the function, in this the right hand side of the differential equation is represented an ML method.

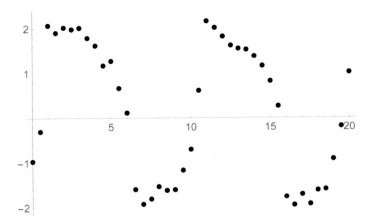

Fig. 15.27 The measured trajectory points

Neural Network can be a proper ML method since it can be differentiated when the activation functions are differentiable. Let us consider a simple deep neural network (Fig. 15.27)

⇒ dropoutNet01 = NetChain[{150, DropoutLayer[0.1], Tanh, 150, Tanh, 1}]

			Input	array
	uninitialized	1	LinearLayer	vector (size: 150)
		2	DropoutLayer	vector (size: 150)
⇐ NetChain [		3	Tanh	vector (size: 150)
		4	LinearLayer	vector (size: 150)
		5	Tanh	vector (size: 150)
		6	LinearLayer	vector (size: 1)
			Output	vector (size: 1)

⇒ dataSVML = Map[#[[1]] → #[[2]]&, dataSV]//N;

⇒ pVm = ListPlot[dataSV, PlotStyle → {Black, PointSize[0.015]}]

The result of the NN approximation,

⇒results5 = NetTrain[dropoutNet01, dataSVML, All,
    MaxTrainingRounds → 20000]

⇐NetTrainResultsObject

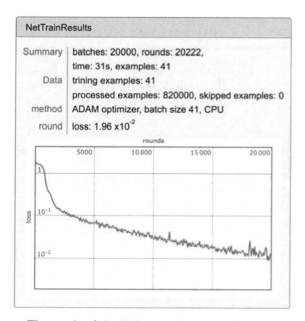

The result of the NN approximation (Fig. 15.28),

$\Rightarrow$ result $=$ results5$[$"TrainedNet"$]$

$\Leftarrow$ NetChain $[$

	Input	real
1	LinearLayer	vector(size: 150)
2	DropoutLayer	vector(size: 150)
3	Tanh	vector(size: 150)
4	LinearLayer	vector(size: 150)
5	Tanh	vector(size: 150)
6	LinearLayer	vector(size: 1)
	Output	scalar

$]$

$\Rightarrow$p11 $=$ Show$[\{$pVm, Plot$[$result$[x], \{x, 0, 22\},$ PlotStyle $\rightarrow \{$Blue$\},$
      PlotRange $\rightarrow$ All$]\}]$

We should like to represent the approximated function in differential equation form,

$$\frac{dy}{dt}(t) = \frac{d}{dt}NN(t),$$

In order to get differential equation form of the approximation, let *differentiate the network*,

$\Rightarrow$ dnet$[$t$_]$: $=$ result$[$t, NetPortGradient$[$"Input"$]]$

We need a pure numerical function (Fig. 15.29),

$\Rightarrow$ Dnet$[$t$_$?NumericQ$]$:= dnet$[$t$]]$

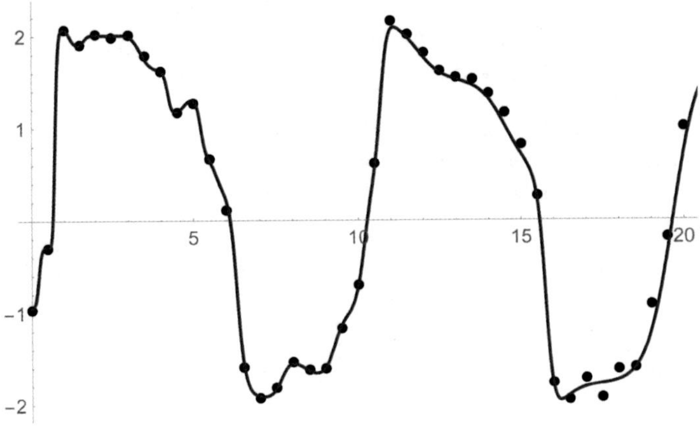

Fig. 15.28 The estimated trajectory by the deep neural network

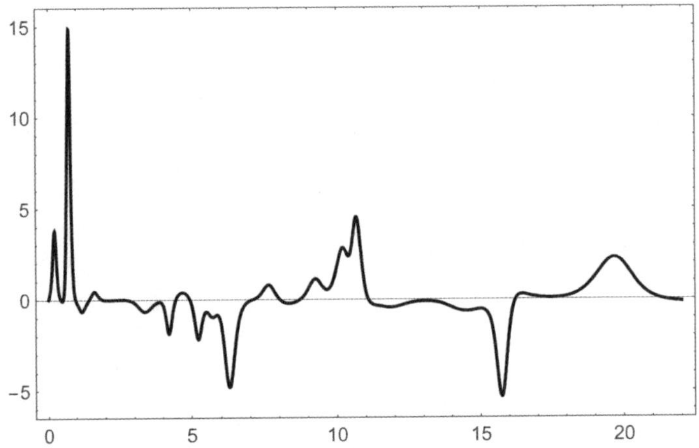

Fig. 15.29 The function approximation as the solution of the neural network differential equation

Let us display this function (Fig. 15.30)

```
⇒Plot[Dnet[t],{t,0,22},Plotstyle⇒Red, Frame⇒True, PlotRange⇒All]
⇒Clear[y]
⇒s = NDSolve[{y′[t] == Dnet[t],y[0] == −1},y,{t,0,22}]//Quiet
⇐{{y → InterpolatingFunction

[ $Failed  Domain: {{2.,22.}}
           Output: scalar  ]}}]
```

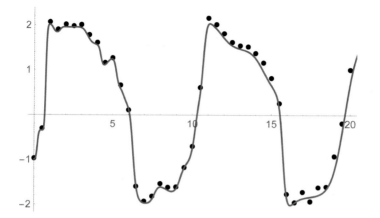

Fig. 15.30 Solution of the neural network differential equation

$$\Rightarrow \text{p2} = \text{Show}[\{\text{pVm}, \text{Plot}[\text{Evaluate}[\text{y}[\text{t}]/.\text{s}], \{\text{t}, 0., 22\},$$
$$\text{PlotRange} \rightarrow \text{All}, \text{Frame} \rightarrow \text{True}, \text{Axes} \rightarrow \text{None}]\}]$$

15.6.2.1   Employing Stochastic Differential Equation Form

In order to get stochastic model for the function approximation we employ Ito form
of the Neural Network Differential Equation. Let define

$$\Rightarrow \text{H}[\text{t}\_] := \text{Dnet}[\text{t}]$$

Considering

$$\Rightarrow \sigma = 0.1; \mu = 0;$$
$$\Rightarrow \text{proc} = \text{ItoProcess}[\text{dy}[\text{t}] == \text{H}[\text{t}]\text{dt} + \text{dw}[\text{t}], \text{y}[\text{t}], \{\text{y}, -1.\}, \text{t},$$
$$\text{w} \approx \textit{WienerProcess}[\mu, \sigma]]$$
$$\Leftarrow \text{ItoProcess}[\{\{\text{Dnet}[\text{t}]\}, \{\{0.1\}\}, \text{y}[\text{t}]\}, \{\{\text{y}\}, \{-1.\}\}, \{\text{t}, 0\}]$$

Simulating the process:

$$\Rightarrow \text{psol} = \text{RandomFunction}[\text{proc}, \{0., 20., 0.1\}, 100]$$

$\Leftarrow$ TemporalData [$\$Failed$  $\begin{array}{l}\text{Time: 0. to 20.}\\ \text{Data points: 20100 Paths: 100}\end{array}$]

Let us display the mean and the standard deviation of the solution (Fig. 15.31),

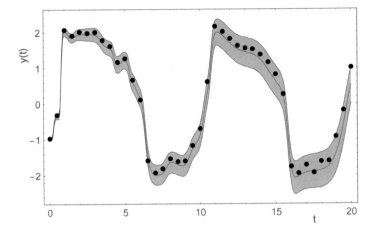

Fig. 15.31 The stochastic solution of the Ito-form of the deep neural network

$\Rightarrow$p1 = Plot[Mean[psol[t]], {t, 0, 20}, PlotStyle $\rightarrow$ {Red, Thick},
    FrameLabel $\rightarrow$ {"t", "y(t)"}, Frame $\rightarrow$ True];]
$\Rightarrow$p2 = Plot[{Mean[psol[t]] + StandardDeviation[psol[t]],
    Mean[psol[t]] $-$ StandardDevia $-$ tion[psol[t]]}, {t, 0, 20},
    Filling $\rightarrow$ {1 $\rightarrow$ {2}}, PlotStyle $\rightarrow$ {{Blue, Thin}, {Blue, Thin}},
    FrameLabel $\rightarrow$ {"t", "y(t)"}, Frame $\rightarrow$ True, Axes $\rightarrow$ None,
    PlotRange $\rightarrow$ All];
$\Rightarrow$Show[{p2, p1, pVm}]

15.6.2.2   Other Hybrid Solution Methods

In the following sections we demonstrate some other techniques.

As first step we approximate the trajectory using a smoothing technique, here a neural network approximation (Fig. 15.32)

$\Rightarrow$pR = Predict[dataSVML, PerformanceGoal $\rightarrow$ "Quality",
    Method $\rightarrow$ {"NeuralNetwork", "NetworkDepth" $\rightarrow$ 5}, TimeGoal $\rightarrow$ 100]

$\Leftarrow$ PredictorFunction [\$Failed   Input type: Numerical   ]
                            Method: NeuralNetwork

$\Rightarrow$ p0 = Show[{Plot[pR[t], {t, 0, 20}], pVm}]

Let employing central differences (Figs. 15.33, 15.34 and 15.35)

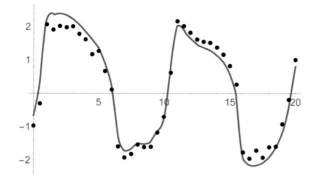

Fig. 15.32 Solution via predict universal function using standard neural network

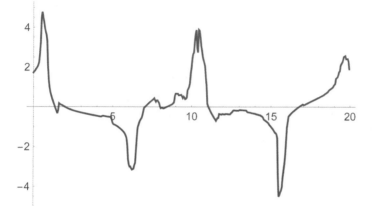

Fig. 15.33 Numerically differentiated trajectory

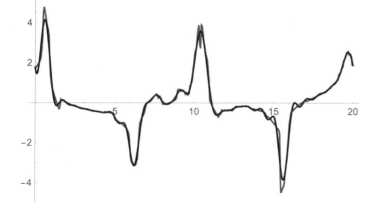

Fig. 15.34 The numerically trajectory can be learned by a Gaussian process

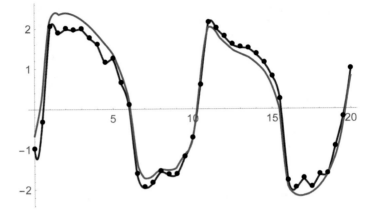

Fig. 15.35 The solution of the differential equation used the Gaussian process as derivative function

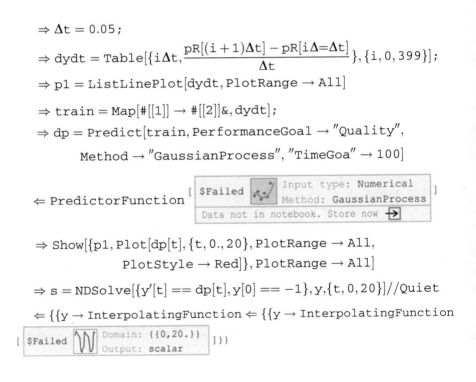

$\Rightarrow \Delta t = 0.05;$

$\Rightarrow$ dydt = Table$\left[\left\{i\Delta t, \dfrac{pR[(i+1)\Delta t] - pR[i\Delta = \Delta t]}{\Delta t}\right\}, \{i, 0, 399\}\right];$

$\Rightarrow$ p1 = ListLinePlot[dydt, PlotRange $\rightarrow$ All]

$\Rightarrow$ train = Map[#[[1]] $\rightarrow$ #[[2]]&, dydt];

$\Rightarrow$ dp = Predict[train, PerformanceGoal $\rightarrow$ "Quality",

      Method $\rightarrow$ "GaussianProcess", "TimeGoa" $\rightarrow$ 100]

$\Leftarrow$ PredictorFunction [$Failed Input type: Numerical Method: GaussianProcess Data not in notebook. Store now]

$\Rightarrow$ Show[{p1, Plot[dp[t], {t, 0., 20}, PlotRange $\rightarrow$ All,

      PlotStyle $\rightarrow$ Red]}, PlotRange $\rightarrow$ All]

$\Rightarrow$ s = NDSolve[{y'[t] == dp[t], y[0] == -1}, y, {t, 0, 20}]//Quiet

$\Leftarrow$ {{y $\rightarrow$ InterpolatingFunction $\Leftarrow$ {{y $\rightarrow$ InterpolatingFunction

[$Failed Domain: {{0, 20.}} Output: scalar]}}

⇒ Show[{Plot[Evaluate[y[t]/.s],{t, 0, 20}, PlotStyle → Red], p0},
PlotRange → All]

Remark Mathematica can solve this problem much more easily via interpolation
and derivation

⇒ pM = Interpolation[dataSV];
⇒ dpM[t_] : = pM'[t]
⇒ s = NDSolve[{y'[t] == dpM[t], y[0] == −1}, y, {t, 0, 20}]//Quiet

⇐ {{y → InterpolatingFunction

[$Failed 〔graph〕 Domain: {{0,20.}}
                     Output: scalar   ]}}

However when the function to be represented by a differential equation is too
noisy this method can fail since it uses interpolation.

Now we can provide the stochastic form of the differential equation. Let

⇒σ = 0.1;
⇒proc = ItoProcess[dy[t] == dpM[t]dt + dw[t], y[t], {y, −1.}, t,
     w ≈ WienerProcess[0, σ]]
⇐ItoProcess[{{InterpolatingFunction

[$Failed 〔graph〕 Domain: {{0,20.}}
                     Output: scalar   ][t]},

{{0.1}}, y[t]}, {{y}, {−1.}}, {t, 0}]

Simulating the process (Fig. 15.36):

⇒ psol = RandomFunction[proc, {0., 20., 0.1}, 100]

⇐ TemporalData [$Failed 〔graph〕 Time: 0. to 20.
                                     Data points: 20100 Paths: 100   ]

⇒ p1M = Plot[Mean[psol[t]], {t, 0, 20}, PlotStyle → {Red, Thick},
   FrameLabel → {"t","y(t)"}, Frame → True];]
⇒ p2M = Plot[{Mean[psol[t]] + StandardDeviation[psol[t]],
   Mean[psol[t]] − StandardDevia − tion[psol[t]]}, {t, 0, 20},
   Filling → {1 → {2}}, PlotStyle → {{Blue, Thin}, {Blue, Thin}},
   FrameLabel → {"t","y(t)"}, Frame → True, Axes → None,
   PlotRange → All];
⇒ Show[{p2M, p1M, pVm}]

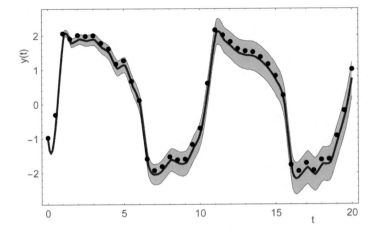

Fig. 15.36 The stochastic form of the solution of the neural differential equation

15.6.3   Comparing the Different Stochastic Modeling Methods

Since the process is stochastic comparing the mean values is not proper. Therefore we compute the log − likelihood function values for the different techniques. For the realization we generated 100 trajectories with step size $\Delta t = 0.1$ for all modeling methods.

The parameter estimation method for the Stochastic Differential Equation model

```
⇒likelihoodSDE =
    Total[
    MapThread[
    Log[PDF[NormalDistribution[Mean[τd[#1]][[1]],
    StandardDeviation[τd[#1]][[1]]], #2]]&, {xoutml, outml}]]
⇒−15.9275
```

for the Machine Learning method

```
⇒ likelihoodML =
  Total[Map[Log[PDF[p[#[[1]],"Distribution"], #[[2]]]]&,
  dataInOut]]
⇒ −9.84123
```

for the Machine Learning Differential Equation method

⇒=

    Total[
    MapThread[
    Log[PDF[NormalDistribution[Mean[psol[#1]],
    StandardDeviation[psol[#1]]],#2]]&,{xoutml,outml}]]
⇒−23.8305
⇒likelihoodML > likelihoodSDE > likelihoodMLDE
⇒True

So the the ML method is better than the scientific model and the MLDE is even better than the ML model.

15.6.4 Image Classification

A variant of Black Hole algorithm has been applied to improve the quality of classification of different of land images employing logistic regression. After dimension reduction via AudioEncoding the hyperparameters maximizing the classification accuracy were computed applying a variant of Black Hole algorithm. The efficiency this optimization method was compared with other global techniques as cxs simulating annealing, differential evaluation and random search method.

15.6.4.1 Vacant Land and Residential Areas

We 16–16 images for vacant and residential lands (Figs. 15.37 and 15.38)

⇒ vacant =

⇒ residental =

Assembling and resizing the images

⇒lands = Join[vacant,residential];
⇒landsReduced = Map[ImageResize[#,{256,256}]&,lands];

Employing AutoEncoder method based on neural network the size of images is reduced to a vector of 2D

⇒reduced = DimensionReduce[landsReduced,2,Method → "AutoEncoder"];
⇒Length[reduced]
⇐32

The two sets of vectors are (Fig. 15.39)

⇒ vacant=

Fig. 15.37 Vacant land areas

⇒ residental=

Fig. 15.38 Residental land areas

Fig. 15.39 Representation of the images after dimension reduction

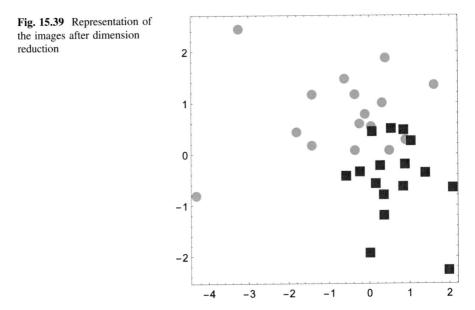

⇒reduced1 = Take[reduced, {1, 16}];
⇒reduced2 = Take[reduced, {17, 32}];
⇒ p0 = ListPlot[{reduced1, reduced2}, PlotStyle → {Green, Red},
      Frame → True, Axes → None, PlotMarkers → {Automatic, Medium},
      AspectRatio → 1]

15.6.4.2   Classification

LogisticRegression models the log probabilities of each class with a linear combination of numerical features.

$$x = \{x_1, x_2, \ldots, x_n\}, \quad \log(P(class = k|x)) \propto x.\theta^{(k)}, \quad \text{where } \theta^{(k)} = \{\theta_1, \theta_2, \ldots, \theta_m\},$$

corresponds to the parameters for class k. The estimation of the parameter matrix θ = $\{\theta^{(1)}, \theta^{(2)}, \ldots, \theta^{(nclass)}\}$ is done by minimizing the loss function.

$$\sum_{i=1}^{m} -\log(P_\theta(class = y_i|x_i)) + \lambda_1 \sum_{i=1}^{n}|\theta_i| + \frac{\lambda_2}{2} \sum_{i=1}^{n} \theta_i^2.$$

Our aim to find the optimal parameters λ_1 and λ_2 which maximize the accuracy of the classification!

⇒ label = Join[Table[0, 16], Table[1, 16]]
⇐ {0, 0, 0, 0, 0, 0, 0, 0, 0, 0, 0, 0, 0, 0, 0, 0, 1, 1, 1, 1, 1, 1, 1, 1, 1, 1, 1, 1, 1, 1, 1, 1}
⇒ class = MapThread[#1 → #2&, {reduced, label}];
..

As illustration let

⇒ λ1 = 0.8; λ2 = 0.3;
⇒ c = Classify[class, Method → {"LogisticRegression",
"L1Regularization" → λ1, "L2Regularization" → λ2},
    "PerformanceGoal" → Quality]//Quiet;

The membership values of three random elements from the 32 ones

⇒ c[reduced[[1]], "Probabilities"]
⇐ <|0 → 0.971337, 1 → 0.0286632| >
⇒ c[reduced[[13]], Probabilities]
⇐ <|0 → 0.696855, 1 → 0.303145| >
⇒ c[reduced[[24]], "Probabilities"]
⇐ <|0 → 0.325858, 1 → 0.674142| >

Let us display the result ⇒ r1 = Transpose[reduced]; (Fig. 15.40)

Fig. 15.40 Result of
classification in case
of $\lambda_1 = 0.8$ and $\lambda_2 = 0.3$

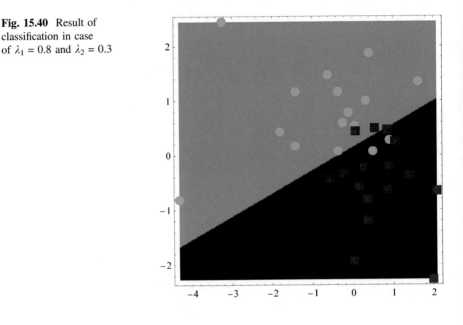

$\Rightarrow \mathtt{r1 = Transpose[reduced]};$

$\Rightarrow \mathtt{u1 = Min[r1[[1]]]}$

$\Leftarrow -4.30843$

$\Rightarrow \mathtt{u2 = Max[r1[[1]]]}$

$\Leftarrow 2.08151$

$\Rightarrow \mathtt{v1 = Min[r1[[2]]]}$

$\Leftarrow -2.271$

$\Rightarrow \mathtt{v2 = Max[r1[[2]]]}$

$\Leftarrow 2.44575$

$\Rightarrow \mathtt{Show[\{DensityPlot[c[\{u,v\}],\{u,u1,u2\},\{v,v1,v2\},}$
$\quad \mathtt{ColorFunction \rightarrow "CMYKColors", PlotPoints \rightarrow 100], p0\}]}$

15.6.4.3 Quality of the Classification

$\Rightarrow \mathtt{cm = ClassifierMeasurements[c, class]}$

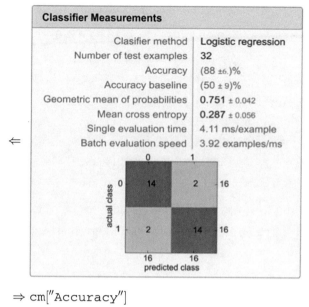

$\Rightarrow \texttt{cm["Accuracy"]}$

$\Leftarrow 0.875$

15.6.4.4   Optimization

The objective function

```
⇒G[{11_,12_}] := Module[{λ1,λ2,c},
    λ1 = 11;λ2 = 12;
    c = ClassifierMeasurements[
    Classify[class,
        Method → {"LogisticRegression","L1Regularization" → 11,
            "L2Regularization" → 12}, PerformanceGoal → "Quality"],
        class];
    c["Accuracy"]]
```

For example

$\Rightarrow \texttt{G[\{0.8, 0.3\}]}$

$\Leftarrow 0.875$

Preparation of proper form of the objective

$\Rightarrow$F = Flatten[Table[{{0.1i, 0.1j}, G[{0.1i, 0.1j}]}, {i, 0, 10},
  {j, 0, 10}], 1];

$\Rightarrow$f = Interpolation[F]

$\Leftarrow$ InterpolatingFunction

[$FAILED 〰 Domain: {{0.,1.}, {0.,1.}}
                 Output: scalar   ]

$\Rightarrow$ H[{u_, v_}] : = f[u, v]

$\Rightarrow$ ContourPlot[H[{x, y}], {x, 0, 1}, {y, 0, 1}, FrameLabel $\rightarrow$ {"λ_1''", "λ_2''"}]

$\Rightarrow$ Plot3D[H[{x, y}], {x, 0, 1}, {y, 0, 1}]

These figs show clearly that global technique is needed! (Figs. 15.41 and 15.42)

15.6.4.5   Simulated Annealing

Let us employ built in function

$\Rightarrow$s = AbsoluteTiming[NMaximize[{H[{u, v}], 0 $\leq$ u $\leq$ 1, 0 $\leq$ v $\leq$ 1},
    {u, v}, Method $\rightarrow$ "SimulatedAnnealing"]]

$\Leftarrow$ {0.263502, {0.910157, {u $\rightarrow$ 0.249465, v $\rightarrow$ 0.45779}}}

Running time

$\Rightarrow$AbsoluteTiming[
    sol = Reap[NMaximize[{H[{u, v}], 0 $\leq$ u $\leq$ 1, 0 $\leq$ v $\leq$ 1}, {u, v},
    EvaluationMonitorⁿSow[{u, v}],
    Method $\rightarrow$ "SimulatedAnnealing"]];]

$\Leftarrow$ {0.0610449, Null}

Function value and the solution

$\Rightarrow$sol[[1]]

$\Leftarrow$ {0.910157, {u $\rightarrow$ 0.249465, v $\rightarrow$ 0.45779}}

Number of iterations

$\Rightarrow${hist} = sol[[2]];

$\Rightarrow$Length[hist]

$\Leftarrow$5.4

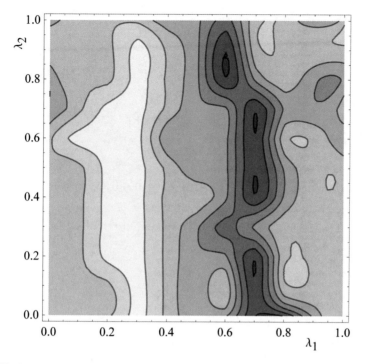

Fig. 15.41 Contour plot of the objective

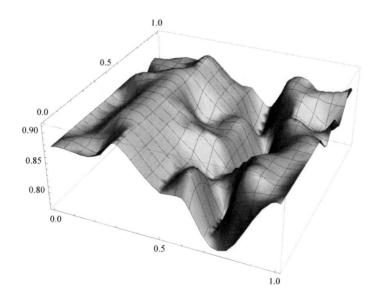

Fig. 15.42 3D plot of the objective

15.6.4.6   Differential Evolution

Let us employ built in function

```
⇒s = AbsoluteTiming[NMaximize[{H[{u,v}], 0 ≤ u ≤ 1, 0 ≤ v ≤ 1},
      {u,v}, Method → "DifferentialEvolution"]]
⇐{0.360245, {0.911362, {u → 0.243052, v → 0.242265}}}
```

Running time

```
⇒AbsoluteTiming[
    sol = Reap[NMaximize[{H[{u,v}], 0 ≤ u ≤ 1, 0 ≤ v ≤ 1}, {u,v},
    EvaluationMonitor : → Sow[{u,v}],
    Method → "DifferentialEvolution"]]; ]
⇐{0.341148, Null}
```

Function value and the solution

```
⇒ sol[[1]]
⇐ {0.911362, {u → 0.243052, v → 0.242265}}
```

Number of iterations

```
⇒ {hist} = sol[[2]];
⇒ Length[hist]
⇐ 1429
```

15.6.4.7   Random Search

Let us employ built in function

```
⇒s = AbsoluteTiming[NMaximize[{H[{u,v}], 0 ≤ u ≤ 1, 0 ≤ v ≤ 1},
      {u,v}, Method → "RandomSearch"]]
⇐{0.660443, {0.911362, {u → 0.243052, v → 0.242265}}}
```

Running time

```
⇒AbsoluteTiming[sol = Reap[NMaximize[{H[{u,v}], 0 ≤ u ≤ 1, 0 ≤ v ≤ 1}, {u,v},
      EvaluationMonitor : → Sow[{u,v}], Method → "RandomSearch"]]; ]
⇐7{0.288491, Null}
```

Function value and the solution

$\Rightarrow$ sol[[1]]

$\Leftarrow \{0.911362, \{u \rightarrow 0.243052, v \rightarrow 0.242265\}\}$

Number of iterations

$\Rightarrow \{hist\} = sol[[2]];$

$\Rightarrow$ Length[hist]

$\Leftarrow 93$

15.6.4.8   Black Hole

$\Rightarrow s = $ AbsoluteTiming[BlackHole02[H, 0, 0, 1, 1, 50, 25]];

Running time

$\Rightarrow s[[1]]$

$\Leftarrow 0.0552989$

Function value and the solution Table 15.7 (Fig. 15.43)

Table 15.7   Result of the different methods

Method	Solution	Function value	Number of iteration	Time (s)
Simulated annealing	{0.2495, 0.2423}	0.9101	54	0.061
Differential evolution	{0.2431, 0.2423}	0.9114	1429	0.340
Random search	{0.2451, 0.2422}	0.9114	93	0.290
Black hole	{0.2432, 0.2440}	0.9114	25	0.053

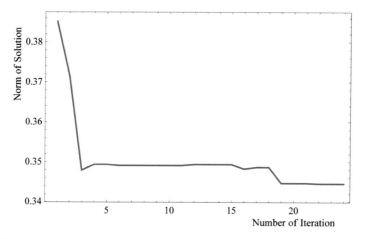

Fig. 15.43   The performance of the BH algorithm

```
⇒Last[s[[2]]]
⇐{0.243246, 0.244008}
⇒H[%]
   ⇐ 0.911361
⇒ListPlot[Map[Norm[#]&, s[[2]]], Joined → True, PlotRange → All,
    ImageSize → 350, Frame → True,
    FrameLabel → {"NumberofIteration", "NormofSolution"}]
```

Now let us carry out the classification with optimal hyper parameters

```
⇒λ1 = 0.2432; λ2 = 0.244;
⇒c =
    Classify[class,
    Method → {"LogisticRegression", "L1Regularization" → λ1,
    "L2Regularization" → λ2}, PerformanceGoal → "Quality"]//
    Quiet;
```

The membership values of three random elements from the 32 ones

```
⇒c[reduced[[1]], "Probabilities"]
⇐<|0 → 0.84648, 1 → 0.15352| >
⇒c[reduced[[13]], "Probabilities"]
⇐<|0 → 0.576428, 1 → 0.423572| >
⇒c[reduced[[24]], "Probabilities"]
⇐<|0 → 0.43054, 1 → 0.56946| >
```

Let us display the result (Fig. 15.44)

```
⇒ r1 = Transpose[reduced];
⇒ u1 = Min[r1[[1]]]
⇐ −4.30843
⇒ u2 = Max[r1[[1]]]
⇐ 2.08151
⇒ v1 = Min[r1[[2]]]
⇐ −2.271
⇒ v2 = Max[r1[[2]]]
⇐ 2.44575
⇒ Show[{DensityPlot[c[{u, v}], {u, u1, u2}, {v, v1, v2},
    ColorFunction → "CMYKColors", PlotPoints → 100], p0}]
```

The quality of the classification

Fig. 15.44 Result of classification in case of $\lambda_1 = 0.2432$ and $\lambda_2 = 0.2440$

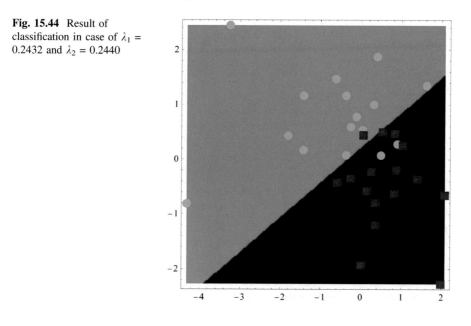

$\Rightarrow \mathtt{cm = ClassifierMeasurements[c, class]}$

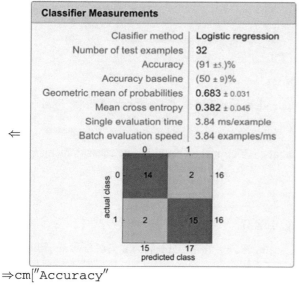

$\Leftarrow$

Classifier Measurements	
Clasifier method	Logistic regression
Number of test examples	32
Accuracy	(91 ±5.)%
Accuracy baseline	(50 ± 9)%
Geometric mean of probabilities	0.683 ± 0.031
Mean cross entropy	0.382 ± 0.045
Single evaluation time	3.84 ms/example
Batch evaluation speed	3.84 examples/ms

$\Rightarrow \mathtt{cm["Accuracy"}$

$\Leftarrow 0.90625$

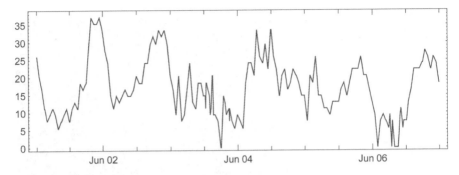

Fig. 15.45 The daily hourly windspeed on the Kennedy Airport

15.6.5   *Regression*

Let us employ the machine learning differential equations technique for estimating wind speed in different days at John F. Kennedy Airport. The wind speeds at John F. Kennedy Airport during summer time for one week (Fig. 15.45)

$$\Rightarrow \text{datas} = \text{WindSpeedData}[''\text{KJFK}'',$$
$$\{\text{DateObject}[\{2013, 6, 1\}], \text{DateObject}[\{2013, 6, 7\}]\}]$$

$\Leftarrow$ TimeSeries [⊞ 〜 Time: 01 Jun 2013 to 06 Jun 2013]
                              Data points: 191

$\Rightarrow$ DateListPlot[datas]

Let us consider the independent variable as the number of days, then

$\Rightarrow$ s = datas[''Path''];

We have 191 measurements, which means roughly one measurements per hour

$\Rightarrow$ n = Length[s]

$\Leftarrow$ 191

The data points are (Fig. 15.46)

$\Rightarrow$ss = MapThread[{#2, QuantityMagnitude[#1[[2]]]}&, {s, Range[n]}];
$\Rightarrow$*zam* = *Interpolation*[ss]

$\Leftarrow$ InterpolatingFunction [$Failed 〜 Domain: {{1, 191.}}]
                                        Output: scalar

$\Rightarrow$pDD1 = ListPlot[ss, PlotStyle $\to$ {Blue, PointSize[0.006]},
       PlotRange $\to$ All, Frame $\to$ True, Axes $\to$ None,
       FrameLabel $\to$ {''days'', ''windspeed''}, AspectRatio $\to$ 1/3]

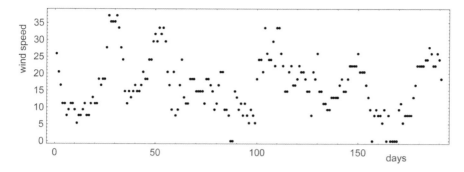

Fig. 15.46 The measured data points

We consider the measurement as a noisy data. Consequently let us apply regression for data via neural network.

Gaussian Process can be a proper ML method since it can be differentiated. Training the process, we employ the model in form

$$w(t + 1) = \text{GaussianProcess}(w(t)),$$

The learning set

```
⇒hu = Transpose[ss][[2]]; train = {};
  Do[AppendTo[train, hu[[i]] → hu[[i + 1]]], {i, 1, 190}]
⇒Short[train, 10]]
⇐{25.9 − > 20.5, 20.5 − > 16.6, 16.6 − > 11.2, 11.2 − > 11.2, 11.2 − > 7.6,
  7.6 − > 9.4, 9.4 − > 11.2, 11.2 − > 11.2, 11.2 − > 9.4, 9.4 − > 5.4, 5.4 − > 7.6,
  7.6 − > 7.6, 7.6 − > 9.4, <<164 > >, 16.6 − > 22.3, 22.3 − > 22.3, 22.3 − > 22.3,
  22.3 − > 22.3, 22.3 − > 24.1, 24.1 − > 24.1, 24.1 − > 27.7, 27.7 − > 25.9,
  25.9 − > 22.3, 22.3 − > 22.3, 22.3 − > 25.9, 25.9 − > 24.1, 24.1 − > 18.4}
```

Now we the teaching process

```
⇒ p = Predict[train, Method → "GaussianProcess",
    PerformanceGoal → "Quality", TimeGoal → 700]
```

⇐ PredictorFunction [$Failed   Input type: Numerical   ]
                                 Method: GaussianProcess

Information about the Gaussian estimator,

```
⇒ Information[p]
```

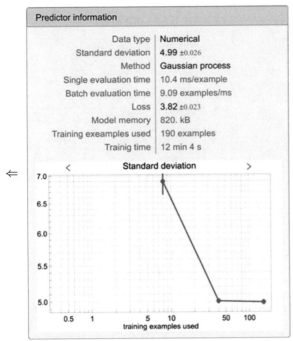

Now we should like to convert the Gaussian estimator in form.

$$w(t) = \mathcal{L}(t),$$

The measured wind velocities $w(t)$,

$\Rightarrow v= \mathtt{Map[\#[[1]]\&, train]};$

The approximated wind velocity estimated by the Gaussian estimator,

$\Rightarrow \varphi= \mathtt{Map[p[\#]\&, v]};$

$\Rightarrow \mathtt{Length[v]}$

$\Leftarrow 190$

The corresponding time points

$\Rightarrow \theta= \mathtt{Range[1, Length[v]]};$

Then the approximation of the Gaussian Process on the time domain using (w_i, t_i) = (θ_i, φ_i) (Fig. 15.47)

$\Rightarrow \mathtt{Show[\{pDD1, ListPlot[Transpose[\{\theta, \varphi\}], Joined \rightarrow True,}$

$\quad \mathtt{PlotStyle \rightarrow \{Thickness[0.001], Black\}]\}]}$

We should like to represent the approximated function in differential equation form,

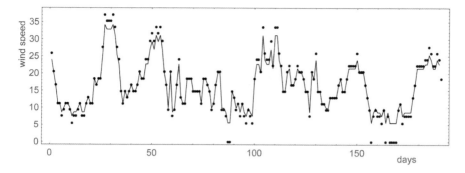

Fig. 15.47 The approximation of the Gaussian process

$$\frac{dy}{dt}(t) = \frac{d}{dt}GP(t)$$

In order to get differential equation form of the approximation, we should *differentiate the Gaussian Process (GP) estimator*. However we have not GP in time domain, therefore we use the interpolation function generated from the corresponding $(w_i, t_i) = (\theta_i, \varphi_i)$ pairs,

$\Rightarrow$ dnet = Interpolation[Transpose[$\{\theta, \varphi\}$]]

$\Leftarrow$ InterpolatingFunction [$\$Failed$ 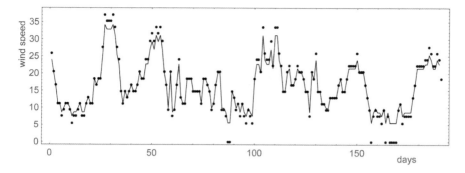 Domain: $\{\{1, 190.\}\}$ Output: scalar]

Now let us derivate this function,

$\Rightarrow$ Dnet[t_?NumericQ] := dnet'[t]

Displaying this function (Fig. 15.48)

$\Rightarrow$ Show[{Plot[Dnet[t], {t, 1, 191}, PlotStyle $\rightarrow$ {Black, Thickness[0.001]}, Frame $\rightarrow$ True, PlotRange $\rightarrow$ All]}, AspectRatio $\rightarrow$ 1/3]

15.6.5.1   The Function Approximation in Form of the Neural Network Differential Equation

Now we can consider the differential equation form of this approximation (Fig. 15.49)

$$\frac{dy}{dt}(t) = \text{Dnet}(t)$$

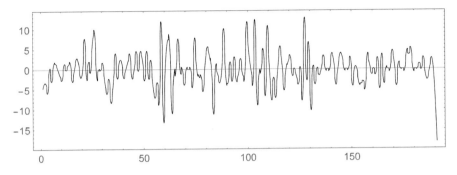

Fig. 15.48 Derivative function of the Gaussian process

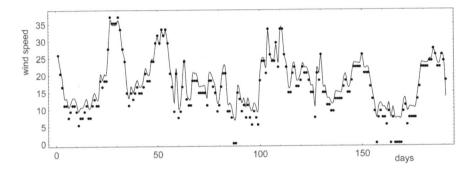

Fig. 15.49 Solution of the neural network differential equation

$\Rightarrow$ s = NDSolve$[\{y'[t] == Dnet[t], y[0] == -1\}, y, \{t, 0, 22\}]//$Quiet

$\Leftarrow \{\{y \rightarrow$ InterpolatingFunction

[$\$$Failed 〰 | Domain: $\{\{1, 191.\}\}$ | Output: scalar] $\}\}$

$\Rightarrow$p2h = Show$[\{$pDD1, Plot$[$Evaluate$[y[t]/.s], \{t, 1., 191\}$, PlotRange $\rightarrow$ All,

15.6.5.2   Employing Stochastic Differential Equation Form

In order to get stochastic model for the function approximation we employ Ito form of the Neural Network Differential Equation. Let define

$\Rightarrow$ H$[t\_]$:= Dnet$[t]$

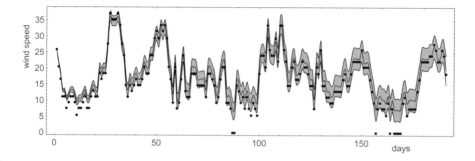

Fig. 15.50 The stochastic solution of the Ito form of the machine learning (Gaussian process) differential equation

Considering

⇒ σ= 0.3 ; μ= 0 ;

⇒ proc = ItoProcess[dy[t] == H[t]dt + dw[t], y[t], {y, 25.9}, t,
    w ≈ WienerProcess[μ,σ]]

⇐ ItoProcess[{{Dnet[t]}, {{0.3}}, y[t]}, {{y}, {25.9}}, {t, 0}]90

Simulating the process:

⇒ psol = RandomFunction[proc, {1., 190., 0.01}, 300]

⇐ TemporalData [$Failed 〰〰 Time: 1. to 191.
                                Data points: 5670300 Paths: 300]
                  Data not in notebook. Store now →

Let us display the mean and the standard deviation of the solution (Fig. 15.50),

⇒ p1 = Plot[Mean[psol[t]], {t, 1, 191}, PlotStyle → {Red,
    Thickness[0.001]}, FrameLabel → {"t","y(t)"}, Frame → True];

⇒ p2 = Plot[{Mean[psol[t]] + StandardDeviation[psol[t]],
    Mean[psol[t]] − StandardDeviation[psol[t]]}, {t, 1, 191},
    Filling → {1 → {2}}, PlotStyle → {{Blue, Thickness[0.001]},
    {Blue, Thickness[0.001]}}, FrameLabel → {"t","y(t)"},
    Frame → True, Axes → None, PlotRange → All];

⇒ Show[{pDD1, p2, p1}, AspectRatio → 1/3.2]

This figure shows that we have quite good approximation, only some low values of velocity are overestimated.

Chapter 16
Parallel Computation

16.1 Introduction

Although in the earlier chapters parallel computations in symbolic as well as in numeric form have been carried out in many times, in this chapter a more systematic overview is given of this topic. On the one hand, wide spread multicore computers in the market place provide ample proof that we are in the multicore era. In years to come probably the number of cores packed into a single chip will represent the increasing computational power rather than the clock frequency of the processor, which has perhaps reached its limit, and will likely stay below 4 GHz for a while. However, nowadays there processors running over 5 GHz, too.

On the other hand, popular software systems like MATLAB *Maple*, and *Mathematica* have extended their capabilities with simple instructions and functions providing a comfortable realization of parallel computation on multicore desktop (or even laptop) computers.

Consequently, parallel computing can be utilized by not only experts but engineers and scientists having no special knowledge in this area. However, despite the easily accessible hardware and the simple, user friendly software, there are some pitfalls and tricks, which are good to know in order to avoid unexpected negative effects, as well as to exploit the advantages of a multicore system.

Using illustrative examples, important features of parallel computation will be demonstrated. We show how some frequently employed algorithms in geosciences can be efficiently evaluated in parallel. We use *Mathematica*, but other systems could be used for some of these examples. As usual, we will show some of the *Mathematica* commands but not the more detailed ones. In most of these examples, the time consumed is displaced first.

Table 16.1 Upper bound on speedup, $f = 0.99$

Basic core equivalents	Base Amdahl	Symmetric	Asymmetric	Dynamic
16	14	14	14	< 16
64	39	39	49	< 60
256	72	80	166	< 223
1024	91	161	531	< 782

16.2   Amdahl's Law

Multicore processors are processing systems composed of two or more independent "cores" or CPUs. The amount of performance gained by the use of a multicore processor strongly depends on the software algorithms and implementation. In particular, the possible gains are limited by the fraction of the software that can be "parallelized" to run on multiple cores simultaneously; this effect is described by *Amdahl's law*. The modern version of this law states that if one enhances a fraction f of a computation by speedup S, then the overall speedup S_a is,

$$S_a(f, S) = \frac{1}{(1 - f) + \frac{f}{S}}.$$

Four decades ago, Amdahl assumed that a fraction f of a program's execution time was infinitely parallelizable, with no overhead, while the remaining fraction, $1 - f$, was totally sequential. He noted that the speedup on n processors is governed by

$$S_a(f, n) = \frac{1}{(1 - f) + \frac{f}{n}}.$$

By this law, in the best case, so-called embarrassingly parallel problems $f \approx 1$ one may realize speedup factors near the number of cores. Many typical applications, however, do not realize such large speedup factors, and thus the parallelization of software is a significant on-going topic of research.

In the near future, more efficient core design, more complex trade-offs between cores and cache capacity, and highly tuned parallel scientific and database codes may lead to considerable improvement. Table 16.1 shows the speedup limit in the case of $f = 0.99$ for different hardware designs: symmetric, asymmetric and dynamic multi-cores.

16.3   Implicit and Explicit Parallelism

On multicore machines, as many independent processes can be executed in parallel as there are cores in the machine. However, the so-called "multi-threading execution model" allows further parallelization, since within the context of a single

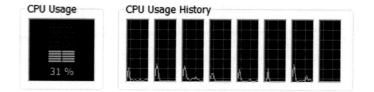

Fig. 16.1 Implicit parallelism at low computation load

process running on one core, several tasks sharing the same resources can be executed concurrently.

Some of the programming systems (e.g. *Matlab*) can exploit multicore and multi-threading automatically, partly or fully, without any special directives of the programming language. This characteristic of a programming language is called *implicit parallelism*. The implicit parallelism may support only certain statements and may be efficient only in case of large computational loads. *Mathematica* also automatically supports the parallelization of some operations. Already Version 11.1 (2017) added automatic multi-threading when computations are performed on multicore computers. To illustrate implicit threading in *Mathematica*, let us compute the greatest singular value of a random $n \times n$ matrix with $n = 1200$.

$\Rightarrow$ f[n_] := Max[SingularValueList[Table[RandomReal[],{n},{n}]]]

The time elapsed and result (there is really no value to return) is (Fig. 16.1):

$\Leftarrow$ {0.430452,Null}

The maximum of the average CPU usage is about 30%.

Increasing the load, for example in case of $n = 5000$, the CPU usage will automatically increase. The computation time yields (Fig. 16.2),

$\Leftarrow$ {8.00045,Null}

The maximum of the average CPU usage is about 50%.

Parallelism provided by the system and controlled by the user, namely using or not using parallel execution, is called *explicit parallelism*. This is a feature that allows or forces the programmer to annotate his program to indicate which parts

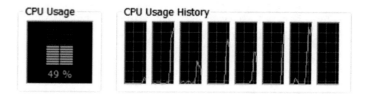

Fig. 16.2 Implicit parallelism at higher computation load

should be executed as independent parallel tasks. This is obviously more work for the programmer than a system with implicit parallelism (where the system decides automatically which parts to run in parallel) but may provide higher performance. *Mathematica* has many functions supporting explicit parallelism and we will use some of them in the following examples.

16.4 Dispatching Tasks

In this section we will illustrate how important can be the proper size of a task dispatched to the threads in order to avoid overloading the "master," resulting in "idle workers". We shall see that improper size of the tasks can increase the running time of parallel evaluation compared to the non-parallel case.

Let us consider the computation of the value of π with a Monte-Carlo method. The standard technique is to generate random numbers from the uniform distribution $[0, 1]$ as pair of coordinates (x_i, y_i) of points in plane. The approximate value of π can be computed as,

$$\pi \approx 4 \frac{N_p}{N},$$

where Np is the number of the points having distance from the origin less or equal with 1, and N is the number of the all randomly generated points.

The Mathematica function to compute this simulation could be

$\Rightarrow$ PI[n_] := 4./nLength[Select[Table[{RandomReal[], RandomReal[]},

$\{n\}], \sqrt{\#[[1]]^2 + \#[[2]]^2} \le 1 \&]]$

If we carry out the computation for 10 million random points, we get (again with the timing given first)

$\Leftarrow$ $\{31.9762, 3.14168\}$

To check the CPU usage on a Windows system, open *Windows Task Manager-Performance*. (Similar utilities exist on other platforms.) Average CPU Usage is 12–13%. This means that practically only one core computation power was exploited. Of course the other cores were not completely idle, because of implicit threading. Now let us try to use explicit parallel mode using a modified statement, with **ParallelTable** instead of **Table**,

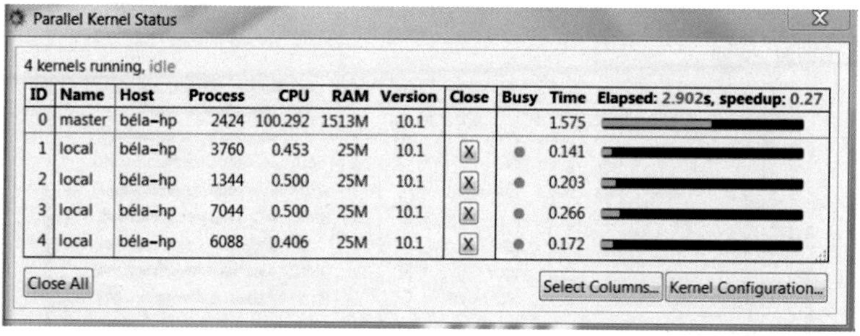

Fig. 16.3 Naive parallelism

$\Rightarrow$ PIParallelNaive[n_]:= 4./n Length[Select[ParallelTable[

{RandomReal[], RandomReal[]},{n}],$\sqrt{\#[[1]]^2 + \#[[2]]^2} \le 1\&$]]

If we invoke this on $n = 10,000,000$ we get

$\Leftarrow$ {35.6434, 3.14332}

No improvement happened, it even became somewhat worse! It is clear that the master was deeply engaged with dispatching tasks, since the workers completed their small tasks very quickly and were queuing at the front of the master and asking for a new task. So the master was overloaded; it is the bottleneck of the process (Fig. 16.3).

Now let us divide the total work into two portions and assign them to the workers. We should slightly modify our function,

$\Rightarrow$ 4Length[Select[Table[RandomReal[], RandomReal[],

{n}],$\sqrt{\#[[1]]^2 + \#[[2]]^2} \le 1\&$]]//N

Now the number of the available threads is eight. Every core represents two threads.

$\Rightarrow$ LaunchKernels[2$ProcessorCount]//Quiet;

This function will be available for all workers. We formulate the parallel evaluation as (Fig. 16.4).

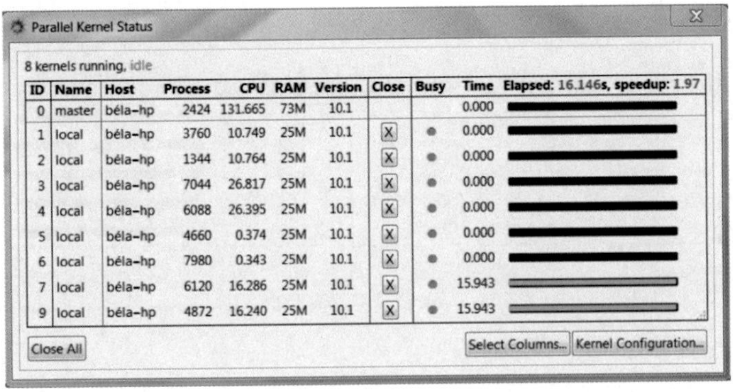

Fig. 16.4 Two workers are active

$$\Rightarrow \texttt{t} = \texttt{AbsoluteTime[]};$$
$$\Rightarrow \texttt{job1} = \texttt{ParallelSubmit[PIParallel[5000000]]};$$
$$\Rightarrow \texttt{job2} = \texttt{ParallelSubmit[PIParallel[5000000]]};$$
$$\Rightarrow \{\texttt{a1}, \texttt{a2}\} = \texttt{WaitAll[}\{\texttt{job1}, \texttt{job2}\}];$$
$$\Rightarrow \texttt{pi} = (\texttt{a1} + \texttt{a2})/10000000;$$
$$\Rightarrow \texttt{time2} = \texttt{AbsoluteTime[]} - \texttt{t}$$
$$\Leftarrow 16.1616284$$
$$\Rightarrow \texttt{pi}$$
$$\Leftarrow 3.14196$$

Now the average CPU Usage is 25%. The problem is that only two workers were on duty, the others were idle. Let us generalize the function for m sub-tasks:

$$\Rightarrow \texttt{PIParallelReasonable[n\_,m\_]} := \texttt{Apply[Plus,}$$
$$\texttt{WaitAll[Map[ParallelSubmit[PIParallel[\#]]\&, Table[n/m,\{m\}]]]]/n}$$

With $m = 4$ we get

$$\Leftarrow \{10.5612, 3.14105\}$$

Average CPU Usage is 50%. There is certainly improvement, but even now half of the workers remained idle. Let us use all the threads, i.e., $m = 8$:

$$\Leftarrow \{9.34044, 3.14238\}$$

We get the best result when the total work is partitioned into as many parts as we have threads. The average processor usage is 100% and all threads are working (Fig. 16.5).

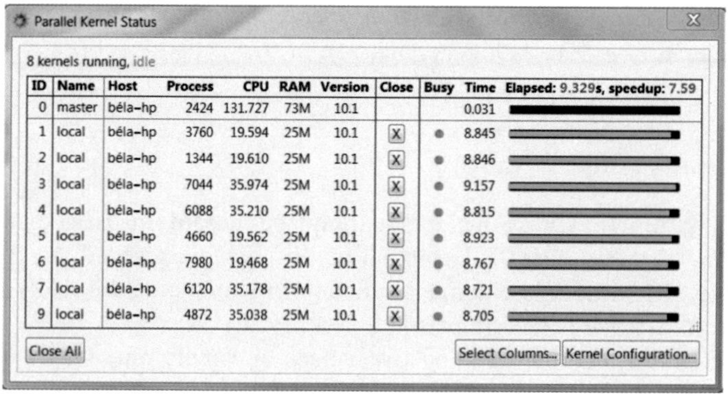
Parallel Kernel Status

8 kernels running, idle

ID	Name	Host	Process	CPU	RAM	Version	Close	Busy	Time	Elapsed: 9.329s, speedup: 7.59
0	master	béla-hp	2424	131.727	73M	10.1			0.031	
1	local	béla-hp	3760	19.594	25M	10.1	X	•	8.845	
2	local	béla-hp	1344	19.610	25M	10.1	X	•	8.846	
3	local	béla-hp	7044	35.974	25M	10.1	X	•	9.157	
4	local	béla-hp	6088	35.210	25M	10.1	X	•	8.815	
5	local	béla-hp	4660	19.562	25M	10.1	X	•	8.923	
6	local	béla-hp	7980	19.468	25M	10.1	X	•	8.767	
7	local	béla-hp	6120	35.178	25M	10.1	X	•	8.721	
9	local	béla-hp	4872	35.038	25M	10.1	X	•	8.705	

Close All | Select Columns... | Kernel Configuration...

Fig. 16.5 All workers are active

The efficiency of the parallel computation can be measured as

$$\eta = \frac{\textit{single core mode time}}{\textit{multi core mode time}} \frac{100}{\textit{number of threads}} \%.$$

In our case it is about (32.5/9.3) × (100/8) = 40%

16.5 Balancing Loads

It is also important to distribute tasks among the workers (threads) uniformly. However this is not always an easy job. To illustrate this situation let us consider the Monte-Carlo analysis of the stability of an infinite slope.

Infinite Slope Stability via Monte-Carlo Analysis

We are going to compute the safety factor for an infinite slope subject to a given ground acceleration value occurring over a specified time period. The safety factor is the ratio of resisting forces to driving forces. A slope will be stable if the resisting forces exceed the driving forces, and the factor of safety is greater than 1. We will use a simplified model of the infinite slope model. Let ϕ be the angle of internal friction (representing the frictional component of soil shear strength), β be the slope angle in degrees, and $0 \leq H \leq 1$ is the dimensionless height of the phreatic surface above the base of the slide mass; Haneberg (2004).

The pseudo static factor of safety of an infinite slope subjected to seismic acceleration is,

```
⇒ SeismicsFS[f_,β_,H_,Cs_] :=
  (((1 - H/2)Cos[β] - CsSin[β])Tan[f])/(Sin[β] + CsCos[β])
```

in which *Cs* is a coefficient of seismic acceleration given in terms of the gravitational acceleration *g*. This is a generalization of the static model (*Cs* = 0),

⇒ SeismicsFS[f,β,H,0]
⇐ (1 − H/2) Cot[β] Tan[φ]

According to the USGS national earthquake hazard maps, a peak ground acceleration of 0.12 g has 0.10 probability of being exceeded in 50 years in Socorro, New Mexico. So we define a function for computing the safety factor at a fixed *Cs* = 0.12 value called Slope[n].

We will censor the offensive negative values by simply removing them. Let us try *n* = 80 000 trials:

⇐ {22.573240, Null}

Here is the histogram showing the *Monte Carlo* simulation results (Fig. 16.6). The mean value is 1.01779. The deviation is 0.184506. Now let us carry out this simulation in parallel.

⇒ LaunchKernels[2$ProcessorCount]//Quiet;
⇒ DistributeDefinitions[Slope, SeismicsFS];

First we distribute the tasks in equal portions among the eight threads (Fig. 16.7).

⇒ n = Table[10000,{8}]
⇐ {10000,10000,10000,10000,10000,10000,10000,10000}
⇒ r1 = Flatten[ParallelMap[Slope[#]&,n]]; //AbsoluteTiming
⇐ {2.605205, Null}

Fig. 16.6 The histogram of the distribution of the safety factor in case Cs = 0.12, employing 80,000 samples

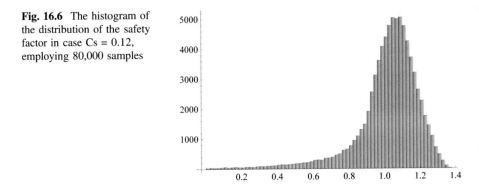

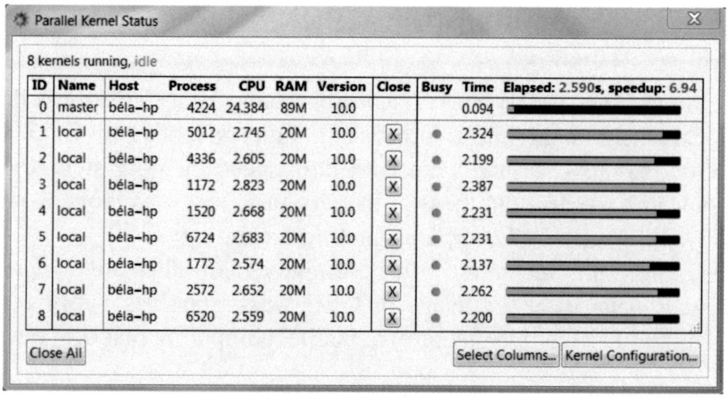
Parallel Kernel Status

8 kernels running, idle

ID	Name	Host	Process	CPU	RAM	Version	Close	Busy	Time	Elapsed: 2.590s, speedup: 6.94
0	master	béla-hp	4224	24.384	89M	10.0			0.094	
1	local	béla-hp	5012	2.745	20M	10.0	X	•	2.324	
2	local	béla-hp	4336	2.605	20M	10.0	X	•	2.199	
3	local	béla-hp	1172	2.823	20M	10.0	X	•	2.387	
4	local	béla-hp	1520	2.668	20M	10.0	X	•	2.231	
5	local	béla-hp	6724	2.683	20M	10.0	X	•	2.231	
6	local	béla-hp	1772	2.574	20M	10.0	X	•	2.137	
7	local	béla-hp	2572	2.652	20M	10.0	X	•	2.262	
8	local	béla-hp	6520	2.559	20M	10.0	X	•	2.200	

Close All | Select Columns... | Kernel Configuration...

Fig. 16.7 The threads have got the same size of load and they were used nearly equally

Now we distribute the tasks in an unequal way:

```
⇒ n ={1000,2000,5000,8000,11000,15000,18000,20000};
```

The running time is longer and the load of the threads are unbalanced (Fig. 16.8).

```
⇒ r1 = Flatten[ParallelMap[Slope[#]&,n]]; //AbsoluteTiming
⇐ {3.806407, Null}
```

That is why it is important to distribute the tasks into as equal loads as possible!

Basically, there are two types of parallel computation, namely *task parallel* and *data parallel*. In reality these two types turn up in mixed form.

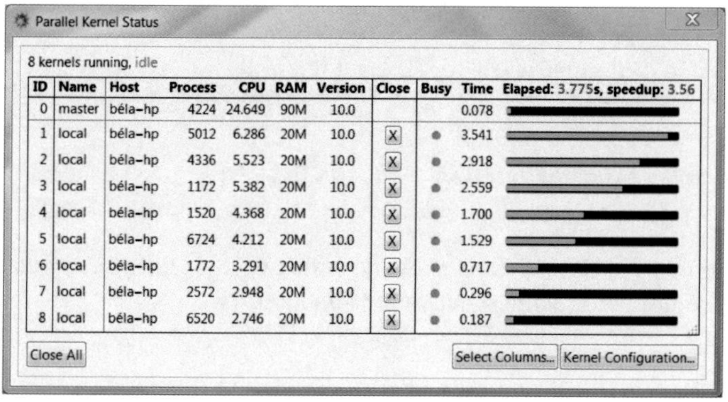
Parallel Kernel Status

8 kernels running, idle

ID	Name	Host	Process	CPU	RAM	Version	Close	Busy	Time	Elapsed: 3.775s, speedup: 3.56
0	master	béla-hp	4224	24.649	90M	10.0			0.078	
1	local	béla-hp	5012	6.286	20M	10.0	X	•	3.541	
2	local	béla-hp	4336	5.523	20M	10.0	X	•	2.918	
3	local	béla-hp	1172	5.382	20M	10.0	X	•	2.559	
4	local	béla-hp	1520	4.368	20M	10.0	X	•	1.700	
5	local	béla-hp	6724	4.212	20M	10.0	X	•	1.529	
6	local	béla-hp	1772	3.291	20M	10.0	X	•	0.717	
7	local	béla-hp	2572	2.948	20M	10.0	X	•	0.296	
8	local	béla-hp	6520	2.746	20M	10.0	X	•	0.187	

Close All | Select Columns... | Kernel Configuration...

Fig. 16.8 The balance of load of the different threads

16.6 Parallel Computing with GPU

Graphics Process Unit (GPU) alias Visual Processing Unit (VPU) was designed to carry out visualization an image processing independently on CPU. So called dedicated GPU has its own hardware with high speed memory storage and multi-core processor unit to carry out these tasks very fast. This development was mainly motivated by the demand of computer gaming technology.

However, today this technique can be employed in other areas of computing, too. This generalization of the usage of GPU was strongly supported by CUDA (Compute Unified Device Architecture) a parallel computing platform developed by NVIDIA.

The most simple way to utilize this feature is to parametrize special built-in functions employing GPU. Some computational systems like Mathematica provides easy access to this computational ability even for standard users.

16.6.1 Neural Network Computing with GPU

The algorithms of training and evaluating of neural networks are mainly based on matrix operations which can be carried out effectively by using CPU architecture.

Let us illustrate this operation first by a function approximation application.

16.6.1.1 Function Approximation

Assuming Gaussian $N(0, 0.2)$ noise, discrete data points of function

$$y(x) = \exp\left(-x^2\right)$$

can be generated in range $x \in [-5, 5]$, see Fig. 16.9

```
⇒ data = Table[x → Exp[−x²] +
    RandomVariate[NormalDistribution[0,.2]],{x, − 5, 5,.1}];
⇒ plot = ListPlot[List@@@data, PlotStyle → Red, PlotRange → All]
```

Let us employ a feedforward network with two hidden layers containing 150 nodes each employing Tanh(x) activation function,

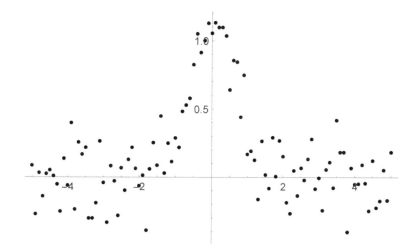

Fig. 16.9 Noisy data points

The structure of the regression neural network

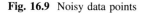

		scalar
	Input	vector (size: 1)
1	LinearLayer	vector (size: 150)
2	Tanh	vector (size: 150)
⟸ NetChain[3	LinearLayer	vector (size: 150)]
4	Tanh	vector (size: 150)
5	LinearLayer	vector (size: 1)
	Output	scalar
		(uninitialized)

⇒ net = NetChain[{150, Tanh, 150, Tanh, 1}, "Input" → "Scalar",
  "Output" → "Scalar"]

In order to avoid over learning the 20% of the data points are used as validation and 80% as training set. First we carry out the training with CPU.

⇒ AbsoluteTiming[
  net2 = NetTrain[net, data, ValidationSet → Scaled[0.2]];]
⟸ {24.0289, Null}

Let us approximate the data with the network function, see Fig. 16.10

⇒ Show[Plot[net2[x], {x, −5, 5}], plot]

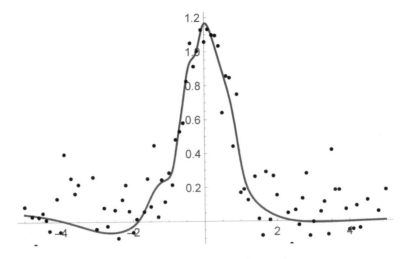

Fig. 16.10 Noisy data points approximated by the neural network

Then we repeat the computation with GPU unit (in our case GeForce GT 630), using parametrization (TargetDevice→"GPU"),

⇒ AbsoluteTiming[net2 = NetTrain[
  net, data, ValidationSet → Scaled[0.2], TargetDevice→"GPU"];]
⇐ {11.2258, Null}

The running time is considerably shorter and even the result is much more realistic, see Fig. 16.11.

⇒ Show[Plot[net2[x], {x, −5, 5}], plot]

Of course this result can strongly depend on the hardware of the GPU.

16.6.1.2   Classification Problem via Deep Learning

Our second example is an image classification problem, which can be solved via deep learning technique employing convolutional neural network. We use the CIFAR-10 database of labeled images of ten classes: airplane, automobile, bird, cat, deer, dog, frog, horse, ship and truck. There are available 50,000 colored RGB images of size 32 × 32. Let us download the images,

⇒ obj = ResourceObject["CIFAR − 10"];
⇒ trainingData = ResourceData[obj, "TrainingData"];

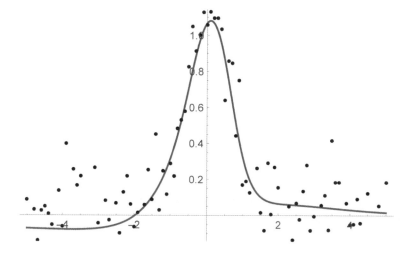

Fig. 16.11 Noisy data points approximated by the neural network

As illustration here are ten randomly selected images,

⇒ RandomSample[trainingData, 10]

Extract the unique classes,

⇒ classes = Union@Values[trainingData]
⇐ {airplane, automobile, bird, cat, deer, dog, frog, horse, ship, truck}

The total number of the images

⇒ Length[trainingData]
⇐ 50000

Actually, we employ the last 1000 images as testing set, therefore the training set,

$\Rightarrow$ actualTrainingSet = RandomSample[trainingData, 49000];

and the testing set is,

$\Rightarrow$ actualTestingSet = Complement[trainingData, actualTrainingSet];

The structure of the convolutional network can be seen in below mentioned table. The network has a 3 dimensional tensor since one image has 32×32 pixel size and three RGB colors, consequently the input size is $3 \times 32 \times 32$. The output is a ten dimensional vector corresponding the ten different classes. The i-th element of the output vector shows the probability of the association of the input image with the i-th class.

Let us define the network,

$\Rightarrow$ lenet = NetChain[
   {ConvolutionLayer[20, 5], Ramp, PoolingLayer[2, 2],
   ConvolutionLayer[50, 5], Ramp, PoolingLayer[2, 2],
   FlattenLayer[], 500, Ramp, 10, SoftmaxLayer[]},
   "Output" $\rightarrow$ NetDecoder[{"Class", classes}],
   "Input" $\rightarrow$ NetEncoder[{"Image", {32, 32}}]
   ]

The structure of the convolutional neural network

			image
		Input	3–tensor (size: $3 \times 32 \times 32$)
	1	ConvolutionLayer	3–tensor (size: $20 \times 28 \times 28$)
	2	Ramp	3–tensor (size: $20 \times 28 \times 28$)
	3	PoolingLayer	3–tensor (size: $20 \times 14 \times 14$)
	4	ConvolutionLayer	3–tensor (size: $50 \times 10 \times 10$)
	5	Ramp	3–tensor (size: $50 \times 10 \times 10$)
$\Leftarrow$ NetChain [	6	PoolingLayer	3–tensor (size: $50 \times 5 \times 5$)]
	7	FlattenLayer	vector (size: 1250)
	8	LinearLayer	vector (size: 500)
	9	Ramp	vector (size: 500)
	10	LinearLayer	vector (size: 10)
	11	SoftmaxLayer	vector (size: 10)
		Output	class
			(uninitialized)

For validation set we use the 20% of the training set. First let us carry out the training with CPU.

⇒ AbsoluteTiming[
  trained = NetTrain[lenet, actualTrainingSet,
  ValidationSet → Scaled[0.2], MaxTrainingRounds → 5];]
⇐ {1538.4, Null}

Let us pick up one image from the testing set randomly

⇒ images = RandomSample[Keys[actualTestingSet], 1]

{ }

Using our trained network we can classify this image,

⇒ trained[images, "Probabilities"]
⇐ {(|airplane → 0.0000499916,
  automobile → 0.497519, bird → 4.10959 × 10⁻⁶, cat → 0.000122994,
  deer → 3.23379 × 10⁻⁷, dog → 0.000022407, frog → 4.71051 × 10⁻⁶,
  horse → 0.0000508431, ship → 0.0149342, truck → 0.487292|)}

Our classifier identifies this object as an automobile with the highest probability (0.497519), however gives a similar chance for being classified as a truck (0.487292).

Let us check more test images

⇒ images = RandomSample[Keys[actualTestingSet], 10]

⇒ trained[images]
⇐ {cat, dog, airplane, automobile, dog, dog, cat, frog, frog, frog}

Now we carry out the training with GPU

```
⇒ AbsoluteTiming[
   trained = NetTrain[lenet, actualTrainingSet,
   ValidationSet → Scaled[0.2], MaxTrainingRounds → 5,
   TargetDevice→"GPU"]; ]
   ⇐ {128.438, Null}
```

We can reach more than ten times speed up. Considering these two neural network examples, one can see that increasing task size results increasing speed-up. Checking the test image.

⇒ trained{ { } \}{,}"Probabilities"]

```
⇐ {(|airplane → 0.0034147,
    automobile → 0.579832, bird → 0.000021244, cat → 0.0000830502,
    deer → 0.0000600025, dog → 0.0000127508, frog → 6.02948 × 10⁻⁷,
    horse → 0.00050736, ship → 0.00266856, truck → 0.4134|)}
```

Remark Images here with improved resolution are illustration however the computation has been carried out with the original images of the CIFAR-10 database.

16.6.2 Image Processing with GPU

The parametrization technique is simple and effective but it is constrained to a limited number of built in functions. Extension can be achieved by using CUDA platform providing functions can be written in C -type compiler languages. These functions then will be linked to a computing system like *Mathematica* via CUDALink. Therefore this method is less efficient than the parametrization technique. Sometimes, surprisingly, if the CPU of our computer is much efficient than its GPU, then the considered function could be executed faster on CPU than on GPU. Here we illustrate this situation in order the call attention to the pitfalls of the GPU application, too.
Let us load CUDA platform,

⇒ Needs["CUDALink`"]

Checking whether our system can support this platform,

$\Rightarrow$ CUDAQ[]

$\Leftarrow$ True

Let us get information about our GPU system,

$\Rightarrow$ CUDAInformation[]

$\Leftarrow$ {1 → {Name → GeForceGT630, ClockRate → 875500,
   ComputeCapabilities → 3., GPUOverlap − > 1,
   MaximumBlockDimensions → {1024, 1024, 64},
   MaximumGridDimensions → {2147483647, 65535, 65535},
   MaximumThreadsPerBlock → 1024,
   MaximumSharedMemoryPerBlock → 49152,
   TotalConstantMemory → 65536, WarpSize → 32,
   MaximumPitch → 2147483647, MaximumRegistersPerBlock − > 65536,
   TextureAlignment → 512, MultiprocessorCount → 1,
   CoreCount → 32, CanMapHostMemory → True,
   ComputeMode → Default, Texture1DWidth → 65536,
   Texture2DWidth → 65536, Texture2DHeight → 65536,
   Texture3DWidth → 4096, Texture3DHeight → 4096,
   Texture3DDepth → 4096, Texture2DArrayWidth → 16384,
   Texture2DArrayHeight → 16384, Texture2DArraySlices → 2048,
   SurfaceAlignment → 512, ConcurrentKernels → True,
   ECCEnabled → False, TCCEnabled → False,
   TotalMemory → 2147483648}}

The task is to convolve an RGB image, see Fig. 16.12

$\Rightarrow$ im

Our kernel is an edge detection filter

$$\Rightarrow M1 = \begin{pmatrix} 0 & 1 & 0 \\ 1 & -4 & 1 \\ 0 & 1 & 0 \end{pmatrix};$$

Employing CPU, we get.

$\Rightarrow$ AbsoluteTiming[ImageConvolve[im, M1]]

$\Leftarrow$ {0.0464621,} (Fig. 16.13)

Fig. 16.12 Budapest panorama with the Danube

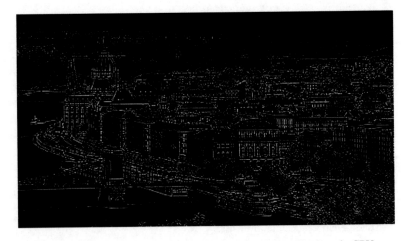

Fig. 16.13 Budapest panorama with the Danube after convolution filtering via CPU

Now let us employ GPU via CUDA function.

⇒ Needs["CUDALink‘"]
⇒ AbsoluteTiming[CUDAImageConvolve[im, M1]]
⇐ {0.155772,} (Fig. 16.14)

The time is somewhat longer, but the quality of the result is definitely better. The explanation is in Table 16.2. This table shows that our system has a strong CPU,

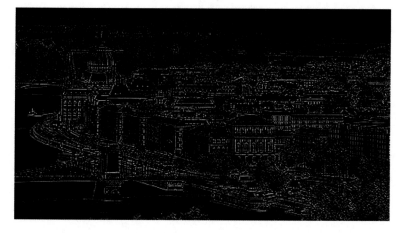

Fig. 16.14 Budapest panorama with the Danube after convolution filtering via CPU

Table 16.2 Details of the effectivity of the different system components

Component	What is rated	Subscore	Base score
Processor	Calculations per second	7.8	
Memory (RAM)	Memory operations per second	7.9	
Graphics	Desktop performance for Windows Aero	6.8	
Gaming graphics	3D business and gaming graphics performance	6.8	Determined by lowest subscore
Primary hard disk	Disk data transfer rate	7.9	

however its average performance index is low, since the low efficiency of the GPU subsystem is the bottle neck.

It means that when we have a corresponding built-in function and small task, in addition our GPU is not strong comparing to the CPU then using the built-in function with CPU can be better than employing CUDALink function.

16.7 Applications

The type of parallel jobs can be *task parallel* (embarrassingly parallel) and *data parallel*. In case of task parallel, multiple workers work on different parts of the problem without communication between each other. The result is independent of the execution order. Let us see a geodesics example.

16.7.1 3D Ranging Using the Dixon Resultant

Suppose three stations are given with known coordinates (x_i, y_i, z_i) and their distances, s_i from a point in 3D space. We are looking for the coordinates of the unknown point (x_0, y_0, z_0) (Fig. 16.15).

The corresponding distance observation equations,

$$\Rightarrow e1 = (x_1-x0)^2 + (y_1-y0)^2 + (z_1-z0)^2 - s_1^2 // \text{Expand}$$
$$\Leftarrow x0^2 + y0^2 + z0^2 - s_1^2 - 2x0x_1 + x_1^2 - 2y0y_1 + y_1^2 - 2z0z_1 + z_1^2$$
$$\Rightarrow e2 = (x_2-x0)^2 + (y_2-y0)^2 + (z_2-z0)^2 - s_2^2 // \text{Expand}$$
$$\Leftarrow x0^2 + y0^2 + z0^2 - s_2^2 - 2x0x_2 + x_2^2 - 2y0y_2 + y_2^2 - 2z0z_2 + z_2^2$$
$$\Rightarrow e3 = (x_3-x0)^2 + (y_3-y0)^2 + (z_3-z0)^2 - s_3^2 // \text{Expand}$$
$$\Leftarrow x0^2 + y0^2 + z0^2 - s_3^2 - 2x0x_3 + x_3^2 - 2y0y_3 + y_3^2 - 2z0z_3 + z_3^2$$

Instead of trying to solve the original system directly, first we shall transform this system into a linear system of two equations containing $z0$ as parameter. With simple subtractions, we get

$$\Rightarrow q = -\{e1 - e2, e2 - e3\} // \text{Expand}$$
$$\Leftarrow \{s_1^2 - s_2^2 + 2x0x_1 - x_1^2 - 2x0x_2 + x_2^2 + 2y0y_1 - y_1^2 -$$
$$2y0y_2 + y_2^2 + 2z0z_1 - z_1^2 - 2z0z_2 + z_2^2, s_2^2 - s_3^2 + 2x0x_2 - x_2^2 -$$
$$2x0x_3 + x_3^2 + 2y0y_2 - y_2^2 - 2y0y_3 + y_3^2 + 2z0z_2 - z_2^2 - 2z0z_3 + z_3^2\}$$

Now, we will write the equations in the following form,

$$\Rightarrow eq_1 = \alpha_1 x0 + \beta_1 y0 + \delta_1 z0 + \varepsilon_1$$
$$\Leftarrow x0\,\alpha_1 + y0\,\beta_1 + z0\,\delta_1 + \varepsilon_1$$

Fig. 16.15 Spacial ranging 3 point problem

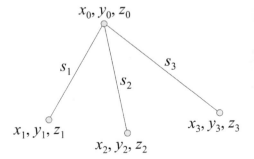

and

$$\Rightarrow \text{eq}_2 = \text{eq}_1 /. \{1 \to 2\}$$
$$\Leftarrow \text{x0}\, \alpha_2 + \text{y0}\, \beta_2 + \text{z0}\, \delta_2 + \varepsilon_2$$

The corresponding coefficients α_i, β_i, δ_i are

$$\Rightarrow \text{coeffs0} = \text{Map}[(\{\text{Coefficient}[\#,\text{x0}], \text{Coefficient}[\#,\text{y0}],$$
$$\text{Coefficient}[\#,\text{z0}]\}//\text{Factor})\&, q]$$
$$\Leftarrow \{\{2(x_1 - x_2), 2(y_1 - y_2), 2(z_1 - z_2)\}, \{2(x_2 - x_3), 2(y_2 - y_3), 2(z_2 - z_3)\}\}$$

and the constant values, ε_i

$$\Rightarrow \text{coeffs1} =$$
$$\text{Table}[\{q[[i]] - \text{coeffs0}[[i]].\{\text{x0},\text{y0},\text{z0}\}//\text{Simplify}\}, \{i,1,2\}]$$
$$\Leftarrow \{\{s_1^2 - s_2^2 - x_1^2 + x_2^2 - y_1^2 + y_2^2 - z_1^2 + z_2^2\}\{s_2^2 - s_3^2 - x_2^2 + x_3^2 - y_2^2 + y_3^2 - z_2^2 + z_3^2\}\}$$

then unify them,

$$\Rightarrow \text{coeffs} = \text{Table}[\text{Union}[\text{coeffs0}[[i]], \text{coeffs1}[[i]]], \{i,1,2\}]$$
$$\Leftarrow \{\{2(x_1 - x_2), 2(y_1 - y_2), 2(z_1 - z_2), s_1^2 - s_2^2 - x_1^2 + x_2^2 - y_1^2 + y_2^2 - z_1^2 + z_2^2\},$$
$$\{2(x_2 - x_3), 2(y_2 - y_3), 2(z_2 - z_3), s_2^2 - s_3^2 - x_2^2 + x_3^2 - y_2^2 + y_3^2 - z_2^2 + z_3^2\}\}$$

These expressions can be assigned to the coefficients in rule form,

$$\Rightarrow \text{coeffsn} = \text{Flatten}[\text{Table}[\text{Inner}[\#1 \to \#2\&, \{\alpha_i, \beta_i, \delta_i, \varepsilon_i\}],$$
$$\text{coeffs}[[i]], \text{List}], \{i,1,2\}]]$$
$$\Leftarrow \{\alpha_1 \to 2(X_1 - X_2), \beta_1 \to 2(Y_1 - Y_2), \delta_1 \to 2(Z_1 - Z_2),$$
$$\varepsilon_1 \to S_1^2 - S_2^2 - X_1^2 + X_2^2 - Y_1^2 + Y_2^2 - Z_1^2 + Z_2^2, \alpha_2 \to 2(X_2 - X_3),$$
$$\beta_2 \to 2(Y_2 - Y_3), \delta_2 \to 2(Z_2 - Z_3),$$
$$\varepsilon_2 \to S_2^2 - S_3^2 - X_2^2 + X_3^2 - Y_2^2 + Y_3^2 - Z_2^2 + Z_3^2\}$$

The symbolic solution of the *3D Ranging Problem* can be solved via symbolic computation, employing the Dixon resultant.

The sequential solution in *Mathematica* is the following:

$$\Rightarrow \ll \text{Resultant }'\text{Dixon}'$$

Now, we can eliminate $y0$ and $z0$ simultaneously,

```
⇒ AbsoluteTiming[solx0D =
  DixonResultant[eq₁, eq₂, e3},{y0,z0},{Y0,Z0}]//Simplify;]
⇐ {1.13882,Null}
⇒ solx0D
```

$$
\begin{aligned}
\Leftarrow (-\beta_2\delta_1 + \beta_1\delta_2)(&x0^2\beta_2^2\delta_1^2 - s_3^2\beta_2^2\delta_1^2 - 2x0x_3\beta_2^2\delta_1^2 + x_3^2\beta_2^2\delta_1^2 + y_3^2\beta_2^2\delta_1^2 + z_3^2\beta_2^2\delta_1^2 \\
&+ x0^2\alpha_2^2(\beta_1^2 + \delta_1^2) - 2x0^2\beta_1\beta_2\delta_1\delta_2 + 2s_3^2\beta_1\beta_2\delta_1\delta_2 + 4x0x_3\beta_1\beta_2\delta_1\delta_2 - 2x_3^2\beta_1\beta_2\delta_1\delta_2 \\
&- 2y_3^2\beta_1\beta_2\delta_1\delta_2 - 2z_3^2\beta_1\beta_2\delta_1\delta_2 + x0^2\beta_1^2\delta_2^2 - s_3^2\beta_1^2\delta_2^2 - 2x0x_3\beta_1^2\delta_2^2 + x_3^2\beta_1^2\delta_2^2 \\
&+ y_3^2\beta_1^2\delta_2^2 + z_3^2\beta_1^2\delta_2^2 + x0^2\alpha_1^2(\beta_2^2 + \delta_2^2) + 2z_3\beta_2^2\delta_1\epsilon_1 - 2z_3\beta_1\beta_2\delta_2\epsilon_1 - 2y_3\beta_2\delta_1\delta_2\epsilon_1 \\
&+ 2y_3\beta_1\delta_2^2\epsilon_1 + \beta_2^2\epsilon_1^2 + \delta_2^2\epsilon_1^2 - 2z_3\beta_1\beta_2\delta_1\epsilon_2 + 2y_3\beta_2\delta_1^2\epsilon_2 + 2z_3\beta_1^2\delta_2\epsilon_2 - 2y_3\beta_1\delta_1\delta_2\epsilon_2 \\
&- 2\beta_1\beta_2\epsilon_1\epsilon_2 - 2\delta_1\delta_2\epsilon_1\epsilon_2 + \beta_1^2\epsilon_2^2 + \delta_1^2\epsilon_2^2 - 2x0\alpha_2(-y_3\beta_2\delta_1^2 + y_3\beta_1\delta_1\delta_2 + z_3\beta_1(\beta_2\delta_1 \\
&- \beta_1\delta_2) + x0\alpha_1(\beta_1\beta_2 + \delta_1\delta_2) + \beta_1\beta_2\epsilon_1 + \delta_1\delta_2\epsilon_1 - \beta_1^2\epsilon_2 - \delta_1^2\epsilon_2) \\
&+ 2x0\alpha_1(z_3\beta_2(\beta_2\delta_1 - \beta_1\delta_2) + y_3\delta_2(-\beta_2\delta_1 + \beta_1\delta_2) + \beta_2^2\epsilon_1 + \delta_2^2\epsilon_1 - \beta_1\beta_2\epsilon_2 - \delta_1\delta_2\epsilon_2))
\end{aligned}
$$

This is a second order polynomial for $x0$,

```
⇒ Exponent[solx0D,x0]
⇐ 2
```

Similarly, eliminating $x0$ and $z0$ simultaneously,

```
⇒ Clear[X0]
⇒ AbsoluteTiming[soly0D =
  DixonResultant[{eq₁,eq₂,e3},{x0,Z0},{X0,Z0}]//Simplify;]
⇐ {1.13374,Null}
⇒ soly0D
```

$$
\begin{aligned}
\Leftarrow -(\alpha_2\delta_1 - \alpha_1\delta_2)(&\alpha_2^2(y0^2\beta_1^2 + (y0^2 - s_3^2 + x_3^2 - 2y0y_3 + y_3^2 + z_3^2)\delta_1^2 + 2z_3\delta_1\epsilon_1 \\
&+ \epsilon_1^2 + 2yy0\beta_1(z_3\delta_1 + \epsilon_1)) + 2x_3\alpha_1\delta_2(-y0\beta_2\delta_1 + y0\beta_1\delta_2 + \delta_2\epsilon_1 - \delta_1\epsilon_2) + (y0\beta_2\delta_1 \\
&- y0\beta_1\delta_2 - \delta_2\epsilon_1 + \delta_1\epsilon_2)^2 + \alpha_1^2(y0^2\beta_2^2 + (y0^2 - s_3^2 + x_3^2 - 2y0y_3 + y_3^2 + z_3^2)\delta_2^2 \\
&+ 2z_3\delta_2\epsilon_2 + \epsilon_2^2 + 2y0\beta_2(z_3\delta_2 + \epsilon_2)) - 2\alpha_2(x_3\delta_1(-y0\beta_2\delta_1 + y0\beta_1\delta_2 + \delta_2\epsilon_1 - \delta_1\epsilon_2) \\
&+ \alpha_1(y0^2\delta_1\delta_2 - s_3^2\delta_1\delta_2 + x_3^2\delta_1\delta_2 - 2y0y_3\delta_1\delta_2 + y_3^2\delta_1\delta_2 + z_3^2\delta_1\delta_2 + y0\beta_2\epsilon_1 \\
&+ \epsilon_1\epsilon_2 + y0\beta_1(y0\beta_2 + z_3\delta_2 + \epsilon_2) + z_3(y0\beta_2\delta_1 + \delta_2\epsilon_1 + \delta_1\epsilon_2))))
\end{aligned}
$$

This is a second order polynomial for $y0$,

```
⇒ Exponent[soly0D,y0]
⇐ 2
```

To get $z0$, we can eliminate $x0$ and $y0$ simultaneously,

```
⇒ AbsoluteTiming[solz0D =
  DixonResultant[{eq₁,eq₂,e3},{x0,y0},{X0,Y0}]//Simplify; ]
⇐ {1.09857,Null}
⇒ solz0D
```

$$\Leftarrow -(\alpha_2\beta_1 - \alpha_1\beta_2)(\alpha_2^2((z0^2 - s_3^2 + x_3^2 + y_3^2 - 2z0z_3 + z_3^2)\beta_1^2 + 2y_3\beta_1(z0\delta_1 + \epsilon_1)$$
$$+ (z0\delta_1 + \epsilon_1)^2) + 2x_3\alpha_1\beta_2(\beta_2(z0\delta_1 + \epsilon_1) - \beta_1(z0\delta_2 + \epsilon_2)) + (\beta_2(z0\delta_1 + \epsilon_1)$$
$$- \beta_1(z0\delta_2 + \epsilon_2))^2 + \alpha_1^2((z0^2 - s_3^2 + x_3^2 + y_3^2 - 2z0z_3 + z_3^2)\beta_2^2 + 2y_3\beta_2(z0\delta_2 + \epsilon_2)$$
$$+ (z0\delta_2 + \epsilon_2)^2) - 2\alpha_2(x_3\beta_1(\beta_2(z0\delta_1 + \epsilon_1) - \beta_1(z0\delta_2 + \epsilon_2)) + \alpha_1((z0\delta_1 + \epsilon_1)(y_3\beta_2$$
$$+ z0\delta_2 + \epsilon_2) + \beta_1((z0^2 - s_3^2 + x_3^2 + y_3^2 - 2z0z_3 + z_3^2)\beta_2 + y_3(z0\delta_2 + \epsilon_2)))))$$

This is a second order polynomial for $y0$,

```
⇒ Exponent[solz0D, z0]
⇐ 2
```

These computations can be evaluated in parallel too.

```
⇒ xyz ={{y0,z0},{x0,z0},{x0,y0}}; XYZ ={{Y0,Z0},{X0,Z0},{X0,Z0}};
⇒ sys ={eq₁,eq₂,e3};
⇒ DistributeDefinitions[DixonResultant,xyz,XYZ,sys];
⇒ {solxPD,solyPD,solzPD}= Map[Simplify[#]&,
  Parallelize[MapThread[DixonResultant[sys,#1,#2]&,{xyz,XYZ}]]];
  //AbsoluteTiming
⇐ {1.64605,Null}
```

The results are the same,

```
⇒ solx0D − solxPD//Simplify
⇐ 0
⇒ soly0D − solyPD//Simplify
⇐ 0
⇒ solz0D − solzPD//Simplify
⇐ 0
```

The load is too small, however; the net win is $(1.14 + 1.13 + 1.1)/1.65 = 2.042$.
So using parallel computation even for a symbolic case can reduce the running time by half!

16.7.2   *Reducing Colors via Color Approximation*

In some problems data being analyzed is too much for one processor. In this case each worker operates on part of the data, and they may or may not communicate with each other. To illustrate, let's consider an example in digital image processing.

On systems with 24-bit color displays, true color images can display up to 16,777,216 (i.e., 2^{24}) colors. On systems with lower screen bit depths, true color images are still displayed reasonably well, using color approximation. Color approximation is the process by which the software chooses replacement colors when direct matches cannot be found. One of the methods to carry out such color approximation is *color quantization*.

An important concept in image quantization is the RGB color cube. The RGB color cube is a three-dimensional array of all the colors that are defined for a particular data type (Fig. 16.16).

Quantization involves dividing the RGB color cube into smaller boxes, and then mapping all colors that fall within each box to the color value at the center of that box.

If the actual image is big in size, one may divide the image into parts, and this quantization process can be carried out in parallel on these image segments. Let us consider an airborne digital photo of Boston.

$\Rightarrow$ pBoston $=$ (Fig. 16.17)

Its size is

$\Rightarrow$ ImageDimensions[pBoston]
$\Leftarrow \{4481, 2881\}$

Fig. 16.16   RGB color cube

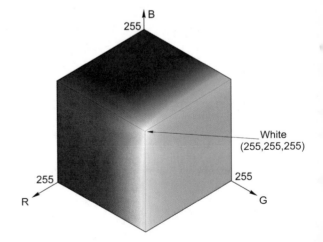

Fig. 16.17 Digital photo of Boston

So the number of pixels is their product, 12,909,761.

Let us divide the RGB cube into 10 "rectangular cuboids" (rectangular solids), and carry out the color reduction by color quantization,

> ⇒ pQ = ColorQuantize[pBoston, 10]; //Timing
> ⇐ {7.69085, Null}

The result is (Fig. 16.18).

Now let us divide the picture into eight parts (Fig. 16.19),

> ⇒ 4481/4
> ⇐ 1120.25
> ⇒ 2881/2
> ⇐ 1440.5
> ⇒ pP = ImagePartition[pBoston, {1120, 1440}]; //Timing
> ⇐ {0.0624004, Null}
> ⇒ pPF = Flatten[pP];
> ⇒ Magnify[pPF, 0.2]

Fig. 16.18 Digital photo of Boston after color reduction

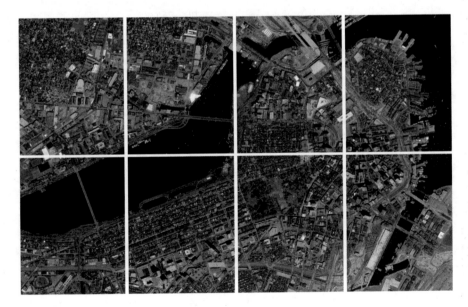

Fig. 16.19 The partitioned image of Boston

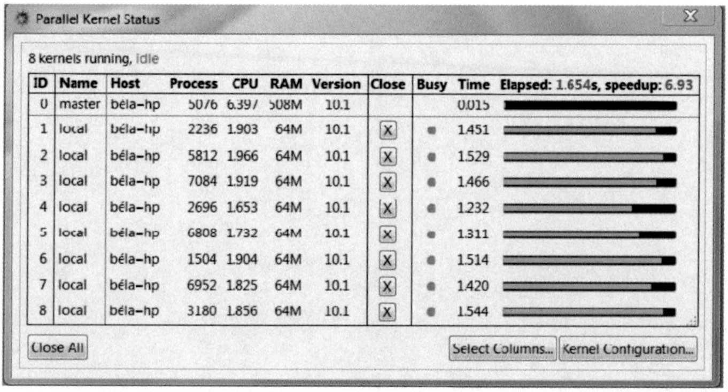

Fig. 16.20 Status of the kernels in case of parallel evaluation of color quantization using **ParallelSubmit**

Now let us compute the color approximation of each sub-image simultaneously in parallel, using **ParallelSubmit** (Fig. 16.20).

$\Rightarrow$ LaunchKernels[2$ProcessorCount]//Quiet;

$\Rightarrow$ DistributeDefinitions[ColorQuantize,pPF];

$\Rightarrow$ pS = WaitAll[Map[ParallelSubmit[ColorQuantize[#,10]]&,pPF]]; //
AbsoluteTiming

$\Leftarrow$ {1.66377,Null}

$\Rightarrow$ Dimensions[pS]

$\Leftarrow$ {8}

We put the parts together,

$\Rightarrow$ pR = ImageAssemble[Partition[pS,4]]; //AbsoluteTiming

$\Leftarrow$ {0.0961564,Null}

$\Rightarrow$ ImageDimensions[pR]

$\Leftarrow$ {4480,2880}

The result is (Fig. 16.21).

The actual net win in running time, 5.865/(0.031 + 1.664 + 0.096) = 3.27471.

The parallelization reduced the running time by a factor of 3.3, which means down to 30% of the traditional computation.

Fig. 16.21 Digital photo of Boston after color reduction employing parallel computation

16.8 Problem

16.8.1 Photogrammetric Positioning by Gauss–Jacobi Method

The relation between the coordinates of a ground point (X_i, Y_i, Z_i) and the coordinates of the corresponding point on the photo plane (x_i, y_i) can be represented by the following linear transformation

$$\begin{pmatrix} x_i - \eta_0 \\ y_i - \xi_0 \\ -f \end{pmatrix} = k_i R \begin{pmatrix} X_i - X_0 \\ Y_i - Y_0 \\ Z_i - Z_0 \end{pmatrix}$$

where

- η_0, ξ_0 are the coordinates of the perspective center on the photo plane,
- f the focal length,
- k_i the scaling factor,
- R the rotation matrix,
- X_0, Y_0, Z_0 the coordinates of the perspective center in the ground system (Fig. 16.22).

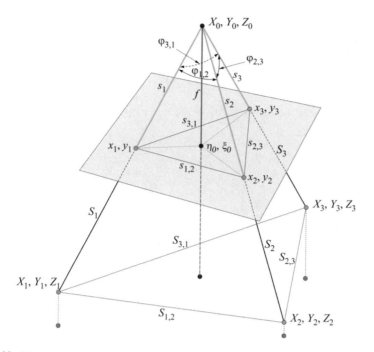

Fig. 16.22 Photogrammetric 3D resection

In general the focal length f is known, and the parameters of the transformation should be determined on the basis of the coordinates of 3 corresponding plane-ground points.

The problem with this representation of the transformation is that the scaling factors k_i differ for the different points, therefore it is reasonable to eliminate them. Let us express the equations of the transformation for the 3 different coordinates,

$$x_i - \eta_0 = k_i \left(R_{1,1}(X_i - X_0) + R_{1,2}(Y_i - Y_0) + R_{1,3}(Z_i - Z_0) \right),$$
$$y_i - \xi_0 = k_i \left(R_{21}(X_i - X_0) + R_{22}(Y_i - Y_0) + R_{23}(Z_i - Z_0) \right),$$
$$-f = k_i \left(R_{21}(X_i - X_0) + R_{22}(Y_i - Y_0) + R_{23}(Z_i - Z_0) \right).$$

Let us divide the first and second equation by the third one and rearrange them and introducing $r_{ij} = R_{i,j}$ we get

$$x_i = \eta_0 - f \frac{(r_{11}(X_i - X_0) + r_{12}(Y_i - Y_0) + r_{13}(Z_i - Z_0))}{(r_{31}(X_i - X_0) + r_{32}(Y_i - Y_0) + r_{33}(Z_i - Z_0))},$$

$$y_i = \xi_0 - f \frac{(r_{21}(X_i - X_0) + r_{22}(Y_i - Y_0) + r_{23}(Z_i - Z_0))}{(r_{31}(X_i - X_0) + r_{32}(Y_i - Y_0) + r_{33}(Z_i - Z_0))}.$$

Next, we express the elements of the rotation matrix with the elements of the skew matrix,

$\Rightarrow$ Clear$\left[''\text{Global } '*'' \right]$

$$\Rightarrow S = \begin{pmatrix} 0 & -c & b \\ c & 0 & -a \\ -b & a & 0 \end{pmatrix};$$

The rotation matrix is

$$\mathbf{R} = (I_3 - S)^{-1}(I_3 + S),$$

where I_3 is a 3×3 identity matrix.

$\Rightarrow$ I$_3$= IdentityMatrix$[3]$;

$\Rightarrow$ R = Inverse$[(I_3 - S)].(I_3 + S)$//Simplify; MatrixForm$[R]$

$$\Leftarrow \begin{pmatrix} \frac{1+a^2-b^2-c^2}{1+a^2+b^2+c^2} & \frac{2ab-2c}{1+a^2+b^2+c^2} & \frac{2(b+ac)}{1+a^2+b^2+c^2} \\ \frac{2(ab+c)}{1+a^2+b^2+c^2} & \frac{1-a^2+b^2-c^2}{1+a^2+b^2+c^2} & -\frac{2(a-bc)}{1+a^2+b^2+c^2} \\ \frac{2(-b+ac)}{1+a^2+b^2+c^2} & \frac{2(a+bc)}{1+a^2+b^2+c^2} & \frac{1-a^2-b^2+c^2}{1+a^2+b^2+c^2} \end{pmatrix}$$

The numerator of the elements of the rotation matrix,

$\Rightarrow$ Table$[r_{i,j}=$ Numerator$[R[[i,j]]], \{i,1,3\}, \{j,1,3\}]$

$\Leftarrow \{\{1+a^2-b^2-c^2, 2ab-2c, 2(b+ac)\},$
$\quad \{2(ab+c), 1-a^2+b^2-c^2, -2(a-bc)\},$
$\quad \{2(-b+ac), 2(a+bc), 1-a^2-b^2+c^2\}\}$

Considering six corresponding points on the photo plane and on the ground,

$\Rightarrow$ n = 6;

$\Rightarrow$ Clear$[X]$

the prototypes of the model equations are

$\Rightarrow e_i = (x_i - \eta 0)(r_{3,1}(X_i - X0) + r_{3,2}(Y_i - Y0) + r_{3,3}(Z_i - Z0)) +$
$f(r_{1,1}(X_i - X0) + r_{1,2}(Y_i - Y0) + r_{1,3}(Z_i - Z0))//\text{Expand}$

$\Rightarrow -fX0 - a^2 fX0 + b^2 fX0 + c^2 fX0 - 2abfY0 + 2cfY0 - 2bfZ0 - 2acfZ0 - 2bX0\eta 0 +$
$2acX0\eta 0 + 2aY0\eta 0 + 2bcY0\eta 0 + Z0\eta 0 - a^2 Z0\eta 0 - b^2 Z0\eta 0 + c^2 Z0\eta 0 + 2bX0x_i -$
$2acX0x_i - 2aY0x_i - 2bcY0x_i - Z0x_i + a^2 Z0x_i + b^2 Z0x_i - c^2 Z0x_i + fX_i + a^2 fX_i -$
$b^2 fX_i - c^2 fX_i + 2b\eta 0X_i - 2ac\eta 0X_i - 2bx_iX_i + 2acx_iX_i + 2abfY_i - 2cfY_i -$
$2a\eta 0Y_i - 2bc\eta 0Y_i + 2ax_iY_i + 2bcx_iY_i + 2bfZ_i + 2acfZ_i - \eta 0Z_i + a^2 \eta 0Z_i +$
$b^2 \eta 0Z_i - c^2 \eta 0Z_i + x_iZ_i - a^2 x_iZ_i - b^2 x_iZ_i + c^2 x_iZ_i$

and

$\Rightarrow e_{i+n} = (y_i - \xi 0)(r_{3,1}(X_i - X0) + r_{3,2}(Y_i - Y0) + r_{3,3}(Z_i - Z0))$
$+ f(r_{2,1}(X_i - X0) + r_{2,2}(Y_i - Y0) + r_{2,3}(Z_i - Z0))//\text{Expand}$

$\Leftarrow -2abfX0 - 2cfX0 - fY0 + a^2 fY0 - b^2 fY0 + c^2 fY0 + 2afZ0 - 2bcfZ0 - 2bX0\xi 0$
$+ 2acX0\xi 0 + 2aY0\xi 0 + 2bcY0\xi 0 + Z0\xi 0 - a^2 Z0\xi 0 - b^2 Z0\xi 0 + c^2 Z0\xi 0 + 2abfX_i$
$+ 2cfX_i + 2b\xi 0X_i - 2ac\xi 0X_i + 2bX0y_i - 2acX0y_i - 2aY0y_i - 2bcY0y_i - Z0y_i$
$+ a^2 Z0y_i + b^2 Z0y_i - c^2 Z0y_i - 2bX_iy_i + 2acX_iy_i + fY_i - a^2 fY_i + b^2 fY_i - c^2 fY_i$
$- 2a\xi 0Y_i - 2bc\xi 0Y_i + 2ay_iY_i + 2bcy_iY_i - 2afZ_i + 2bcfZ_i - \xi 0Z_i + a^2 \xi 0Z_i$
$+ b^2 \xi 0Z_i - c^2 \xi 0Z_i + y_iZ_i - a^2 y_iZ_i - b^2 y_iZ_i + c^2 y_iZ_i$

$\Rightarrow \text{sysn} = \text{Table}[\{e_i, e_{i+n}\}/.i^R j, \{j, 1, n\}]//\text{Flatten};$

$\Rightarrow \text{Short}[\text{sysn}, 10]$

$\Leftarrow \{-fX0 - a^2 fX0 + b^2 fX0 + c^2 fX0 - 2abfY0 + 2cfY0 - 2bfZ0 - 2acfZ0 - 2bX0\eta 0$
$+ 2acX0\eta 0 + 2aY0\eta 0 + 2bcY0\eta 0 + Z0\eta 0 - a^2 Z0\eta 0 - b^2 Z0\eta 0 + c^2 Z0\eta 0 + 2bX0x_1$
$- 2acX0x_1 - 2aY0x_1 - 2bcY0x_1 - Z0x_1 + <<10>> + 2b\eta 0X_1 - 2ac\eta 0X_1 - 2bx_1X_1$
$+ 2acx_1X_1 + 2abfY_1 - 2cfY_1 - 2a\eta 0Y_1 - 2bc\eta 0Y_1 + 2ax_1Y_1 + 2bcx_1Y_1 + 2bfZ_1$
$+ 2acfZ_1 - \eta 0Z_1 + a^2 \eta 0Z_1 + b^2 \eta 0Z_1 - c^2 \eta 0Z_1 + x_1Z_1 - a^2 x_1Z_1 - b^2 x_1Z_1 + c^2 x_1Z_1,$
$<<1>>, <<8>>, <<1>>, <<1>>\}$

Let us consider the following data: the ground coordinates, (X_i, Y_i, Z_i).
The coordinate values of the ground points,

$\Rightarrow \text{dataXYZ} =$
$\{X_1 \to -460., Y_1 \to -920, Z_1 \to -153, X_2 \to 460., Y_2 \to -920., Z_2 \to 0.,$
$X_3 \to -460., Y_3 \to 0, Z_3 \to 0, X_4 \to 460., Y_4 \to 0, Z_4 \to 153,$
$X_5 \to -460., Y_5 \to 920, Z_5 \to -153., X_6 \to 460., Y_6 \to 920, Z_6 \to 0\};$

and the image coordinates,

```
⇒ dataxy =
   {x₁ → 18996.171, y₁ → −64147.679, x₂ → 113471.749, y₂ → −73694.266,
    x₃ → 16504.609, y₃ → 16331.649, x₄ → 128830.826, y₄ → 21085.172,
    x₅ → 13716.588, y₅ → 106386.802, x₆ → 120577.473, y₆ → 128214.823};
```

In our case the focal length $f = 153\,000$.
Then the equations in numerical form,

```
⇒ sysnN = sysn/.dataXYZ/.dataxy/.f → 153000;
⇒ Short[sysnN, 20]
```
$$\Leftarrow \{-7.32864 \times 10^7 - 3.4953 \times 10^7 a - 6.74736 \times 10^7 a^2 - 2.93415 \times 10^7 b$$
$$- 281520000ab + 7.32864 \times 10^7 b^2 + 281520000c - 6.42945 \times 10^7 ac$$
$$- 3.4953 \times 10^7 bc + 6.74736 \times 10^7 c^2 - 153000X0 - 153000a^2 X0$$
$$+ 37992.3bX0 + 153000b^2 X0 - 37992.3acX0 + 153000c^2 X0$$
$$- 37992.3aY0 - 306000abY0 + 306000cY0 - 37992.3bcY0$$
$$- 18996.2Z0 + 18996.2a^2 Z0 - 306000bZ0 + 18996.2bη0 - 153a^2η0$$
$$- 920.bη0 - 153b^2η0 + 920.acη0 + 1840bcη0 + 153c^2η0 - 2bX0η0$$
$$+ 2acX0η0 + 2aY0η0 + 2bcY0η0 + Z0η0 - a^2 Z0η0 - b^2 Z0η0 + c^2 Z0η0,^2$$
$$- 1.30945 \times 10^8 + 1.6485 \times 10^8 a + <\,<57>\,> + c^2 Z0ξ0, <\,<1>\,>,$$
$$<\,<6>\,>, <\,<1>\,>, <\,<1>\,>, 140760000 + 2.35915 * 10^8 a$$
$$- 140760000a^2 - 1.17958 \times 10^8 b + 1.4076 \times 10^8 ab + 140760000b^2$$
$$1.4076 \times 10^8 c + 1.17958 \times 10^8 ac + <\,<38>\,> + 2acX0ξ0 +$$
$$+ 2aY0ξ0 + 2bcY0ξ0 + Z0ξ0 - a^2 Z0ξ0 - b^2 Z0ξ0 + c^2 Z0ξ0\}$$
```
⇒ Length[%]
⇐ 12
```

This means we have twelve equations, but only eight unknown variables (a, b, c, $X0$, $Y0$, $Z0$, $η0$, $ξ0$), consequently our system is overdetermined.

```
⇒ χ={a, b, c, X0, Y0, Z0, η0, ξ0};
```

Let us employ the *Gauss-Jacobi combinatorial algorithm*. We have 6 corresponding points and every four points represent four equations,

```
⇒ n = 6; m = 4;
```

The number of the combinations,

$\Rightarrow$ mn = Binomial[n, m]

$\Leftarrow$ 15

And the actual combinations,

$\Rightarrow$ qs = Partition[Map[#&, Flatten[Subsets[Range[n], {m}]]], m]

$\Leftarrow$ {{1, 2, 3, 4}, {1, 2, 3, 5}, {1, 2, 3, 6}, {1, 2, 4, 5}, {1, 2, 4, 6},
{1, 2, 5, 6}, {1, 3, 4, 5}, {1, 3, 4, 6}, {1, 3, 5, 6}, {1, 4, 5, 6},
{2, 3, 4, 5}, {2, 3, 4, 6}, {2, 3, 5, 6}, {2, 4, 5, 6}, {3, 4, 5, 6}}

We use data preparation, namely we apply a mask {1, 2, 3, 4} for each combination,

$\Rightarrow$ datapXYZ =

Table[Map[Select[dataXYZ, MemberQ[qs[[i]], #[[1, 2]]]&]/.
{#[[1]] $\rightarrow$ 1, #[[2]] $\rightarrow$ 2, #[[3]] $\rightarrow$ 3, #[[4]] $\rightarrow$ 4}&,
{qs[[i]]}], {i, 1, mn}];

For example, for the subset {1, 2, 3, 6},

$\Rightarrow$ datapXYZ[[3]]

$\Leftarrow$ {{$X_1 \rightarrow -460., Y_1 \rightarrow -920., Z_1 \rightarrow -153., X_2 \rightarrow 460., Y_1 \rightarrow -920.,$
$Z_1 \rightarrow 0., X_3 \rightarrow -460., Y_3 \rightarrow 0, Z_3 \rightarrow 0, X_4 \rightarrow 460., Y_4 \rightarrow 920, Z_4 \rightarrow 0$}}

similarly,

$\Rightarrow$ datapxy =

Table[Map[Select[dataxy, MemberQ[qs[[i]], #[[1, 2]]]&]/.
{#[[1]] $\rightarrow$ 1, #[[2]] $\rightarrow$ 2, #[[3]] $\rightarrow$ 3, #[[4]] $\rightarrow$ 4}&,
{qs[[i]]}], {i, 1, mn}];

and

$\Rightarrow$ datapXYZ[[3]]

$\Leftarrow$ {{$x_1 \rightarrow 18996.171, y_1 \rightarrow -64147.679, x_2 \rightarrow 113471.749, y_2 \rightarrow -73694.266,$
$x_3 \rightarrow 16504.609, y_3 \rightarrow 16331.649, x_4 \rightarrow 120577.473, y_4 \rightarrow 128214.823$}}

To solve the square subsets, we need the equations of the determined problem, namely the $n = 4$ points problem,

$$\Rightarrow e_{i+4} = (y_i - \xi 0)(r_{3,1}(X_i - X0) + r_{3,2}(Y_i - Y0) + r_{3,3}(Z_i - Z0))$$
$$+ f(r_{2,1}(X_i - X0) + r_{2,2}(Y_i - Y0) + r_{2,3}(Z_i - Z0)) // \text{Expand}$$

$$\Leftarrow -2abfX0 - 2cfX0 - fY0 + a^2fY0 - b^2fY0 + c^2fY0 + 2afZ0 - 2bcfZ0 - 2bX0\xi0$$
$$+ 2acX0\xi0 + 2fY0\xi0 + 2bcY0\xi0 + Z0\xi0 - a^2Z0\xi0 - b^2Z0\xi0 + c^2Z0\xi0 + 2abfX_i$$
$$+ 2cfX_i + 2b\xi0X_i - 2ac\xi0X_i + 2bX0y_i - 2acX0y_i - 2aaY0y_i - 2bcY0y_i - Z0y_i$$
$$+ a^2Z0y_i + b^2Z0y_i - c^2Z0y_i - 2bX_iy_i + 2acX_iy_i + fY_i - a^2fY_i + b^2fY_i - c^2fY_i$$
$$- 2a\xi0Y_i - 2bc\xi0Y_i + 2ay_iY_i + 2bcy_iY_i - 2afZ_i + 2Z_i - \xi0bcfZ_i$$
$$+ a^2\xi0Z_i + b^2\xi0Z_i - c^2\xi0Z_i + y_iZ_i - a^2y_iZ_i - b^2y_iZ_i + c^2y_iZ_i$$

$$\Rightarrow \text{sysn} = \text{Table}[\{e_i, e_{i+4}\}/.i \to j, \{j, 1, 4\}] // \text{Flatten};$$

$$\Rightarrow \text{Short}[\text{sysn}, 10]$$

$$\Leftarrow \{-fX0 - a^2fX0 + b^2fX0 + c^2fX0 - 2abfY0 + 2cfY0 - 2bfZ0 - 2acfZ0 - 2bX0\eta0$$
$$+ 2acX0\eta0 + 2aY0\eta0 + 2bcY0\eta0 + Z0\eta0 - a^2Z0\eta0 - b^2Z0\eta0 + c^2Z0\eta0 + 2bX0x_1$$
$$- 2acX0x_1 - 2aY0x_1 - 2bcY0x_1 - Z0x_1 + <<10>> + 2b\eta0X_1 - 2ac\eta0X_1 - 2bx_1X_1$$
$$+ 2acx_1X_1 + 2abfY_1 - 2cfY_1 - 2a\eta0Y_1 - 2bc\eta0Y_1 + 2ax_1Y_1 + 2bcx_1Y_1 + 2bfZ_1$$
$$+ 2acfZ_1 - \eta0Z_1 + a^2\eta0Z_1 + b^2\eta0Z_1 - c^2\eta0Z_1 + x_1Z_1 - a^2x_1Z_1 - b^2x_1Z_1 + c^2x_1Z_1,$$
$$<<1>>, <<4>>, <<1>>, <<1>>\}$$

Then the solution of the subsets,

$$\Rightarrow \text{AbsoluteTiming}[\text{solGJ} = \text{MapThread}[$$
$$\text{NSolve}[\text{sys}/.\text{Flatten}[\#1]/.\text{Flatten}[\#2]/.f \to 153000, \chi]\&,$$
$$\{\text{datapXYZ}, \text{datapxy}\}];]$$
$$\Leftarrow \{73.866130, \text{Null}\}$$

This computation takes a long time, therefore it is reasonable to employ parallel computation. Let us define all variables involved in the computation as parallel variables,

$$\Rightarrow \text{LaunchKernels}[2 \ \$\text{ProcessorCount}]//\text{Quiet};$$
$$\Rightarrow \text{DistributeDefinitions}[\text{sys}, f, \text{datapXYZ}, \text{datapxy}, \chi]$$
$$\Leftarrow \{\text{sys}, \text{datapXYZ}, \text{datapxy}, \chi\}$$

Then the parallel computation could be applied straight away, but here the application of the function **Parallelize** is reasonable,

$$\Rightarrow \text{AbsoluteTiming}[\text{solGJ} = \text{Parallelize}[\text{MapThread}[$$
$$\text{NSolve}[\text{sys}/.\text{Flatten}[\#1]/.\text{Flatten}[\#2]/.f \to 153000, \chi]\&,$$
$$\{\text{datapXYZ}, \text{datapxy}\}]];]$$
$$\Leftarrow \{22.198839, \text{Null}\}$$

Now the running time is considerably shorter (Fig. 16.23).

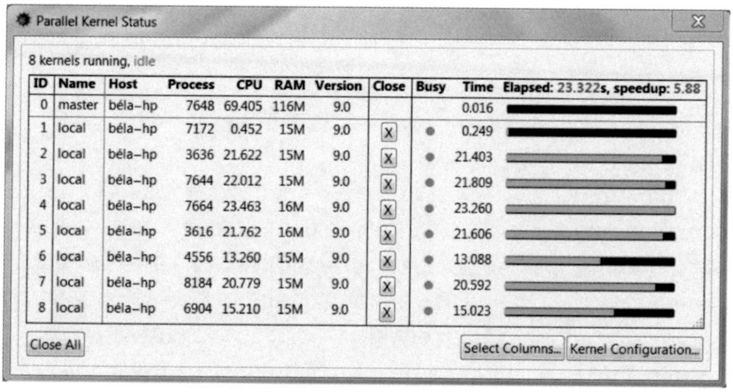

Fig. 16.23 Status of the kernels in case of parallel evaluation of the Gauss–Jacobi algorithm using parallelize

It can be seen that only seven threads were activated and their tasks needed different running times! Indeed, the solutions have different types and different lengths, for example

$$\Rightarrow \text{solGJ}[[1]]$$
$$\Leftarrow \{\{a \rightarrow 212.292, b \rightarrow 213.217, c \rightarrow -22.6786, X0 \rightarrow 5220.9,$$
$$Y0 \rightarrow 5503.69, Z0 \rightarrow 330.045, \eta 0 \rightarrow -528188., \xi 0 \rightarrow -550286.\},$$
$$\{a \rightarrow 2.23966, b \rightarrow -2.23331, c \rightarrow 0.0123486, X0 \rightarrow 4738.68,$$
$$Y0 \rightarrow 5021.47, Z0 \rightarrow 3229.65, \eta 0 \rightarrow -528188., \xi 0 \rightarrow -550286.\},$$
$$\{a \rightarrow 11.63, b \rightarrow 11.8209, c \rightarrow -52.4359, X0 \rightarrow 22.2177,$$
$$Y0 \rightarrow 482.218, Z0 \rightarrow -1369.61, \eta 0 \rightarrow 0.0120128, \xi 0 \rightarrow 0.0154296\}$$

In order to select the admissible solutions from a subset, first we compute the minimum of the residual errors for each subset solution,

$$\Rightarrow \text{obj} = \text{Apply}[\text{Plus}, \text{Map}[\#^2 \&, \text{sys}]];$$
$$\Rightarrow \text{objmin} = \text{Map}[\text{Min}[\text{Abs}[\#]] \&,$$
$$\text{Table}[\text{obj}/.\text{dataXYZ}/.\text{dataxy}/.\text{solGJ}[[i,j]]/.f \rightarrow 153000,$$
$$\{i, 1, \text{mn}\}, \{j, 1, \text{Length}[\text{solGJ}[[i]]]\}]];$$

then we select the solutions belonging to these minimums,

```
⇒ solGJS = Table[Select[solGJ[[i]],
  Abs[obj/.dataXYZ/.dataxy/.#/.f → 153000] ==
  objmin[[i]]&],{i,1,mn}]
⇐ {{{a → 0.0520854, b → −0.0546015, c → 0.0183141, X0 → −460.,
    Y0 → 0.000080254, Z0 → 1530., η0 → 0.0120128, ξ0 → 0.0154296}},
  {{a → 0.0520853, b → −0.0546014, c → 0.0183141, X0 → −460.,
    Y0 → 0.000179128, Z0 → 1530., η0 → 0.0160779, ξ0 → 0.0396817}},
  {{a → 0.0520854, b → −0.0546015, c → 0.0183141, X0 → −460.,
    Y0 → −0.000023012, Z0 → 1530., η0 → 0.0090842, ξ0 → 0.0047149}},
  {{a → 0.0520854, b → −0.0546015, c → 0.0183141, X0 → −460.,
    Y0 → 0.00002137, Z0 → 1530., η0 → −0.00226708, ξ0 → 0.0052136}},
  {{a → 0.0520854, b → −0.0546014, c → 0.0183141, X0 → −460.,
    Y0 → 0.00015106, Z0 → 1530., η0 → 0.0305765, ξ0 → 0.0287886}},
  {{a → 0.0520854, b → −0.0546015, c → 0.0183141, X0 → −460.,
    Y0 → −0.000054, Z0 → 1530., η0 → −0.001947, ξ0 → −0.00323999}},
  {{a → 0.0520854, b → −0.0546015, c → 0.0183141, X0 → −460.,
    Y0 → 0.000093005, Z0 → 1530., η0 → 0.0021706, ξ0 → 0.0273753}},
  {{a → 0.0520854, b → −0.0546014, c → 0.0183141, X0 → −460.,
    Y0 → 0.000115357, Z0 → 1530., η0 → 0.0132181, ξ0 → 0.0155434}},
  {{a → 0.0520854, b → −0.0546015, c → 0.0183141, X0 → −460.,
    Y0 → −0.00001176, Z0 → 1530., η0 → −0.0094104, ξ0 → 0.0126929}},
  {{a → 0.0520854, b → −0.0546015, c → 0.0183141, X0 → −460.,
    Y0 → 0., Z0 → 1530., η0 → 0.00455397, ξ0 → 0.0057907}},
  {{a → 0.0520854, b → −0.0546015, c → 0.0183141, X0 → −460.,
    Y0 → 0.0000418975, Z0 → 1530., η0 → −0.0122799, ξ0 → 0.019443}},
  {{a → 0.0520854, b → −0.0546014, c → 0.0183141, X0 → −460.,
    Y0 → 0.0000871336, Z0 → 1530., η0 → 0.0177359, ξ0 → 0.0148248}},
  {{a → 0.0520854, b → −0.0546015, c → 0.0183141, X0 → −460.,
    Y0 → −0.000432414, Z0 → 1530., η0 → −0.0230733, ξ0 → −0.03286}},
  {{a → 0.0520855, b → −0.0546011, c → 0.0183142, X0 → −460.,
    Y0 → −0.0010589, Z0 → 1530., η0 → 0.0951372, ξ0 → −0.183949}},
  {{a → 0.0520854, b → −0.0546015, c → 0.0183141, X0 → −460.,
    Y0 → 0.00024182, Z0 → 1530., η0 → −0.00770075, ξ0 → 0.0280156}}}
```

The average values of the subset solutions is

```
⇒ solGJ = MapThread[#1 → #2&, {χ, Map[Mean[#]&,
  Transpose[Table[Map[#[[2]]&,
  Flatten[solGJS[[i]],1]],{i,1,mn}]]]}]
⇐ {a → 0.0520854, b → −0.0546014, c → 0.0183141, X0 → −460.,
  Y0 → −0.0000379376, Z0 → 1530., η0 → 0.0095926, ξ0 → −0.000169176}
```

The actual net win in running time 73.9/22.2 = 3.33.

It goes without saying that these time values can change machine to machine, hardware to hardware therefore the results provide only qualitative information about the effect of the parallel execution technique.

Reference

Haneberg WC (2004) Computational geosciences with Mathematica. Springer, Berlin, Heidelberg

Printed in the United States
by Baker & Taylor Publisher Services